AF323765

Vibrational Dynamics of Molecules

Vibrational Dynamics of Molecules

Editor

Joel M Bowman
Emory University, USA

NEW JERSEY · LONDON · SINGAPORE · BEIJING · SHANGHAI · HONG KONG · TAIPEI · CHENNAI · TOKYO

Published by

World Scientific Publishing Co. Pte. Ltd.

5 Toh Tuck Link, Singapore 596224

USA office: 27 Warren Street, Suite 401-402, Hackensack, NJ 07601

UK office: 57 Shelton Street, Covent Garden, London WC2H 9HE

Library of Congress Cataloging-in-Publication Data

Names: Bowman, J. M. (Joel M.), editor.

Title: Vibrational dynamics of molecules / editor, Joel M. Bowman.

Description: New Jersey : World Scientific, [2022] | Includes bibliographical references and index.

Identifiers: LCCN 2021054883 | ISBN 9789811237904 (hardcover) |

ISBN 9789811237911 (ebook for institutions) | ISBN 9789811237928 (ebook for individuals)

Subjects: LCSH: Vibrational spectra. | Molecular dynamics. | Chemistry, Physical and theoretical.

Classification: LCC QD96.V53 V493 2022 | DDC 539/.6--dc23/eng20220405

LC record available at https://lccn.loc.gov/2021054883

British Library Cataloguing-in-Publication Data

A catalogue record for this book is available from the British Library.

For any available supplementary material, please visit
https://www.worldscientific.com/worldscibooks/10.1142/12305#t=suppl

Desk Editor: Shaun Tan Yi Jie

Typeset by Stallion Press
Email: enquiries@stallionpress.com

Preface

The title of this book, "Vibrational Dynamics of Molecules", is a general one. This was intentional, as the vibrational dynamics of molecules, ranging from isolated molecules (polyatomics to biomolecules), molecular clusters, to finally the condensed phase is a subject of central interest to both experimental and theoretical/computational chemists. The book presents modern theoretical and computational approaches that span this large landscape from the leading groups in the world.

The ultimate theoretical approach is solving the Schrödinger equation for the nuclear motion "exactly". This is of course a major challenge in two respects. First, the equation itself is a partial differential equation in high mathematical dimensionality. Second, the equation requires knowledge of the high-dimensional potential that describes the interaction of all the atoms in the molecule(s). This is the well-known Born-Oppenheimer potential, which is obtained rigorously from the variation of the electronic energy with respect to nuclear motion. This variation of electronic energies is obtained from solving the electronic Schrödinger equation, which shares some things in common mathematically with the nuclear Schrödinger equation, but which also has crucial differences. Perhaps the biggest difference is the nature of the interactions. In the electronic Schrödinger equation these are the familiar two-body Coulomb interactions. By contrast, in the nuclear Schrödinger equation the interactions are not of a universal type nor are they two-body; they are many-body, albeit bound by the number of vibrational degrees of freedom.

This difference between the potentials in these equations notwithstanding, the methods used to solve the nuclear Schrödinger equation borrow heavily from methods, developed over many years, to solve the electronic Schrödinger equation. This is seen in the first four chapters of the book, "Vibrational Configuration Interaction Theory" (Schröder and Rauhut), "Vibrational Coupled Cluster Theory" (Christiansen), "Tensor Network States for Vibrational Spectroscopy" (Glaser, Baiardi, and Reiher), and "Diffusion Monte Carlo Approaches for Studying Large Amplitude Vibrational Motions in Molecules and Clusters" (Finney, DiRisio, and McCoy). The chapter "Collocation Methods for Computing Vibrational Spectra" (Carrington) explores a very different approach to solve the nuclear Schrödinger equation by using sparse and scattered grid points where the potential is given.

Of course, solving the nuclear Schrödinger equation is basically a means to an end, namely addressing interesting and important applications that are often driven by experiments. This is the focus of the chapters "Vibration-rotation-tunneling levels and spectra of Van der Waals molecules" (van der Avoird), "Vibrational and Rovibrational Spectroscopy Applied to Astrochemistry" (Fortenberry and Lee), "MULTIMODE, the n-mode Representation of the Potential and Illustrations to IR Spectra of Glycine and Two Protonated Water Clusters" (Yu, Qu, Houston, Conte, Nandi, and Bowman), and "Vibrational Spectra of Flexible Systems using the MCTDH approach" (Meyer, Schröder and Vendrell).

Nuclei are thousands of times more massive than electrons and so treating the nuclear motion classically is commonly done, especially in modeling large molecules such as biomolecules and molecules in the condensed phase. Through the use of time-dependent correlation functions, vibrational spectroscopy can also be approached using classical dynamics. Indeed, since solving the classical equations of motion is much easier than solving the nuclear Schrödinger equation, the classical approach can be applied to large molecules and molecules in the condensed phase. This is the subject of the chapter "Direct Dynamics for Vibrational Spectroscopy: from Large

Molecules in the Gas Phase to the Condensed Phase" (Bougueroua, Chantitch, Chen, Pezzotti, and Gaigeot).

The one atom that is problematic to treat classically is the lightest and most ubiquitous in the universe, namely hydrogen. H atoms are present in all biomolecules and of course in the most important liquid, water. So, finding an approach to describe the motion of H atoms for the many instances where a full quantum approach is not feasible and yet where classical mechanics is suspect has been a major focus of theoretical chemistry. Broadly speaking, these are known as semi-classical approaches. A very powerful one, especially for vibrational dynamics, is the subject of the chapter "Vibrational Dynamics for Molecular and Supra-Molecular Systems" (Conte and Ceotto).

The final chapter, "Introduction to vibropolaritons: spectroscopy, relaxation and chemical reactions" (Ribeiro and Yuen-Zhou), introduces readers to a new research area in which nuclear motion is strongly coupled to electromagnetic fields in a cavity. This is an area where the methods reviewed in the earlier chapters will likely play an important role, as this new field rapidly develops.

J. M. Bowman

About the Editor

Dr. Joel M. Bowman is the Samuel Candler Dobbs Professor of Theoretical Chemistry at Emory University, USA, where he held the Department Chair position twice (1990–1993, 2003–2006). He is an elected Fellow of the American Physical Society (since 1990), American Association for the Advancement of Science (since 2005), and International Academy of Quantum Molecular Science (since 2013). He is a recipient of the Alexander von Humboldt Research Award in 2018 and the Dudley Herschbach Prize for Theoretical Chemistry in 2013. He is the author of more than 500 publications, and has a h-index of 86 as of January 2022. He is Editor of *Spectrochimica Acta A* and a member of the Editorial Boards of *Chemical Physics*, *Advances in Physical Chemistry* and the *International Journal of Quantum Chemistry*. He holds a PhD in Chemistry from the California Institute of Technology, USA.

Dr. Bowman is widely considered as one of the founding fathers of theoretical reaction dynamics. He has made significant contributions in the theory and computation of many aspects of chemical reaction dynamics and molecular vibrations. Notable among these were the development of *ab initio* potential energy surfaces in high dimensionality using permutationally invariant fitting bases. Examples include the reactions $X + CH_4 \rightarrow HX + CH_3$, $X = H$, $O(^3P)$, F, Cl and intersystem crossing in $O + C_2H_4$. Potentials for H_5^+, CH_5^+, $H_5O_2^+$, etc., have led to the most rigorous analyses of this complex cations. The approach has also resulted in the most accurate *ab initio* potential and dipole moment for water, built from 1, 2, 3-body high-level electronic energies and precisely fit. In addition, he developed the vibrational self-consistent field and virtual state CI approaches to coupled molecular vibrations. Subsequently the efficient and accurate n-mode representation of the potential was developed and incorporated into the code MULTIMODE. This code has been used in many applications ranging from the rovibrational spectroscopy of polyatomic molecules to the vibrational dynamics of molecular clusters, including water clusters and hydrated ions. He also discovered roaming dynamics, and developed powerful methods to combine aspects of transition state theory with reduced dimensionality quantum scattering treatment of reaction dynamics, among which J-shifting has been widely used.

Contents

xi

Contents

4. Diffusion Monte Carlo Approaches for Studying Large Amplitude Vibrational Motions in Molecules and Clusters — 145

Jacob M. Finney, Ryan J. DiRisio, and Anne B. McCoy

5. Collocation Methods for Computing Vibrational Spectra   174

Tucker Carrington

6. Vibration-Rotation-Tunneling Levels and Spectra of Van der Waals Molecules   194

Ad van der Avoird

7.   Vibrational and Rovibrational Spectroscopy Applied to Astrochemistry   235

Ryan C. Fortenberry and Timothy J. Lee

8. MULTIMODE, The n-Mode Representation of the Potential and Illustrations to IR Spectra of Glycine and Two Protonated Water Clusters

296

Qi Yu, Chen Qu, Paul L. Houston,
Riccardo Conte, Apurba Nandi, and Joel M. Bowman

9. Vibrational Spectra of Flexible Systems using the MCTDH Approach   340

H.-D. Meyer, M. Schröder and O. Vendrell

10. Semiclassical Vibrational Dynamics for Molecular and Supra-Molecular Systems   378

Riccardo Conte and Michele Ceotto

11. Direct Dynamics for Vibrational Spectroscopy: From Large Molecules in the Gas Phase to the Condensed Phase

416

Sana Bougueroua, Vladimir Chantitch, Wanlin Chen, Simone Pezzotti, and Marie-Pierre Gaigeot

https://doi.org/10.1142/9789811237911_0001

Chapter 1

Vibrational Configuration Interaction Theory

Benjamin Schröder and Guntram Rauhut*

*Institute for Theoretical Chemistry,
University of Stuttgart, 70569 Stuttgart, Germany*
**rauhut@theochem.uni-stuttgart.de*

1.1. Introduction

The time-independent variational calculation of transition energies for non-rotating molecules is an attractive route for investigating vibrational spectra as it is not bound to a specific Hamiltonian or a certain set of coordinates and, perhaps most importantly, will be exact in the limit.[1,2] In contrast to other approaches it can easily deal with high state densities and the associated resonances, which occur quite frequently for larger molecules. At least on paper the underlying theory of an associated finite-basis approach is fairly simple: take a Hamiltonian and a meaningful basis set, set up the Hamiltonian matrix and diagonalize it, that's about it. However, in practice, it is much more intricate for technical reasons and usually a number of approximations and truncated expansions need to be employed to limit the rapidly exceeding computational effort.

Attempts to determine accurate vibrational frequencies and wavefunctions this way date back to the very beginnings of computational vibrational spectroscopy,[3,4] but it is due to the enormous computational power and the high accuracy of electronic structure methods for representing the potential energy surface, PES, nowadays that vibrational structure methods attracted broad interest.

Among these, the workhorse of the developed methods is vibrational configuration interaction theory,[5,6] VCI, often based on Hartree products generated from a preceding vibrational self-consistent field calculation, VSCF.[7,8] However, computational implementations of VCI theory show a certain flexibility with respect to technical details, which are usually employed to accelerate these calculations. It is not the objective of this contribution to provide an overview about all variants of VCI theory,[5,6,9–25] but to outline general aspects, which hold for most of these refined methods. Moreover, strong analogies to configuration interaction theory in electronic structure theory do exist, but usually these hold only for some general aspects rather than the Hamiltonian-bound peculiarities. In all that follows, we limit the discussion to approaches based on the Eckart-Watson Hamiltonian,[26] which makes use of rectilinear normal coordinates or any unitarily transformed sets thereof. Moreover, we will not consider grid-based approaches, *e.g.*, collocation methods (see Chapter 5), and thus assume the PES being represented by any analytical functions, *e.g.*, polynomials or B-splines, in a sum-of-products (SOP) form. As a complete representation of the PES is only feasible for the smallest systems, usually expansions of it are employed, *e.g.*, Taylor or n-mode expansions,[1,9] which can conveniently be truncated at a given order. Note that both expansions are related to each other, but usually the faster converging n-mode expansion is used, while an early truncated Taylor expansion in normal coordinate space, *i.e.*, a quartic force field (QFF), is not able to provide the high accuracy being expected from VCI calculations, but usually is followed by second-order vibrational perturbation theory, VPT2.[27–30]

1.2. General Aspects

Formally, the main aspects of VCI theory are the build-up of the Hamiltonian matrix and its diagonalization or, more precisely, the determination of individual eigenpairs of it. Once an SOP representation of the PES is used, the multidimensional wavefunction, Ψ_v, of the vibrational state v can be expressed by simple Hartree products, Φ^I, of one-mode wavefunctions, φ_i^I, belonging to single coordinates, q_i, *i.e.*,

$$\Psi_v(\mathbf{q}) = \sum_I C_{vI} \Phi^I(\mathbf{q}) = \sum_I C_{vI} \prod_i \varphi_i^I(q_i). \qquad (1.1)$$

Due to the analogs of the Slater–Condon rules for Hartree products, the Hamiltonian matrix has a band structure with respect to the excitation levels within the Hartree products and the width of this band is controlled by the order of the expansion of the PES. The contribution of the PES to a single matrix element of the Hamiltonian can be written as

$$H_{IJ}^{\text{(PES)}} = \sum_d \sum_{i_1,\dots,i_d} \sum_{r_1,\dots,r_d} p_{r_1\dots r_d}^{(i_1\dots i_d)}$$

$$\times \left(\prod_{k\in\{1,\dots,d\}} \left\langle \varphi_{i_k}^I \left| q_{i_k}^{r_k} \right| \varphi_{i_k}^J \right\rangle \right) \prod_{l\notin\{1,\dots,d\}} \delta_{n_{i_l}^I n_{i_l}^J}. \qquad (1.2)$$

Within this equation, d denotes the individual orders of the underlying n-mode expansion and $p_{r_1\dots r_d}^{(i_1\dots i_d)}$ are the cocfficients arising from a fit of the PES to an SOP representation.[31] As the 1D integrals $Q_{kr}^{IJ} = \left\langle \varphi_{i_k}^I \left| q_{i_k}^{r_k} \right| \varphi_{i_k}^J \right\rangle$ depend on just one coordinate, they can be easily evaluated and stored in memory, despite their formal dependence on the indices I and J referring to Hartree products. The Kronecker delta at the end of this equation arises from the overlap integrals of the orthonormal modals and $n_{i_l}^I$ denotes an element of the quantum number vector (QNV), *i.e.*, a pointer to the employed modal of mode i_l within the Hartree product I or J. This gives rise to the analogs of the Slater–Condon rules. For better readability, we will simplify our notation in the following and denote modes by the indices $i, j, k, \dots$ rather than $i_1, i_2, i_3, \dots$. It is rather obvious that the CPU time for evaluating a single matrix element rises rapidly with the size of the molecules and the order of the PES expansion, $d_{\max}$. These demands can be diminished by simple contraction schemes,[32] *e.g.*,

$$Y_{r_i\dots r_j}^{(i\dots j,k),IJ} = \sum_{r_k} p_{r_i\dots r_j,r_k}^{(i\dots j,k)} Q_{kr}^{IJ}. \qquad (1.3)$$

The contracted integrals Y request more memory, but still can be stored without problems and the number of sums over the basis functions of the PES (*e.g.*, polynomials), *i.e.*, $r_i \ldots r_k$, is reduced by one. Typically this leads to speed-ups by factors between 2 and 10 depending on the representation of the PES.[32] Further speed-ups can be obtained by exploiting molecular point group symmetry, which affects both, the coefficients of the SOP representation of the PES and the corresponding integrals Q_{ir}^{IJ}.

Another aspect concerns the storage of the Hamiltonian matrix. In principle, direct CI techniques are available,[10] which avoid a storage of the Hamiltonian matrix completely, but on cost of higher CPU time demands. These approaches shall not be considered here as the storage of this matrix usually is not a problem even for large initial correlation spaces with millions of Hartree products. As will be outlined in detail below, the dimensionality of the matrix can significantly be reduced by configuration selection schemes, *i.e.*, only those Hartree products will be considered, which show non-negligible contributions to the vibrational state of interest. Typically this reduces the matrix by several orders of magnitude. Besides that, its symmetric structure due to the Hermitian Watson operator and its block diagonality arising from symmetry-adapted coordinates lead to a further reduction of the memory demands. However, even when accounting for all these aspects does the Hamiltonian matrix still remain very sparse. Consequently, a storage of it in a sparse matrix format, *i.e.*, position and value, leads to significant savings. As a consequence, Hamiltonian matrices for initial correlation spaces of up to 10^{10} Hartree products can be stored without problems on customary workstations with no more than 16 GByte of memory. However, the actual demands are very difficult to estimate and may vary substantially from molecule to molecule or even in dependence on the vibrational state of interest.

An example shall be given for the Hamiltonian matrices being used for the calculation of the 9^1 (B_{2u}) state and its overtone 9^2 (A_g), respectively, of ethylene (*cf.* Table 1.1). The PES has been represented by an n-mode expansion up to fourth order and the initial correlation space comprised up to sixtuple excitations. Note that the

Table 1.1. Memory requirements (in MByte) for the Hamiltonian matrices of the 9^1 (B_{2u}) and 9^2 (A_g) states of ethylene in dependence on different properties and techniques.

	9^1 (B_{2u})	9^2 (A_g)
Initial space	3,203,570,683	3,203,570,683
+ Hermiticity	1,601,785,419	1,601,785,419
+ Block diagonality	27,694,143	27,443,488
+ Configuration selection	69	2,043
+ Sparsity	12	172

sparsity recognition threshold for the elements of the Hamiltonian matrix within Table 1.1 has been set to 10^{-10} a.u., which means that Hamiltonian matrix elements below an absolute value of about $0.00002\,\mathrm{cm}^{-1}$ are neglected. In both cases the savings are tremendous and the final memory requirements are very modest. This example demonstrates the importance to exploit all different techniques and in particular the selection of important Hartree products.

1.3. Basis Functions

Historically, the one-mode wavefunctions (modals), φ_i, for generating the Hartree products, Φ^I, and needed for the evaluation of the Hamiltonian matrix were chosen to be harmonic oscillator (HO) functions, which is kind of a natural choice and is still in use in modern work.[14,21,24,25,33] These functions show several appealing features: (1) they have a simple analytical form, (2) they show symmetry, and (3) they are derived from a spectroscopically motivated approximation. As a consequence, many of the integrals Q_{ir}^{IJ} vanish for symmetry reasons once the individual 1D potential functions (*e.g.*, monomials) show also linear or point symmetry. As outlined above, this can be exploited to accelerate the calculation of the Hamiltonian matrix elements. For many applications, this choice is sufficient and ultimately will yield good results. However, once anharmonicity effects become strong, *e.g.*, in dissociation potentials, quartic potentials, double-well potentials, *etc.*, an HO basis will require huge correlation spaces or lead to unconverged results.

Table 1.2.   Convergence of the zero point vibrational energy of nitrosamine in dependence of the type of the correlating functions within a configuration-selective VCI approach.

	VSCF modals		HO functions	
N_{CF}^a	E_0	N_{sel}^b	E_0	N_{sel}^b
3	7035.9	1,896	7218.3	64,101
4	7023.8	3,129	7111.6	339,188
5	7020.8	3,900	7104.1	$>500,000^c$
6	7018.9	4,699	7064.9	$>500,000^c$
7	7017.9	5,694	7060.7	$>500,000^c$
8	7016.8	6,195	7042.8	$>500,000^c$
9	7016.7	6,554	7040.9	$>500,000^c$
10	7016.5	8,272	7031,9	$>500,000^c$
11	7016.5	13,548	7030,7	$>500,000^c$

[a]Number of correlating functions per mode.
[b]Number of selected Hartree products.
[c]Not fully converged.

An example is given for the vibrational ground state energy of nitrosamine, a simple pentatomic molecule with strong quartic contributions to the PES. Table 1.2 shows that anharmonic VSCF modals lead to faster convergence with respect to the size of the correlation space and are in general better suited for VCI calculations. As mentioned above, these are usually obtained from an optimization within vibrational self-consistent field (VSCF) theory.[7] As these functions do neither show point nor line symmetry, the Q_{ir}^{IJ} integrals need to be considered in full, unless q_i belongs to a symmetrical mode. The subsequent VCI calculations for individual states can be based on VSCF modals obtained for the vibrational ground state or on state-specific modals, *i.e.*, the VSCF equations are solved for the reference Hartree product of the state of interest.[34] Usually state-specific modals outperform ground state-based modals and may be the only meaningful choice in special situations, but they are associated with two problems: (1) In a state-specific framework an individual VCI calculation will be performed for each state. Consequently, the final VCI eigenvectors need not be orthogonal. However, in practice this effect is negligible, in particular in VCI calculations close to the exact full VCI (FVCI) limit. (2) In case of resonances, *e.g.*, Fermi or Darling-Dennison resonances, one Hartree product may be preferred over the other, which leads to an unbalanced calculation. Likewise, in calculations for non-Abelian molecules, it is preferable to base the calculations of degenerate modes on equal footing. These problems can be lifted by modals obtained from configuration averaged VSCF (ca-VSCF) theory.[35] In the basis of the primitive functions μ and ν, which build up the individual modals, the energy expression of ca-VSCF, to be minimized under the orthonormality constraint, is given in Eq. (1.4).

$$
\begin{aligned}
E_{\text{CA}} &= \frac{1}{N} \sum_I E_I \\
&= \sum_i \sum_{a_i} \gamma_{a_i} \sum_{\mu\nu} c^{a_i}_{\mu i} c^{a_i}_{\nu i} \left[-\frac{1}{2} \tilde{T}^i_{\mu\nu} + \sum_{r_i} \tilde{Q}^{ir}_{\mu\nu} \left[p^{(i)}_{r_i} \right. \right. \\
&\quad \left. \left. + \sum_j \sum_{b_j} \gamma_{b_j} \sum_{\rho\sigma} c^{b_j}_{\rho j} c^{b_j}_{\sigma j} \sum_{r_j} \tilde{Q}^{jr}_{\rho\sigma} \left[\frac{1}{2} p^{(ij)}_{r_i r_j} + \cdots \right] \right] \right].
\end{aligned}
\tag{1.4}
$$

Here we have introduced fractional quantum numbers γ_{a_i}. Note that $\gamma_{a_i} = n_{a_i}/N$, with n_{a_i} being the number of configurations in which the modal a_i of mode i is used. $p^{(i)}_{r_i}$ and $p^{(ij)}_{r_i r_j}$ are the coefficients of the expansion of the PES, while $\tilde{T}^i_{\mu\nu}$ and $\tilde{Q}^{ir}_{\mu\nu}$ are the usual integrals over the kinetic and PES operators. The tilde has been used to distinguish these integrals from the corresponding contracted quantities in the modal basis. In case of a single Hartree product, Eq. (1.4) reduces to the energy expectation value being used in conventional VSCF theory ($\gamma_{a_i} = 1$). In practice, the effects due to a ca-VSCF optimization of the modals relative to a VSCF calculation with just one Hartree product are fairly small, but state-specific modals, in general, lead to larger leading VCI coefficients and thus a more secure assignment of the vibrational state.[34] In contrast to molecular orbitals in electronic structure theory, modals show a significantly reduced flexibility arising from their lower dimensionality and thus a lesser impact on the final results.

1.4. Completeness Issues

Due to the curse of dimensionality and the representation of the wavefunction by a linear combination of Hartree products, expansions need to be introduced into VCI theory, which must be controlled with respect to convergence in order to yield physically meaningful and quantitatively correct vibrational state energies. The simple but overall important question is: When is it safe to truncate these expansions and what means safe?

1.4.1. *Completeness of the Watson Hamiltonian*

The most obvious aspect in that respect concerns the order of the PES expansion. As the evaluation of the multidimensional PES is the time-determining step within vibrational structure calculations, the order of truncation usually controls the accuracy of the calculation (besides the level of electronic structure theory). Let us assume that the PES has been evaluated by high-level coupled-cluster methods in combination with a decent basis set. Let us further assume that the molecule is of semi-rigid nature and that the minimum of interest on the PES is energetically well separated from any other minima. In that case, the mean absolute deviation (MAD) of the fundamental vibrational frequencies usually is in the range of 1–$5\,\mathrm{cm}^{-1}$ with respect to experimental gas phase data once the n-mode expansion of the PES has been truncated after fourth order. In almost all cases the maximum deviations arise from CH-stretching modes, which are already in the region of high state density. In that range of accuracy it is not clear at all if an improvement of the electronic structure calculations is more important than the inclusion of PES terms of even higher order or vice versa. Once lower accuracy is acceptable or the level of electronic structure theory remains limited, *e.g.*, density functional theory, or the molecule contains less than three hydrogen atoms, a truncation of the n-mode expansion after third order may be sufficient. However, n-mode expansions of the PES, which have been truncated too early, tend to introduce artifical resonances in the subsequent VCI calculations, which are resolved upon inclusion of higher order terms.

Moreover, it is very important to keep the vibrational structure calculations in balance with the level of electronic structure theory. As semi-diagonal or complete normal coordinate QFFs can be obtained cheaply in comparison to a fourth order n-mode expansion, many authors simply combine such QFFs with VCI approaches.[14,21,24,25] In that case the latter calculations run extremely fast, because the number of integrals to be evaluated and Hartree products to be considered is much smaller. However, the accuracy remains limited and usually these calculations compete with results from VPT2 theory, which often then appears to be the preferable choice (see Chapter 7).

Table 1.3 shows the mean absolute and maximum deviations of VCI fundamental frequencies with respect to experimental values for a small set of molecules of increasing size. All PESs have been obtained from electronic structure calculations at the level of explicitly correlated coupled-cluster theory. While the results including 3-mode coupling terms (3D) show relatively large errors and which are in the range of VPT2 results, the inclusion of 4-mode couplings (4D) results in significant improvements in almost all cases. Again, the large maximum deviations in the 3D calculations arise exclusively from the CH or BH stretching modes. As the calculation of a PES is rather costly at the coupled-cluster level,

Table 1.3. Mean absolute (MAD) and maximum (MAX) deviations of VCI fundamental frequencies (in cm^{-1}) with respect to experimental results in dependence on the expansion of the PES.

	2D		3D		4D	
Molecule	MAD	MAX	MAD	MAX	MAD	MAX
H_2CO	13.1	51.5	2.4	7.4	1.4	3.2
CH_2F_2	11.3	79.5	2.0	7.9	1.6	4.3
CHFClBr	1.5	4.0	1.1	3.5	1.1	3.2
C_2H_4	10.7	27.4	10.6	34.0	2.7	5.9
C_2H_3F	16.9	88.5	3.2	10.2	3.0	5.9
C_3H_4	24.5	107.5	4.7	10.8	2.6	6.3
B_2H_6	21.1	81.5	10.8	28.8	4.6	11.7
C_2H_6	16.9	40.4	8.7	33.8	1.5	3.6

one is tempted to use electronic structure methods of lower quality, *e.g.*, density functional theory (DFT). However, as DFT has a larger intrinsic error bar, the accuracy of the 4D results in Table 1.3 cannot be achieved by pure DFT potentials, but mixed multi-level potentials are a remedy to that,[36,37] *i.e.*, the individual terms of the n-mode expansion are evaluated at different levels of electronic structure theory. In these calculations, the 4D calculations are usually not the time-limiting step, while coupled-cluster accuracy can be retained.

A further aspect concerns the completeness with respect to an expansion of the μ-tensor within the vibrational angular momentum (VAM) terms and the Watson correction term. An n-mode expansion of the μ-tensor and a subsequent transformation of it by means of Kronecker product fitting[31] to an analytic SOP representation is computationally cheap and can be performed up to high orders. For the Watson correction term, *i.e.*, $-\frac{1}{8}\sum_\alpha \mu_{\alpha\alpha}$, the expansion usually is determined up to the same order as the corresponding n-mode expansion of the PES. It can then simply be added as a mass-dependent pseudopotential to the PES and thus be absorbed into the coefficients of the SOP representation of the PES. Once the PES has been expanded up to 3-mode or 4-mode coupling terms, the μ-tensor contributions can be considered to be converged. Moreover, as the Watson correction term usually leads to a shift of the absolute state energies that is comparable for a large range of vibrational states, its impact on transition energies remains modest in most cases. The situation alters completely for the VAM terms, which can formally be written as

$$
H_{IJ}^{(\mathrm{VAM})} = -\frac{1}{2}\sum_{\alpha\beta}\sum_{ijkl}\zeta_{ij}^{\alpha}\zeta_{kl}^{\beta}\left\langle \Phi^I(\mathbf{q})\left| q_i\frac{\partial}{\partial q_j}\mu_{\alpha\beta}(\mathbf{q})q_k\frac{\partial}{\partial q_l}\right|\Phi^J(\mathbf{q})\right\rangle,
$$

$$(1.5)$$

where ζ_{ij}^{α} denote the well-known skew-symmetric Coriolis ζ-constants with $\zeta_{ij}^{\alpha} = -\zeta_{ji}^{\alpha}$ and $\zeta_{ii}^{\alpha} = 0$. It is obvious that an SOP representation of an n-mode expansion of the μ-tensor leads to the multiplication of different 1D integrals, which are simple to evaluate. However, as the VAM operator changes its form in dependence on the equality of the mode indices, the summation of these products can be rather demanding. Moreover, the well-known commutator relationship $[\pi_\alpha, \mu_{\alpha\beta}] = 0$ does not hold for the individual terms of the n-mode expansion as this would break the hermiticity. It is thus apparent that Eq. (1.5) must be unrolled. For example, for the constant contribution, $\mu_{\alpha\beta}^0$, the term for the diagonal elements of the Hamiltonian matrix can be written as

$$-\frac{1}{2}\sum_{\alpha\beta}\mu_{\alpha\beta}^0\sum_i\sum_{j\neq i}\zeta_{ij}^\alpha\left[\zeta_{ij}^\beta\left(\frac{1}{4}+Q_{i2}^{II}\right)+\sum_{k\neq\{i,j\}}\zeta_{kj}^\beta Q_{i1}^{II}Q_{k1}^{II}T_j^{II}\right].$$

$$(1.6)$$

Clearly, this expression is much cheaper to evaluate as the number of loops as occurring in Eq. (1.5) has been reduced. Likewise expressions can be derived for the non-diagonal elements of the Hamiltonian matrix, which reduce the formal $\mathcal{O}(N^4)$ scaling to $\mathcal{O}(N^k)$, with k varying between 0 and 3 in dependence on the number of modes with different quantum numbers in their associated modals within the configurations on the lhs and rhs of the Hamiltonian matrix elements.[38] Consequently, the computational cost for adding the zeroth order terms to the Hamiltonian matrix is very modest, but increases rapidly with respect to the order of the expansion. However, for many applications it is fully sufficient to add zeroth order terms to the full Hamiltonian matrix and up to first order terms to the diagonal elements.[39] For example, for the fundamental antisymmetric stretching mode ν_3 and the bending mode ν_1 of water, the 0D VAM contributions to the transition energy amount to more than $10\,\mathrm{cm}^{-1}$ and can thus not be neglected, while the 1D VAM terms contribute less than $1\,\mathrm{cm}^{-1}$.[40] Consequently, only for highly accurate calculations, high excitations or systems with large amplitude motions this stringent approximation needs to be scrutinized. Note that second-order vibrational perturbation theory, VPT2, inherently is restricted to 0D VAM contributions which are included in the off-diagonal anharmonicity constants x_{ij}. For water the only non-vanishing VAM contribution occurs for x_{13} and leads to a correction of more than $10\,\mathrm{cm}^{-1}$ in the ν_1 and ν_3 fundamental, in close agreement with the previously quoted VCI results.

1.4.2. *Completeness of the correlation space*

The exponential increase of the correlation space with respect to the number of Hartree products or symmetry-adapted configurations is a long-standing problem and closely related to the comparable situation in electronic structure theory. Usually the correlation space is expanded in terms of excitation levels, *i.e.*, single, double, triple excitations and so on.[10,34] It is well-known from electronic structure theory that for accurate calculations the excitation level needs to exceed the order of the Hamiltonian, which due to the two-electron potential is of second order. This can roughly be transferred as a rule of thumb to VCI theory: The excitation level should at least be one order higher than the order of the PES representation. For accurate calculations, which include 4-mode coupling terms in the potential, the inclusion of sixtuple excitations, *i.e.*, VCI[6], was found to have non-negligible effects once results close to the full VCI are aimed at. This, at the same time, diminishes any size-extensivity problems arising from the truncated VCI expansion. The number of Hartree products, P, for a non-symmetric system is given as

$$P = \sum_{d}^{D} \binom{N}{d} L^d, \tag{1.7}$$

where D denotes the maximal order of the expansion, N is the number of modes and L is the number of modals per mode. Obviously L must be kept as small as possible, which can be achieved by optimized modals (see above). In order to control the VCI space in the input stream by some keywords, the parameters D, L, and the sum of quantum numbers, S, for a given Hartree product are used in many implementations. Values between 12 and 15 were found to be useful for S, but must be correlated with respect to D. While D and S are fairly unproblematic, L depends on the system. In many cases values of about 6–7 are meaningful, but when the high-lying correlating modals are no longer bound by the potential, an increase of this number may not necessarily lead to improved VCI results. In order to reduce the size of the correlation space, molecular symmetry needs to be exploited as much as possible. However, even then the correlation space for systems with less than 10 atoms may easily exceed many millions of configurations, which leads to an intractable problem if no further techniques are applied. A reduction of the order of the expansion D, as has been done in old-fashioned CISD calculations in electronic structure calculations, must be considered outdated, as many alternatives exist to overcome this problem (see below).

1.5. Configuration Selection

The simple idea of configuration selection consists in the reduction of the correlation space to those Hartree products, which are relevant for a given vibrational state. In order to keep this number as small as possible, individual VCI calculations need to be performed for all states to be determined, making use of eigenvalue solvers for single states rather than diagonalizations of the entire matrices. The challenge of all selection techniques is a reliable prediction of relevant Hartree products prior to their explicit use within the VCI calculation. Several routes have been proposed for doing so.[13, 15–20, 22, 23] Two major branches can be distinguished: (1) the first one starts with a kernel (typically the state-specific VSCF configuration) and a criterion is used to generate new configurations out of this, which will then be tested upon importance and be added to the kernel of the new iteration once it cannot be neglected. This leads to a steady increase of the correlations space until convergence being tested by intermediate eigenvalue determinations is achieved. (2) The second family of approaches avoids a criterion for the generation of new Hartree products, but starts with a large initial correlation space (see above) and simply uses a selection criterion to determine important configurations out of these. Batch algorithms and binary encoding can be used to diminish the memory requirements arising from very long configuration lists. Again, intermediate eigenvalue calculations

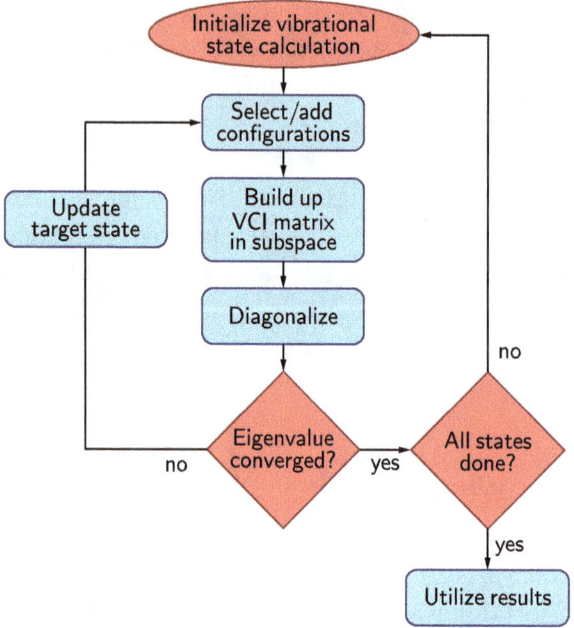

Figure 1.1.   Flow chart of an iterative configuration-selective VCI approach.

are needed within this iterative procedure, which is sketched in Figure 1.1. The latter approach, which is based on the early work of Malrieu[41] in electronic structure theory and which has been revived by Evangelista and co-workers,[42] shall exemplarily be discussed in some detail.

The criterion for selection can be based on energy contributions for a test configuration as obtained from a VMP2-based criterion or 2×2 VCI matrices, $e.g.$,

$$\Delta \varepsilon_K = \frac{1}{2} \left(\pm \Delta_K \mp \sqrt{\Delta_K^2 + 4U_K^2} \right), \qquad (1.8)$$

with

$$\Delta_K = \sum_{I,J \in \{\mathcal{A}\}}^{N_{\mathrm{sel}}} C_I C_J \langle \Phi^I | \hat{H} | \Phi^J \rangle - \langle \Phi^K | \hat{H} | \Phi^K \rangle, \qquad (1.9)$$

and

$$U_K = \sum_{I \in \{\mathcal{A}\}}^{N_{\text{sel}}} C_I \langle \Phi^I | \hat{H} | \Phi^K \rangle. \tag{1.10}$$

Essentially, the number of selected configuration, N_{sel}, increases from iteration to iteration and the criterion is based on the direct interaction of the test configuration Φ^K with respect to the VCI target state in the subspace $\{\mathcal{A}\}$ of the preceding iteration step. An energy threshold for the energy correction to the target state, $\Delta\varepsilon_K$ (*cf.* Eq. (1.8)), being in the range of $10^{-10}\,E_h$, is employed to decide about the importance of the test configuration.

As this or the related VMP2-based selection criterion is rather costly to evaluate and may dominate the calculation of state energies, additional techniques are usually employed to diminish the computational costs. (a) Once a new set of configurations $\{\mathcal{A}\}$ has been selected from the initial space, a VCI matrix in this subspace will be set up and the eigenpair of interest has to be determined. Based on the obtained eigenvector, the importance of a configuration in $\{\mathcal{A}\}$ can be checked and thrown out of the subspace to be considered in the selection process if the VCI coefficient was found to be too small. This cleaning process reduces the subspace and leads to accelerations in the next iteration. (b) It is also meaningless to test each unselected configuration individually. As many integral evaluations can be avoided on the basis of the Slater–Condon rules, it is much more efficient to base the Slater–Condon type rules on classes of configurations, *i.e.*, batches of configurations with excitation patterns within the same set of modes.[43] Hence, the concept of classes introduces jumps in the very long list of unselected configurations, which results in speed-ups within the configuration selection process by about an order of magnitude. (c) Once a VMP2-like selection criterion is used instead of the 2×2 VCI criterion displayed above, an unnormalized vector of the VMP2-like wavefunction within the space of the unselected Hartree products is a byproduct of this calculation. Significant elements of this vector, *i.e.*, corresponding to the newly selected configurations in a particular iteration step, can be used to eliminate the intermediate diagonalization step (*cf.* Figure 1.1), which is only meant to update the wavefunction for the reference state.[43] Consequently, the wavefunction for state v in the $(a+1)$th iteration step can be written in the following form:

$$\Psi_v^{(a+1)} = \sum_{K \in \{\mathcal{A}\}} C_{vK}^{(a)} \Phi^K + \sum_{J \neq K} \frac{\sum_{K \in \{\mathcal{A}\}} C_{vK}^{(a)} \langle \Phi^K | H' | \Phi^J \rangle}{\epsilon_v^{(a)} - \epsilon_J} \Phi^J$$

$$(1.11)$$

$$= \Psi_v^{(a)} + \sum_{J \neq K} C_{vK,J}^{(a+1)} \Phi^J, \tag{1.12}$$

where the summation over J refers to configurations selected in the $(a+1)$th iteration. As VMP2 does not properly deal with resonances, this simplification is only used once the iterations approach convergence, *i.e.*, all relevant resonances have already been accounted for at the VCI level in the previous iterations. At the switch from pure VCI wavefunctions for the target state to a mixed VCI/VMP2 target state, the VCI coefficients are normalized, while the additional coefficients of the VMP2-like wavefunction are not. As a consequence, this resulting reference wavefunction in the $(a+1)$th iteration is unnormalized. However, these effects were found to be very small and usually the final impact of this approximation is far less than one wavenumber for an arbitrary vibrational state, but the acceleration gained usually exceeds a factor of two for the entire computation time of high-lying vibrations.

While for fundamental vibrational states up to $2{,}000\,\mathrm{cm}^{-1}$ convergence of the configuration space usually can be achieved within 6–8 iterations, CH-stretching modes may easily require twice as many or even more iteration steps. Due to the steady increase of the Hamiltonian matrix, for which the eigenpair has to be determined, the computational effort increases with each iteration step, which renders the calculation of CH-stretching modes comparably costly. However, most fundamental states and low-lying combination bands can be described by no more than 10^5 configurations, which usually is a substantial reduction with respect to the initial configuration space consisting of millions or billions of configurations. Figure 1.2

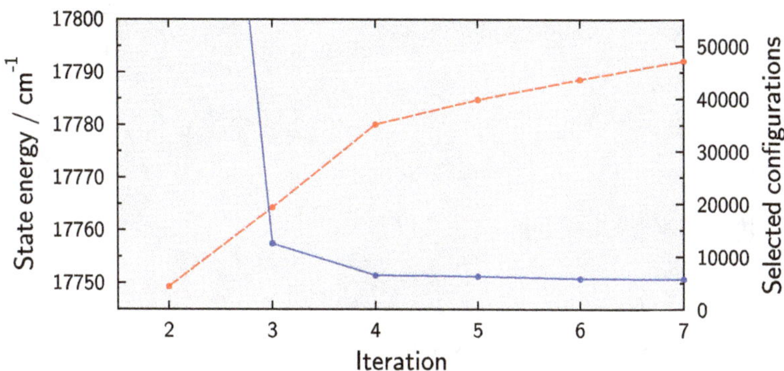

Figure 1.2. Vibrational state energy (blue) of the 13^1 CH-stretching mode of C_2H_5F and number of selected configurations (red) in dependence on the iteration step within a configuration-selective VCI calculation.

shows an example for the convergence of a configuration-selective VCI calculation. The initial space for this calculation comprised $4.3 \cdot 10^7$ Hartree products.

1.6. Eigenpair Determination and the Assignment of States

Within VCI theory vibrational quantum numbers are not good quantum numbers and thus the identity of the target state may cause an additional problem.[44–46] Once the leading VCI coefficient is very low, *e.g.*, <0.7, a wrong state may be picked, which usually is associated with poor convergence in iterative eigenvalue solvers. For that reason, the chosen eigenvalue solver needs to be able to handle such situations. As a full diagonalization of the Hamiltonian matrix is only possible for the smallest systems, this usually isn't a viable route. Likewise, eigenvalue determinations, which require the knowledge of all eigenpairs below the target state, are also problematic as, for example, CH-stretching modes already lie high up in the eigenvalue spectrum, where the state density may be very high already. Consequently, solvers are needed, which can handle a single interior eigenstate or a bunch of coupling states.[1] The Lanczos or Jacobi–Davidson eigenvalue solvers are qualified in that respect, but may fail in difficult situations and usually they

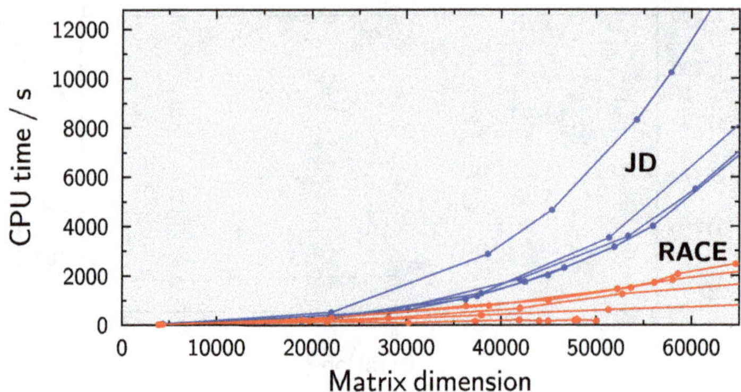

Figure 1.3. CPU times for the eigenvalue determination of the CH-stretching modes of C_2H_5F in dependence on the size of the matrix and the eigenvalue solver used.

are outperformed by the iterative residuum-based RACE algorithm, which has been designed for the situations in vibrational structure calculations.[47] The latter method is not only more reliable, but very often also faster than the Jacobi–Davidson algorithm as is shown in Figure 1.3. It is important to note that a failure of the eigenvalue solver in iterative configuration-selective VCI calculations inevitably leads to the determination of incomplete correlation spaces and thus inaccurate results. Therefore, a reliable start vector must be provided to the eigenvalue solver which, for molecules belonging to non-Abelian point groups and a correlation basis consisting of individual Hartree products rather than symmetry-adapted configurations, requires more than one entry for the VSCF or harmonic reference Hartree product.[46,48] Once the state of interest does not have a unique state identity, *e.g.*, due to strong resonances, multi-VCI approaches should be employed, *i.e.*, configuration-selective VCI approaches, which select the configurations for a set of coupling states, supported by the eigenvalue solver, which needs to provide all solutions.

A complete assignment of a vibrational state provides also the irreducible representation (irrep) of it. For Abelian molecules this is usually no problem as this information can be obtained from the symmetry properties of the underlying normal coordinates — as long as canonical normal coordinates are used (see below). However, for overtones and combination bands of non-Abelian molecules the situation alters, but the problem can be solved by using the character projection operator

$$\hat{P}^{(\mu)} = \frac{d_\mu}{g} \sum_R \chi^{(\mu)}(R)^* \hat{O}_R. \qquad (1.13)$$

The dimension of the irrep $\Gamma^{(\mu)}$ is denoted with d_μ, $\chi^{(\mu)}(R)$ is the character of the representation matrix $\mathbf{D}^{(\mu)}(R)$ and $\hat{O}_R$ is a symmetry operator of the point group. Since the irrep of $|\Psi_v\rangle$ is unique, there are two possible cases

$$\hat{P}^{(\mu)}|\Psi_v\rangle = \begin{cases} |\Psi_v\rangle \\ 0 \end{cases}, \qquad (1.14)$$

and thus the irrep of any state can be determined by screening of all possible irreps within the point group.

1.7. Unitarily Transformed Normal Coordinates

In the last decade, several research groups investigated the performance of vibrational structure calculations based on unitarily transformed normal coordinates, *i.e.*, localized or optimized normal coordinates $\tilde{\mathbf{q}}$.[49–53]

$$\tilde{\mathbf{q}} = \mathbf{U}\mathbf{q}. \qquad (1.15)$$

Different techniques have been devised to obtain these alternative sets of rectilinear coordinates and much work has been devoted to study the impact on the construction of the multidimensional PES. However, the consequences on the subsequent VCI calculations are also substantial and experience with such calculations is still limited. In the following we will focus specifically on localized normal coordinates (LNCs), but most likely similar results will hold for most other unitarily transformed sets. Localization of normal coordinates can easily be performed following the Pipek–Mezey localization scheme[54,55] within electronic structure theory, in which a delocalization measure of a vector is used to minimize the spatial extent of it. It is important to note that a full localization usually leads to poor results, while constraint localizations show the desired behaviour.[52] As a consequence, the definition of the unitary transformation matrix, $\mathbf{U}$, is a matter of debate. LNCs are not symmetry-adapted, but can introduce permutational symmetry with respect to the coordinates. Consequently, the symmetry blocking of the Hamiltonian matrix is lost, which results in larger (initial) correlation spaces and thus larger memory demands. However, with respect to CPU times, the configuration selection may overcompensate the effects of a larger correlation space and the performance may suffer less than expected. Actually, the opposite is observed in many cases and exemplary results are shown in Table 1.4 for the four CH-stretching modes of ethylene.[56] Due to the much smaller number of configurations needed in the calculations with canonical normal coordinates (NCs), the LNC calculations are much slower and any benefits in an improved calculation of the PESs go along with a worse performance in the VCI calculations. Similar trends have been observed for other molecules, but there are also few cases in which the LNC calculation shows a slightly better performance.[56] Consequently, the actual performance of the VCI calculations is difficult to predict, but the overall trend for small molecules tends to a slightly worse performance of the LNC calculations.[48]

Table 1.4. Number of Hartree products selected in a configuration-selective VCI approach for the four CH-stretching modes of ethylene in dependence on the set of coordinates.

	$N_{\text{sel}}(\text{NC})^{\text{a}}$	$N_{\text{sel}}(\text{LNC})^{\text{a}}$
ν_{11}	5,571	14,085
ν_1	8,570	16,589
ν_5	5,420	13,057
ν_9	5,064	11,547

[a]Number of selected Hartree products.
Note: The initial correlation space comprised up to quintuple excitations.

Table 1.5.　Vibrational VSCF and VCI frequencies and VCI coefficients of the CH-stretching modes of ethylene based on a PES expanded in LNCs.

Mode	Sym	VSCF	VCI	C_{11}^{a}	C_1^{a}	C_5^{a}	C_9^{a}
ν_{11}	B_{3u}	3023.3	2984.2	-0.44	0.44	0.44	-0.44
ν_1	A_g	3023.3	3023.6	0.44	0.44	0.44	0.44
ν_5	B_{1g}	3023.3	3078.6	0.47	-0.47	0.48	-0.47
ν_9	B_{2u}	3023.3	3099.5	0.48	0.48	-0.48	-0.48

[a]The coefficients C_i in one row denote the entries of the VCI vector of the respective CH-stretching vibration and refer to the VSCF Hartree products.

Another aspect concerns the state assignment. Due to the permutational symmetry of the LNCs in calculations of symmetric molecules, VCI states dominated by these coordinates lose their state identity, *i.e.*, the underlying VSCF calculations yield identical results. This, again, is shown for the four CH-stretching modes of ethylene in Table 1.5. Each VCI state is represented equally by all four Hartree products, weighted by the coefficients C_{11}, C_1, C_5 and C_9 and an unambiguous assignment on these grounds is a tedious and crror-prone task. This problem can be circumvented by evaluating the overlap integral, I_{ovlp}, of the VCI wavefunction with the corresponding multidimensional HO wavefunction.[44, 46]

$$I_{\mathrm{ovlp}} = \left\langle \prod_j \phi_j^{\mathrm{HO}}(q_j) | \sum_I C_I \Phi^I(\mathbf{Uq}) \right\rangle. \qquad (1.16)$$

However, this may be a high-dimensional integral, whose repeated evaluation is costly. Remember that the state identity is also needed for the start vector of the eigenvalue solver. Luckily, its evaluation to very high precision is not needed and it can efficiently be determined by sparse grid methods, *e.g.*, Smolyak quadrature.[46] Based on this approach, the assignment of states within the LNC framework is unproblematic, even for non-Abelian molecules, unless the multidimensional harmonic wavefunction is a poor representation of the exact one as, for example, occurring in weakly bound molecular clusters.

1.8. Incremental VCI Approaches, iVCI

Much of the increase in the accuracy of PESs is due to the exploitation of parallel computing capabilities that enable high-levels of *ab initio* calculations to be combined with up to 4-body terms of the PES expansion. At the same time, many VCI implementations that follow such highly parallelized PES calculations show only little or no parallelization for technical reasons. A common approach then is to use an embarassingly parallel implementation that distributes the vibrational states to be calculated[19, 20] (*cf.* Figure 1.1). However, such an approach is only efficient when the number of states is larger than the number of available computing units and comes with rather high memory demands. Inspired by a recent revival of CI calculations in electronic structure theory that employ a many-body expansion of the correlation energy[57–59] an incremental implementation of VCI (iVCI) has been developed[60, 61] that allows for a highly efficient parallelization combined with low memory costs and thus this approach in principle is perfectly suited for many-core architectures.

Within iVCI the correlation energy is subject to a many-body expansion and the energy E_v^{iVCI} of a vibrational state v is given by

$$E_v^{\text{iVCI}} = E_v^{\text{VSCF}} + \sum_{o=1}^{n_{\text{b}}} \sum_{[\lambda]_o \in S_o[\Lambda]_{n_{\text{b}}}} \Delta\varepsilon_v^{[\lambda]_o}, \qquad (1.17)$$

where E_v^{VSCF} is the VSCF energy and $[\Lambda]_{n_{\text{b}}} = [1, 2, \ldots, n_{\text{b}}]$ is the complete set of n_{b} body indices. The action of the S_o operator on the latter is to generate all unique (ordered) subtuples of length o, *e.g.*, $S_2[1, 2, 3]_3 = \{[1, 2]_2, [1, 3]_2, [2, 3]_2\}$. An arbitrary correlation energy increment $\Delta\varepsilon_v^{[\lambda]_o}$ is given by

$$\Delta\varepsilon_v^{[\lambda]_o} = E_v^{[\lambda]_o} - E_v^{\text{VSCF}} - \sum_{p=1}^{o-1} \sum_{[\kappa]_p \in S_p[\lambda]_o} \Delta\varepsilon_v^{[\kappa]_p}. \qquad (1.18)$$

The main task of iVCI thus consists in determining the state energies $E_v^{[\lambda]_o}$ in the configurational subspace spanned by the respective bodies.

From an algorithmic standpoint iVCI can be thought of as a shell that surrounds a conventional VCI implementation mainly connected via the configuration generator. Therefore, many of the previously outlined techniques can be combined with iVCI, especially the configuration selection and the RACE eigenvalue solver. The question then is how to define a body and different avenues have been investigated:

- The most simple definition is based on the normal coordinate index i of the VSCF modals $\varphi_i^{n_i}$ and a body b_r then comprises all modals $\{\varphi_r^{n_r}\}$ except for the modal present in the reference configuration. The subspaces spanned by such bodies contain only configurations excited in the active modes with respect to the reference configuration. Considering Eq. (1.7) the subspaces are limited in both $N = 0$ and D with $D \leq N$, *i.e.*, a three-body increment with body indices $[\lambda]_3 = [r, s, t]$ can contain only up to 3-fold excitations in the three active modes q_r, q_s and q_t.

- A second definition also uses the VSCF modals but employs the difference of the excitation level n_i with respect to the one used in the reference configuration. For example, given the reference configuration $\Phi^I = \varphi_i^1 \prod_{j \neq i} \varphi_j^0$ the first body b_1 would comprise the modals $\{\varphi_i^0, \varphi_i^2\}$ as well as the modals φ_j^1 with $j \neq i$. Such bodies then span correlation subspaces that are complete with respect to D, S and N in Eq. (1.7) but limit the value of L.

- The third definition constructs the bodies in an entirely different fashion by directly partitioning the complete space of Hartree products based on an energy criterion. To this end, the energy range covered by the correlation space is divided into n_b initially equal energy windows and all Hartree products with a summed modal energy that are within such a window are combined to form the respective body.

Each body definition has its respective strengths and weaknesses. For example, mode-based bodies can be related to the underlying n-mode expansion of the potential and thus provide physical interpretability. However, they show rather complicated convergence patterns preventing an early truncation of the expansion.[60] Considering level-based bodies one observes a better convergence but the limitations discussed for L in Eq. (1.7) now apply also to the number of bodies yielding many-body expansions that do not scale with the size of the molecule, *i.e.*, the number of vibrational modes.[61] Energy-based bodies offer a maximum of flexibility compared to the previous two definitions since the number of bodies that are employed is not limited and can be chosen almost arbitrarily. However, this definition often leads to one-body increments that comprise almost all of the relevant configuration space for low-lying vibrational states.[61] While this results in fast convergence, higher-order terms will involve diagonalization of rather large VCI matrices which might not be required at all (see below). The following discussion will focus exemplarily on the energy-based bodies.

The frequent occurrence of vibrational resonances poses an important issue in iVCI calculations, since strongly coupled configurations that belong to subspaces generated by different bodies will lead to substantial high-order corrections and thus may require long expansions. To resolve this, the explicit account of resonances in the construction of bodies is required. The problem of resonances is well known in perturbational approaches and much research has been dedicated to defining a reliable estimator.[30,62–66] A rather cheap and at the same time easily tunable approach, suggested by Krasnoshchekov *et al.*,[62] is used in iVCI:

$$\chi = \left| \frac{P_{12}}{E_1 - E_2} \right|, \qquad (1.19)$$

where P_{12} is the VPT2 coupling matrix element (*i.e.*, Fermi and Darling–Dennison coupling[66]) between interacting vibrational states with diagonal (deperturbed) VPT2 energies E_1 and E_2. Given the generated list of resonances based on a cut-off value for χ the Hartree products contained in a specific energy-based body can be checked for possible resonances and the energy windows adjusted accordingly. This frequently leads to a reduced number of bodies compared to the initial partitioning of the correlation space as well as differing numbers of bodies for individual states due to the use of state-specific modals, but to a significantly improved convergence behaviour (see below).

The contribution of individual increments to the correlation energy will vary by several orders of magnitude within a given order and thus a fraction of them may be screened out prior to their calculation. This observation is also exploited within the construction of n-mode potentials and a simple screening criterion was successfully introduced.[56] In iVCI a similar approach is being used, where the importance of an increment with indices $[\lambda]_o$ is checked against all increments $S_{o-1}[\lambda]_o$ of the previous order:

$$\max_{[\kappa]_{o-1} \in S_{o-1}[\lambda]_o} |\Delta \varepsilon_v^{[\kappa]_{o-1}}| > T_o, \tag{1.20}$$

where the threshold function T_o is defined according to

$$T_o = \begin{cases} 0 & \text{for } o < o_{\text{init}} \\ T_{\text{init}} \cdot f_{\text{rel}}^{o-o_{\text{init}}} & \text{for } o \geq o_{\text{init}}. \end{cases} \tag{1.21}$$

The pre-screening will ensure that the expansion automatically terminate at an order $o_{\max} < n_b$ once any increment has been skipped. As discussed above, such a scheme is especially beneficial for the energy-based bodies where fast convergence is observed.

As an example, results for the CH-stretching fundamentals in ethylene employing configuration-selective iVCI with up to sixtuple excitations and 12 initial energy-based bodies are displayed in Figure 1.4. Panel (a) depicts the convergence with bodies that have not been adapted for resonances, which lead to sizable contributions of varying sign up to 6-body terms. Upon inclusion of resonances ($\chi \geq 0.375$) the expansion essentially converges already in the third order (panel (d)). The middle panels (b and e) compare the theoretical limit of increments that can be screened based on a constant threshold of $0.001\,\text{cm}^{-1}$ (straight lines) and the actually screened increments (dashed lines) based on the threshold function in Eq. (1.21). Finally, the pre-screened results in panels (c) and (f) show that the screening algorithm terminates the expansion at an order that would be expected based on the unscreened results. At the same time, the difference of iVCI with respect to conventional VCI is

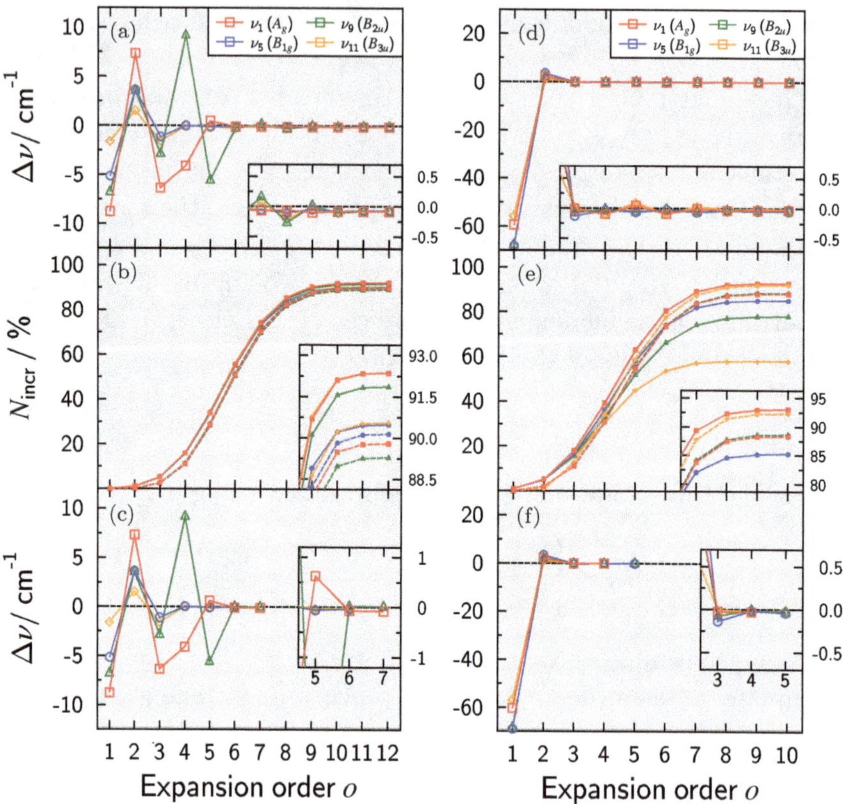

Figure 1.4.   Performance of iVCI for the CH-stretching fundamentals in ethylene. The left panel (a)–(c) depicts results obtained with energy-based bodies and on the right (d)–(f) results employing energy-based bodies optimized for resonances. The top two graphs (a & d) depict the convergence without pre-screening employed and the bottom two graphs (c & f) depict results with pre-screening. The middle row (b & e) compares cumulative relative numbers of increments with absolute energy below $0.001\,\mathrm{cm}^{-1}$ in the unscreened results (straight lines) and the actual number of screened increments (dashed lines).

minuscule compared to the residual error with respect to experiment due to the employed PES which is about 2–$5\,\mathrm{cm}^{-1}$.[56] Consequently, iVCI constitutes an interesting route to exploit modern computing architectures in an efficient manner and is a subject of current research.

1.9. Infrared Intensities

In order to guide the analysis of experimental infrared (IR) spectra and to allow for a direct comparison with these, the explicit calculation of intensities is mandatory.[67,68] Double harmonic calculations often do not suffice for this task as the transition dipole moments vanish for all but the fundamental vibrational bands and intensity borrowing in case of resonances cannot be accounted for. The integrated absorption coefficient can be written as

$$A = \frac{2\pi^2}{3} \frac{N_A}{\epsilon_0 h^2 c^2} \left| \langle \Psi'' | \boldsymbol{\mu} | \Psi' \rangle \right|^2 (E' - E'')\Delta n(T). \tag{1.22}$$

Herein the double prime denotes the reference state and a single prime the final state. $\Delta n(T)$ depends on the difference in the fraction of molecules in the respective states at a given temperature T. In the limit of $T \to 0$ this quantity will be 1. Apparently this equation necessitates the calculation of a dipole moment surface (DMS) and thus the accuracy of the IR intensities depends on both, the quality of the vibrational wavefunction and that of the DMS. As the quantum chemical calculation of a PES inevitably is based on energy points being obtained from calculations relying on Hartree–Fock or density functional theory, a low-level DMS can readily be determined in the course of the PES evaluation without additional computational cost. However, an accurate calculation of the DMS by correlated wavefunction methods, *e.g.*, coupled-cluster, requests the solution of the coupled-perturbed equations, which approximately double the CPU time demands within the construction of the surfaces. As experimentally determined intensities depend sensitively on the experimental conditions, very often a qualitative assignment instead of a quantitative one is provided, *e.g.*, medium, weak, *etc.* In that case, low-level Hartree–Fock DMSs may already suffice, while sensitive rovibrational calculations (see below) certainly request a higher level of *ab initio* calculations.[32] Table 1.6 shows a comparison of the intensities of some selected IR-active vibrational bands of ethylene in dependence of the electronic structure levels of the DMSs.

Table 1.6. VCI infrared intensities (km/mol) of selected IR-active modes of ethylene in dependence on the computational level.

Mode	Sym	GS-based[a]		State-specific[a]	
		RHF[b]	CCSD(T)[b]	RHF[b]	CCSD(T)[b]
ν_{12}	B_{3u}	14.4	9.5	14.4	9.5
ν_{11}	B_{3u}	17.4	13.8	17.1	13.6
ν_9	B_{2u}	26.0	19.6	27.6	20.7
ν_7	B_{1u}	117.1	86.7	116.8	86.5
$\nu_2 + \nu_{12}$	B_{3u}	3.4	2.2	3.6	2.4
$\nu_6 + \nu_{10}$	B_{1u}	1.0	0.8	1.0	0.8

[a]Ground state-based vs. state-specific VCI calculations.
[b]Electronic structure level of the DMS.

As the wavefunctions of the reference and final vibrational states, *i.e.*, Ψ'' and Ψ', can be retrieved from independent VCI matrices being evaluated from different sets of modals, *i.e.*, state-specific calculations, these may not necessarily be orthogonal. In consequence, mode-dependent overlap integrals $\bar{S}_k^{IJ}$ need to be evaluated, where the bar denotes the dependence from both sets of modals. The vibrational transition dipole moments can then be calculated according to

$$\langle \Psi'' | \mu_\alpha | \Psi' \rangle = p_\alpha^0 \sum_{IJ} C_I'' C_J' \prod_k \bar{S}_k^{IJ}$$

$$+ \sum_{IJ} C_I'' C_J' \sum_i \sum_r p_{\alpha i}^r \bar{Q}_{ir}^{IJ} \prod_{k \neq i} \bar{S}_k^{IJ} + \cdots . \quad (1.23)$$

Within this equation an n-mode expansion of the DMS has been assumed being represented by the coefficients p. The integrals $\bar{Q}$ correspond to the integrals in Eq. (1.2), *i.e.*, $\bar{Q}_{ir}^{IJ} = \langle \varphi_i^{I''} | q_i^r | \varphi_i^{J'} \rangle$. As the storage of these integrals does not constitute a computational bottleneck, the evaluation of state-specific IR intensities usually does not cause any problems and deviations with respect to strictly orthogonal ground state-based IR intensities were found to be very small. Note that the correlation spaces of ground state-based and state-specific calculations are not directly comparable unless they are complete. As a consequence of that, IR intensities based on ground state-based or state-specific VCI differ slightly. With that, the calculation of IR intensities based on low-level DMSs simply is a byproduct of any VCI calculation, while the highly accurate determination of intensities may be a very demanding task.

1.10. Rovibrational CI Theory, RVCI

The techniques outlined so far apply all to the case of a non-rotating $(J = 0)$ molecule. In contrast, gas-phase IR spectra of small- to medium-sized molecules show a resolvable rotational fine structure of the individual vibrational transitions. A first step towards a description of the molecular rotation can be achieved by averaging the μ-tensor over converged VCI wavefunctions[69] and accounting for the missing Coriolis contribution.[70] This approach usually leads to rotational constants that agree to within a few tenths of a percent or better with experiment,[67,69,70] depending on the level of *ab initio* theory used for the PES. However, this only provides an effective view of rovibrational spectra and no information regarding intensities of rovibrational transitions can be extracted. This requires a complete variational treatment of the rovibrational problem and very advanced programs to perform such calculations do exist.[10,71-76] The following outlines an approach to rovibrational calculations based on VCI theory.[77]

The line intensity of a rovibrational transition can be calculated according to Eq. (1.22) with E'' and E' now referring to rovibrational term energies. The thermal population difference is given by

$$\Delta n(T) = g_{\text{ns}} \frac{e^{-E''/k_{\text{B}}T} \left(1 - e^{-(E'-E'')/k_{\text{B}}T}\right)}{Z(T)}, \tag{1.24}$$

where $Z(T)$ is the partition function for the chosen temperature and g_{ns} is the nuclear spin statistical weight of the lower state which can be obtained via the Landau–Lifshitz formula.[78] The dependence of the rotational partition function on the temperature and the rotational quantum number J for ethylene is shown in Figure 1.5.

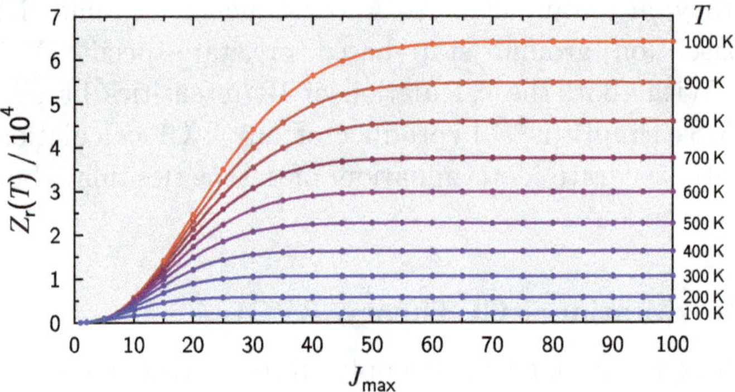

Figure 1.5.   Convergence of the rotational partition function $Z_r(T)$ for ethylene with an increasing rotational quantum number J_{max} at different temperatures.

The transition dipole moments $\langle\Psi''|\boldsymbol{\mu}|\Psi'\rangle$ are now calculated using the rovibrational wavefunctions of the involved states. Similar to the solution of the vibrational problem in VCI, the rotation vibration configuration interaction (RVCI) employs a variational *ansatz* for the wavefunction

$$|\Psi_J\rangle = \sum_v \sum_{k=-J}^{J} C_{vJk}\,|\Psi_v\rangle\,|Jkm\rangle, \qquad (1.25)$$

where $|\Psi_v\rangle$ are VCI wavefunctions. Eq. (1.25) employs the most common choice of rigid rotor functions $|Jkm\rangle$[79] as the rotational basis functions but also their Wang-type linear combinations[80] can be used. In the absence of an external field the Hamiltonian matrix elements do not depend on the quantum number m and rovibrational states therefore are $(2J+1)$-fold degenerate. Within RVCI a further optimization of the rotational basis is possible by solving the rotational problem for the vibrational ground state only. The resulting eigenfunctions then provide a better suited rotational basis for the molecule under study termed molecule specific rotational basis, MSRB,[77] which diminishes assignment problems that are prevalent in rovibrational calculations. Setting up the Hamiltonian matrix using the full Eckart–Watson operator in the chosen basis and diagonalizing it yields the desired term energies and wavefunctions. The use of compact VCI wavefunctions leads to comparably small

RVCI matrices with up to a few thousand basis functions for typical values of J and therefore only moderate memory requirements for the calculation.

The computational bottleneck in RVCI is the calculation of the squared rovibrational transition dipole moments

$$|\langle \Psi''_J |\boldsymbol{\mu}| \Psi'_J \rangle|^2 = (2J''+1)(2J'+1) \left| \sum_{\sigma=-1}^{1} (-1)^\sigma \sum_{v''v'} \langle \Psi''_v |\mu_{\mathrm{m}}^{(1,\sigma)}| \Psi'_v \rangle \right.$$

$$\left. \times \sum_{k''k'} (-1)^{k''} (C''_{vJk})^* C'_{vJk} \begin{pmatrix} J'' & 1 & J' \\ k'' & \sigma & -k' \end{pmatrix} \right|^2, \qquad (1.26)$$

where the $(2J+1)$ prefactors arise from the m-degeneracy and the last summation in Eq. (1.26) involves Wigner 3-j symbols.[79] The vibrational dipole moment integrals in spherical tensor form $\langle \Psi''_v |\mu_{\mathrm{m}}^{(1,\sigma)}| \Psi'_v \rangle$ are obtained from the molecule-fixed Cartesian components in Eq. (1.23)[81] and can be stored following the preceding VCI calculation. Even then, Eq. (1.26) has to be evaluated for a large number of transitions and involves nested summations over rovibrational basis functions in the initial and final state wavefunction. Therefore, selection rules and molecular symmetry need to be exploited as much as possible to limit the computational work.

Figure 1.6 shows an example IR stick spectrum at $300\,\mathrm{K}$ calculated for ethylene in the region of the CH-stretching fundamentals.[82,83] This spectrum has been obtained from an RVCI calculation that includes 50 vibrational states including the fundamentals, first overtones and binary combination states and a DMS that includes up to 3-mode coupling. The rotational kinetic energy term employs a μ-tensor expanded up to 2D terms and for the Coriolis coupling term a constant μ-tensor has been found sufficient. The largest RVCI matrix for $J_{\mathrm{max}} = 35$ was of dimension 3,550 which should be compared to $3 \cdot 10^{17}$, being the corresponding size in the primitive basis set used to construct the VSCF modals. For comparison experimental data from the HITRAN database[84] are depicted. The RVCI spectrum consists of about 18,000 rovibrational transitions in the depicted spectral range compared to about 6,900 lines which are

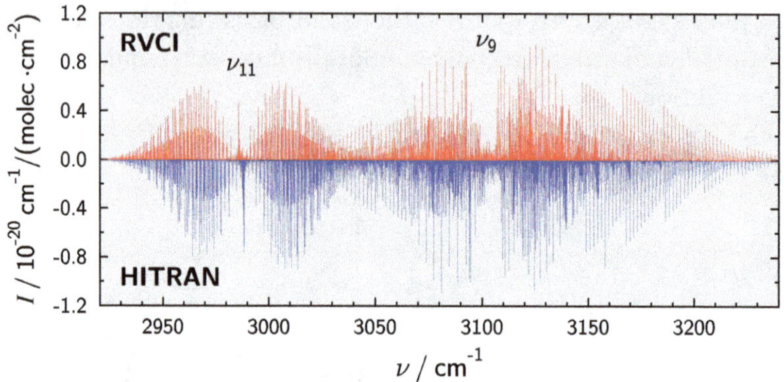

Figure 1.6. Comparison of the computed RVCI infrared spectrum of C_2H_4 in the region of the CH-stretching fundamentals with data taken from the HITRAN database.[84]

part of the HITRAN database. Aside from a small constant frequency shift due to the error in the VCI band centers both spectra show excellent agreement with respect to the overall band shapes. The RVCI calculation required only 2 h of computation time including the VCI, the RVCI wavefunction and term energy as well as the final transition moment calculations. Clearly, the almost black-box nature of RVCI allows for a fully automated accurate calculation of rovibrational spectra at low cost as a routine application for small molecules.

1.11. Summary

VCI theory has reached a state of maturity and can routinely be applied in a black-box type manner to small molecules of up to 10 atoms. However, even calculations for systems of up to 20 atoms are feasible and have already been performed, but on the cost of long computation times. It is the interplay of many different tricks and techniques which render VCI theory an attractive route for the highly accurate calculation of (ro)vibrational spectra. Consequently, there is a high need for new efficient approximations and concepts, shifting the limits of VCI theory to even larger molecules. One of them is iVCI theory, which is perfectly adapted to modern multi-core architectures. It is the open-ended accuracy of VCI theory and the fact that it is not limited to a single local minimum of the PES which renders this approach highly attractive. This allows for the pure *ab initio* calculation of reliable rovibrational spectra of molecules with more than 5 atoms, which became feasible only within the last years.[85] In any case, it must be kept in mind that in almost all applications, the VCI calculation is not the bottleneck of a spectroscopical study, but the determination of the multidimensional potential energy surface at high levels of electronic structure theory. All developments and calculations presented in this work have been implemented into and performed with the MOLPRO package of *ab initio* programs and are thus available to the scientific community.[86]

Acknowledgments

Continuous support by the Deutsche Forschungsgemeinschaft (DFG) is kindly acknowledged. Parts of this contribution have been realized within DFG project Ra 656/26-1. We are grateful for stimulating discussions with T. Mathea, S. Erfort and M. Tschöpe.

References

1. Bowman, J. M.; Carrington, T.; Meyer, H.-D. Variational quantum approaches for computing vibrational energies of polyatomic molecules. *Mol. Phys.* **2008**, *106*(16–18), 2145–2182.
2. Christiansen, O. Vibrational structure theory: new vibrational wave function methods for calculation of anharmonic vibrational energies and vibrational contributions to molecular properties. *Phys. Chem. Chem. Phys.* **2007**, *9*, 2942–2953.
3. Carney, G. D.; Sprandel, L. L.; Kern, C. W. Variational Approaches to Vibration-Rotation Spectroscopy for Polyatomic Molecules. In *Advances in Chemical Physics*, Vol. 37; Prigogine, I., Rice, S. A., Eds.; John Wiley & Sons, Inc., 1978; pp. 305–379.
4. Whitehead, R.; Handy, N. Variational calculation of vibration-rotation energy levels for triatomic molecules. *J. Mol. Spectrosc.* **1975**, *55*(1), 356–373.
5. Bowman, J. M.; Christoffel, K.; Tobin, F. Application of SCF-CI theory to vibrational motion in polyatomic molecules. *J. Phys. Chem.* **1979**, *83*, 905–912.

6. Christoffel, K. M.; Bowman, J. M. Investigations of self-consistent field, SCF CI and virtual state configuration-interaction vibrational energies for a model 3-mode system. *Chem. Phys. Lett.* **1982**, *85*, 220–224.

7. Bowman, J. M. Self-consistent field energies and wavefunctions for coupled oscillators. *J. Chem. Phys.* **1978**, *68*, 608–610 .

8. Gerber, R. B.; Ratner, M. A. Self-Consistent-Field Methods for Vibrational Excitations in Polyatomic Systems. In *Advances in Chemical Physics*, Vol. 70; Prigogine, I., Rice, S. A., Eds.; John Wiley & Sons, Inc., 1988; pp. 97–132.

9. Carter, S.; Culik, S. J.; Bowman, J. M. Vibrational self-consistent field method for many-mode systems: a new approach and application to the vibrations of CO adsorbed on Cu(100). *J. Chem. Phys.* **1997**, *107*(24), 10458–10469.

10. Carter, S.; Bowman, J. M.; Handy, N. C. Extensions and tests of MULTIMODE: a code to obtain accurate vibration/rotation energies of many-mode molecules. *Theor. Chem. Acc.* **1998**, *100*, 191–198.

11. Cassam-Chenaï, P.; Liévin, J. The VMFCI method: a flexible tool for solving the molecular vibration problem. *J. Comput. Chem.* **2006**, *27*(5), 627–640.

12. Benoit, D. M. Efficient correlation-corrected vibrational self-consistent field computation of OH-stretch frequencies using a low-scaling algorithm. *J. Chem. Phys.* **2006**, *125*(24), 244110.

13. Sibaev, M.; Crittenden, D. L. Balancing accuracy and efficiency in selecting vibrational configuration interaction basis states using vibrational perturbation theory. *J. Chem. Phys.* **2016**, *145*(6), 064106.

14. Sibaev, M.; Crittenden, D. L. PyVCI: a flexible open-source code for calculating accurate molecular infrared spectra. *Comput. Phys. Commun.* **2016**, *203*, 290–297.

15. Scribano, Y.; Benoit, D. M. Iterative active-space selection for vibrational configuration interaction calculations using a reduced-coupling VSCF basis. *Chem. Phys. Lett.* **2008**, *458*, 384.

16. Carbonniere, P.; Dargelos, A.; Pouchan, C. The VCI-P code: an iterative variation-perturbation scheme for efficient computations of anharmonic vibrational levels and IR intensities of polyatomic molecules. *Theor. Chem. Acc.* **2010**, *125*(3–6), 543–554.

17. Baraille, I.; Larrieu, C.; Dargelos, A.; Chaillet, M. Calculation of non-fundamental IR frequencies and intensities at the anharmonic level. I. The overtone, combination and difference bands of diazomethane, H_2CN_2. *Chem. Phys.* **2001**, *273*, 91.

18. Odunlami, M.; Bris, V. L.; Begue, D.; Baraille, I.; Coulaud, O. A-VCI: a flexible method to efficiently compute vibrational spectra. *J. Chem. Phys.* **2017**, *146*, 214108.

19. Gohaud, N.; Begue, D.; Darrigan, C.; Pouchan, C. New parallel software (P_Anhar) for anharmonic vibrational calculations: application to $(CH_3Li)_2$. *J. Comput. Chem.* **2005**, *26*(7), 743–754.

20. Neff, M.; Rauhut, G. Toward large scale vibrational configuration interaction calculations. *J. Chem. Phys.* **2009**, *131*, 124129.

21. Fetherolf, J. H.; Berkelbach, T. C. Vibrational heat-bath configuration interaction. *J. Chem. Phys.* **2021**, *154*(7), 074104.

22. Garnier, R.; Odunlami, M.; Bris, V. L.; Begue, D.; Baraille, I.; Coulaud, O. Adaptive vibrational configuration interaction (A-VCI): a posteriori error estimation to efficiently compute anharmonic IR spectra. *J. Chem. Phys.* **2016**, *144*, 204123.

23. Garnier, R. Dual vibration configuration interaction (DVCI). An efficient factorization of molecular Hamiltonian for high performance infrared spectrum computation. *Comput. Phys. Comm.* **2019**, *234*, 263–277.

24. Lesko, E.; Ardiansyah, M.; Brorsen, K. R. Vibrational adaptive sampling configuration interaction. *J. Chem. Phys.* **2019**, *151*(16), 164103.

25. Bhatty, A. U.; Brorsen, K. R. An alternative formulation of vibrational heat-bath configuration interaction. *Mol. Phys.* **2021**, *119*, e1936250.

26. Watson, J. K. G. Simplification of the molecular vibration-rotation Hamiltonian. *Mol. Phys.* **1968**, *15*(5), 479–490.

27. Nielsen, H. H. The Vibration-rotation Energies of Molecules and their Spectra in the Infra-red. In *Handbuch der Physik*, Vol. 37, Part I; Flügge, S., Ed.; Springer, 1959; pp. 173–313.

28. Amat, G.; Nielsen, H. H.; Tarrago, G. *Rotation-Vibration of Polyatomic Molecules*. Dekker: New York, 1971.

29. Clabo, D. A.; Allen, W. D.; Remington, R. B.; Yamaguchi, Y.; Schaefer, H. F. A systematic study of molecular vibrational anharmonicity and vibration-rotation interaction by self-consistent-field higher-derivative methods — asymmetric-top molecules. *Chem. Phys.* **1988**, *123*, 187–239.

30. Piccardo, M.; Bloino, J.; Barone, V. Generalized vibrational perturbation theory for rotovibrational energies of linear, symmetric and asymmetric tops: theory, approximations, and automated approaches to deal with medium-to-large molecular systems. *Int. J. Quant. Chem.* **2015**, *115*(15), 948–982.

31. Ziegler, B.; Rauhut, G. Efficient generation of sum-of-products representations of high-dimensional potential energy surfaces based on multimode expansions. *J. Chem. Phys.* **2016**, *144*, 114114.

32. Oschetzki, D.; Neff, M.; Meier, P.; Pfeiffer, F.; Rauhut, G. Selected aspects concerning the efficient calculation of vibrational spectra beyond the harmonic approximation. *Croat. Chem. Acta* **2012**, *85*, 379.

33. Schröder, B.; Sebald, P. High-level theoretical rovibrational spectroscopy beyond fc-CCSD(T): the C_3 molecule. *J. Chem. Phys.* **2016**, *144*(4), 044307.

34. Christiansen, O.; Luis, J. M. Beyond vibrational self-consistent-field methods: benchmark calculations for the fundamental vibrations of ethylene. *Int. J. Quantum Chem.* **2005**, *104*(5), 667–680.

35. Meisner, J.; Hallmen, P. P.; Kästner, J.; Rauhut, G. Vibrational analysis of methyl cation — rare gas atom complexes: CH_3^+-Rg (Rg = He, Ne, Ar, Kr). *J. Chem. Phys.* **2019**, *150*(8), 084306.

36. Pflüger, K.; Paulus, M.; Jagiella, S.; Burkert, T.; Rauhut, G. Multi-level vibrational SCF calculations and FTIR measurements on furazan. *Theor. Chem. Acc.* **2005**, *114*, 327–332.

37. Yagi, K.; Hirata, S.; Hirao, K. Multiresolution potential energy surfaces for vibrational state calculations. *Theor. Chim. Acta* **2007**, *118*, 681–691.

38. Mathea, T.; Rauhut, G. Advances in vibrational configuration interaction calculations — Part 1: efficient calculation of vibrational angular momentum terms. *J. Comput. Chem.* **2021**, *42*(32), 2321–2333.

39. Neff, M.; Hrenar, T.; Oschetzki, D.; Rauhut, G. Convergence of vibrational angular momentum terms within the Watson Hamiltonian. *J. Chem. Phys.* **2011**, *134*(6), 064105.

40. Carbonniere, P.; Barone, V. Coriolis couplings in variational computations of vibrational spectra beyond the harmonic approximation: implementation and validation. *Chem. Phys. Lett.* **2004**, *392*(4), 365–371.

41. Huron, B.; Malrieu, J. P.; Rancurel, P. Iterative perturbation calculations of ground and excited-state energies from multiconfigurational zeroth-order wavefunctions. *J. Chem. Phys.* **1973**, *58*, 5745.

42. Schriber, J. B.; Evangelista, F. A. Adaptive configuration interaction for computing challenging electronic excited states with tunable accuracy. *J. Chem. Theory Comp.* **2017**, *13*(11), 5354–5366.

43. Mathea, T.; Petrenko, T.; Rauhut, G. Advances in vibrational configuration interaction calculations — Part 2: fast screening of the correlation space. *J. Comput. Chem.* **2022**, *43*(1), 6–18.

44. Mátyus, E.; Fábri, C.; Szidarovszky, T.; Czakó, G.; Allen, W. D.; Császár, A. G. Assigning quantum labels to variationally computed rotational-vibrational eigenstates of polyatomic molecules. *J. Chem. Phys.* **2010**, *133*(3), 034113.

45. Šmydke, J.; Császár, A. G. Understanding the structure of complex multidimensional wave functions. A case study of excited vibrational states of ammonia. *J. Chem. Phys.* **2021**, *154*(14), 144306.

46. Mathea, T.; Rauhut, G. Assignment of vibrational states within configuration interaction calculations. *J. Chem. Phys.* **2020**, *152*(19), 194112.

47. Petrenko, T.; Rauhut, G. A new efficient method for the calculation of interior eigenpairs and its application to vibrational structure problems. *J. Chem. Phys.* **2017**, *146*, 124101.

48. Mathea, T.; Petrenko, T.; Rauhut, G. VCI calculations based on canonical and localized normal coordinates for non-abelian molecules: accurate assignment of the vibrational overtones of allene. *J. Phys. Chem. A* **2021**, *125*(4), 990–998.

49. Yagi, K.; Keçeli, M.; Hirata, S. Optimized coordinates for anharmonic vibrational structure theories. *J. Chem. Phys.* **2012**, *137*(20), 204118.

50. Yagi, K.; Otaki, H. Vibrational quasi-degenerate perturbation theory with optimized coordinates: applications to ethylene and trans-1,3-butadiene. *J. Chem. Phys.* **2014**, *140*(8), 084113.

51. Thomsen, B.; Yagi, K.; Christiansen, O. Optimized coordinates in vibrational coupled cluster calculations. *J. Chem. Phys.* **2014**, *140*(15), 154102.

52. Panek, P. T.; Jacob, C. R. On the benefits of localized modes in anharmonic vibrational calculations for small molecules. *J. Chem. Phys.* **2016**, *144*(16), 164111.

53. Klinting, E. L.; König, C.; Christiansen, O. Hybrid optimized and localized vibrational coordinates. *J. Phys. Chem. A* **2015**, *119*(44), 11007–11021.

54. Pipek, J.; Mezey, P. G. A fast intrinsic localization procedure applicable for ab initio and semiempirical linear combination of atomic orbital wave functions. *J. Chem. Phys.* **1989**, *90*(9), 4916–4926.

55. Rauhut, G. Configuration selection as a route towards efficient vibrational configuration interaction calculations. *J. Chem. Phys.* **2007**, *127*(18), 184109.

56. Ziegler, B.; Rauhut, G. Localized normal coordinates in accurate vibrational structure calculations: benchmarks for small molecules. *J. Chem. Theory Comput.* **2019**, *15*(7), 4187–4196.

57. Boschen, J. S.; Theis, D.; Ruedenberg, K.; Windus, T. L. Correlation energy extrapolation by many-body expansion. *J. Phys. Chem. A* **2017**, *121*, 836.

58. Zimmerman, P. M. Incremental full configuration interaction. *J. Chem. Phys.* **2017**, *146*, 104102.

59. Eriksen, J. J.; Gauss, J. Incremental treatments of the full configuration interaction problem. *WIREs Comput. Mol. Sci.* **2021**, *11*(5), e1525.

60. Schröder, B.; Rauhut, G. Incremental vibrational configuration interaction theory, iVCI: implementation and benchmark calculations. *J. Chem. Phys.* **2021**, *154*(12), 124114.

61. Schröder, B.; Rauhut, G. Comparison of body definitions within incremental vibrational configuration interaction theory, iVCI. Manuscript in preparation.

62. Krasnoshchekov, S. V.; Isayeva, E. V.; Stepanov, N. F. Criteria for first- and second-order vibrational resonances and correct evaluation of the Darling–Dennison resonance coefficients using the canonical Van Vleck perturbation theory. *J. Chem. Phys.* **2014**, *141*(23), 234114.

63. Krasnoshchekov, S. V.; Dobrolyubov, E. O.; Syzgantseva, M. A.; Palvelev, R. V. Rigorous vibrational fermi resonance criterion revealed: two different approaches yield the same result. *Mol. Phys.* **2020**, *118*(11), e1743887.

64. Martin, J. M. L.; Lee, T. J.; Taylor, P. R.; Franois, J. The anharmonic force field of ethylene, C_2H_4, by means of accurate *ab initio* calculations. *J. Chem. Phys.* **1995**, *103*(7), 2589–2602.

65. Davisson, J. L.; Brinkmann, N. R.; Polik, W. F. Accurate and efficient calculation of excited vibrational states from quartic potential energy surfaces. *Mol. Phys.* **2012**, *110*(19–20), 2587–2598.

66. Rosnik, A. M.; Polik, W. F. VPT2+K spectroscopic constants and matrix elements of the transformed vibrational Hamiltonian of a polyatomic molecule with resonances using Van Vleck perturbation theory. *Mol. Phys.* **2014**, *112*(2), 261–300.

67. Dinu, D. F.; Ziegler, B.; Podewitz, M.; Liedl, K. R.; Loerting, T.; Grothe, H.; Rauhut, G. The interplay of VSCF/VCI calculations and matrix-isolation IR spectroscopy — Mid infrared spectrum of CH_3CH_2F and CD_3CD_2F. *J. Mol. Spectrosc.* **2020**, *367*, 111224.

68. Oschetzki, D.; Zeng, X.; Beckers, H.; Banert, K.; Rauhut, G. Azidoacetylen — interpretation of gas phase infrared spectra based on high-level vibrational configuration interaction calculations. *Phys. Chem. Chem. Phys.* **2013**, *15*, 6719–6725.

69. Czakó, G.; Mátyus, E.; Császár, A. G. Bridging theory with experiment: a benchmark study of thermally averaged structural and effective

spectroscopic parameters of the water molecule. *J. Phys. Chem. A* **2009**, *113*(43), 11665–11678.

70. Rauhut, G. Anharmonic Franck–Condon factors for the $\tilde{X}^2B_1 \leftarrow \tilde{X}^1A_1$ photoionization of ketene. *J. Phys. Chem. A* **2015**, *119*(41), 10264–10271.

71. Carter, S.; Sharma, A. R.; Bowman, J. M.; Rosmus, P.; Tarroni, R. Calculations of rovibrational energies and dipole transition intensities for polyatomic molecules using MULTIMODE. *J. Chem. Phys.* **2009**, *131*(22), 224106.

72. Bowman, J. M.; Carter, S.; Huang, X. MULTIMODE: a code to calculate rovibrational energies of polyatomic molecules. *Int. Rev. Phys. Chem.* **2003**, *22*(3), 533–549.

73. Yurchenko, S. N.; Thiel, W.; Jensen, P. Theoretical ROVibrational Energies (TROVE): a robust numerical approach to the calculation of rovibrational energies for polyatomic molecules. *J. Mol. Spectrosc.* **2007**, *245*(2), 126–140.

74. Mátyus, E.; Czakó, G.; Császár, A. G. Toward black-box-type full- and reduced-dimensional variational (ro)vibrational computations. *J. Chem. Phys.* **2009**, *130*(13), 134112.

75. Rey, M.; Nikitin, A. V.; Tyuterev, V. G. Complete nuclear motion Hamiltonian in the irreducible normal mode tensor operator formalism for the methane molecule. *J. Chem. Phys.* **2012**, *136*(24), 244106.

76. Nikitin, A.; Rey, M.; Champion, J.; Tyuterev, V. Extension of the MIRS computer package for the modeling of molecular spectra: from effective to full ab initio ro-vibrational Hamiltonians in irreducible tensor form. *J. Quant. Spectrosc. Radiat. Transf.* **2012**, *113*(11), 1034–1042.

77. Erfort, S.; Tschöpe, M.; Rauhut, G. Toward a fully automated calculation of rovibrational infrared intensities for semi-rigid polyatomic molecules. *J. Chem. Phys.* **2020**, *152*(24), 244104.

78. Landau, L.; Lifshitz, E. *Quantum Mechanics*, 3rd ed. Pergamon: Oxford, 1977.

79. Zare, R. N. *Angular Momentum: Understanding Spatial Aspects in Chemistry and Physics.* John Wiley & Sons, Inc.: New York, 1988.

80. Wang, S. C. On the asymmetrical top in quantum mechanics. *Phys. Rev.* **1929**, *34*, 243–252.

81. Bunker, P. R.; Jensen, P. *Molecular Symmetry and Spectroscopy.* NRC Research Press: Ottawa, 1998.

82. Carter, S.; Sharma, A. R.; Bowman, J. M. First-principles calculations of rovibrational energies, dipole transition intensities and partition function for ethylene using MULTIMODE. *J. Chem. Phys.* **2012**, *137*(15), 154301.

83. Mant, B. P.; Yachmenev, A.; Tennyson, J.; Yurchenko, S. N. ExoMol molecular line lists — XXVII. Spectra of C_2H_4. *MNRAS* **2018**, *478*(3), 3220–3232.

84. Gordon, I.; Rothman, L.; Hill, C.; *et al.* The HITRAN2016 molecular spectroscopic database. *J. Quant. Spectrosc. Radiat. Transf.* **2017**, *203*, 3–69.

85. Tschöpe, M.; Schröder, B.; Erfort, S.; Rauhut, G. High-level rovibrational calculations on ketenimine. *Front. Chem.* **2021**, *8*, 1222.

86. Werner, H.-J.; Knowles, P. J.; Manby, F. R.; *et al.* The Molpro quantum chemistry package. *J. Chem. Phys.* **2020**, *152*, 144107.

Chapter 2

Vibrational Coupled Cluster Theory

Ove Christiansen

Department of Chemistry, University of Aarhus,
Langelandsgade 140, DK-8000 Aarhus C, Denmark
ove@chem.au.dk

2.1. Introduction

Coupled cluster theory has shown to be an attractive approach for computing accurate energies and wave functions in different quantum mechanical contexts. The strength of coupled cluster theory is the ability to provide approximate but accurate solutions to correlated problems. This chapter gives an introduction to the theoretical background for vibrational coupled cluster (VCC) theory and VCC response theory. Methods for computing anharmonic vibrational states and spectra are described together with key aspects for understanding the computational choices and the input/output of such computations. Finally, recent work aimed at describing time-dependent wave packet dynamics using time-dependent vibrational coupled cluster (TDVCC) and time-dependent vibrational coupled cluster with time-dependent modals (TDMVCC) will be briefly discussed.

2.2. Second Quantization for Many-Mode Systems

We consider a system with M degrees of freedom, called modes. The modes are described by some coordinates, q_m. One-mode functions, modals, are written as $\phi_{s^m}^m(q_m)$ for the modal indexed by s^m for

41

mode m. For a start we assume the basis is orthonormal

$$\langle \phi_{r^m}^m(q_m) | \phi_{s^m}^m(q_m) \rangle = \delta_{r^m s^m}, \tag{2.1}$$

and that there are N^m modals per mode. We shall later describe procedures for obtaining good basis sets for VCC computations, but at this stage it is only important to note that it is a general basis (as opposed to a special choice like a harmonic oscillator basis). To form a basis of M-mode states we consider Hartree products (HPs) constructed as products of the above one-mode wave functions

$$|\Phi_{\mathbf{s}}(\mathbf{q})\rangle = \prod_{m=1}^{M} \phi_{s^m}^m(q_m) = \phi_{s^1}^1(q_1)\phi_{s^2}^2(q_2)\cdots\phi_{s^M}^M(q_M), \tag{2.2}$$

where $\mathbf{s}$ is an index vector in the M-dimensional space. From the orthonormality of the one-mode functions follows orthonormality of the M-mode HPs

$$\langle \Phi_{\mathbf{s}} | \Phi_{\mathbf{r}} \rangle = \delta_{\mathbf{sr}} = \prod_{m=1}^{M} \delta_{s^m r^m}. \tag{2.3}$$

A general wave function in the M-mode space can now be written in the basis of HPs as

$$|\Psi(\mathbf{q})\rangle = \sum_{\mathbf{s}}^{\mathbf{N}} c_{\mathbf{s}} |\Phi_{\mathbf{s}}\rangle$$

$$= \sum_{s^1}^{N^1} \sum_{s^2}^{N^2} \cdots \sum_{s^M}^{N^M} c_{s^1 s^2 \ldots s^M} |\Phi_{s^1 s^2 \ldots s^M}(q_1, q_2, \ldots, q_M)\rangle, \tag{2.4}$$

with summation limits given by the vector of modal basis sizes, $\mathbf{N} = (N^1, N^2, \ldots, N^M)$. The exact wave function to the vibrational problem (the problem of treating N distinguishable particles in a potential) is obtainable in the limiting case of (i) a complete one-mode basis and (ii) an untruncated M-mode function in Eq. (2.4). However, the complexity related to such numerical exact quantum dynamics for many mode systems is seen already in the number of C coefficients in the above sum. Assume for simplicity that all modes are treated with a basis of the same size, N_b. Then the number of coefficients are $N_b^M = \exp(M \ln(N_b))$. This exponential scaling will make it computationally prohibitive to construct complete wave functions in the M-mode space except for small M. It is therefore decisive to develop wave function approximations that provide high accuracy with as few wave function variables as possible and in a reasonable computational time. This is where the interest in VCC arises. VCC is best formulated in terms of second quantization and following Ref. 1 we will show how such a formulation of many-mode dynamics can be constructed.

We now wish to construct a second quantization formulation that stands in one-to-one correspondence with the first quantization formulation, such that all computed quantities will be equivalent. In one-to-one correspondence to HPs we introduce occupation number vectors (ONVs). An ONV is a vector of integers, $k_{r^m}^m$. For every modal, indexed by r^m, belonging to every mode, indexed by m, $k_{r^m}^m$ specifies the occupation, and the ONV can be written as

$$|\mathbf{k}\rangle = |\{k_1^1, k_2^1, \ldots, k_{N^1}^1\}, \ldots, \{k_1^m, k_2^m, \ldots, k_{N^m}^m\},$$
$$\ldots, \{k_1^M, k_2^M, \ldots, k_{N^M}^M\}\rangle. \tag{2.5}$$

The special ONV with all elements equal to zero is denoted the vacuum state, $|\text{vac}\rangle$. Covering and extending the HP overlaps in Eq. (2.3) the inner product between two ONVs is defined by

$$\langle \mathbf{k} | \mathbf{l} \rangle = \prod_{m=1}^{M} \prod_{r^m=1}^{N^m} \delta_{k_{r^m}^m l_{r^m}^m}. \tag{2.6}$$

Having introduced the space we now turn to introduce operators beginning with creation $(a_{r^m}^{m\dagger})$ and annihilation operators $(a_{r^m}^m)$ for modal r^m of mode m, defined through

$$a_{r^m}^{m\dagger} |\ldots, k_{r^m}^m, \ldots\rangle = \sqrt{k_{r^m}^m + 1} |\ldots, k_{r^m}^m + 1, \ldots\rangle, \tag{2.7}$$

$$a_{r^m}^m |\ldots, k_{r^m}^m, \ldots\rangle = \sqrt{k_{r^m}^m} |\ldots, k_{r^m}^m - 1, \ldots\rangle. \tag{2.8}$$

The annihilation operator $a_{r^m}^m$ gives zero when applied to an ONV with $k_{r^m}^m = 0$, and for example, $a_{r^m}^m |\text{vac}\rangle = 0$. The following commutator relations can be derived:[1]

$$[a_{r^m}^{m\dagger}, a_{s^{m\prime}}^{m\prime\dagger}] = [a_{r^m}^m, a_{s^{m\prime}}^{m\prime}] = 0, \tag{2.9}$$

$$[a_{r^m}^m, a_{s^{m\prime}}^{m\prime\dagger}] = \delta_{mm\prime} \delta_{r^m s^m}. \tag{2.10}$$

The ONV space is a large index space where typically the physical interesting states will be in the subspace corresponding to M-mode HPs, *i.e.*, those ONVs where only one $k_{i^m}^m$ is equal to 1 and all others modals 0 for each of the M modes. In the context of VCC we will often choose a particular ONV and generate the rest of the physically relevant M-mode basis from this reference ONV. Accordingly we define a reference described in terms of an index vector $\mathbf{i} = (i^1, i^2, \ldots, i^M)$,

$$|\Phi\rangle = \prod_{m=1}^{M} a_{i^m}^{m\dagger} |\text{vac}\rangle . \tag{2.11}$$

Ordering the indices from 0, to $N^m - 1$, for each mode and considering the special case of the harmonic oscillator basis the reader may find comfort in finding the ground state of an M-dimensional harmonic oscillator to be represented in terms of a zero vector, $\mathbf{i} = (0, 0, \ldots, 0)$. However, the choice of reference is general and covers any other index vector for any type of one-mode basis set, including anharmonic ones. We use the convention that indices i^m denote modals occupied in the reference while a^m, b^m, c^m denote unoccupied modals, and r^m, s^m general or unspecified modals. To generate other M-mode basis states, the excitation operators

$$\tau_{\mathbf{a^m}}^{\mathbf{m}} = \prod_{m \in \mathbf{m}} a_{a^m}^{m\dagger} a_{i^m}^m, \tag{2.12}$$

are introduced, indexed by the modes excited and the modals occupied in the resulting state. Here the so-called mode combination (MC) $\mathbf{m}$ is the set of modes which are promoted out of the reference ONV. The vector $\mathbf{a^m} = (a^{m_1}, a^{m_2}, \ldots)$ specifies what modals the operator excites to for the particular modes in play. For the case of one- and two-mode excitation operators we have, for example, MCs with only one or two modes respectively:

$$\tau_{a^{m_1}}^{m_1} = a_{a^{m_1}}^{m_1\dagger} a_{i^{m_1}}^{m_1}, \tag{2.13}$$

$$\tau_{a^{m_1} a^{m_2}}^{m_1, m_2} = a_{a^{m_1}}^{m_1\dagger} a_{i^{m_1}}^{m_1} a_{a^{m_2}}^{m_2\dagger} a_{i^{m_2}}^{m_2}. \tag{2.14}$$

As there is only one occupied modal for each mode, the corresponding i^m index is almost always omitted in this case as the reference state is usually clear from the context. We will often use a compound index specifying just some excitation, either completely general, μ, general for a given MC, $\mu^{\boldsymbol{m}}$, or general for a given excitation level j, μ_j. Following from Eqs. (2.9)–(2.10) and Eq. (2.7) the excitation operators satisfy a few simple relations that are extremely important for the efficient formulation of VCC namely

$$[\tau_\mu, \tau_\nu] = 0, \tag{2.15}$$

$$\langle \Phi | \tau_\nu = 0 \iff \tau_\mu^\dagger | \Phi \rangle = 0. \tag{2.16}$$

It is important to note the difference between the type of second quantization presented here and the ladder operators of the harmonic oscillator. As stated from the outset, the present formulation is completely general with respect to basis and indeed later we shall typically not use harmonic oscillator basis sets, though it is fully possible. The ladder operators refer to harmonic oscillators, and they work in a different way. For example, a step-down operator, as the name indicates, takes a harmonic oscillator state $|3\rangle_{\mathrm{ho}}$ to $|2\rangle_{\mathrm{ho}}$ while the same operator applied to $|2\rangle_{\mathrm{ho}}$ takes it to $|1\rangle_{\mathrm{ho}}$. Here $|n\rangle_{\mathrm{ho}}$ are harmonic oscillator states, *i.e.*, they are not directly ONV kets, but the relation for the given basis is simple (*i.e.*, $|1\rangle_{\mathrm{ho}} = |0, 1, 0, \ldots\rangle_{\mathrm{ONV}}$, *etc.*) The present many-mode second quantization, when applied to a single mode and skipping the mode index, would work with annihilation operators for each quantum level, *i.e.*, $a_2 |2\rangle_{\mathrm{ho}} = a_2 |0, 0, 1, \ldots\rangle_{\mathrm{ONV}} = |\mathrm{vac}\rangle$ while $a_2 |1\rangle_{\mathrm{ho}} = 0$, and similarly $a_1 |2\rangle_{\mathrm{ho}} = 0$ while $a_1 |1\rangle_{\mathrm{ho}} = |\mathrm{vac}\rangle$. Do note, however, that the step down effect for a given state can be achieved by a combination of creation and annihilation operators. For example, $a_1^\dagger a_2 |2\rangle = |1\rangle$ but the same operator applied to any basis state other than $|2\rangle$ gives zero, and it is thus very far from being a step-down operator. Rather it is just an example of a transfer operator, $E^m_{rm\,sm} = a^{m\dagger}_{rm}a^m_{sm}$, and the excitation operators above are just another specialization.

Let us finally also dwell briefly on the MC notation and a touch of set logic, because this is a convenient notation we shall use throughout. In addition to MCs we also frequently use mode-combination ranges (MCRs), or in other words, families of MCs meaning MCRs are just sets of sets. For example, if there are three modes in the system (1,2,3) we have the complete MCR of all possibly non-equivalent and non-empty sets as $\{\{1\}, \{2\}, \{3\}, \{1,2\}, \{1,3\}, \{2,3\}, \{1,2,3\}\}$. We will frequently talk of the MCR for an operator O (Hamiltonian, cluster operator, *etc.*) as MCR$[O]$ by which we mean the set of included MCs for the operator. We sometimes let the number of indices of an MC be indicated by a subscript (if no confusion can occur), *i.e.*, $\mathbf{m}_k = \{m_1, m_2, \ldots, m_k\}$. Then MCR$[O]$ can accordingly be seen as the union of MCR$[O_k]$ for the different included k-mode MCs. Typically an MC refers to a physical mode-coupling (in the Hamiltonian or the wave function) of the modes in the MC. Neglecting MCs with more than n indices is a common approximation. Including only up to two mode couplings in our example thus excludes $\{1,2,3\}$. In this case only one MC was neglected, but from basic combinatorics the number of k-mode MCs for an M-mode system is $\binom{M}{k}$ and the simplication of including only up to n-mode MCs versus all MCs can therefore be very significant when $n \ll M$.

2.3. The Hamiltonian

Consider a sum of products form of the vibrational Hamiltonian

$$H = \sum_t c_t \prod_{m \in \mathbf{m}^t} h^{mo^{tm}} = \sum_t c_t \prod_{m \in \mathbf{m}^t} \sum_{r^m, r^m} h^{mo^{tm}}_{r^m s^m} a^{m\dagger}_{r^m} a^m_{s^m}. \quad (2.17)$$

Here $\mathbf{m}^t$ is the MC inluding only active modes for the term t, and o^{tm} is used to index the one-mode operators for each mode. The expression after the first equality covers in principle both first and second quantization. The final expression is a second quantization representation that ensures a one-to-one correspondence between all

M-mode integrals in first and second quantization with one-mode integrals $h_{r^m s^m}^{m o^{tm}} = \int \phi_{r^m}^m(q_m) h^{m o^{tm}} \phi_{s^m}^m(q_m) dq_m$. Matrix elements between two HPs/ONVs are obtained as a sum of products of 1D integrals for the active modes

$$\langle \Phi_{\mathbf{r}} | H | \Phi_{\mathbf{s}} \rangle = \sum_t c_t \prod_{m \in \mathbf{m}^t} h_{r^m s^m}^{m o^{tm}} \prod_{m \notin \mathbf{m}^t} \delta_{r^m s^m}. \qquad (2.18)$$

The sum of products Hamiltonian is very useful in practice, and current implementations of VCC employ this format to achieve efficiency. Second quantization analogues of any other operator can equally well be developed using similar procedures.[1] Thus, there is no theoretical issue in employing such operators with VCC, only significant practical issues. The potential energy surface (PES) part of the Hamiltonian is not known exactly and have to be computed and represented approximately in any case. This suggests using the convenient and flexible sum of products form from the outset. The sum of products form naturally covers Taylor expansions where the $h^{m o^{tm}}$ operators from the PES are polynomials q_m^j. A more general PES will have to involve computation of a grid of electronic structure points followed in most cases with fitting or interpolation. Consider the important case where we write a PES expanded in mode couplings as

$$V(\mathbf{q}) = \sum_k \sum_{\mathbf{m}_k \in \mathrm{MCR}[V_k]} \bar{V}^{\mathbf{m}_k}(q_{m_1}, q_{m_2}, \ldots, q_{m_k}). \qquad (2.19)$$

Here MCR[V] is the set of MCs included in the potential and divided into subsets MCR[V_k]. The $\bar{V}^{\mathbf{m}_k}$ potentials are k-dimensional sub-potentials for the k modes in $\mathbf{m}_k$ and can be constructed from (but is not directly equal to) cuts of the PES for these modes.[2–5] Including all possible MCs with up to at most n modes, *i.e.*, including all $\mathbf{m}_k$ with $k \leq n$ we have a so-called n-mode representation.[2,3,5] Each of the coupling terms must in turn be represented in some way, either through a grid or by fitting it to a basis. We have, for example, developed an approach where a set of analytical functions, a fit-basis $\{h^{m,o^m}; o^m = 1, 2, \ldots, O^m\}$, can be provided on input for each mode and all mode-couplings fitted accordingly to[4]

$$\bar{V}^{\mathbf{m}_k}(q_{m_1}, q_{m_2}, \ldots, q_{m_k}) = \sum_{\mathbf{o}_k^{\mathbf{m}_k}} c_{\mathbf{o}_k^{\mathbf{m}_k}}^{\mathbf{m}_k} \prod_{m \in \mathbf{m}_k} h^{m,o^m}. \tag{2.20}$$

Similar expansions are possible for other operators including (parts of) the kinetic energy operator, if it is not already in a sufficiently simple approximate form. The outcome of all this is a Hamiltonian that can be written as

$$H = \sum_{k=0}^{n} \sum_{\mathbf{m}_k} H^{\mathbf{m}_k} = \sum_{k=0}^{n} \sum_{\mathbf{m}_k \in \mathrm{MCR}[H_k]} \sum_{\mathbf{o}_k^{\mathbf{m}_k}} c_{\mathbf{o}_k^{\mathbf{m}_k}}^{\mathbf{m}_k} \prod_{m \in \mathbf{m}_k} h^{m,o^m}, \tag{2.21}$$

which is approximate if $n < M$. The contribution from $k = 0$ is only a constant and is for many cases unimportant, at least for the wave function determination. The Hamiltonian is now given in terms of a limited set of one-mode operators (h^{m,o^m}) and the coefficients, $c_{\mathbf{o}_k^{\mathbf{m}_k}}^{\mathbf{m}_k}$. The key characteristics of the operator's complexity is given by n and the number of one-mode operators for each mode, for simplicity, denoted O if they are similar. In the context of implementation the sum of products form is attractive as this form allows us to work with the Hamiltonian in the various steps of our wave function computation through the action of one-mode operators separately. The PES in the form of Eq. (2.21) can be computed by automatic procedures employing the adaptive density guided approach (ADGA)[4,6,7] for automatic, accurate and economical grid determination, double-incremental approach for large systems achieving linear scaling total PES construction cost,[5,8] and Gaussian progress regression for boosting efficiency,[9,10] all in a form ready for VCC computation.

2.4. Vibrational Self-consistent Field Theory

In the previous section, we discussed a general basis and reference state without being concrete on possible choices. An attractive basis and reference state for a VCC computation can be obtained from a vibrational self-consistent field (VSCF) computation. The VSCF method itself is historic[11–13] and we focus only on the essentials for understanding the use of VSCF for providing a basis for the following VCC.

In VSCF we choose a single configuration ansatz for the wave function and optimize the modals entering into the wave function, all in analogy to Hartree–Fock theory for electrons. The VSCF state can in second quantization (SQ) be written exactly as the reference ket in Eq. (2.11). The VSCF state is obtained by variationally optimizing the VSCF energy, $E = \langle \Phi | H | \Phi \rangle$, with respect to the M modals given by the occupation vector $\mathbf{i}$. The occupied modals for the optimal state are eigenfunctions of the mean field operators[14]

$$F^m_{r^m s^m} = \left\langle \Phi \left| \left[\left[a^m_{r^m}, H \right], a^{m\dagger}_{s^m} \right] \right| \Phi \right\rangle.$$
(2.22)

Consider now the case where we expand all VSCF modals in a set of primitive basis functions, $\xi^m_{\nu^m}$,

$$\phi^m_{r^m}(q_m) = \sum_{\nu^m} C^m_{\nu^m r^m} \xi^m_{\nu^m}.$$
(2.23)

The set of harmonic oscillator functions is one obvious possibility for a primitive basis, but there are many others. We have found that B-splines[15] are convenient in our VSCF and VCC computations because they are flexible, local and allow accurate numerical computation of all types of $h^{m,o^{tm}}_{\mu^m \nu^m} = \langle \xi^m_{\mu^m} | h^{m,o^{tm}} | \xi^m_{\nu^m} \rangle$ integrals needed.[16] These features are highly useful in particular for adaptive PES construction. For example, the local nature of B-spline functions means that one can increase the number of basis functions in some domain of space without necessarily expanding the domain covered, unlike what would be the case for a harmonic oscillator basis. Thereby we can increase accuracy while avoiding using the PES in regions where it is undetermined/problematic due to limited data/grid points.

We express the VSCF eigenvalue equation in the primitive basis $\xi^m_{\nu^m}$ as

$$\mathbf{F}^{m,\mathrm{prim}} \mathbf{C}^m = \mathbf{S}^m \mathbf{C}^m \epsilon^m.$$
(2.24)

where $\mathbf{F}^{m,\text{prim}}$ is the matrix representation of the mean field operator in the primitive basis and $S^m_{\mu^m \nu^m} = \langle \xi^m_{\mu^m} | \xi^m_{\nu^m} \rangle$ is the overlap matrix. Solving this set of equations self-consistently for all the occupied modals of all modes we obtain the VSCF modals. By computing all eigenvalues and eigenvectors of Eq. (2.24) for all modes we obtain a full set of VSCF modal energies $\epsilon^m_{q^m}$ (the diagonal elements of the diagonal matrices $\boldsymbol{\epsilon}^m$) and VSCF modals (from Eq. (2.23) with the coefficients matrices $\mathbf{C}^m$). If we use N^m_{prim} primitive basis functions we also have N^m_{prim} modals on output while to define the VSCF state we only needed the occupied modal indexed i^m. The other modals can be regarded as virtual modals and are indexed by a^m following the earlier convention for excitations. If there is no interaction between the modes of the system, VSCF is exact, and the only approximation will be due to the primitive basis. Even if the one-mode part is highly anharmonic the VSCF modals can fully describe the one-mode anharmonicity. In addition they are based on a mean-field description of the interaction between modes. Therefore the VSCF modals form an excellent choice of basis functions and are much more attractive than, for example, harmonic oscillator functions. It is a computationally simple task to transform all h^{m,o^m} one-mode integrals from the primitive basis to the final VSCF basis. A fairly large primitive basis can be used for the VSCF part, since VSCF computations are extremely fast and there is no need to be restrictive in this step. The basis for the correlated VCC computation can subsequently be restricted to the N^m energetically lowest VSCF modals of a ground state VSCF computation. Thus, N^m_{prim} can be of order 100 or more (often determined by some automatic procedure) without any problems, while a limited modal basis size N^m of order 10 is often sufficient depending of course on the case. This approach thereby combines the attractive features of the primitive basis and the anharmonic VSCF basis.

2.5. Vibrational Coupled Cluster Theory

The VSCF wave function consists of a single HP and to proceed beyond VSCF in accuracy we need to include more. Including more

HPs through a linear variational approach we obtain vibrational configuration interaction (VCI) wave functions.[17–20] In principle we already have such a general linear wave function ansatz in Eq. (2.4) and we now simply choose VSCF modals for constructing the HPs and introduce approximate VCI wave functions by imposing limitations in the sum over configurations. For the purpose of introducing a hierarchy of excitation level VCI and VCC wave functions we write the wave function expansion in terms of a reference state and exitation operators and employing intermediate normalization. This particular way is not completely standard for VCI but useful for comparing wave functions. We write accordingly

$$|\text{VCI}\rangle = \sum_{\mathbf{s}} c_{\mathbf{s}} \Phi_{\mathbf{s}} = (1+C)\,|\Phi\rangle = (1+C_1+C_2+C_3\cdots)\,|\Phi\rangle\,. \quad (2.25)$$

Here the configuration interaction generating operator, C, is introduced, which can be written in various levels of detail

$$C = \sum_{\mu} C_{\mu} \tau_{\mu} = \sum_{\mathbf{m}\in\text{MCR}[C]} C^{\mathbf{m}} = \sum_{k} \sum_{\mathbf{m}_k\in\text{MCR}[C_k]} C^{\mathbf{m}_k}$$

$$= C_1 + C_2 + C_3 + \cdots \quad (2.26)$$

The first expression for C is simply given in terms of general excitations out of the reference, while in the following we use the MC notation where MCR$[C]$ is the set of MCs included in the expansion. In the third form we sum explicitly over excitation levels, and accordingly divide MCR$[C]$ into a union of MCR$[C_k]$ for $k = 1, 2, \ldots, M$. The configuration interaction operator $C^{\mathbf{m}_k}$ excites all k modes in $\mathbf{m}_k$ in all possible ways

$$C^{\mathbf{m}_k} = \sum_{\mathbf{a}^{\mathbf{m}_k}} C_{\mathbf{a}^{\mathbf{m}_k}}^{\mathbf{m}_k} \tau_{\mathbf{a}^{\mathbf{m}_k}}^{\mathbf{m}_k}, \quad (2.27)$$

where $\mathbf{a}^{\mathbf{m}_k} = a^{m_1}, a^{m_2}, \ldots, a^{m_k}$.

Including all possible excitations the full vibrational configuration interaction (FVCI) wave function is obtained, which would be the exact wave function for the given Hamiltonian and one-mode basis, but suffering from an exponential N_b^M scaling of the number of C coefficients with the number of modes. Approximations can be introduced in a systematic way as VCI[n] methods defined by including all excitations exciting at most n-modes at a time, or similarly, including only MCs with at most n modes. The hierarchy VCI[1], VCI[2], VCI[3], ..., VCI[M] converges by definition to FVCI.

We now turn to VCC theory where we employ from an exponential wave function ansatz

$$|\text{VCC}\rangle = \exp(T)\,|\Phi\rangle\,. \tag{2.28}$$

The cluster operator, T, is defined in analogy to above as

$$T = \sum_{\mu} T_{\mu}\tau_{\mu} = \sum_{\mathbf{m}\in\text{MCR}[T]} T^{\mathbf{m}} = \sum_{k}\sum_{\mathbf{m}_k\in\text{MCR}[T_k]} T^{\mathbf{m}_k}$$

$$= T_1 + T_2 + T_3 + \cdots \tag{2.29}$$

$$T^{\mathbf{m}_k} = \sum_{\mathbf{a}^{\mathbf{m}_k}} t^{\mathbf{m}_k}_{\mathbf{a}^{\mathbf{m}_k}} \tau^{\mathbf{m}_k}_{\mathbf{a}^{\mathbf{m}_k}}\,. \tag{2.30}$$

In this expression, MCR[T] is used to specify the MCs included in the cluster expansion. The parameters $t^{\mathbf{m}_k}_{\mathbf{a}^{\mathbf{m}_k}}$ are denoted cluster amplitudes. We now introduce the hierarchy of VCC[n] methods where $n = 1, 2, \ldots, M$. On convergence, *i.e.*, for $n = M$, MCR[T] includes all possible MCs for a given system, and Eq. (2.28) is merely an alternative parameterization of the FVCI wave function which we may denote full vibrational coupled cluster (FVCC). The hierarchy of VCC[n] methods is analogously to CCS, CCSD, CCSDT, *etc.*, in electronic structure theory, but for brevity the [n] notation is often used instead of S,SD,SDT,SDTQ, *etc.*

To do computation with the VCC approach we must determine the cluster amplitudes. We introduce the VCC ansatz into the Schrödinger equation and transform with the operator $\exp(-T)$,

$$\exp(-T)H\exp(T)\,|\Phi\rangle = E\,|\Phi\rangle\,. \tag{2.31}$$

We note that the action of $\exp(\pm T)$ has well defined meaning from the SQ formulation introduced previously. Projection onto the reference state and the set of excitations included in the excitations space, $\langle \mathbf{a^m}| = \langle \Phi| \tau_{\mathbf{a^m}}^{\mathbf{m}}$, gives

$$E_{\mathrm{VCC}} = \langle \Phi| H \exp(T) |\Phi\rangle , \tag{2.32}$$

$$0 = e_{\mathbf{a^m}}^{\mathbf{m}} = \langle \mathbf{a^m}| \exp(-T)H \exp(T) |\Phi\rangle . \tag{2.33}$$

The cluster amplitudes are thus determined from solving Eq. (2.33) and the energy are in turn computed using Eq. (2.32). The amplitude equations are complicated, coupled and highly non-linear equations. These are solved in an iterative manner considering the VCC equations in Eq. (2.33) as an error vector, $\mathbf{e(t)}$, depending on the amplitudes collected in a vector $\mathbf{t}$.

2.5.1. *Iterative solution of the VCC equations*

A first-order Taylor expansion of the $\mathbf{e(t)}$ vector for the amplitudes of a given iteration, $\mathbf{t}^n$, around that of the previous iteration gives

$$\mathbf{e}^{(n)} = \mathbf{e}(\mathbf{t}^{(n)}) = \mathbf{e}(\mathbf{t}^{(n-1)}) + \mathbf{A}\Delta\mathbf{t}^{(n)} + \cdots , \tag{2.34}$$

where the important VCC Jacobian matrix has been introduced

$$A_{\mu\nu} = \frac{\partial e_\mu(t)}{\partial t_\nu} = \langle \mu| \exp(-T)[H, \tau_\nu] \exp(T) |\Phi\rangle . \tag{2.35}$$

Aiming at solving for $\mathbf{e}^{(n)} = \mathbf{0}$ and assuming higher order terms are negligble a set of linear equations for $\Delta\mathbf{t}^{(n)}$ is obtained

$$\mathbf{A}\Delta\mathbf{t}^{(n)} = -\mathbf{e}(\mathbf{t}^{(n-1)}). \tag{2.36}$$

These linear equations can be solved using iterative methods to obtain the amplitude update, $\Delta\mathbf{t}^{(n)}$. This is possible and has been implemented, but this double iterative full Newton approach (iterating in the non-linear equations with iterative solution for the step vector) is comparatively expensive. Instead we briefly exploit a perturbational point of view taking the diagonalized VSCF mean field Hamiltonian as a zeroth-order Hamiltonian

$$F = \sum_m F^m = \sum_m \sum_{r^m s^m} F_{r^m s^m}^m a_{r^m}^{m\dagger} a_{s^m}^m = \sum_m \sum_{r^m} \epsilon_{r^m}^m a_{r^m}^{m\dagger} a_{r^m}^m . \tag{2.37}$$

Note that $F|\Phi\rangle = \sum_m \epsilon_{im}^m |\Phi\rangle$, *i.e.*, the zeroth-order amplitudes are equal to zero. From a little SQ algebra follows that the zeroth-order Jacobian where the Hamiltonian is replaced by F is diagonal and simple to evaluate from VSCF modal energy differences

$$A^{\mathbf{m},\mathbf{m}'(0)}_{\mathbf{a^m b^{m'}}} = \langle \mathbf{a^m} | \exp(-T)[F, \tau^{\mathbf{m}'}_{\mathbf{b^{m'}}}] \exp(T) |\Phi\rangle$$

$$= \delta_{\mathbf{m},\mathbf{m}'} \prod_{m \notin \mathbf{m}} \delta_{a^m b^m} \left(\sum_{m \in \mathbf{m}} (\epsilon_{a^m}^m - \epsilon_{im}^m) \right). \qquad (2.38)$$

With this diagonal approximate Jacobian it is now trivial to find the update vector $\Delta\mathbf{t}^{(n)}$ from Eq. (2.36). Furthermore, acceleration methods such as DIIS[21] or CROP[22] provide additional efficiency gains achieving the solution in fewer steps and thereby fewer error-vector evaluations.[23]

2.5.2. *Implementation and computational scaling*

VCC computations can be carried out with the MidasCpp code.[24] Assuming we have computed a suitable PES, a VCC computation requires us merely to run a VSCF computation with the chosen PES for the given state, choose the number of VSCF modals per mode to use in the VCC computation, and give convergence method and convergence thresholds (suitable defaults exist). Since essentially all computational time lies with the error-vector evaluation, the efficient implementation of this is decisive, but also extremely challenging in general. This is perhaps not surprising in view of the very many years of research invested in implementing electronic structure CC methods at even quite low excitation levels. Note also that the electronic Hamiltonian is limited to one- and two-body terms while here we can have and often will have higher coupling levels in the Hamiltonian.

It is impossible to enter into details here and we refer to the original literature.[25–29] Overall, different routes to implementing VCC exist. One is to derive and code all terms explicitly by hand. This is fully possible for VCC[2] restricted to Hamiltonians with at most two-mode couplings.[28] However, for example, VCC[4] with a Hamiltonian including up to four-mode terms, requires evaluation of 1852 individual terms. Explicit derivation, analysis and implementation by hand is clearly not realistic in general. Very general implementations have been made based on a transforming VCC to extended VCI formats[25] or even representing all operators in full matrix form.[30] These approaches lead to codes that are fairly inefficient, but have the gain of allowing very general and very safe implementations of new methods, and were used, respectively, in the very first implementation of VCC,[25] and for exploring advanced alternative parameterizations such as extended VCC.[30]

The acceptable solution with respect to both efficiency and generality has so far been to write computer code for all of the following steps: (i) automatic derivation of the detailed equations for all contributions, (ii) automatic analysis for identifying intermediates required for achieving low computational scaling, (iii) automatic evaluation of the final expressions.[26,27] The first step employs the second quantization analysis and commutation rules above. In the second step the sum-over-product Hamiltonian in Eq. (2.21) is essential, allowing the individual factors for different modes to be applied separately and making reshuffling of summations possible for obtaining low operation counts. We refer to Ref. 26 for details on the overall scaling propertics with respect to the computational decisive parameters M, N_b and O. In summary, for VCC[2]/H2, VCC[3]/H3, VCC[4]/H4, VCC[5]/H5 where Hn denotes a Hamiltonian with only up to n-mode couplings, we have M^3, M^4, M^6 and M^7 scaling. The scaling with N_b and O is less decisive as N_b and O do not increase with the size of the system, but play a role for the absolute speed of a given computation. For both, the scaling is low order polynomial (combined with the M scaling we have, for example, $M^3 N_b^2$ and $M^2 N_b^3$ for VCC[2]/H2). The number of terms in the Hamiltonian depends on O with a leading term in M and O for an n-mode expansion as $\binom{M}{n} O^n$, but the overall VCC computational cost does not scale directly with the number of terms of the Hamiltonian, as the relevant M and O summations enter the automatically constructed intermediates. Still O^k scaling with some k ($1 \leq k \leq n$) are found. We furthermore note that VCI is also covered with the same implementation, and it was confirmed that VCI[n]/Hm and VCC[n]/Hm have similar leading-order scaling properties, but VCC[n]/Hm has a larger number of terms.

2.5.3. *Approximate VCC models from perturbational arguments*

Besides configuration interaction, perturbation theory obviously also has a long history in the context of application to vibrational problems. While many applications of perturbation theory depart from the harmonic oscillator solution, one can also build a Møller-Plesset perturbation (VMP) theory using VSCF as the zeroth-order state.[31,32] However, as discussed before in the solution of the VCC equations, one can analyse the terms in the VCC equations with respect to expected importance as judged from a perturbational analysis. Before this was just part of developing convenient numerical methods for solving the equations and the results of the VCC[n] computations are independent of any sort of perturbation theory. However, now we shall describe methods which select some terms for inclusion while excluding others, to define iterative VCC methods intermediate between the pure VCC[n] models in accuracy and computational cost.

Consider a Hamiltonian including up to three-mode couplings, $H = H_1 + H_2 + H_3$. We partition this Hamiltonian into a mean field part and a remainder in such a way that the mode-coupling level origins of the terms are kept. Thus, we write the Hamiltonian as

$$H = F^{(0)} + U_2^{(1)} + U_3^{(2)}, \tag{2.39}$$

$$F^{(0)} = H_1 + F_2 + F_3, \tag{2.40}$$

where F_2 and F_3 are the mean-field operator parts corresponding to the two- and three-mode parts of the Hamiltonian, respectively, and correspondingly $U_2^{(1)} = H_2 - F_2$ and $U_3^{(2)} = H_3 - F_3$. An order[33] has been assigned to the U operators. In the VCC[2pt3] model all VCC[2] terms for the given Hamiltonian is retained. Terms involving three-mode excitations are restricted to only those of lowest non-vanishing order in the above order counting with the additional rule that singles excitations are treated as zeroth-order parameters. The latter rule is enforced to obtain good response functions and excitation energies and is implemented in theory and practice by using T_1-similarity transformed Hamiltonians

$$\tilde{H} = \exp(-T_1) H \exp(T_1). \tag{2.41}$$

The VCC[2pt3] equations for two- and three-mode excitations are now

$$e^{2\text{pt}3}_{\mu_2} = e^{\text{VCC}[2]}_{\mu_2} + \langle \mu_2 | \, [\tilde{H}_1 + \tilde{H}_2 + F_3, T_3] \, | \Phi \rangle , \tag{2.42}$$

$$e^{2\text{pt}3}_{\mu_3} = \langle \mu_3 | \, \tilde{H} + [\tilde{H}_1 + \tilde{H}_2, T_2] + [F, T_3] \, | \Phi \rangle , \tag{2.43}$$

leaving out many other terms of the VCC[3] equations. The VCC[3] singles equations are unchanged in VCC[2pt3] and therefore not given. The VCC[2pt3] method scales as M^3 with system size thus matching VCC[2], though more expensive in absolute terms. Most importantly it is a gain compared to the M^4 scaling of VCC[3]. Benchmark calculations[33] clearly showed that the accuracy of VCC[2pt3] is significantly better than VCC[2] and approaching VCC[3] supporting that this method is a useful approach in computations on large systems where VCC[3] is not possible. Similar VCC[3pt4] methods in between VCC[3] and VCC[4] in accuracy and computational cost have also been developed.[34]

2.5.4. *Tensors, tensor decomposition and VCC and VCI compared*

Let us return to the basic VCC parameterization and the summation over $\mathbf{a^m}$ in Eq. (2.30) with summation over all virtual modals for all modes of the given MC. The wave function parameters can be regarded as a set of tensors — one for each MC. Thus, $t^{\mathbf{m}}_{\mathbf{a^m}}$ are the elements of the tensor $\boldsymbol{t^m}$ with dimensionality $\dim(\mathbf{m})$. For example, for $\mathbf{m} = (m_1, m_2, m_3)$ the amplitude constitutes a 3-way tensor, and the summation runs over all $a^{m_1}, a^{m_2}, a^{m_3}$, *i.e.*, over all $N^{\mathbf{m}}_{\text{vir}} = N^{m_1}_{\text{vir}} N^{m_2}_{\text{vir}} N^{m_3}_{\text{vir}}$ individual excitations. The full set of VCC parameters is thus a stack of tensors of different dimensionalities and refering to different MCs (vectors for each mode, matrix for pairs,

3-way tensors for triples, *etc.*). Equivalent stack-of-tensors structures are found for the error vector (Eq. (2.33)) and the various vectors occuring in the VCC response calculations. Truncated VCI can be interpreted similarly, but note that the FVCI wave function can most straightforwardly be seen as one huge tensor, see Eq. (2.4).

Let us now return to the basic VCC ansatz and perform a formal expansion

$$\exp(T)\,|\Phi\rangle = \left(1 + T_1 + \left(T_2 + \frac{1}{2}T_1^2\right)\right.$$
$$\left. + (T_3 + T_1 T_2 + \frac{1}{6}T_1^3) + \cdots\right)|\Phi\rangle. \qquad (2.44)$$

We can now compare this to the VCI expansion of Eq. (2.25) noting that in the FVCI limit the two should be equivalent. We thus learn that the two- and three-mode excited configurations in VCI are generated from a combination of the T_1, T_2, and T_3 cluster operators in VCC, *i.e.*, $C_2 = T_2 + \frac{1}{2}T_1^2$, $C_3 = T_3 + T_1 T_2 + \frac{1}{6}T_1^3$. Coefficients for individual configurations can correspondingly be found. For example,

$$\mathbf{c}^{m_1 m_2 m_3} = \mathbf{t}^{m_1 m_2 m_3} + \mathbf{t}^{m_1}\mathbf{t}^{m_2 m_3} + \mathbf{t}^{m_2}\mathbf{t}^{m_1 m_3}$$
$$+ \mathbf{t}^{m_3}\mathbf{t}^{m_1 m_2} + \mathbf{t}^{m_1}\mathbf{t}^{m_2}\mathbf{t}^{m_3}. \qquad (2.45)$$

Here, products of amplitude tensors should be considered as direct products with the final order of indices defined by the modes of the resulting tensor. In general the $\mathbf{c}$ coefficient tensor for a given MC is

$$\mathbf{c}^{\mathbf{m}} = \sum_{\mathrm{MCR}\in\mathrm{SMCR}[\mathbf{m}]} \prod_{\mathbf{m}_k\in\mathrm{MCR}} \mathbf{t}^{\mathbf{m}_k}. \qquad (2.46)$$

Here SMCR[$\mathbf{m}$] is the set of all partitions of $\mathbf{m}$, *i.e.*, the set of all possible sets of mutually disjoint and non-empty subsets of $\mathbf{m}$ whose union is $\mathbf{m}$. In this manner there is a clear recipe for transforming a VCC wave function into a VCI form, and it is relatively simple to invert this. This underlines that for the exact full expansion FVCI and FVCC are just two different parameterizations of the same. In this way FVCC is an, admittingly very special, decomposition of a

FVCI. For approximate VCC the cluster parameters will of course be different as they are determined through solving the VCC equations and not by decomposing VCI amplitudes. A truncated cluster expansions inherently include higher excitation VCI coefficients as can be seen by casting the VCC back into VCI format. Thus, including only up to two-mode excitations in VCC we still have three-mode couplings as seen from a VCI format, $\mathbf{c}^{m_1 m_2 m_3} = \mathbf{t}^{m_1} \mathbf{t}^{m_2 m_3} + \mathbf{t}^{m_2} \mathbf{t}^{m_1 m_3} + \mathbf{t}^{m_3} \mathbf{t}^{m_1 m_2} + \mathbf{t}^{m_1} \mathbf{t}^{m_2} \mathbf{t}^{m_3}$. This offers one explanation for VCC[n] being a compact wave function representation with higher accuracy than VCI[n] but fewer parameters and lower cost than VCI[n'] for $n' > n$.

The fact that the exponential generates higher order product terms also ensures the celebrated correct separability of VCC (some times described using words as size-consistency/size-extensivity). For a system consisting of non-interacting sub-systems, A and B, with an additively separable Hamiltonian, $\hat{H}_{\text{tot}} = \hat{H}_A + \hat{H}_B$, it is easy to see that the exact energy and the VCC energy is additively separable, $E_{\text{tot}} = E_A + E_B$, while the wave function is multiplicatively separable. Due to the exponential ansatz this is found for both complete and truncated VCC wave functions with an additively separable cluster operator, $T = T^A + T^B$. The proof follows essentially the text book case for electronic structure theory.[35] However, the result is significant since it means that $T^{AB} = 0$ – non-interacting sub-systems do not need VCC parameters to describe their "non-interaction" while in standard VCI $C^{AB} \neq 0$. Therefore, as systems become larger with more and more sub-systems that have only very weak direct interactions in the Hamiltonian, we can have a realistic hope that the cluster amplitudes will tend to become small and eventually zero in the non-interacting limit (unlike for VCI parameter vectors). It would be attractive to have a setup where the effort used for a given mode coupling could be adjusted according to the need. We therefore turn to tensor decomposition of the cluster amplitudes.

In the canonical polyadic (CP) tensor decomposition method a D-dimensional tensor $\mathcal{F}$ is decomposed into a sum of $R^{\mathcal{F}}$ vector outer products

$$F_{i_1 i_2 \dots i_{D^{\mathcal{F}}}} \approx \sum_{r=1}^{R^{\mathcal{F}}} \prod_{k=1}^{D} f_{i_k r}^{k}. \qquad (2.47)$$

Here $R^{\mathcal{F}}$ is denoted the tensor *rank*. Early studies indicated that cluster amplitude tensors could well be decomposed with generally favorable low ranks[36,37] and more favorable ranks for VCC compared to VCI due to the above separability discussion. In the canonical polyadic vibrational coupled cluster (CP-VCC) method[38,39] this is taken to the full consequence such that all tensors of an VCC computation are represented in CP format throughout the whole computation, while all VCC equations are still solved to their usual threshold. Thus, at no time is any full tensor computed. Instead amplitudes, error vectors, and intermediate quantities are represented in the CP format and automatic procedures and tensor recompressions are repeatingly used to find the appropriate ranks and economical CP representations. Many tedius numerical procedures are required for achieving this, but there are key advantages obtained with the final working CP-VCC algorithm. For VCC[n] with $n > 2$ CP-VCC leads to vast reductions in memory while representing essentially the same contents. The automatic rank determination procedures correctly find that while the amplitude tensors for some MCs require a significant rank for accuracy, the cluster tensors for the majority of MCs can actually be kept very small, if not even zero. Zero rank corresponds to the tensor being simply screened away. Due to the additive separability of the cluster operator, the fraction of very small or zero ranks should be expected to increase with system size. Indeed, for CP-VCC[3] computations in the sequence naphtalene, anthracene, tetracene increasing data reduction was found, with almost 3 orders of magnitude reduction in the amount of data necessary to describe the numerically same VCC[3] wave function for tetracene. Even more pronounced cases was found for higher level VCC[n]. Another advantage of the CP format is that the costly tensor contractions of the general VCC algorithms are significantly reduced in cost. Extension of the CP-VCC to VCC response theory has been reported[39] but here there may be even further significant gains by further method development.

2.5.5. *Coordinates and kinetic energy operators*

The choice of coordinates has not been discussed so far for VCC since for the theory and implementation the choice of coordinates is not essential. What is essential in practice is solely that one is able to provide a Hamiltonian in the sum of products form. VCC has been employed with standard normal coordinates using both the simplest possible kinetic energy operator, the exact Watson kinetic operator, and various approximations in between. VCC has also been reported for various other rectilinear coordinates. This includes both optimized coordinates which proved quite favorable in reducing couplings and thereby the size of certain cluster amplitudes.[40] Other coordinates based on localization (localized coordinates[41] or hybrid optimized and localized coordinates (HOLC)[42]) and/or fragmentation (FALCON[43]) have also been employed.[44] Recently VCC computation was also reported using polyspherical coordinates with exact kinetic energy operators.[45] Very encouraging convergence of the VCC[n] hierarchy was obtained, showing promise for future use, in particular if well-defined, systematic and accurate n-mode approximations to the kinetic energy operator can be developed. Altogether, combining VCC with local coordinates of various sorts seems to hold great promise for the future.

2.6. Response Theory

2.6.1. *General aspects of response theory*

Response theory is a very general theory used to identify expressions for evaluating properties, excitation energies and transition probabilities.[46–48] Essentially, one studies the time-dependent response of exact and approximate wave functions to an applied perturbation. For example, applying a frequency-dependent electric field to a system it is well known in exact theory that when the frequency matches an excitation energy of the system we can have a transition. By arguments of analogy, we study what happens to our approximate wave function in a similar formal study and in this way we find expressions for evaluating approximate excitation energies. For an exact state the linear response function (LRF) is

$$\langle\langle X; Y \rangle\rangle_\omega^\gamma = \sum_{k \neq o} \left[\frac{\langle\Psi_0|X|\Psi_k\rangle\langle\Psi_k|Y|\Psi_0\rangle}{\omega + i\gamma - \omega_k} - \frac{\langle\Psi_0|Y|\Psi_k\rangle\langle\Psi_k|X|\Psi_0\rangle}{\omega + i\gamma + \omega_k} \right].$$

$$(2.48)$$

Here the unperturbed Hamiltonian is denoted H_o and it has eigenenergies E_0 for the reference state $|\Psi_0\rangle$ of the response expansion and E_k for excited state $|\Psi_k\rangle$. The excitation energies of the systems are $\hbar\omega_k = (E_k - E_0)$. We assume the unperturbed eigenstates $\{|\Psi_0\rangle, |\Psi_k\rangle\}$ form a complete set of eigenstates. The frequency ω is the external frequency associated with perturbation operator Y. X and Y are general Hermitian operators. The additional γ in the denominator can be denoted a damping term, and plays a role in what we will denote damped response theory but is ignored in many other contexts. We can account phenomenologically for a finite lifetime of the excited states with a separate γ_k for all states. For simplicity we choose one common γ throughout,[48] as this is sufficient for our analysis.

Let us first consider the case where all $\gamma_k = 0$. In this case the LRF has poles when $\omega = \omega_f$ for some state f and the corresponding residue is

$$\lim_{\omega \to \omega_f} (\omega \to \omega_f)\langle\langle X; Y \rangle\rangle_\omega^{\gamma=0} = \langle\Psi_0|X|\Psi_f\rangle\langle\Psi_f|Y|\Psi_0\rangle. \qquad (2.49)$$

Taking $X = Y$ we can obtain the transition probabilities for transitions out of the reference states, $|\langle\Psi_0|Y|\Psi_f\rangle|^2$, from the residues. Transition probabilities between excited states can be obtained from the quadratic response function, as well as the higher order response function allows determination of two- and higher photon absorption. The linear and higher order response functions themselves can be used to compute frequency-dependent polarizabilities and hyperpolarizabilities, and a wealth of other properties. In quantum chemistry, response theory has long been used for electronic properties. However, response theory has now also been introduced for vibrational wave functions[49] including both VSCF and VCI but has especially become a vital part of VCC computations.[27,50–52]

In the above considerations we used $\gamma = 0$ to find excitation energies and subsequently transition properties, and we can easily imagine how one can use these to eventually simulate a spectrum. However, one can also compute the spectrum in other ways that do not first determine eigenstates. For example, through time-dependent theory or through damped response theory. For the damped response case consider $\gamma \neq 0$ and let $X = Y = \mu_\alpha$ which is the α component of the dipole operator. The imaginary part of the LRF can be written as

$$\mathrm{Im}\{\langle\langle\mu_\alpha;\mu_\alpha\rangle\rangle_\omega^\gamma\} = -\sum_{k\neq 0} |\langle\Psi_0|\mu_\alpha|\Psi_k\rangle|^2$$

$$\times\left(\frac{\gamma}{(\omega-\omega_k)^2+\gamma^2} - \frac{\gamma}{(\omega+\omega_k)^2+\gamma^2}\right). \quad (2.50)$$

We can introduce an oscillator strength distribution computed from the LRF that becomes discrete in the $\gamma \to 0$ limit as

$$f(\omega) = -\frac{4\omega}{3}\sum_{\alpha=x,y,z}\lim_{\gamma\to 0}\mathrm{Im}[\langle\langle\mu_\alpha;\mu_\alpha\rangle\rangle_\omega^\gamma]$$

$$= \frac{4\omega}{3}\pi\sum_{k\neq 0}\sum_{\alpha=x,y,z} |\langle\Psi_o|\mu_\alpha|\Psi_k\rangle|^2(\delta(\omega-\omega_k) - \delta(\omega+\omega_k)).$$

$$(2.51)$$

Thereby a link between the LRF and absorption (that is proportional to the oscillator strength) has been established. With a non-zero γ we can from the imaginary part of the damped LRF (Eq. (2.50)) compute directly a spectrum that essentially is a Lorentzian convolution of the discrete spectrum. Why should we want to proceed on this route? Consider a set of similar molecules with similiar groups but of increasing size. As the size of the molecule increases the number of states increases dramatically and there are typically more and more states in a given energy window. Calculating the spectrum directly for a certain frequency interval may then become highly advantageous compared to first computing very many states and subsequently compute their absorption. This is further compounded by the fact that (i) the absorption may be small for a large fraction of the states (ii) the computation of eigenstates can become more and more problematic the more dense the spectrum is. For example, if the excitation is described through diagonalization of some matrix, it should be noted that small changes in the values of the matrix elements may signficantly change the eigenstates. As a result contributions to the absorption from individual roots can in dense spectra be an almost ill-defined research question given the almost guaranteed presence of small errors in the PES, total Hamiltonian, or wave function. It can therefore be attractive to compute the spectra directly and focus on identifying the original sources and motions involved as opposed to individual states. This changes the challenge to computing the complex LRF itself instead of finding its poles.

2.6.2. *Excitation energies from VCC response theory*

The poles of the VCC LRF response function (for $\gamma = 0$) are found at the eigenvalues of the VCC Jacobian $\mathbf{A}$ introduced in Eq. (2.35). Thus VCC excitation energies are found by solving the VCC response eigenvalue equations,[27,50]

$$\mathbf{L}^f \mathbf{A} = \mathbf{L}^f \omega_f \tag{2.52}$$

$$\mathbf{A} \mathbf{R}^f = \omega_f \mathbf{R}^f. \tag{2.53}$$

Note that $\mathbf{A}$ is not symmetric. As a consequence the VCC left and right eigenvectors, $\mathbf{L}^f$ and $\mathbf{R}^f$ are not each others' adjoints. However, they can be required to be bi-orthogonal,

$$\mathbf{L}^f \mathbf{R}^g = \delta_{fg}. \tag{2.54}$$

If we denote the number of VCC cluster amplitudes as N_{amp} the VCC Jacobian is an $N_{\mathrm{amp}} \times N_{\mathrm{amp}}$ matrix. This matrix will be much too large to store and the VCC response eigenvalue problem need to solved by iterative methods.[27,39,53] In reduced-space methods the linear algebraic problem at hand is expressed in a reduced basis and solved in the following way that can be compared directly to program output. Given a basis of trial vectors $\mathbf{b}_i$ we construct the reduced $\mathbf{A}^R$ as $A_{ij}^R = \mathbf{b}_i{}^T \mathbf{A} \mathbf{b}_j = \mathbf{b}_i{}^T \sigma_j$ with $\sigma_j = \mathbf{A}\mathbf{b}_j$. The reduced space problem, $\mathbf{A}^R \mathbf{R}^{f,R} = \omega_f^R \mathbf{R}^{f,R}$, is then solved for the current approximation for the excitation energies ω_f^R, and the reduced space eigenvectors $\mathbf{R}^{f,R}$. From these the current approximation to the full space eigenvectors can be constructed, $\mathbf{R}^{f,app} = \sum_j R_j^{f,R} \mathbf{b}^j$. The quality of the approximate solution for state f can be judged from the residual vector

$$\mathcal{R}_f = (\mathbf{A} - \omega_f^R)\mathbf{R}^{f,app}. \tag{2.55}$$

If the norm of the residual is smaller than a given threshold the excitation is converged, but if not a new trial vector is generated (for example, by using the zeroth-order Jacobian and applying $(\mathbf{A}^{(0)} - \omega_f^R)^{-1}$ to the residual vector and orthonomalizing to previous basis vectors), added to the space, and the iterations continued. One may solve for a given number of the lowest excitation energies or attempt to target specific states such as fundamentals (noting that such targeting may be more difficult and require more advanced options for achieving convergence[53]). In practice the VCC response excitations are done on top of a ground state VCC computation by requesting the desired states and giving a residual convergence threshold, generating a sequence of iterations with excitation energies and residuals. It should be emphasized that in no part of the computation (ground state or response) is any matrix constructed in the full parameter space.

Carrying out the linear transformations for obtaining the σ vectors is the most computationally expensive part of all response computations. The right transformation

$$\sigma_{j,\mu} = \sum_\nu A_{\mu\nu} b_{j,\nu} = \sum_\nu \langle \mu | \exp(-T)[H, \tau_\nu] \exp(T) | \Phi \rangle b_{j,\nu}$$

$$= \langle \mu | \exp(-T)[H, B_j] \exp(T) | \Phi \rangle, \tag{2.56}$$

with $B_j = \sum_\nu b_{j,\nu} \tau_\nu$, has similar challenges and requires the same type of techniques as the computation of error vectors. The computational cost for VCC response excitation energy computations has therefore the same scaling behaviour as the ground state comptutation wrt the key parameters M, N_b and O. For determining excitation energies we only need to solve either the left or the right eigenvalue equations, but not necessarily both. From response theory we obtain also computational tractable expressions for calculation of transition properties such as dipole transition intensities or Raman scattering. In these cases both the right and left eigenvectors, as well as additional sets of equations, must be solved. We refer to the original literature for more detailed discussions.[27,50]

2.6.3. *Spectra from damped VCC response functions*

We now turn to the case of determining IR or Raman spectra by computing the LRF for $\gamma \neq 0$.[51,52,54] To compute, for example, IR spectra with VCC response theory the following type of contributions are needed

$$f(\omega) \leftarrow \langle\langle \mu_\alpha; \mu_\alpha \rangle\rangle_\omega^\gamma \leftarrow \eta^{\mu\alpha}(\mathbf{A} - (\omega - i\gamma)\mathbf{1})^{-1}\xi^{\mu\alpha}, \qquad (2.57)$$

where $\eta^{\mu\alpha}$ and $\xi^{\mu\alpha}$ are some characteristic VCC response vectors.[50,51]

In one approach a Lanzcos algorithm is used to determine successively better approximations to the VCC Jacobian $\mathbf{A}$ in a tri (or band) diagonal form making the evaluation easy. In the case of VCI response theory the standard symmetric Lanzcos algorithm can be used, but since the VCC Jacobian is asymmetric the asymmetric matrix Lanzcos algorithm[55] must be used in the VCC case. A chain of Lanzcos vectors are generated based on start vectors constructed from $\eta^{\mu\alpha}$ and $\xi^{\mu\alpha}$ and thereby tailored the particular LRF. The integer chain length thereby defines a level of approximation for a full spectrum for all frequencies ω. Both VCI and VCC damped response have been implemented in single vector update versions[51,56] and band versions[52] where contributions from several response functions to the spectra are simultaneously computed. In the Lanzcos approach one specifies a chain length and analyzes after the computation the extent of convergence for the features of interest with the possibility of restart and continuation. For low chain lengths the overall spectrum has only a rough similarity with the full spectrum but it improves with increasing chain length and it can be shown mathematically how the LRF contributions converge with increasing chain length.[51,56] Strong features show their presence early, while weak features in the mid-frequency range requires long chain lengths.

In a different approach one solves for the specific frequencies of interest,[54] *i.e.*, one solves the linear response equations $(\mathbf{A} - (\omega - i\gamma)\mathbf{1})^{-1}\xi^{\mu\alpha}$ for given ω and γ. This can be done in a specially designed reduced space approach to cope with the complex $\omega - i\gamma$ term. Convergence is checked for the residual of the damped linear response equations. This approach is useful for focusing on a specific frequency window whatever the strength or frequency of the absorption, but does not provide the more global view of absorption as a function of frequency that the Lanzcos approach generates. The two methods are thus somewhat complementary. It is true for both approaches that the full matrix $\mathbf{A}$ is never explicitly computed or stored, only vectors transformed with $\mathbf{A}$ similar to the above iterative eigenvalue algorithm.

2.6.4. *Excited states by VCC or VCC response theory: Discussion*

With the availability of VCC one can imagine a state-specific approach also for excited states. Thus separate but similar VCC computations are done for both ground and excited states where for each state its own non-linear VCC equations are solved. This determines two different states and total energies, and excitation energies are trivially obtained by subtraction. We denote this a state-specific VCC approach, ss-VCC. Naturally, the quality of the description depends on the state and the level of approximation applied. However, among the critical issues are that (i) non-linear optimization can be very complicated in practice for excited states when there are interferences from other close lying states, (ii) single reference-based VCC may not be adequate, (iii) the computed states are non-orthogonal making it far from trivial to compute transition properties. These aspects usually make it much more attractive to use the response theory approach where (i) excited states are determined from linear eigenvalue equations which are much simpler to solve than non-linear equations, (ii) the response approach can much better allow for describing states that have several leading configurations, (iii) the response approach defines recipies for computations of transition properties.

Let us consider a three-mode system as an example and compare ss-VCC and VCC response theory departing from a VCC ground state. We denote the ground state leading configuration as $[0, 0, 0]$. For an excited state of the form $[0, 0, 1]$ in ss-VCC we target the corresponding HP as the reference for a VCC, *i.e.*, $\mathbf{i} = (0, 0, 1)$ and the excitation space is constructed relative to this reference. The VCC response approach use $\mathbf{i} = (0, 0, 0)$ as reference and determines the cluster amplitudes for this state and subsequently the $\mathbf{R}$ (and/or $\mathbf{L}$) vectors describe the physical excitation as well as the differences in correlations in the excited state compared to the ground state. In CC response theory (using one fixed reference state for simplicity of the argument) we can only hope to describe excited states of a type covered by the excitation space included in VCC from the outset. For example, $[0,0,1]$ and $[0,0,2]$ require only single excitations (not a problem) while the combination bands $[0,1,1]$, $[0,1,2]$ require at least two-mode excitations to be present and $[0,0,1]$ and $[0,0,2]$ are likely to be more accurately described than $[0,1,1]$, $[0,1,2]$. The state $[1,1,1]$ would require a VCC including three-mode excitations and will not at all be present in a VCC[2] response computation. We should thus expect a decaying accuracy of VCC response theory with increasing number of modes excited out of the VCC reference state. In contrast ss-VCC is readily applicable to all those states indicating the generality of this approach. However, the presented analysis is deceptive in its simplicity, and the problems mentioned before makes the practical computation and analysis with ss-VCC far from trivial. For example, if two configurations $[0,0,2]$ and $[0,1,0]$ are almost equally important (a typical resonance case) then standard single reference ss-VCC requires two-mode excitations to excite from $[0,0,2]$ to $[0,1,0]$ and vice versa. This makes two computations using either $[0,0,2]$ or $[0,1,0]$ biased towards either reference and therefore problematic. In VCC response, on the other hand, both count as single excitations compared to the ground state. Thus, VCC response comes out more balanced and would typically give more accurate results with less work. The above raised problem for ss-VCC calls for a multi-configurational treatment. VCC response theory can in many cases accurately handle situations that in the langurage of other modes of computation can be considered multi-reference cases. This is in fact similar to the electronic structure CC response theory.[57] In the absence of other knowledge it is with the currently available methods recommended to employ the VCC response approach, that is usually simpler and often more accurate in practice, and only depart from this when needs develop. The analysis above also shows that it is worth paying attention to the level of excitation in judging the quality of VCC response results and typically given as output from the computation. They can thus be scrutinized and expectations should be that one-mode excitations (fundamentals and overtones) are more accurately described than two-mode combinations, *etc.*[50]

2.7. Time-Dependent Wave Functions

2.7.1. *Time-dependent dynamics with time-independent modals*

We shall end this chapter considering recent work on describing time-dependent vibrational wave packets with VCC methods. We shall first discuss the use of a time-independent basis. A general time-dependent state is in this case written as a linear combination of time-independent HPs (in first quantization or ONVs in second quantization) with time-dependent expansion coefficients

$$|\Psi(\mathbf{q}, t)\rangle = \sum_{\mathbf{s}}^{\mathbf{N}} C_{\mathbf{s}}(t) |\Phi_{\mathbf{s}}(\mathbf{q})\rangle. \tag{2.58}$$

In accord with the earlier parts of this chapter we can denote such an ansatz as a time-dependent vibrational configuration interaction (TDVCI) wave function. The evolution of the $C_{\mathbf{s}}(t)$ coefficient can be obtained by inserting into and projecting onto the time-dependent Schrödinger equation (TDSE) or by the time-dependent variational principle giving

$$i\dot{C}_\mathbf{s}(t) = \sum_\mathbf{r} \langle \Phi_\mathbf{s}(\mathbf{q})| \, H \, |\Phi_\mathbf{r}(\mathbf{q})\rangle \, C_\mathbf{r}(t). \qquad (2.59)$$

If there are no trunctions in the included configurations the time-dependent full vibrational configuration interaction (TDFVCI) wave function is obtained representing (numerically) the exact solution of the TDSE within the space spanned by the selected modals basis. Increasing the latter towards the complete basis set (CBS) limit the (numerical) exact wave function evolution is in principle obtained. However, the dimension of the space grows as N_b^M and TDFVCI is in practice restricted to rather small M and N_b. Approximate TDVCI[n] methods can be introduced where the expansion parameters are included/excluded based on the maximum number of modes excited (n) relative to a selected reference HP. The TDVCI[n] hierarchy converges to TDFVCI for $n \to M$ as depicted in Figure 2.1.

Instead of the linear configuration space parameterization we will now discuss introduction of a TDVCC framework. Similar concepts of space truncation can be used as before. However, significant extensions to the theory are needed, first or all since VCC is not variational and second since our presentation has so far been rather ket-centric. To evaluate anything other than the energy a dual bra state is needed. This is well known in the context of VCC response theory, but the reader was spared of the details. Now we are forced to consider at least the fundamental aspects to be able to give a reasonable impression of the TDVCC theory.

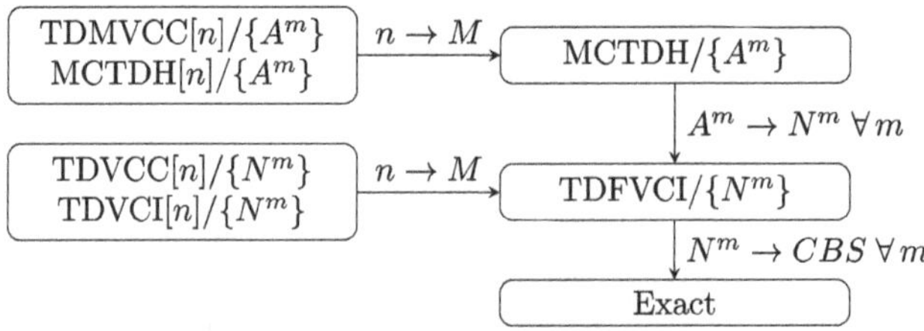

Figure 2.1.   Summary of various limits of different time-dependent wave-function parameterizations.

In place of using variational principles TDVCC equation of motion (EOM) can be derived from a bivariational principle (BVP)[58–61] requiring stationarity with respect to independent variation of both Ψ and Ψ' for the functional

$$\mathcal{I}[\Psi, \Psi'] = \int_{t_0}^{t_1} \left\langle \Psi' \left| \left(i\frac{\partial}{\partial t} - H \right) \right| \Psi \right\rangle dt, \qquad (2.60)$$

with contraints $\delta\Psi'(t_0) = \delta\Psi'(t_1) = \delta\Psi(t_0) = \delta\Psi(t_1) = 0$. This yields the TDSE and its complex conjugate for exact states. The standard time-dependent variational principle is obtained if $\Psi = \Psi'$. However, for TDVCC the latter is not fulfilled and we have to introduce separate TDVCC parameterizations of the ket and bra states. The ket state is

$$|\Psi(t)\rangle = |\text{CC(t)}\rangle = \exp(-i\epsilon(t)) \exp(T(t)) |\Phi\rangle = \exp(-i\epsilon(t)) |\widetilde{\text{CC}}\rangle, \qquad (2.61)$$

where the tilde denotes a time-dependent wave function since the cluster amplitudes t_μ is time dependent in $T(t) = \sum_\mu t_\mu(t)\tau_\mu$. The generalized phase-factor (not necessarily real for TDVCC) $\epsilon(t)$ is also introduced. The bra state $\langle\Psi'|$ is

$$\langle\Psi'(t)| = \langle\Phi| (1 + L(t)) \exp(-T(t)) \exp(i\epsilon'(t)), \qquad (2.62)$$

where the L operator is

$$L(t) = \sum_\mu l_\mu(t)\tau_\mu^\dagger. \qquad (2.63)$$

After some derivations[61] we obtain the EOM for all wave function parameters

$$\dot{t}_\mu = -i\langle\mu| \exp(-T(t))H \exp(T(t))|\Phi\rangle \equiv -ie_\mu, \qquad (2.64)$$

$$\dot{l}_\mu = +i\langle\Phi(1 + L(t)) \exp(-T(t))|[H, \tau_\mu] \exp(T(t))|\Phi\rangle$$

$$\equiv i\left(\eta_\mu + \sum_\nu l_\nu A_{\nu\mu} \right), \qquad (2.65)$$

$$\dot{\epsilon} = \dot{\epsilon}' = \langle\Phi| \exp(-T(t))H \exp(T(t))|\Phi\rangle, \qquad (2.66)$$

where we have introduced the vector $\eta_\mu = \langle\Phi|[\exp(-T(t))H \exp(T(t)), \tau_\mu]|\Phi\rangle$ while $e_\mu = \langle\mu|\exp(-T(t))H\exp(T(t))|\Phi\rangle$, $A_{\mu\nu} = \langle\mu|[\exp(-T(t))H\exp(T(t)), \tau_\nu]|\Phi\rangle$ are respectively the error vector and the error-vector Jacobian matrix previously encountered, only now with complex and time-dependent cluster amplitudes.

The numerical studies confirm that the TDVCC bra and ket states converge to the exact limits as TDVCI wave functions for complete expansions. For truncated expansions the expected gains of approximately implicit inclusion of higher excitations and better separability of computed properties are found.[61] For time-independent modals both specialized two-mode implementations[61] and efficient general coupling level implementations[62] of TDVCC are available, allowing time-dependent simulations with polynomial scaling cost with respect to system size similar to the earlier described, but unlike most methods in this field.

2.7.2. *Time-dependent dynamics with time-dependent modals*

The use of a time-independent basis can be expected to lead to a need for quite large basis sets to be able to describe all important effects at all times. A physically appealing alternative is therefore to let the modals evolve variationally in time with the expectation that thereby a much reduced number of HPs is needed for achieving sufficient accuracy. This is at the core of multi-configuration time-dependent Hartree (MCTDH)[63,64] where modals and configurations are simultaneously evolved in accord with the time-dependent variational principle. Thus the wave function is a linear superposition of HPs constructed using time-dependent modals

$$|\tilde{\Psi}(\mathbf{q}, t)\rangle = \sum_\mathbf{s}^\mathbf{A} \tilde{C}_\mathbf{s}(t)\,|\tilde{\Phi}_\mathbf{s}(\mathbf{q}, t)\rangle = \sum_\mathbf{s}^\mathbf{A} \tilde{C}_\mathbf{s}(t) \prod_{m=1}^{M} \tilde{\phi}_{s^m}^m(q_m, t). \quad (2.67)$$

There is a number of active time-dependent modals for each mode $A^m \leq N^m$, where the TDFVCI limit is obtained if $A^m \to N^m$ for all modes, as also illustrated in Figure 2.1. Due to the modals evolving variationally MCTDH has significant flexibility and MCTDH has shown itself as a versatile method for quantum dynamics computations. Nevertheless, the standard complete active space MCTDH still comes with an exponential scaling, although with a much reduced base compared to TDFVCI.

Recently, MCTDH was reformulated in terms of SQ, MCs and excitation operators and truncated multi-configuration time-dependent Hartree (MCTDH[n]) and multi-reference multi-configuration time-dependent Hartree (MR-MCTDH(n)) methods were introduced,[65,66] where the MCTDH wave function is truncated based on excitation levels and MC considerations. This was done in combination with variational determination of both basis functions, configuration coefficents *and* the so-called constraint operator. The variational constraint operator is important for defining the time evolution of the active modals,[65] and is a non-trivial feature of the MCTDH[n] approach. It is not needed in standard MCTDH which is formally invariant under unitary transformation of active modals, but it is also important in the following.

The MCTDH-based methods are all still based on linear parameterizations in the configuration space. In the very recently proposed[67] TDMVCC[n] method the exponential parameterization of TDVCC is combined with time-dependent modals. The theory is quite involved and will not be repeated here in any detail. The principal theoretical aspects are, (i) the use of the BVP to determine the EOM for all involved parameters, (ii) a systematic biorthogonal second quantization formulation with separately evolving bra and ket modals, (iii) bra and ket states are parameterized in terms of $T(t)$ and $L(t)$ as in TDVCC above, only now with time-dependent bi-orthogonal modals also in the excitation operators, (iv) constraint operators with non-redundant parts variationally determined, and (v) a non-trivial redundancy analysis showing that all one-mode parameters T and L can be excluded when a variational constraint operator is used. The resulting EOM includes equations similar to the above t_μ, l_μ EOM for TDVCC as well as EOM for the bra and ket modals that share similarties with MCTDH theory. As illustrated in Figure 2.1 MCTDH[n] and TDMVCC[n] both converge to the full MCTDH method when $n \to M$, and thereby also to the FVCI limit when

$A^m \to N^m$. Away from these limits MCTDH[n] and TDMVCC[n] give numerically different results from each other and from both MCTDH and TDVCC. TDMVCC[n] presents a highly interesting new hierarchy of quantum-dynamics methods that is expected to provide very compact wave function parameterizations by combining the fast convergence of VCC in configuration space with time-dependent adaptive modals in the spirit of MCTDH. The method is very new and only a pilot TDMVCC implementation has been presented but initial results were very encouraging. It is a topic of future research to develop efficient formulations and implementations of TDMVCC to allow further exploring its potential.

Acknowledgements

OC thanks Independent Research Fund Denmark for support. OC thanks many previous and current students and post docs for contributing to the development of VCC methods, and thanks Denis Artiukhin for commenting on an early draft.

References

1. Christiansen, O. A second quantization formulation of multi-mode dynamics. *J. Chem. Phys.* **2004**, *120*, 2140–2148.
2. Carter, S.; Culik, S. J.; Bowman, J. M. Vibrational self-consistent field method for many-mode systems: A new approach and application to the vibrations of CO adsorbed on CU(100). *J. Chem. Phys.* **1997**, *107*, 10458–10469.
3. Kongsted, J.; Christiansen, O. Automatic generation of force fields and property surfaces for use in variational vibrational calculations of anharmonic vibrational energies and zero-point vibrational averaged properties. *J. Chem. Phys.* **2006**, *125*, 124108–124116.
4. Klinting, E. L.; Thomsen, B.; Godtliebsen, I. H.; Christiansen, O. Employing general fit-bases for construction of potential energy surfaces with an adaptive density-guided approach. *J. Chem. Phys.* **2018**, *148*, 064113.
5. König, C.; Christiansen, O. Linear-scaling generation of potential energy surfaces using a double incremental expansion. *J. Chem. Phys.* **2016**, *145*, 064105.

6. Sparta, M.; Toffoli, D.; Christiansen, O. An adaptive density-guided approach for the generation of potential energy surfaces of polyatomic molecules. *Theor. Chem. Acc.* **2009**, *123*, 413–429.

7. Sparta, M.; Hansen, M. B.; Matito, E.; Toffoli, D.; Christiansen, O. Using electronic energy derivative information in automated potential energy surface construction for vibrational calculations. *J. Chem. Theory Comput.* **2010**, *6*, 3162–3175.

8. Artiukhin, D. G.; Klinting, E. L.; König, C.; Christiansen, O. Adaptive density-guided approach to double incremental potential energy surface construction. *J. Chem. Phys.* **2020**, *152*, 194105.

9. Schmitz, G.; Artiukhin, D. G.; Christiansen, O. Approximate high mode coupling potentials using Gaussian process regression and adaptive density guided sampling. *J. Chem. Phys.* **2019**, *150*, 131102.

10. Schmitz, G.; Klinting, E. L.; Christiansen, O. A Gaussian process regression adaptive density guided approach for potential energy surface construction. *J. Chem. Phys.* **2020**, *153*, 064105.

11. Bowman, J. M. The Self-consistent field approach to polyatomic vibrations. *Acc. Chem. Res.* **1986**, *19*, 202–208.

12. Gerber, R. B.; Ratner, M. A. self-consistent field review. *Adv. Chem. Phys.* **1988**, *70*, 97.

13. Christiansen, O. Vibrational structure theory: New vibrational wave functions methods for calculation of anharmonic vibrational energies and vibrational contributions to molecular properties. *Phys. Chem. Chem. Phys.* **2007**, *9*, 2942–2953.

14. Hansen, M. B.; Sparta, M.; Seidler, P.; Christiansen, O.; Toffoli, D. A new formulation and implementation of vibrational self-consistent field (VSCF) theory. *J. Chem. Theory Comput.* **2010**, *6*, 235–248.

15. de Boor, C. *A Practical Guide to Splines*, Revised ed. Springer-Verlag: New York, 2001.

16. Toffoli, D.; Sparta, M.; Christiansen, O. Accurate multimode vibrational calculations using a B-spline basis: theory, tests and application to dioxirane and diazirinone. *Mol. Phys.* **2011**, *109*, 673–685.

17. Bowman, J. M.; Carter, S.; Huang, X. MULTIMODE: a code to calculate rovibrational energies of polyatomic molecules. *Int. Rev. Phys. Chem.* **2003**, *22*, 533.

18. Rauhut, G. Configuration selection as a route towards efficient vibrational configuration interaction calculations. *J. Chem. Phys.* **2007**, *127*, 184109.

19. Christiansen, O. Vibrational structure theory: new vibrational wave function methods for calculation of anharmonic vibrational energies and vibrational contributions to molecular properties. *Phys. Chem. Chem. Phys.* **2007**, *9*, 2942.

20. Oschetzki, D.; Neff, M.; Meier, P.; Pfeiffer, F.; Rauhut, G. Selected aspects concerning the efficient calculation of vibrational spectra beyond the harmonic approximation. *Croat. Chem. Acta* **2012**, *85*, 379–390.

21. Pulay, P. Convergence acceleration of iterative sequences. the case of scf iteration. *Chem. Phys. Lett.* **1980**, *73*, 393–398.

22. Ettenhuber, P.; Jørgensen, P. Discarding information from previous iterations in an optimal way to solve the coupled cluster amplitude equations. *J. Chem. Theory Comput.* **2015**, *11*, 1518–1524.

23. Madsen, N. K.; Godtliebsen, I. H.; Christiansen, O. Efficient algorithms for solving the non-linear vibrational coupled-cluster equations using full and decomposed tensors. *J. Chem. Phys.* **2017**, *146*, 134110.

24. Christiansen, O.; Artiukhin, D.; Godtliebsen, I. H.; Gras, E. M.; GyHorffy, W.; Hansen, M. B.; Hansen, M. B.; Kongsted, J.; Klinting, E. L.; König, C.; *et al. MidasCpp (Molecular Interactions, dynamics and simulation Chemistry program package in C++)*; 2021; https://midascpp.gitlab.io.

25. Christiansen, O. Vibrational coupled cluster theory. *J. Chem. Phys.* **2004**, *120*, 2149–2159.

26. Seidler, P.; Christiansen, O. Automatic derivation and evaluation of vibrational coupled cluster theory equations. *J. Chem. Phys.* **2009**, *131*, 234109–234115.

27. Seidler, P.; Sparta, M.; Christiansen, O. Vibrational coupled cluster response theory: A general implementation. *J. Chem. Phys.* **2011**, *134*, 054119-1–15.

28. Seidler, P.; Hansen, M. B.; Christiansen, O. Towards fast computations of correlated vibrational wave functions: Vibrational coupled cluster response excitation energies at the two-mode coupling level. *J. Chem. Phys.* **2008**, *128*, 154113-1–12.

29. Seidler, P.; Christiansen, O. Vibrational Coupled Cluster Theory. In *Recent Progress in Coupled Cluster Methods: Theory and Applications*, 1st ed.; Crsky, P., Paldus, J., Pittner, J., Eds.; Springer, 2010; pp. 491–512.

30. Hansen, M. B.; Madsen, N. K.; Christiansen, O. Extended vibrational coupled cluster: Stationary states and dynamics. *J. Chem. Phys.* **2020**, *153*, 044133.

31. Norris, L. S.; Ratner, M. A.; Roitberg, A. E.; Gerber, R. B. Møller-Plesset perturbation theory applied to vibrational problems. *J. Chem. Phys.* **1996**, *105*, 11261–11267.

32. Christiansen, O. Møller-Plesset perturbation theory for vibrational wave functions. *J. Chem. Phys.* **2003**, *119*, 5773–5781.

33. Seidler, P.; Matito, E.; Christiansen, O. Vibrational coupled cluster theory with full two-mode and approximate three-mode couplings: The VCC[2pt3] model. *J. Chem. Phys.* **2009**, *131*, 034115-1–12.

34. Zoccante, A.; Seidler, P.; Hansen, M. B.; Christiansen, O. Approximate inclusion of four-mode couplings in vibrational coupled-cluster theory. *J. Chem. Phys.* **2012**, *136*, 204118–204112.

35. Helgaker, T.; Jørgensen, P.; Olsen, J. *Molecular Electronic-Structure Theory*. Wiley: Chichester, 2000.

36. Godtliebsen, I. H.; Thomsen, B.; Christiansen, O. Tensor decomposition and vibrational coupled cluster theory. *J. Phys. Chem. A* **2013**, *117*, 7267–7279.

37. Godtliebsen, I. H.; Hansen, M. B.; Christiansen, O. Tensor decomposition techniques in the solution of vibrational coupled cluster response theory eigenvalue equations. *J. Chem. Phys.* **2015**, *142*, 024105.

38. Madsen, N. K.; Godtliebsen, I. H.; Losilla, S. A.; Christiansen, O. Tensor-decomposed vibrational coupled-cluster theory: Enabling large-scale, highly accurate vibrational-structure calculations. *J. Chem. Phys.* **2018**, *148*, 024103.

39. Madsen, N. K.; Jensen, R. B.; Christiansen, O. Calculating vibrational excitation energies using tensor-decomposed vibrational coupled-cluster response theory. *J. Chem. Phys.* **2021**, *154*, 054113.

40. Thomsen, B.; Yagi, K.; Christiansen, O. Optimized coordinates in vibrational coupled cluster calculations. *J. Chem. Phys.* **2014**, *140*, 154102-1–15.

41. Jacob, C. R.; Reiher, M. Localizing normal modes in large molecules. *J. Chem. Phys.* **2009**, *130*, 084106.

42. Klinting, E. L.; König, C.; Christiansen, O. Hybrid optimized and localized vibrational coordinates. *J. Phys. Chem. A* **2015**, *119*, 11007–11021.

43. König, C.; Hansen, M. B.; Godtliebsen, I. H.; Christiansen, O. FALCON: A method for flexible adaptation of local coordinates of nuclei. *J. Chem. Phys.* **2016**, *144*, 074108.

44. Klinting, E. L.; Christiansen, O.; Knig, C. Toward accurate theoretical vibrational spectra: a case study for maleimide. *J. Phys. Chem. A* **2020**, *124*, 2616–2627.

45. Klinting, E. L.; Lauvergnat, D.; Christiansen, O. Vibrational coupled cluster computations in polyspherical coordinates with the exact analytical kinetic energy operator. *J. Chem. Theory Comput.* **2020**, *16*, 4505–4520.

46. Olsen, J.; Jørgensen, P. Time-Dependent Response Theory with Applications to Self-Consistent Field and Multiconfigurational Self-Consistent Field Wave Functions. In *Modern Electronic Structure*

78   *O. Christiansen*

      Theory, Vol. 2, Chapter 13; Yarkony, D. R., Ed.; World Scientific, 1995; pp. 857–990.

47. Christiansen, O.; Jørgensen, P.; Hättig, C. Response functions from a Fourier component variational perturbation theory applied to a time-averaged quasienergy. *Int. J. Quantum Chem.* **1998**, *68*, 1–52.

48. Norman, P. A perspective on nonresonant and resonant electronic response theory for time-dependent molecular properties. *Phys. Chem. Chem. Phys.* **2011**, *13*, 20519–20535.

49. Christiansen, O. Response theory for vibrational wave functions. *J. Chem. Phys.* **2005**, *122*, 194105.

50. Seidler, P.; Christiansen, O. Vibrational excitation energies from vibrational coupled cluster response theory. *J. Chem. Phys.* **2007**, *126*, 204101.

51. Thomsen, B.; Hansen, M. B.; Seidler, P.; Christiansen, O. Vibrational absorption spectra from vibrational coupled cluster damped linear response functions calculated using an asymmetric Lanczos algorithm. *J. Chem. Phys.* **2012**, *136*, 124101–124117.

52. Godtliebsen, I. H.; Christiansen, O. A band Lanczos approach for calculation of vibrational coupled cluster response functions: simultaneous calculation of IR and Raman anharmonic spectra for the complex of pyridine and a silver cation. *Phys. Chem. Chem. Phys.* **2013**, *15*, 10035–10048.

53. Györffy, W.; Seidler, P.; Christiansen, O. Solving the eigenvalue equations of correlated vibrational structure methods: preconditioning and targeting strategies. *J. Chem. Phys.* **2009**, *131*, 024108-1–16.

54. Godtliebsen, I. H.; Christiansen, O. Calculating vibrational spectra without determining excited eigenstates: Solving the complex linear equations of damped response theory for vibrational configuration interaction and vibrational coupled cluster states. *J. Chem. Phys.* **2015**, *143*, 134108.

55. Bai, Z.; Demmel, J.; Dongarra, J.; Ruhe, A.; van der Vorst, H., Eds. *Templates for the Solution of Algebraic Eigenvalue Problems: A Practical Guide*. SIAM: Philadelphia, 2000.

56. Seidler, P.; Hansen, M. B.; Györffy, W.; Toffoli, D.; Christiansen, O. Vibrational absorption spectra calculated from vibrational configuration interaction response theory using the Lanczos method. *J. Chem. Phys.* **2010**, *132*, 164105–164115.

57. Sneskov, K.; Christiansen, O. Excited state coupled cluster methods. *Wiley Interdis. Rev.: Comput. Mol. Sci.* **2012**, *2*(4), 566–584.

58. Arponen, J. Variational principles and linked-cluster exp *S* expansions for static and dynamic many-body problems. *Ann. Phys.* **1983**, *151*(2), 311–382.

59. Froelich, P.; Lwdin, P. On the Hartree–Fock scheme for a pair of adjoint operators. *J. Math. Phys.* **1983**, *24*, 88–96.

60. Kvaal, S. Ab initio quantum dynamics using coupled-cluster. *J. Chem. Phys.* **2012**, *136*, 194109.

61. Hansen, M. B.; Madsen, N. K.; Zoccante, A.; Christiansen, O. Time-dependent vibrational coupled cluster theory: theory and implementation at the two-mode coupling level. *J. Chem. Phys.* **2019**, *151*, 154116.

62. Madsen, N. K.; Jensen, A. B.; Hansen, M. B.; Christiansen, O. A general implementation of time-dependent vibrational coupled-cluster theory. *J. Chem. Phys.* **2020**, *153*, 234109.

63. Meyer, H.-D.; Manthe, U.; Cederbaum, L. S. The multi-configurational time-dependent Hartree approach. *Chem. Phys. Lett.* **1990**, *165*, 73–78.

64. Beck, M. H.; Jäckle, A.; Worth, G. A.; Meyer, H.-D. The multiconfiguration time-dependent Hartree method: a highly efficient algorithm for propagating wavepackets. *Phys. Rep.* **2000**, *324*, 1–105.

65. Madsen, N. K.; Hansen, M. B.; Worth, G. A.; Christiansn, O. Systematic and variational truncation of the configuration space in the multiconfiguration time-dependent Hartree method: The MCTDH[n] hierarchy. *J. Chem. Phys.* **2020**, *152*, 084101.

66. Madsen, N. K.; Hansen, M. B.; Worth, G. A.; Christiansen, O. MR-MCTDH[n]: Flexible configuration spaces and nonadiabatic dynamics within the MCTDII[n] framework. *J. Chem. Theory Comput.* **2020**, *16*, 4087–4097.

67. Madsen, N. K.; Hansen, M. B.; Christiansen, O.; Zoccante, A. Time-dependent vibrational coupled cluster with variationally optimized time-dependent basis sets. *J. Chem. Phys.* **2020**, *153*, 174108.

https://doi.org/10.1142/9789811237911_0003

Chapter 3

Tensor Network States for Vibrational Spectroscopy

Nina Glaser, Alberto Baiardi, and Markus Reiher[*]

*Laboratorium für Physikalische Chemie, ETH Zürich,
Vladimir-Prelog-Weg 2, 8093 Zürich, Switzerland*
**markus.reiher@phys.chem.ethz.ch*

3.1. Introduction

The exact solution of vibrational and electronic quantum many-body problems is obtained by solving the full configuration interaction (CI) eigenvalue problem.[1] However, the practical application of full CI is limited due to its exponentially growing computational cost. This changed in the last decade, which has witnessed an impressive development of modern full CI schemes.[2,3] In this context, methods based on so-called tensor network states have made it possible to calculate accurate energies for systems with up to 100 single-particle basis functions.[4–17] These methods tame the exponential scaling of full CI by efficiently compressing the many-body wave function representation by tensor factorization and by optimizing the wave function with algorithms that are tailored to each specific factorization. The potential of these methods is determined by a complex interplay between two components: representation effectiveness and availability of an efficient optimization algorithm. The ideal tensor network state should provide a compact description of a quantum many-body state, especially in the presence of strong correlation, where other simplified methods, based either on truncated CI or on low-order perturbation theory, fail. At the same time, it should support efficient optimization algorithms. Finding the optimal balance between these two components is a key challenge for the development of tensor-based methods.

The development of tensor-based quantum chemical methods has not followed a straight path and a variety of different algorithms have been designed for time-independent and time-dependent, electronic and vibrational many-body chemical problems. The most popular tensor factorization for nuclear quantum dynamics is the multi-layer multi-configurational time-dependent Hartree (ML-MCTDH) algorithm.[18–22] ML-MCTDH factorizes the full CI tensor based on the so-called hierarchical Tucker format[23] and solves the underlying time-dependent (TD) Schrödinger equation with the multi-layer generalization of the MCTDH method.[24] The ML-MCTDH accuracy relies heavily on the possibility of partitioning the quantum system into subsets of strongly interacting particles. It is, therefore, maximally efficient when applied to system-bath Hamiltonians, even though it has also been successfully applied to complex molecular systems.[25–27] The imaginary-time formulation of ML-MCTDH enables the optimization of the ground state of many-body Hamiltonians.[28] For vibrational calculations, Carrington and co-workers proposed the canonical polyadic factorization[29, 30] and its hierarchical variant[31] to represent anharmonic vibrational wave functions for molecules with more than 30 modes. The same factorization has also been combined with other wave function *ansätze* that do not rely on the full CI parameterization. For instance, Christiansen and co-workers applied tensor decomposition algorithms to compactly encode vibrational coupled cluster wave functions.[32, 33]

The development of tensor network states for electronic-structure calculations has focused on a third class of tensor factorizations. The reference method for electronic problems has been the density matrix renormalization group (DMRG) algorithm,[4, 5] which relies on the so-called tensor train (TT) factorization.[34] It is, in fact, possible to demonstrate that the TT factorization, also known as the matrix product state (MPS) representation in physics,[35] efficiently encodes the ground state wave function of short-ranged Hamiltonians.[36] While DMRG has originally been devised for one-dimensional spin chains in condensed matter physics, it has later been applied to Hamiltonians that have no inherently (pseudo-)one-dimensional structure, such as molecular electronic-structure problems.[37–42] It turned out that DMRG can also efficiently encode the electronic wave function of strongly correlated systems in transition metal complexes and clusters.[43] Since then, it has become one of the reference methods for large-scale CI electronic-structure calculations.[6–10, 12–17, 44–46]

DMRG has recently been generalized to vibrational problems and was applied to anharmonic vibrational-structure calculations.[47, 48] The resulting theory, abbreviated vDMRG, delivers accurate anharmonic energies for molecules with more than 30 modes, *i.e.*, for system sizes that are at the current limits of alternative tensor-based methods. Also other methods originally designed for large-scale electronic full CI calculation have been applied to vibrational problems, including multidimensional DMRG generalizations[49] and selected CI[50, 51] algorithms, which are as accurate as traditional state-of-the-art vibrational full CI algorithms. The success of tensor-based electronic-structure methods extended to vibrational problems suggests that they can further expand the range of applicability of anharmonic vibrational-structure calculations. In this respect, DMRG is the natural starting point for developing tensor-based vibrational-structure algorithms. It is, in fact, the most popular tensor-based method for quantum chemical applications due to its numerical stability and its flexibility, and represents the foundation of many other more complex tensor-based algorithms.

In this chapter, we introduce the tensor network state framework with a particular focus on the DMRG algorithm. We first review the theoretical foundations of DMRG. Then, we discuss the optimization of ground and excited state anharmonic vibrational wave functions with DMRG. Afterwards, we elaborate on extensions of DMRG to solve the nuclear time-dependent Schrödinger equation, leading to the time-dependent DMRG (TD-DMRG) method. Finally, we explain the main shortcomings and limitations of the current vDMRG variants, and discuss possible strategies to overcome them.

3.2. Tensor Decompositions and Tensor Network States

The description of quantum many-body systems with tensor networks has received an increasing amount of attention in recent years.[52] Tensor decomposition can be understood as a fragmentation scheme for factorizing the wave function into smaller pieces. The parametrization of the many-body wave function is decomposed into a multitude of tensors, which are connected according to a specific pattern reflecting the entanglement structure of the system. The resulting wave function representation is known as a tensor network state.

3.2.1. *Diagrammatic notation of tensor networks*

Tensors are multidimensional arrays of (real or complex) numbers, with their rank corresponding to the number of indices required to specify their elements. Hence, a tensor of rank zero is a scalar, rank one corresponds to a vector, rank two denotes a matrix, and so forth. Tensors are conveniently represented with a graphical notation,[53] where the tensor core is represented by a solid shape, and its indices are depicted by lines, as shown in Figure 3.1.

A core attribute of tensors is that high-rank tensors can be decomposed into a multitude of lower-rank tensors connected in a network. Mathematically, such a tensor network is a set of tensors where the indices are contracted according to a particular pattern, with index contraction referring to summing over all possible values of the repeated indices of a set of tensors. Contractions are graphically represented by joining index lines in tensor network diagrams, as illustrated in Figure 3.2. A variety of different tensor

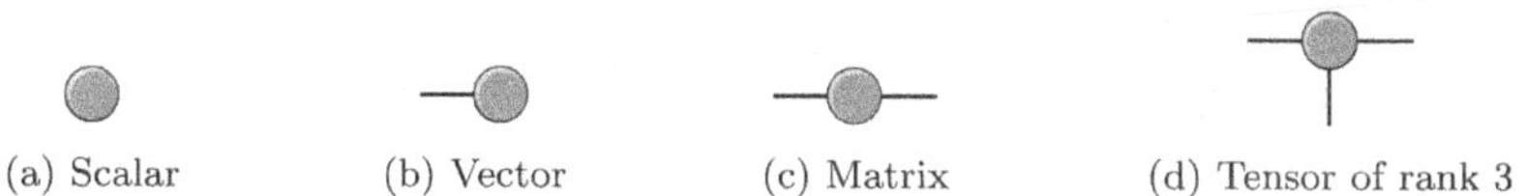

(a) Scalar   (b) Vector   (c) Matrix   (d) Tensor of rank 3

Figure 3.1. Diagrammatic notation of tensors up to rank 3.

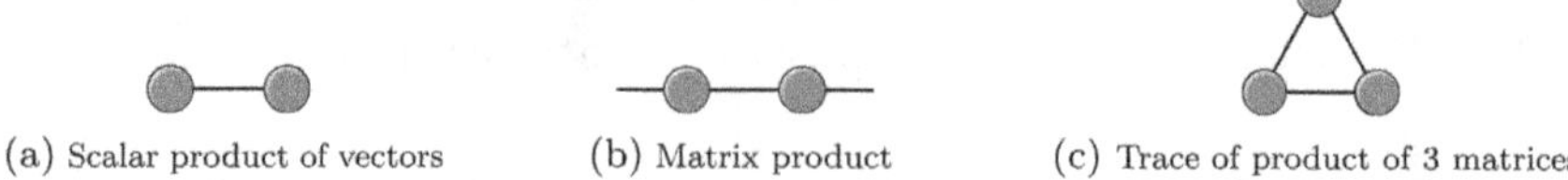

Figure 3.2.   Tensor network diagrams of different index contractions.

network topologies can be realized when decomposing a large-rank tensor. Ideally, the chosen network should provide the most compact representation of the original tensor.

3.2.2. *Tensor networks for quantum states*

Tensor networks can be employed to parametrize quantum many-body states. An arbitrary many-body system with L basis functions is described by the full CI wave function, which can be written as

$$|\Psi\rangle = \sum_{\sigma_1,\sigma_2,\ldots,\sigma_L} c_{\sigma_1,\sigma_2,\ldots,\sigma_L} |\sigma_1\rangle \otimes |\sigma_2\rangle \otimes \cdots \otimes |\sigma_L\rangle, \tag{3.1}$$

$$= \sum_{\sigma} c_{\sigma} |\sigma\rangle, \tag{3.2}$$

where $|\sigma_i\rangle$ represents the possible occupations of the ith basis function, $\otimes$ denotes the tensor product, and $|\sigma\rangle$ is the occupation number vector (ONV). The CI coefficients c_{σ} can be collected in a single large tensor $\mathbf{C}$ of rank L, as visualized in Figure 3.3(a). A compact representation of the tensor $\mathbf{C}$ is obtained by replacing it with a network of smaller tensors. For this purpose, the chosen tensor network should be flexible enough to capture the relevant correlation and its topology should reflect the entanglement structure of the underlying many-body wave function. We review four of the most

Figure 3.3.   (a) Tensor diagram of the CI coefficient tensor and (b) of the corresponding matrix product state for a 5-site system. In both diagrams, the open indices correspond to the physical degrees of freedom of the local Hilbert spaces spanned by $|\sigma_i\rangle$.

common tensor network types in quantum chemistry, each of them corresponding to a different factorization of the CI tensor. We start with the MPS and demonstrate how the full CI wave function can be decomposed into an MPS, before briefly introducing tree tensor network states (TTNS), projected entangled pair states (PEPS), and complete graph tensor network states (CGTNS).

One of the simplest and most common tensor decompositions of a many-body wave function is the tensor train (TT) factorization,[54] which encodes the wave function as an MPS. In an MPS, the rank-L full CI tensor $\mathbf{C}$ is factorized as the product of L rank-3 tensors, one per single-particle basis function of the many-body system,

$$|\Psi\rangle = \sum_{\sigma_1\ldots\sigma_L} \sum_{a_1\ldots a_{L-1}} M^{\sigma_1}_{1a_1} M^{\sigma_2}_{a_1 a_2} \cdots M^{\sigma_L}_{a_{L-1}1} |\sigma_1\rangle \otimes |\sigma_2\rangle \otimes \cdots \otimes |\sigma_L\rangle,$$

$$(3.3)$$

$$= \sum_{\sigma} \mathbf{M}^{\sigma_1} \mathbf{M}^{\sigma_2} \cdots \mathbf{M}^{\sigma_L} |\sigma\rangle, \qquad (3.4)$$

where the CI coefficients c_σ are obtained as the product of a set of matrices $\mathbf{M}^{\sigma_l}$. The diagrammatic notation of an MPS is given in Figure 3.3(b).

The exact MPS representation of a full CI wave function is constructed by reshaping the CI tensor $\mathbf{C}$ of rank L into a matrix $\mathbf{M}_{\sigma_1,(\sigma_2\ldots\sigma_L)}$ with dimension $N \times N^{L-1}$, where N is the number of possible single-particle states, and then calculating the singular value decomposition (SVD) of the resulting matrix. Any general $p \times n$ matrix $\mathbf{M}$ can be factorized as

$$\mathbf{M} = \mathbf{U}\mathbf{S}\mathbf{V}^\dagger, \qquad (3.5)$$

where matrices $\mathbf{U}$ and $\mathbf{V}$ have orthogonal columns, and $\mathbf{S}$ is a diagonal matrix with real non-negative elements, the singular values. The dimensions of the decomposed matrices are $p \times \min(p, n)$ for $\mathbf{U}$, $\min(p, n) \times n$ for $\mathbf{V}$, and $\min(p, n) \times \min(p, n)$ for $\mathbf{S}$. By repeatedly applying SVD to the reshaped matrices of the CI coefficients, as illustrated in Figure 3.4, the CI wave function can be decomposed into a product of L matrices, hence resulting in an MPS parametrization. The linear arrangement of the tensors is referred to as the MPS

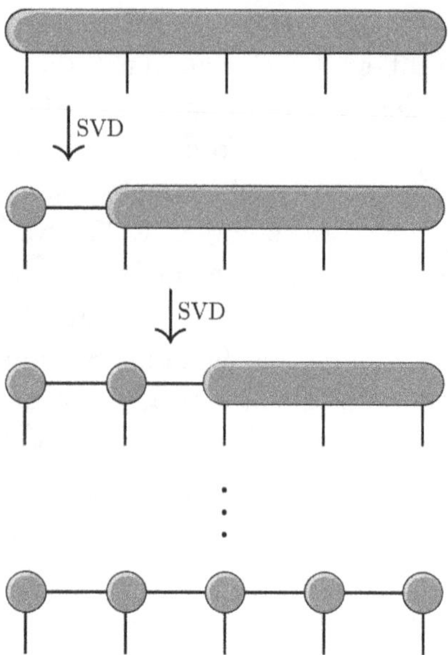

Figure 3.4.   Graphical representation of the decomposition of the CI tensor into an MPS by a sequence of singular value decomposition steps.

lattice, with each tensor being attributed to a specific site on the lattice.

The exact decomposition of a full CI state results in an MPS with matrix dimensions $1 \times N$, $N \times N^2$, ..., $N^{\frac{L}{2}-1} \times N^{\frac{L}{2}}$, $N^{\frac{L}{2}} \times N^{\frac{L}{2}-1}$, ..., $N^2 \times N$, $N \times 1$ for $\mathbf{M}^{\sigma_1}$ to $\mathbf{M}^{\sigma_L}$, respectively. Hence, the number of entries still scales exponentially with L, as for the original CI tensor. The key advantage of the MPS representation is that each SVD step (see Eq. (3.5)) can be truncated to control the size of the factorization. This is achieved by limiting the dimension of the matrices $\mathbf{M}^{\sigma_i}$ to some maximum size of $m \times m$, where m is referred to as the bond dimension. In practice, this truncation step is realized by setting all but the m largest diagonal elements in $\mathbf{S}$ to zero, and retaining only the m first columns of $\mathbf{U}$ and $\mathbf{V}$. The error introduced in each truncation step can be straightforwardly evaluated based on the sum of the squared values of the discarded singular values.[55]

The bond dimension m is the key parameter of the MPS wave function representation. A maximal reduction of the bond dimension to $m = 1$ results in an ordinary product state, which corresponds to the mean-field reference that neglects all correlation within the system. The MPS representation can be enhanced by increasing the bond dimension, with more correlated systems requiring higher values of m to accurately encode many-particle correlation effects. The resulting MPS wave function of bond dimension m scales polynomially, while also accounting for correlation between the different sites, therefore providing a compact parametrization of the full CI problem. The efficiency of the MPS compression is ensured by the area law,[36] which entails that for short-ranged one-dimensional Hamiltonians the bond dimension m required to represent the ground state with a given accuracy is independent of the system size L, therefore taming the exponential scaling of the full CI wave function. While most Hamiltonians encountered in quantum chemistry do not satisfy the area law conditions, the wave functions can nevertheless be efficiently parametrized as MPS. It has been shown that even though the MPS truncation introduces an approximation of the full CI wave function for long-range Hamiltonians, MPS with significantly reduced bond dimensions can accurately represent a broad range of systems.[6–10,12–17,43–45] Hence, the MPS decomposition provides one of the most powerful representations of molecular wave functions in the presence of strong correlation effects.

While the MPS parameterization yields extremely compact wave function representations for quasi-monodimensional Hamiltonians, other tensor network geometries might encode the entanglement structure of certain systems more efficiently. Two of the simplest generalizations of MPS are TTNSs[49,56–59] and PEPSs.[60,61]

While MPSs encode a one-dimensional correlation structure, TTNS employ tree-like connections, whereas PEPS generalize the MPS structure to two spatial dimensions. Another commonly employed tensor factorization in quantum chemistry is the CGTNS,[62–64] where each site is connected to every other site, as illustrated in Figure 3.5.

Compared to MPS, TTNSs allow for a more flexible representation of the wave function entanglement, while PEPSs are tailored for two-dimensional quantum systems, whereas CGTNS account for

                    N. Glaser, A. Baiardi, & M. Reiher

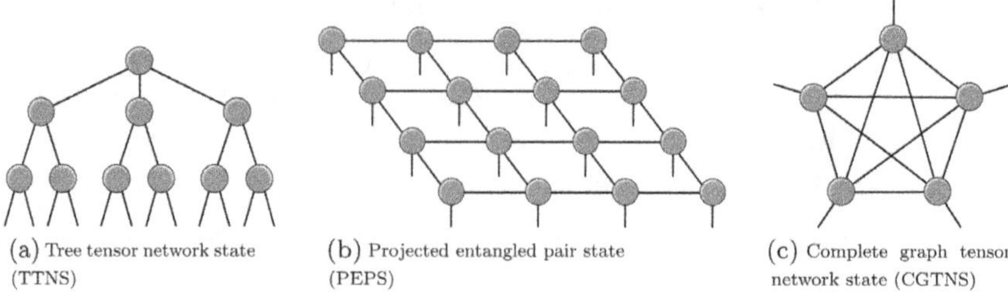

Figure 3.5.   Tensor network diagrams of MPS generalizations.

all interactions between any pair of sites. While these geometries can lead to more compact wave function parametrizations, more complex connection graphs also result in higher computational costs for optimizing their parameters. Hence, there is often an implicit trade-off between tensor network state flexibility and the computational scaling, because the numerical optimization of the tensor network becomes less efficient for more complex tensor factorizations.

One of the most commonly applied tensor network-based algorithms is the density matrix renormalization group (DMRG), which is based on an efficient deterministic variational optimization of matrix product states. Tensor network states with a more complex structure than MPS are often variationally optimized with Monte Carlo-based optimization algorithms.[65] The development of increasingly efficient sampling schemes[66] is constantly enhancing the efficiency of these stochastic optimization algorithms.[62, 64] However, they all share a common limitation, namely that the wave function is optimized in the full CI space. Therefore, they do not fully exploit the compact structure of the underlying tensor factorization. A notable exception are TTNSs, for which a DMRG-like optimization strategy has been devised.[57] Moreover, the recently developed algorithms to solve the time-dependent Schrödinger equation for TTNSs[67, 68] have paved the route towards the optimization of TTNSs with imaginary-time propagation algorithms. Although these developments could set the ground for new efficient tensor-based methods in the near future, DMRG currently provides the optimal balance between the complexity of the tensor factorization and of its optimization.

Therefore, we will focus on DMRG-based algorithms for the remainder of this chapter.

3.3. MPS/MPO Formulation of the Density Matrix Renormalization Group

Finding the optimal tensor network state representation of a quantum many-body wave function is a very challenging task that requires solving complex non-linear optimization problems. For wave functions encoded as MPS, an efficient optimization technique is offered by the DMRG algorithm. DMRG was first introduced in solid state physics in 1992 by White[4] as a method to calculate ground state wave functions of one-dimensional quantum lattices. In its original formulation, DMRG was not contrived as a tensor network approach, but relied on the theoretical framework of the numerical renormalization group.[69] The tensor network-based formulation of DMRG was developed later, when it was discovered that the entire problem can be reformulated in terms of matrix product parametrizations.[35,70] This second formulation proved especially useful for quantum chemical applications, as it straightforwardly supports arbitrarily complex Hamiltonians. Therefore, this MPS-based formulation of the DMRG algorithm will be the main focus for the remainder of this chapter.

3.3.1. *Variational optimization in the MPS/MPO framework*

As a variational method, DMRG optimizes the MPS parameters by minimizing the expectation value of the Hamiltonian $\langle\Psi|\hat{\mathcal{H}}|\Psi\rangle$ for normalized wave functions Ψ. In practice, the Lagrangian $\mathcal{L}$

$$\mathcal{L} = \langle\Psi|\hat{\mathcal{H}}|\Psi\rangle - \lambda\left(\langle\Psi|\Psi\rangle - 1\right) \tag{3.6}$$

is minimized, with λ being the Lagrangian multiplier enforcing the normalization constraint. The expectation value of the Hamiltonian is conveniently minimized by expressing the Hamiltonian operator in a form analogous to the MPS, known as a matrix product operator

(MPO). A general operator $\hat{\mathcal{W}}$ of the form

$$\hat{\mathcal{W}} = \sum_{\boldsymbol{\sigma},\boldsymbol{\sigma}'} w_{\boldsymbol{\sigma},\boldsymbol{\sigma}'} |\boldsymbol{\sigma}\rangle\langle\boldsymbol{\sigma}'|, \tag{3.7}$$

can be exactly expressed as an MPO

$$\hat{\mathcal{W}} = \sum_{\boldsymbol{\sigma},\boldsymbol{\sigma}'} \sum_{b_1 \ldots b_{L-1}} W_{1b_1}^{\sigma_1\sigma_1'} W_{b_1 b_2}^{\sigma_2\sigma_2'} \cdots W_{b_{L-1}1}^{\sigma_L\sigma_L'} |\boldsymbol{\sigma}\rangle\langle\boldsymbol{\sigma}'|. \tag{3.8}$$

By introducing

$$\hat{W}_{b_{l-1}b_l}^{[l]} = \sum_{\sigma_l,\sigma_l'} W_{b_{l-1}b_l}^{\sigma_l\sigma_l'} |\sigma_l\rangle\langle\sigma_l'|, \tag{3.9}$$

we can simplify the MPO expression to

$$\hat{\mathcal{W}} = \sum_{b_1 \ldots b_{L-1}} \hat{W}_{1b_1}^{[1]} \hat{W}_{b_1 b_2}^{[2]} \cdots \hat{W}_{b_{L-1}1}^{[L]} \tag{3.10}$$

$$= \hat{\mathbf{W}}^{[1]} \hat{\mathbf{W}}^{[2]} \cdots \hat{\mathbf{W}}^{[L]}. \tag{3.11}$$

The matrices $\hat{\mathbf{W}}^{[l]}$ are operator-valued matrices that collect the elementary operators acting on site l. In the MPS/MPO framework, the calculation of both the overlap and the expectation value can be interpreted in terms of index contractions of tensor networks. The corresponding tensor network diagrams are shown in Figure 3.6.

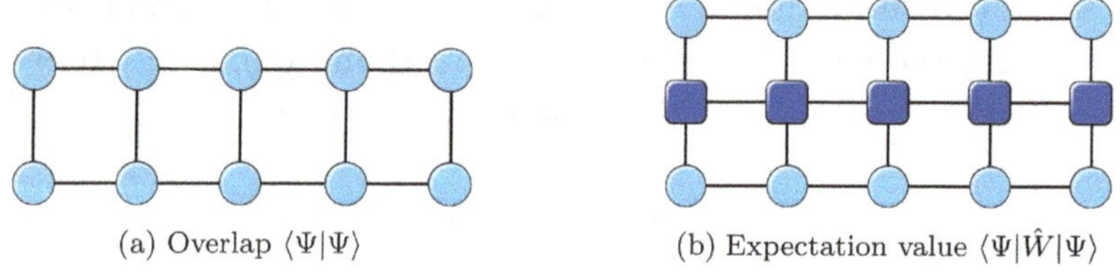

(a) Overlap $\langle\Psi|\Psi\rangle$ (b) Expectation value $\langle\Psi|\hat{W}|\Psi\rangle$

Figure 3.6. (a) Tensor network diagram representing the calculation of the overlap $\langle\Psi|\Psi\rangle$ and (b) the expectation values of the Hamiltonian, expressed as an MPO, over an MPS. In the right panel, the dark blue squares represent the matrix product operator $\hat{W}$, while the light blue circles denote the MPS tensors of the wave function Ψ.

The bond dimensions of an MPO are uniquely determined by the definition of the operator in the given basis set. As discussed, for instance, in Ref. 71, the maximum value of the MPO bond dimension increases with the number of long-range interaction terms. For example, for two-body operators, the MPO bond dimension grows as $\mathcal{O}(L^3)$, and for three-body operators it grows as $\mathcal{O}(L^5)$. Larger MPO bond dimensions not only increase the memory requirements of a calculation, but also the computational cost of contracting the MPO with the MPS to evaluate the expectation value. The calculation time can be decreased by compressing the MPO. Analogous to the MPS compression, the MPO bond dimension can be reduced by successively applying truncated SVDs up to a given target accuracy.[72,73]

3.3.2. *Canonical form*

While the minimization of the Lagrangian is, in principle, possible for any form of the MPS, the corresponding equations can be expressed in a much more convenient form if the MPS is brought into the so-called canonical form. This transformation lays the foundation for an efficient optimization with the DMRG algorithm, as described in Section 3.3.4.

The MPS representation of a given wave function is not unique, since an identity matrix $\mathbf{I} = \mathbf{X}\mathbf{X}^{-1}$ can be inserted between any two matrices $\mathbf{M}^{\sigma_l}$ and $\mathbf{M}^{\sigma_{l+1}}$ on the MPS lattice. Hence, the overall MPS will not change if the matrix $\mathbf{M}^{\sigma_l}$ is replaced by $\mathbf{M}^{\sigma_l}\mathbf{X}$ while $\mathbf{M}^{\sigma_{l+1}}$ is replaced by $\mathbf{X}^{-1}\mathbf{M}^{\sigma_{l+1}}$. By virtue of this gauge freedom, the MPS can be brought into the canonical form. In the so-called left-canonical form

$$|\Psi\rangle = \sum_{\sigma} \mathbf{A}^{\sigma_1} \mathbf{A}^{\sigma_2} \cdots \mathbf{A}^{\sigma_L} |\sigma\rangle \qquad (3.12)$$

all matrices $\mathbf{A}^{\sigma_l}$ are left-normalized, meaning that

$$\sum_{i=1}^{N} \mathbf{A}^{\sigma_i\dagger} \mathbf{A}^{\sigma_i} = \mathbf{I}, \qquad (3.13)$$

where the sum runs over the local basis at the site l. Alternatively, the MPS can be brought into the right-canonical form where

$$|\Psi\rangle = \sum_{\sigma} \mathbf{B}^{\sigma_1}\mathbf{B}^{\sigma_2}\cdots\mathbf{B}^{\sigma_L}|\sigma\rangle, \qquad (3.14)$$

with right-normalized matrices $\mathbf{B}^{\sigma_l}$ such that

$$\sum_{i=1}^{N} \mathbf{B}^{\sigma_i}\mathbf{B}^{\sigma_i\dagger} = \mathbf{I}. \qquad (3.15)$$

The tensor associated with site l, $\mathbf{M}^{\sigma_l}$ with elements $M^{\sigma_l}_{a_{l-1}a_l}$, can be left-normalized by reshaping it into a single matrix with elements $M_{(\sigma_l a_{l-1}),a_l}$. SVD is then applied to the resulting matrix with

$$M_{(\sigma_l a_{l-1}),a_l} = \sum_{s_l} U_{(\sigma_l a_{l-1}),s_l} S_{s_l,s_l} V^{\dagger}_{s_l,a_l}. \qquad (3.16)$$

The matrix $U_{(\sigma_l a_{l-1}),s_l}$ can be reshaped back into a tensor with elements $A^{\sigma_l}_{a_{l-1}s_l}$ obeying the left-normalization condition. The remaining components of the SVD, *i.e.*, $\mathbf{SV}^{\dagger}$, are multiplied into the matrices at the next site to update the tensor $\mathbf{M}^{\sigma_{l+1}}$. The left-canonical form of an MPS is obtained with sequential SVDs starting from the left, until the end of the lattice is reached. Equivalently, the right-canonical form is obtained with sequential SVDs starting from the right, at each step replacing the tensor $\mathbf{M}^{\sigma_l}$ by the reshaped matrix $\mathbf{V}^{\dagger}$ and updating the tensor to the left by multiplying $\mathbf{US}$ into the matrices at site $(l-1)$.

There exists also a so-called mixed-canonical form[8,46]

$$|\Psi\rangle = \sum_{\sigma} \mathbf{A}^{\sigma_1}\cdots\mathbf{A}^{\sigma_{l-1}}\mathbf{M}^{\sigma_l}\mathbf{B}^{\sigma_{l+1}}\cdots\mathbf{B}^{\sigma_L}|\sigma\rangle, \qquad (3.17)$$

where all matrices to the left of an arbitrarily chosen site l are left-normalized, while all remaining matrices to the right are right-normalized, as illustrated in Figure 3.7. This form is of special importance in DMRG, because it facilitates the efficient sweep-based optimization of the MPS, which will be introduced in Section 3.3.4.

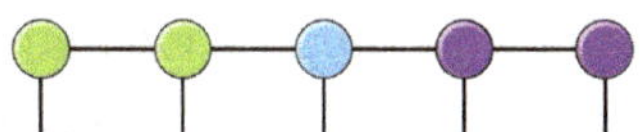

Figure 3.7.   5-site MPS in mixed canonical form. The light green circles represent left-normalized MPS tensors **A**, while the purple ones denote right-normalized tensors **B**. The light blue circle in the center represents an unnormalized MPS tensor **M**.

3.3.3. *Expectation values*

A general expectation value of an operator in MPO form over an MPS can be written as[8,46,71]

$$\langle \Psi | \hat{\mathcal{W}} | \bar{\Psi} \rangle = \sum_{\substack{a_1,\ldots,a_{L-1} \\ a'_1,\ldots,a'_{L-1} \\ \boldsymbol{\sigma\sigma'}}} \left(M^{\sigma_1}_{1a_1} \cdots M^{\sigma_L}_{a_{L-1}1} \right)^* \sum_{b_1,\ldots,b_{L-1}} W^{\sigma_1\sigma'_1}_{1b_1}$$

$$\cdots W^{\sigma_L\sigma'_L}_{b_{L-1}1} \left(\bar{M}^{\sigma'_1}_{1a'_1} \cdots \bar{M}^{\sigma'_L}_{a'_{L-1}1} \right). \tag{3.18}$$

The order in which the sums in Eq. (3.18) are evaluated determines the computational cost of the expectation value evaluation. For instance, if all sums over the a and b indices are calculated first, and then the sum over the σ indices is evaluated, the wave function and operators will first be converted into the CI format, and only afterwards the expectation value will be calculated. In this way, no computational advantage has been gained because the compact form of the MPS wave function is not exploited. By contrast, it is highly efficient to regroup the terms entering Eq. (3.18) as

$$\langle \Psi | \hat{\mathcal{W}} | \bar{\Psi} \rangle = \sum_{\substack{a_{L-1}a'_{L-1}b_{L-1} \\ \sigma_L\sigma'_L}} M^{\sigma_L*}_{a_{L-1}1} W^{\sigma_L\sigma'_L}_{b_{L-1}1} \left(\cdots \left(\sum_{\substack{a_1a'_1b_1 \\ \sigma_2\sigma'_2}} M^{\sigma_2*}_{a_1a_2} W^{\sigma_2\sigma'_2}_{b_1b_2} \right.\right.$$

$$\times \left. \left(\sum_{\sigma_1\sigma'_1} M^{\sigma_1*}_{1a_1} W^{\sigma_1\sigma'_1}_{1b_1} \bar{M}^{\sigma'_1}_{1a'_1} \right) \bar{M}^{\sigma'_2}_{a'_1a'_2} \cdots \right) \bar{M}^{\sigma'_L}_{a'_{L-1}1}.$$

$$\tag{3.19}$$

The expectation value can then be evaluated by first calculating the expression in the innermost parentheses, which corresponds to a contraction of the tensors associated with the first site. The parentheses can then be evaluated sequentially until the outermost part has been reached. This procedure can be converted into a set of recursive equations,

$$\mathbb{L}^{b_0}_{a_0 a'_0} = 1, \tag{3.20}$$

$$\mathbb{L}^{b_1}_{a_1 a'_1} = \sum_{\sigma_1 \sigma'_1} M^{\sigma_1 *}_{1 a_1} W^{\sigma_1 \sigma'_1}_{1 b_1} \bar{M}^{\sigma'_1}_{1 a'_1}, \tag{3.21}$$

$$\vdots$$

$$\mathbb{L}^{b_l}_{a_l a'_l} = \sum_{\substack{a_{l-1} a'_{l-1} b_{l-1} \\ \sigma_l \sigma'_l}} M^{\sigma_l *}_{a_{l-1} a_l} W^{\sigma_l \sigma'_l}_{b_{l-1} b_l} \mathbb{L}^{b_{l-1}}_{a_{l-1} a'_{l-1}} \bar{M}^{\sigma'_l}_{a'_{l-1} a'_l}. \tag{3.22}$$

At the last site we eventually obtain

$$\mathbb{L}^{b_L}_{a_L a'_L} = \langle \Psi | \hat{\mathcal{W}} | \bar{\Psi} \rangle. \tag{3.23}$$

The tensor $\mathbb{L}^{b_l}_{a_l a'_l}$ is referred to as the left boundary at site l.[8,46,71] The right boundary $\mathbb{R}^{b_{l-1}}_{a'_{l-1} a_{l-1}}$ can be defined analogously as

$$\mathbb{R}^{b_{l-1}}_{a'_{l-1} a_{l-1}} = \sum_{\substack{a_l a'_l b_l \\ \sigma_l \sigma'_l}} \bar{M}^{\sigma'_l}_{a'_{l-1} a'_l} W^{\sigma_l \sigma'_l}_{b_{l-1} b_l} \mathbb{R}^{b_l}_{a'_l a_l} M^{\sigma_l *}_{a_{l-1} a_l}. \tag{3.24}$$

Note that a naive evaluation of Eq. (3.21) scales as $\mathcal{O}(m^4 \cdot b_l^2 \cdot N^2)$ for site l, and therefore the cost of each contraction step is independent of L. Accordingly, the overall scaling with L is linear, and not exponential as for full CI.

With the definition of the left and right boundaries, the expectation value $\langle \Psi | \hat{\mathcal{W}} | \bar{\Psi} \rangle$ can be calculated at any site on the lattice according to

$$\langle \Psi | \hat{\mathcal{W}} | \bar{\Psi} \rangle = \sum_{a_l a'_l b_l} \mathbb{L}^{b_l}_{a_l a'_l} \mathbb{R}^{b_l}_{a'_l a_l}. \tag{3.25}$$

It should be noted that the left or right boundaries will not change if the tensor network is modified further to the right or to the left of site l, respectively. Therefore, the boundaries can be stored for a given site, and only need to be updated once an MPS tensor which is contained within that boundary is changed. This property allows for an efficient sweep-based optimization, as described in the next Section.

3.3.4. *MPS optimization*

In a variational DMRG calculation, the expectation value $\langle \Psi | \hat{\mathcal{H}} | \Psi \rangle$ is optimized with respect to the tensor entries of site l. The expectation value can be evaluated at site l based on the left and right boundaries formed from left- and right-normalized matrices (see Figure 3.8(b)) as

$$\langle \Psi | \hat{\mathcal{H}} | \Psi \rangle = \sum_{\substack{a_{l-1} a_l \sigma_l \\ a'_{l-1} a'_l \sigma'_l \\ b_{l-1} b_l}} M^{\sigma_l *}_{a_{l-1} a_l} W^{\sigma_l \sigma'_l}_{b_{l-1} b_l} \mathbb{L}^{b_{l-1}}_{a_{l-1} a'_{l-1}} M^{\sigma'_l}_{a'_{l-1} a'_l} \mathbb{R}^{b_l}_{a'_l a_l}. \tag{3.26}$$

To find the optimal MPS parameters for a given site l, the gradient of the Lagrangian in Eq. (3.6) calculated with respect to the tensor coefficient $M^{\sigma_l *}_{a_{l-1} a_l}$ is set to zero. This results in the eigenvalue equation

$$\sum_{\substack{a'_{l-1} a'_l \sigma'_l \\ b_{l-1} b_l}} W^{\sigma_l \sigma'_l}_{b_{l-1} b_l} \mathbb{L}^{b_{l-1}}_{a_{l-1} a'_{l-1}} M^{\sigma'_l}_{a'_{l-1} a'_l} \mathbb{R}^{b_l}_{a'_l a_l} = \lambda M^{\sigma_l}_{a_{l-1} a_l}, \tag{3.27}$$

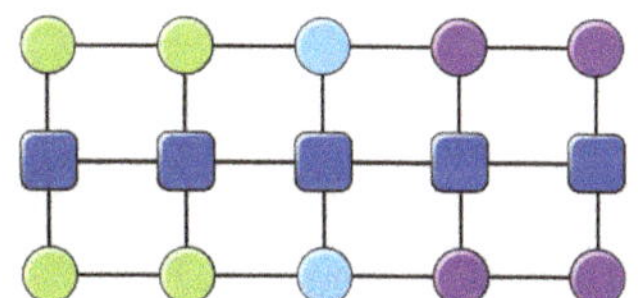

(a) Expectation value without boundaries

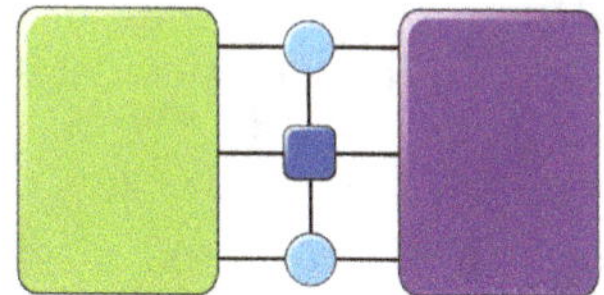

(b) Expectation value with boundaries

Figure 3.8. Tensor network diagrams of the expectation value $\langle \Psi | \hat{H} | \Psi \rangle$: (a) displays the full tensor network contraction according to Eq. (3.18), while in (b) left and right boundaries are shown as light green and purple boxes containing the left and right contracted portions of the network according to Eq. (3.26), respectively.

which can be interpreted as a conventional eigenvalue problem $\boldsymbol{H}\boldsymbol{v} = \lambda\boldsymbol{v}$, where $\boldsymbol{v}$ is a one-dimensional vector with the elements $v_{a_{l-1}a_l\sigma_l} = M^{\sigma_l}_{a_{l-1}a_l}$, and the matrix $\boldsymbol{H}$ is defined as[46]

$$H_{(a_{l-1}a_l\sigma_l),(a'_{l-1}a'_l\sigma'_l)} = \sum_{b_{l-1}b_l} \mathbb{L}^{b_{l-1}}_{a_{l-1}a'_{l-1}} W^{\sigma_l\sigma'_l}_{b_{l-1}b_l} \mathbb{R}^{b_l}_{a'_la_l}. \tag{3.28}$$

Eq. (3.27) can be solved with an iterative eigenvalue solver such as the Jacobi–Davidson algorithm.[74,75] The solution to Eq. (3.27) for every coefficient $M^{\sigma_l}_{a_{l-1}a_l}$ returns the optimal MPS tensor $\boldsymbol{M}^{\sigma_l}$ for a given site l. Hence, to fully optimize an MPS, the equation must be solved for all sites of the DMRG lattice in a sequential manner. The optimization is then repeated iteratively until self-consistency between all tensors is reached.

DMRG optimizes the MPS sequentially with a sweeping procedure, as moving from one site to the next enables an optimal recycling of the boundaries. First, a guess MPS is constructed and brought into the right-canonical form, and the right boundary $\mathbb{R}$ is calculated for every site l on the lattice. Then, a so-called forward sweep is conducted, where for every site from $l{=}1$ to $l{=}(L-1)$

1. the eigenvalue problem of Eq. (3.27) is solved to obtain the optimal tensor $\mathbf{M}^{\sigma_l}$,
2. the obtained $\mathbf{M}^{\sigma_l}$ tensor is left-normalized (see Eq. (3.13)) to obtain $\mathbf{A}^{\sigma_l}$ from $\mathbf{U}$,
3. the MPS coefficients for the next site are updated by multiplying $\mathbf{S}\mathbf{V}^\dagger$ into $\mathbf{M}^{\sigma_{l+1}}$,
4. the left boundary $\mathbb{L}^{b_l}_{a_la'_l}$ is updated, and the next site to the right on the lattice is selected, as illustrated in Figure 3.9.

Once the whole lattice has been traversed from the left, the procedure is reversed at site $l = L$. A backward sweep is conducted, where for every site from $l = L$ to $l = 2$

1. the eigenvalue problem of Eq. (3.27) is solved to obtain the optimal tensor $\mathbf{M}^{\sigma_l}$,
2. the obtained $\mathbf{M}^{\sigma_l}$ is right-normalized by performing SVD to obtain $\mathbf{B}^{\sigma_l}$ from $\mathbf{V}^\dagger$,

3. the MPS coefficients for the neighboring site are updated by multiplying $\mathbf{US}$ into $\mathbf{M}^{\sigma_{l-1}}$,

4. the right boundary $\mathbb{R}^{b_l}_{a'_l a_l}$ is updated, and the next site to the left on the lattice is selected.

After the entire lattice has been traversed forth and back, the MPS parameters are optimized further by performing subsequent sweeps until convergence is reached.

The sweeping procedure described above optimizes the tensor entries for a single site at a time. Alternatively, two sites can be optimized simultaneously with the so-called two-site DMRG variant by introducing two-site tensors for both the MPS and the MPO.[8] In this case, the two-site tensor

$$T^{\sigma_i,\sigma_{i+1}}_{a_{i-1},a_{i+1}} = \sum_{a_i} M^{\sigma_i}_{a_{i-1},a_i} M^{\sigma_{i+1}}_{a_i,a_{i+1}}, \qquad (3.29)$$

is constructed and optimized at each microiteration. The optimized MPS tensors are then obtained from the SVD of the optimized $\mathbf{T}^{\sigma_i,\sigma_{i+1}}$ tensor, reshaped as $T_{(a_{i-1}\sigma_i),(a_i\sigma_{i+1})}$. The two-site algorithm enhances the DMRG convergence and reduces the risk of getting stuck in local minima. Additionally, it allows for an adaptive setting of the bond dimension, meaning that m can be changed at every microiteration step. In fact, the number of non-zero singular values of $\mathbf{T}^{\sigma_i,\sigma_{i+1}}$ may change during the optimization, and,

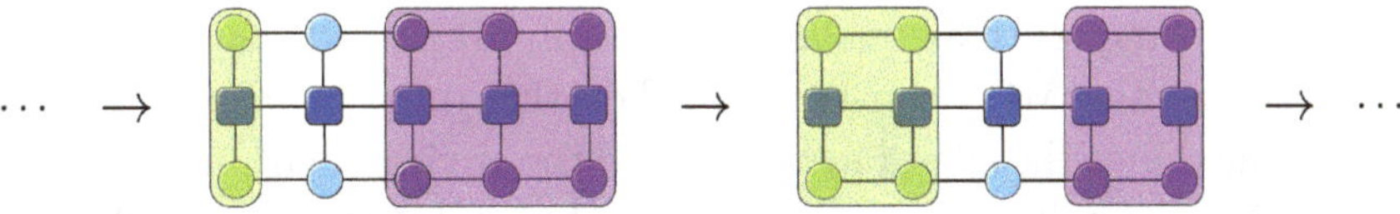

Figure 3.9.   Graphical illustration of two steps of the forward sweep. The shaded boxes represent the left and right boundaries containing the contracted portions of the network in green and purple, respectively. After solving the eigenvalue problem for site l, $\mathbf{M}^{\sigma_l}$ is left-normalized (hence, turning green), which entails that the tensor for site $(l+1)$ is also updated and no longer right-normalized (hence, turning blue). The left boundary on site l is updated and the optimization progresses to site $(l+1)$.

consequently, so does the bond dimension required for a reliable MPS approximation of the wave function. While being more stable, two-site DMRG is computationally less efficient than the single-site algorithm as it optimizes larger tensors at each step of the sweeping procedure. For vibrational applications, the single-site algorithm usually achieves satisfactory convergence[48] and the two-site optimization is applied only to cases where convergence issues are observed.

3.3.5. *Site selection and ordering with quantum information measures*

By truncating the MPS bond dimension m, DMRG tames the exponential scaling of the full CI problem, but, at the same time, it limits the extent to which correlation can be encoded in the wave function. This trade-off can be counteracted by an optimal arrangement of the sites on the DMRG lattice. Ideally, a tensor network is constructed in such a way that it directly encodes the correlation structure of the system which it represents. For an MPS, this translates into placing strongly correlated sites in close vicinity to each other. Due to the sequential optimization of the MPS with the sweeping procedure, the convergence of DMRG is greatly affected by the ordering of the sites.

For an optimal ordering of the sites on the DMRG lattice, the correlation structure of the system at hand has to be determined. Fortunately, DMRG enables the characterization of the wave function structure with measures from quantum information theory. These measures can quantify correlation effects in MPSs and can be utilized to optimize the MPS construction. For this purpose, the system is partitioned into the site of interest i and the environment, in which all other sites are collected. The one-site reduced density matrix (RDM) can then be constructed to calculate the von-Neumann-type entropy as[76]

$$s(1)_i = - \sum_{\alpha=1}^{N_i} \omega_{\alpha,i} \ln \omega_{\alpha,i}, \qquad (3.30)$$

which is referred to as the single-site entropy of the site i. The sum runs over all eigenvalues $\omega_{\alpha,i}$ of the one-site RDM, with N_i being the size of the local basis of site i. Hence, the single-site entropy quantifies how much site i contributes to the deviation of the many-body wave function from a pure state (*i.e.*, one that populates only one of the possible states of that site). At the same time, the single-site entropy also provides a measure for the entanglement of site i with all other sites, therefore characterizing the multi-configurational character of site i.

In an analogous manner, the two-site entropy $s(2)_{ij}$ can be defined for a pair of sites i and j. In this case, the system is defined by sites i and j, and a two-site reduced density matrix is constructed. The two-site entropy is then calculated as

$$s(2)_{ij} = -\sum_{\alpha=1}^{N_i N_j} \omega_{\alpha,ij} \ln \omega_{\alpha,ij}, \qquad (3.31)$$

where the sum runs over all $N_i N_j$ eigenvalues $\omega_{\alpha,ij}$ of the two-site RDM, where N_i and N_j are the sizes of the local bases for sites i and j, respectively. The two-site entropy describes how much the combined state of sites i and j is affected by the environment.

We define the two-site entropy reduced by the single-site entropies as the mutual information I_{ij} between two sites i and j,

$$I_{ij} = \frac{1}{2}\left(s(1)_i + s(1)_j - s(2)_{ij}\right)\left(1 - \delta_{ij}\right), \qquad (3.32)$$

which is a measure for how strongly the states of sites i and j are mutually entangled.

Based on the von-Neumann-type entropies and the mutual information, the ordering of the sites on the lattice can be optimized and the sites to be included in the MPS can be selected.[77] As with other quantum chemical methods, states which are not strongly correlated do not need to be explicitly included in the wave function expansion. Hence, both site selection and ordering on the DMRG lattice are greatly facilitated by these measures from quantum information theory. In practice, the measures are obtained from a partially converged DMRG calculation with a small maximum bond dimension m, because the one- and two-site entropies converge much faster with respect to the bond dimension than the energy. The results from such a calculation are then leveraged for optimizing the MPS site selection and ordering for a subsequent accurate and fully converged DMRG calculation with a sufficiently large bond dimension.

3.4. DMRG for Vibrational Problems

As tensor-based methods are powerful tools for a variety of quantum chemical problems, we introduced tensor networks and the DMRG optimization algorithm in a general manner in the previous Sections. In this Section, we will now focus on the vibrational DMRG (vDMRG) theory.[47, 48, 78]

3.4.1. *Vibrational Hamiltonians*

Within the Born–Oppenheimer approximation, molecular vibrations are determined by the potential energy surface (PES) $\mathcal{V}(R)$, which is obtained from the solution of the electronic Schrödinger equation at different nuclear configurations R. The PES is, in general, not known exactly and is usually approximated either with a Taylor series expansion around a reference geometry, or with a so-called n-mode expansion.[79–82] For a vDMRG calculation, the Hamiltonian and, therefore, the PES must be expressed in a second-quantized form obtained by projection onto a finite modal basis set. Depending on the choice for the PES approximation and for the underlying local modal basis, different forms of the second-quantized vibrational Hamiltonian are obtained. Two quantization choices, namely the canonical quantization[83] and the n-mode[84] representation, are discussed in the context of the vDMRG method in this Section.

3.4.1.1. *Canonical quantization*

The canonical quantization is a convenient framework to encode a PES that is expressed as a Taylor series around a reference geometry in terms of the dimensionless Cartesian normal modes Q_i. The Taylor series expansion of the PES around a stationary point $\mathbf{Q}_e$ is given, in atomic units, by

$$\mathcal{V} = \frac{1}{2} \sum_{i=1}^{M} \left(\frac{\partial^2 V}{\partial Q_i^2} \right)_{\mathbf{Q}_e} Q_i^2$$

$$+ \frac{1}{6} \sum_{ijk} \left(\frac{\partial^3 V}{\partial Q_i \partial Q_j \partial Q_k} \right)_{\mathbf{Q}_e} Q_i Q_j Q_k + \cdots \qquad (3.33)$$

$$= \frac{1}{2} \sum_{i=1}^{M} \omega_i^2 Q_i^2 + \frac{1}{6} \sum_{ijk} k_{ijk} Q_i Q_j Q_k + \cdots , \qquad (3.34)$$

where we omitted the energy at the reference point and note that all first derivatives and the off-diagonal second derivatives vanish by definition of $\mathbf{Q}_e$. The harmonic potential is obtained by retaining only the second-order term of Eq. (3.34). As the Schrödinger equation can be solved exactly for the harmonic oscillator Hamiltonian, the corresponding eigenfunctions can be taken as a reference modal basis to expand the PES. The second-quantization representation of the resulting Hamiltonian is obtained by expressing the position operators Q_i in terms of canonical quantization creation and annihilation operators as

$$Q_i = \frac{1}{\sqrt{2}} \left(\hat{b}_i^+ + \hat{b}_i \right). \qquad (3.35)$$

Hence, a pair of second-quantization operators $\hat{b}_i^+$ and $\hat{b}_i$ is introduced for each mode i, whose definition implies that the vibrational wave function is expanded in terms of the harmonic oscillator eigenfunctions. The canonical quantization also determines the mapping of the Hamiltonian to the vDMRG lattice, namely that each site corresponds to one of the M vibrational normal modes, as visualized in Figure 3.10. As harmonic modes are bosons, the canonical operators fulfill bosonic commutation relations and their application to a given occupation number vector results in:

$$\hat{b}_i^+ |n_1, \ldots, n_i, \ldots, n_M\rangle = \sqrt{n_i + 1} |n_1, \ldots, n_i + 1, \ldots, n_M\rangle, \qquad (3.36)$$

$$\hat{b}_i |n_1, \ldots, n_i, \ldots, n_M\rangle = (1 - \delta_{n_i,0}) \sqrt{n_i} |n_1, \ldots, n_i - 1, \ldots, n_M\rangle. \qquad (3.37)$$

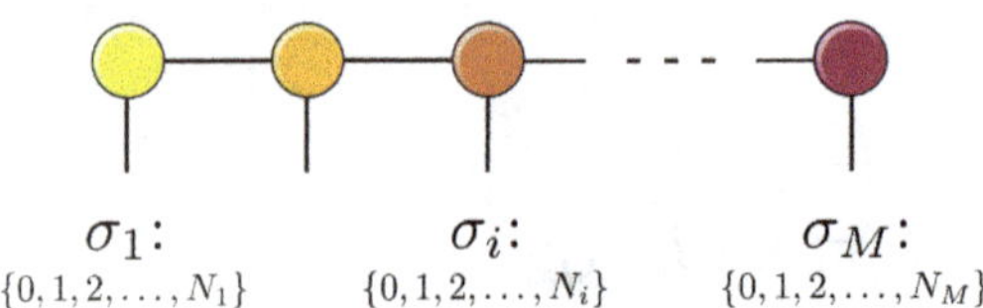

Figure 3.10. Vibrational MPS in the canonical quantization basis. Each site corresponds to a certain vibrational mode, as highlighted by the different colors. For each vibrational mode i, the local basis consists of all vibrational excitations of that mode up to the defined upper bound N_i.

The dimension of the local basis and, hence, the maximum occupation of each site is, in principle, unbounded. In practice, however, a threshold N_i is set to limit the size of the local basis at site i, as highly excited basis functions usually have a negligible effect on the anharmonic vibrational wave function.

While the canonical quantization provides an efficient description of weakly anharmonic systems, it suffers from two severe limitations in the case of strong anharmonicity. First, the Taylor series expansion of the PES does not accurately parametrize potentials without a clear single reference geometry, such as double wells. Second, the harmonic oscillator eigenfunctions as a basis set do not represent strongly anharmonic wave functions accurately. These drawbacks can be overcome with the n-mode quantization, which allows for a more flexible representation of the vibrational Hamiltonian.

3.4.1.2. *n-Mode quantization*

In the n-mode picture, the PES is approximated with a many-body expansion in which terms are grouped together based on the number of modes that are coupled together.[79–82] The potential is written as

$$\mathcal{V} = \sum_{i=1}^{M} \mathcal{V}^{[i]}(Q_i) + \sum_{i<j} \mathcal{V}^{[ij]}(Q_i, Q_j)$$

$$+ \sum_{i<j<k} \mathcal{V}^{[ijk]}(Q_i, Q_j, Q_k) + \cdots, \tag{3.38}$$

where M is the number of modes. The one-body term $\mathcal{V}^{[i]}(Q_i)$ contains the variation of the potential upon change of the i-th coordinate Q_i. Analogously, the two-body term $\mathcal{V}^{[ij]}(Q_i, Q_j)$ represents the variation of the PES for the simultaneous change of two coordinates, Q_i and Q_j and, therefore, increases with the degree of coupling between modes i and j. In contrast to the electronic case, interactions are not limited to one- and two-body terms, as higher-order coupling terms may be relevant. In practice, however, it is often sufficient to only include low-order terms as they contain most of the correlation between the modes.

The n-mode expansion of the PES supports a second-quantization form that is based on generic basis sets. This allows one to choose a basis set that is optimized to yield compact CI wave functions, such as modals obtained from a vibrational self-consistent field (VSCF) calculation.[85, 86] In the n-mode picture, a pair of creation and annihilation operators is introduced for each of the modal basis functions k_i of each mode i, as opposed to the canonical quantization, where only a single pair of operators is required per mode. Hence, an occupation number vector is given as $|\mathbf{n}\rangle = |n_1^1, \ldots, n_{N_1}^1, \ldots, n_1^i, \ldots, n_{N_i}^i, \ldots, n_1^M, \ldots, n_{N_M}^M\rangle$, where $n_{k_i}^i$ represents the occupation of the k_i-th basis function associated with the i-th mode. The corresponding creation $\hat{a}_{k_i}^+$ and annihilation $\hat{a}_{k_i}$ operators are defined as

$$\hat{a}_{k_i}^+|\mathbf{n}\rangle = |\mathbf{n} + \mathbf{1}_{k_i}^i\rangle \delta_{0, n_{k_i}^i}, \tag{3.39}$$

$$\hat{a}_{k_i}|\mathbf{n}\rangle = |\mathbf{n} - \mathbf{1}_{k_i}^i\rangle \delta_{1, n_{k_i}^i}, \tag{3.40}$$

where $|\mathbf{1}_{k_i}^i\rangle$ is the ONV with zero entries except $n_{k_i}^i = 1$.

The second-quantized n-mode potential results in[84]

$$\mathcal{V} = \sum_{i} \sum_{k_i, h_i} \langle k_i|\mathcal{V}_i(Q_i)|h_i\rangle a_{k_i}^+ a_{h_i}$$
$$+ \sum_{i,j} \sum_{k_i, h_i} \sum_{k_j, h_j} \langle k_i k_j|\mathcal{V}_{ij}(Q_i, Q_j)|h_i h_j\rangle a_{k_i}^+ a_{k_j}^+ a_{h_j} a_{h_i} + \cdots.$$
$$\tag{3.41}$$

As each mode is described by a single basis state in each many-body basis function, exactly one of the corresponding modals is occupied in an ONV, whereas all other modals of that mode are unoccupied. Each individual site can only be in one of two states: either it is occupied, if the mode is represented by that specific modal, or it is unoccupied, if the mode is in another basis state. Therefore, the creation and annihilation operators $a_{k_i}^+$ and a_{h_i} always occur in pairs for a given mode i in Eq. (3.41), as mode i is excited from modal h_i to modal k_i.

In a vDMRG calculation in the n-mode picture, a vibrational mode is no longer mapped to a single site on the lattice, but to a number N_i of sites, one for each local basis function. Therefore, in the n-mode quantization the lattice size L no longer corresponds to the number of modes, but to the total number of modals of all modes. The n-mode lattice is graphically represented in Figure 3.11. For a given vibrational problem with M modes, each with N_i basis functions, the canonical quantization results in a significantly smaller vDMRG lattice with size $L=M$, whereas the n-mode lattice has size $L = \sum_i N_i$. The increase in computational cost with the lattice size L is counterbalanced by the smaller local basis in the n-mode picture, which decreases the cost of a single microiteration of the vDMRG algorithm. In general, the n-mode quantization is advantageous in case of strong anharmonicity, as it allows for a more flexible description of the vibrational states and is able to accurately parametrize strongly coupled and highly anharmonic modes.

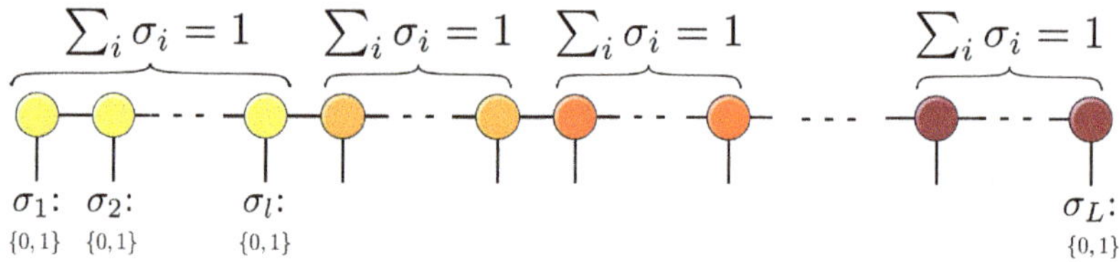

Figure 3.11. Vibrational MPS in the n-mode quantization. Each vibrational mode is mapped to a number of sites, displayed here as circles of the same color. Each site corresponds to one of the modals of that specific mode. Each mode is described by only one modal for any given ONV, so that exactly one of their modals has to be in an occupied state, with all others being unoccupied. Hence, the local basis of each site is either unoccupied ($\sigma_i = 0$) or occupied ($\sigma_i = 1$), with the sum over all sites belonging to the same mode being 1.

3.4.2. *Initial guess and bond dimension*

A guess MPS has to be chosen as a starting point for the optimization to initiate a vDMRG calculation. While the DMRG algorithm may converge independently of the initial guess MPS, the convergence rate to the global minimum can be enhanced significantly by appropriately choosing the starting guess. In practice, the MPS is commonly initialized with the mean-field reference state. For the ground state, this corresponds to all modes being in the lowest-energy basis state in the canonical quantization, while in the n-mode picture each mode is initialized in the lowest-energy modal.

The most vital parameter of a vDMRG calculation is the bond dimension m, as it controls the trade-off between computational complexity and the degree of approximation of the full CI wave function. The minimal bond dimension required to accurately represent a many-body wave function as an MPS depends on the extent of its entanglement. As the correlation within a system is usually not known prior to the calculation, the energy convergence with respect to the bond dimension has to be monitored. In practice, this is done by performing several vDMRG calculations with different bond dimensions and extrapolating the energy of the optimized MPS. By virtue of the variational principle, a lower final energy consistently indicates a better approximation to the ground state wave function. Therefore, a vDMRG calculation will be converged if the energy does not change when m is increased. In contrast to electronic-structure DMRG calculations, converged vDMRG results can be obtained already with relatively small bond dimensions, as demonstrated in Section 3.4.3.

3.4.3. *Example — Anharmonic ZPVE of ethylene*

vDMRG can be straightforwardly applied to calculate the anharmonic zero point vibrational energy (ZPVE) of molecular systems. As discussed above, vDMRG convergence must be monitored with respect to both the number of sweeps and the bond dimension. To illustrate this, we apply vDMRG to ethylene with the sixth-order Taylor-expanded PES taken from Fetherolf and Berkelbach[51] that

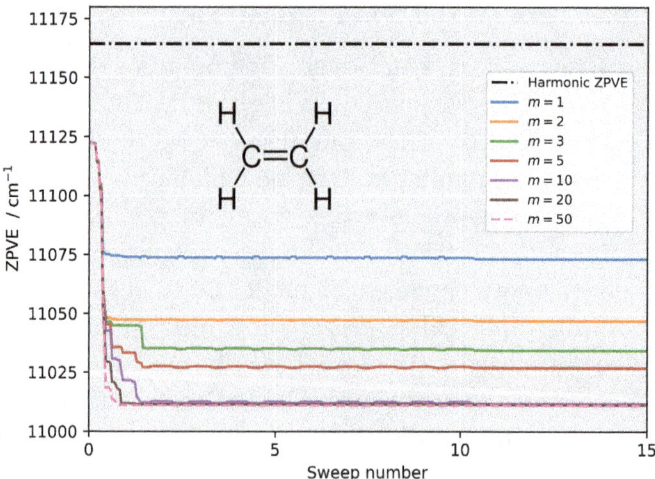

Figure 3.12. vDMRG energy convergence with respect to the number of sweeps of the ground state of ethylene for a range of bond dimensions m.

we encode as an MPO based on the canonical quantization picture. For all examples, we rely on the single-site DMRG optimization algorithm implemented in the `QCMaquis` software module.[87]

All calculations reported in Figure 3.12 are initialized in the harmonic ground state. Their initial energy, the anharmonic ZPVE of the harmonic ground state wave function, is significantly lower than the harmonic reference ZPVE. During the vDMRG calculation, the vibrational energy decreases rapidly during the first few sweeps as the MPS is optimized. For $m = 1$ (blue line in Figure 3.12), the calculation converges to the anharmonic mean-field wave function. By increasing the bond dimension m, the MPS can account for more correlations between the different modes and the anharmonic ZPVE decreases accordingly. No significant change in the final energy is observed for bond dimensions higher than $m = 20$, indicating convergence of the vDMRG calculations with respect to m. For the first 10 sweeps, small oscillations are observed in the energy convergence. This phenomenon is observed because, during the first 10 sweeps, the MPS is perturbed based on the subspace expansion algorithm[88] in order to avoid converging to local minima.

As can be seen in Figure 3.12, the ZPVE of ethylene decreases significantly if anharmonic effects are taken into account. vDMRG converges rapidly, both with the number of sweeps as well as with the bond dimension m, so that for ethylene the energies are converged to sub-cm^{-1} accuracy already after a few sweeps.

3.5. Excited State DMRG

The DMRG algorithm is based on the variational principle and is, therefore, tailored to the optimization of ground state wave functions. When applied to vibrational problems, vDMRG returns the anharmonic ZPVE, E_0. The quantities of interest in vibrational spectroscopy are, however, not the ZPVEs, but the transition frequencies ν_k, which can be calculated from the excited state energies E_k as

$$h\nu_k = E_k - E_0. \tag{3.42}$$

Therefore, in order to calculate ν_k, excited state vDMRG algorithms are required. Various approaches have been developed to target excited state solutions of the time-independent Schrödinger equation with DMRG. In this section, we will introduce the most promising ones for quantum chemical calculations. While calculating excited state energies and wave functions is the key to simulate vibrational spectra in the frequency domain, the very same quantities can also be calculated with an entirely different approach based on the solution of the time-dependent Schrödinger equation with a vDMRG-based algorithm that we will review in the subsequent Section. We note that although many of these methods have originally been developed for vibrational problems, they can be straightforwardly applied to electronic-structure problems as well.

3.5.1. *Excited state targeting by constrained optimization with vDMRG[ortho]*

The simplest extension of the regular vDMRG algorithm for vibrational spectroscopy targets excited states based on a constrained optimization.[48] As all non-degenerate eigenstates of a Hermitian operator are mutually orthogonal, excited states can still be targeted with a variational calculation provided that the optimization is restricted to the variational space orthogonal to all lower-energy states. In the vDMRG[ortho] method,[35,71,89] excited states are calculated sequentially by optimizing the current state under the constraint that the MPS is orthogonal to the previously calculated MPSs of all lower-lying states.

While the vDMRG[ortho] method allows for a straightforward calculation of low-energy excited states, there are significant drawbacks if it is applied to higher-lying excited states. As the algorithm requires all lower-lying states as input, not only the computational effort increases, but also the error which is introduced in the calculation. In fact, the constrained optimization becomes increasingly unstable as the number of constraints grows, especially for small bond dimensions. Possible remedies for this instability are to increase the bond dimension or to track a specified state during the calculation with an overlap criterion, as introduced in the next section.

3.5.2. *Enhancing convergence by root homing: vDMRG[maxO]*

Another extension of vDMRG resting on the variational energy minimization is the vDMRG[maxO] method.[78] In this variant, the MPS optimization is steered by a maximum overlap criterion. During the DMRG sweeping procedure, Eq. (3.27) yields multiple eigenpairs for each microiteration step. In standard DMRG, the tensor with the lowest energy is selected to propagate the ground state wave function. This selection scheme can be adapted to follow a different eigenpair based on the maximum overlap criterion instead of the minimal energy criterion, as illustrated in Figure 3.13. This procedure is known as root homing,[90] as a specific root of the eigenvalue equation is selected.

For a vDMRG[maxO] calculation, a reference MPS has to be chosen with respect to which the overlap is calculated. The predetermined reference MPS is often selected from the eigenstates of the non-interacting Hamiltonian, corresponding to the excited state mean-field reference. Alternatively, the reference MPS can be

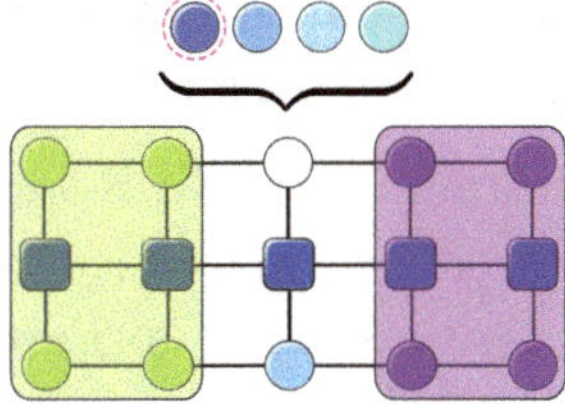

Figure 3.13.   Graphical illustration of the root homing procedure. At each microiteration step of the optimization procedure, multiple eigenpairs of Eq. (3.27) are obtained, as illustrated by the four circles. These are sorted according to their overlap with the reference MPS, as visualized by the darker to lighter blue shading of the tensors. In a vDMRG[maxO] calculation, the tensor which results in the largest overlap with the reference MPS, highlighted by the red contour, is selected and propagated during the subsequent MPS optimization.

obtained from a preliminary vDMRG calculation with a smaller bond dimension m. To allow for an efficient overlap calculation, partial overlaps are constructed as illustrated in Figure 3.14(b), stored for each site, and updated during the sweeping procedure.

The vDMRG[maxO] method can be applied to directly target excited states by choosing a reference MPS which approximates the state of interest. vDMRG[maxO] can also be combined with other excited state methods to ensure convergence to the desired state even in regions with a high density of states. Particularly appealing is the combination of the maximum overlap criterion with the vDMRG[S&I] method which is introduced in the next Section.

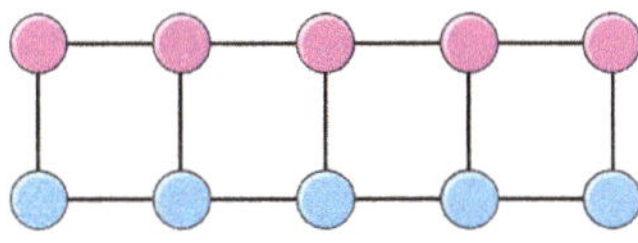

(a) Overlap without boundaries

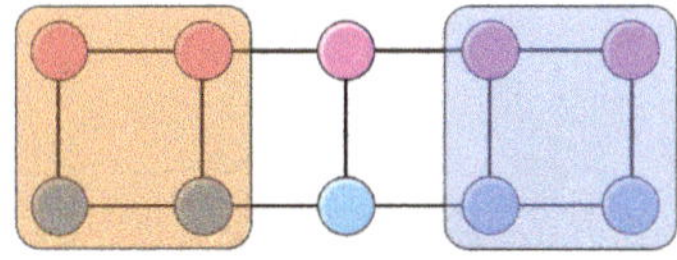

(b) Overlap with boundaries

Figure 3.14.   Tensor network diagrams of the overlap between two different MPSs $\langle \Phi | \Psi \rangle$. (a) The overlap of two different MPSs, as highlighted by their different coloring, is evaluated with a naive contraction of the entire network. (b) The overlap evaluation can be greatly facilitated by calculating and storing the boundaries, which collect the precontracted portions of the network, as illustrated by the shaded boxes. In vDMRG[ortho], the overlap boundaries are stored and updated for every site l on the lattice in analogy to the expectation value boundaries discussed in Section 3.3.3.

3.5.3. *Auxiliary operator-based algorithms: vDMRG[S&I] and vDMRG[f]*

As an alternative to constraining the MPS optimization, the vDMRG calculation can be steered by mapping the Hamiltonian to an auxiliary operator in order to directly optimize an excited state with the ground state DMRG algorithm. For instance, the vDMRG[S&I] approach applies conventional vDMRG to the shifted-and-inverted auxiliary Hamiltonian, Ω_ω[78]

$$\Omega_\omega = (\omega - \mathcal{H})^{-1}, \tag{3.43}$$

where ω is an energy shift parameter. The smallest eigenvalue of Ω_ω corresponds to the first eigenvalue of $\mathcal{H}$ with an energy greater than ω. Hence, a ground state optimization applied to Ω_ω yields the excited state directly above ω, as illustrated in Figure 3.15.

The vDMRG[S&I] approach requires an energy shift ω as an input parameter so that the algorithm can then optimize a specific excited state. Since the energy of the target state is, in general, not known *a priori*, a guess based on experimental values or results from previous calculations has to be made. This becomes an issue in regions with a high density of states, as the energy shift ω

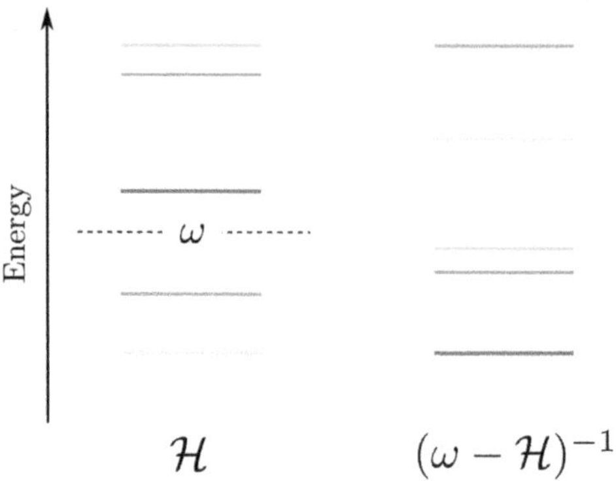

Figure 3.15.   Illustration of the energy eigenvalues of the shifted and inverted Hamiltonian Ω_ω. The red state, which is the first with an energy above the shift ω, is mapped to the ground state of the auxiliary operator $\Omega_\omega = (\omega - \mathcal{H})^{-1}$ and can, therefore, be targeted by variational optimization with vDMRG[S&I].

must be chosen with high accuracy to ensure that vDMRG[S&I] converges to the desired eigenstate. Fortunately, the vDMRG[S&I] method can be made more robust by combining it with the maximum overlap criterion. However, another major issue remains. Encoding the auxiliary operator Ω_ω as an MPO is a challenging numerical problem because, even though $\mathcal{H}$ might be represented as a compact MPO, its inverse may be encoded by an arbitrarily complex MPO representation. Algorithms that approximate the inverse of an MPO have been developed,[34,91] but they have not been applied to quantum chemical problems so far. The MPO inversion can be avoided by applying the S&I transformation to the local, site-centered operator defined by Eq. (3.28)[89,92] based on the theory of harmonic Ritz values.[75] However, if the S&I transformation is applied after the eigenvalue equation is projected onto a given site, the method's variationality and, therefore, its stability is no longer ensured.[47] For this reason, vDMRG[S&I] may not converge if applied to optimize high-energy states, even if it is combined with vDMRG[maxO].

The S&I transformation is not the only one fulfilling the requirement that the lowest-energy eigenfunction of the corresponding auxiliary operator is the excited state with energy closest to the shift parameter ω. For instance, the folded operator which is defined as

$$\Theta_\omega = (\omega - \mathcal{H})^2, \tag{3.44}$$

also satisfies this condition, and is the auxiliary operator employed in the folded vDMRG method (vDMRG[f]). On the one hand, vDMRG[f] has the key advantage that the operator defined in Eq. (3.44) can be straightforwardly encoded as an MPO. On the other hand, evaluating the expectation value in vDMRG[f] is computationally expensive due to the need of contracting the MPS twice with the MPO to account for the square in Eq. (3.44), as illustrated in Figure 3.16. The key advantage is that vDMRG[f] allows for a more stable optimization than vDMRG[S&I] since it preserves the method's variationality and, therefore, ensures a smooth convergence to the target state for any energy range.

                    N. Glaser, A. Baiardi, & M. Reiher

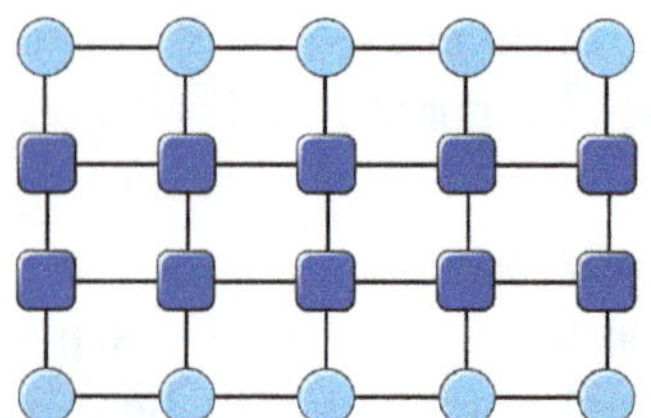

Figure 3.16. Tensor network diagram for the expectation value of the folded Hamiltonian $\langle\Psi|\Theta_\omega|\Psi\rangle$. Note that the MPO has to be applied twice onto the MPS to account for the squared operator in Eq. (3.44).

3.5.4. *Inverse power iteration with MPS: The vDMRG[IPI] algorithm*

An appealing alternative to variational excited state vDMRG variants are methods relying on projection-based algorithms, such as the vDMRG[IPI] approach[47] that combines DMRG with the inverse power iteration (IPI) method.[93] In IPI, the shift-and-invert operator is repeatedly applied onto a guess wave function,

$$|\Psi_k\rangle = (\mathcal{H} - \omega)^{-1}|\Psi_{k-1}\rangle, \tag{3.45}$$

where k labels the iteration step in this recursive equation. Starting from an initial guess $|\Psi_0\rangle$, the wave function $|\Psi_k\rangle$ converges with increasing k towards the eigenfunction of the Hamiltonian with the energy closest to the shift ω. The explicit inversion of the Hamiltonian can be avoided by solving the linear system

$$\Gamma_\omega|\Psi_k\rangle = |\Psi_{k-1}\rangle \tag{3.46}$$

for a normalized wave function $|\Psi_k\rangle$, where we define $\Gamma_\omega = (\mathcal{H} - \omega)$. In vDMRG[IPI], Γ_ω is encoded as an MPO and the wave function $|\Psi_k\rangle$ is approximated as an MPS with a fixed bond dimension m. Unfortunately, the application of an MPO onto an MPS increases the bond dimension of the latter.[8] Consequently, Eq. (3.46) can only be solved approximately with DMRG by calculating the optimal MPS of a given bond dimension m with the least squares method, therefore minimizing the functional

$$O_\omega\left[\Psi_k\right] = \left\|\Gamma_\omega|\Psi_k\rangle - |\Psi_{k-1}\rangle\right\|^2$$
$$= \langle\Psi_k|\Gamma_\omega^2|\Psi_k\rangle - 2\,\mathrm{Re}\left(\langle\Psi_k|\Gamma_\omega|\Psi_{k-1}\rangle\right) + \langle\Psi_{k-1}|\Psi_{k-1}\rangle.$$

$$(3.47)$$

Provided that the initial guess MPS is an adequate approximation of the targeted eigenstate, the minimum of the functional $O_\omega\left[\Psi_k\right]$ is also the minimum of the functional $\tilde{O}_\omega\left[\Psi_k\right]$ defined as[94]

$$\tilde{O}_\omega\left[\Psi_k\right] = \langle\Psi_k|\Gamma_\omega|\Psi_k\rangle - 2\langle\Psi_{k-1}|\Psi_k\rangle. \qquad (3.48)$$

Thus, in practice the functional $\tilde{O}_\omega\left[\Psi_k\right]$ is minimized with the alternating least squares algorithm by optimizing one site at a time in a sweep-like fashion. The tensor $M^{\sigma_l,(k)}_{a_{l-1},a_l}$ defining the MPS representation of $|\Psi_k\rangle$ that minimizes Eq. (3.48) for a given site l is obtained by solving the following linear system[47,94]:

$$\left(\boldsymbol{H} - \omega\boldsymbol{I}\right)\boldsymbol{v}^{(k)} = \boldsymbol{b}^{(k)} \qquad (3.49)$$

where $\boldsymbol{H}$ is defined in Eq. (3.28), $\boldsymbol{I}$ is the identity, and $\boldsymbol{v}^{(k)}$ collects the entries of the tensor $M^{\sigma_l,(k)}_{a_{l-1},a_l}$. Moreover, $\boldsymbol{b}^{(k)}$ is defined as

$$b^{(k)}_{a_{l-1}a_l\sigma_l} = \sum_{a'_{l-1},a'_l} \mathbb{A}^{(k)}_{a_{l-1},a'_{l-1}} M^{\sigma_l,(k-1)*}_{a'_{l-1},a'_l} \mathbb{B}^{(k)}_{a_l,a'_l}, \qquad (3.50)$$

where $\mathbb{A}^{(k)}$ and $\mathbb{B}^{(k)}$ are defined recursively, in analogy to the boundaries $\mathbb{L}$ and $\mathbb{R}$, as

$$\mathbb{A}^{(k)}_{a_l,a'_l} = \sum_{a_{l-1},a'_{l-1}} \sum_{\sigma_l} M^{\sigma_l,(k-1)*}_{a'_{l-1},a'_l} M^{\sigma_l,(k)}_{a_{l-1},a_l} \mathbb{A}^{(k)}_{a_{l-1},a'_{l-1}}, \qquad (3.51)$$

$$\mathbb{B}^{(k)}_{a_l,a'_l} = \sum_{a_{l+1},a'_{l+1}} \sum_{\sigma_{l+1}} M^{\sigma_{l+1},(k-1)*}_{a'_l,a'_{l+1}} M^{\sigma_{l+1},(k)}_{a_l,a_{l+1}} \mathbb{B}^{(k)}_{a_{l+1},a'_{l+1}}, \qquad (3.52)$$

with $\mathbb{A}^{(k)}_{a_0,a'_0} = 1$ and $\mathbb{B}^{(k)}_{a_L,a'_L} = 1$.

For vDMRG[IPI] calculations, both an energy shift ω and a starting guess $|\Psi_0\rangle$ must be specified. While the algorithm is rather robust with respect to the accuracy of the energy guess ω, the choice of the initial state is crucial to ensure stable convergence to the correct eigenstate. An effective guess can be obtained, for weakly anharmonic states, from the corresponding mean-field vibrational wave function. Therefore, in practice, the wave function is initialized with the nth mode in the first excited mean-field state for the calculation of the nth fundamental frequency, while all other modes are initialized in the vibrational ground state.

3.5.5. *Towards large-scale excited state DMRG: vDMRG[FEAST]*

A common issue of all vDMRG excited state methods that rely on an energy shift to target a specific state is their sensitivity towards the shift parameter ω, especially within highly dense energy ranges. To accurately choose ω *a priori* can be a tremendous challenge, and therefore, an automated setting of the energy shift would greatly facilitate excited state calculations. A promising method in this respect is the vDMRG[FEAST] approach,[94] where the FEAST algorithm[95] is combined with vDMRG. vDMRG[FEAST] is a particularly robust and efficient method to calculate excited state wave functions encoded as MPS due to three key advantages. Firstly, the energy shift parameter ω is determined automatically within the FEAST algorithm and therefore need not be set externally. Secondly, linear equations are solved, as in DMRG[IPI], instead of more involved eigenvalue problems. Finally, the linear systems to be solved in vDMRG[FEAST] are independent of one another and, therefore, the algorithm can be straightforwardly parallelized.

The FEAST algorithm is an iterative subspace diagonalization algorithm that can be applied to solve generalized eigenvalue problems.[95] FEAST relies on a simultaneous optimization of all eigenfunctions with an energy within a given target interval by applying a projection operator onto a set of guess wave functions. Let $I_E = [E_{\min}, E_{\max}]$ be an energy interval that contains M eigenvalues of the Hamiltonian $\mathcal{H}$ with eigenfunctions $I_M = \{\Psi^{(1)}, \ldots, \Psi^{(M)}\}$. By leveraging the Cauchy integral theorem, the projector $\mathcal{P}_M$ onto the space spanned by the I_M functions can be expressed as a complex contour integral with

$$\mathcal{P}_M = \sum_{i=1}^{M} |\Psi^{(i)}\rangle\langle\Psi^{(i)}| = \frac{1}{2\pi i} \oint_{\mathcal{C}} (z - \mathcal{H})^{-1} dz. \qquad (3.53)$$

Starting from a linearly independent set of M guess states $\{\Phi_{\text{guess}}^{(1)}, \ldots, \Phi_{\text{guess}}^{(M)}\}$, a basis of the subspace spanned by I_M can be defined as $S_M := \{\mathcal{P}_M \Phi_{\text{guess}}^{(1)}, \ldots, \mathcal{P}_M \Phi_{\text{guess}}^{(M)}\}$. The eigenfunctions within the interval I_E are then obtained by solving the original eigenvalue problem in the subspace spanned by S_M. Calculating the projector defined in Eq. (3.53) requires both inverting the Hamiltonian and evaluating exactly the contour integral, resulting in a computational problem which is as complex as the original eigenvalue equation itself. This issue is circumvented by approximating the complex contour integral with an N_{p}-point numerical quadrature, as illustrated in Figure 3.17. With this approximation, the elements of S_M are expressed as

$$\mathcal{P}_M \Phi_{\text{guess}}^{(i)} \approx \frac{1}{2\pi i} \sum_{k=1}^{N_{\text{p}}} w_k (z_k - \mathcal{H})^{-1} \Phi_{\text{guess}}^{(i)} = \frac{1}{2\pi i} \sum_{k=1}^{N_{\text{p}}} w_k \Phi^{(i,k)}. \quad (3.54)$$

For a given quadrature node k, the quadrature weight w_k and the complex energy shift z_k are determined by the numerical integration algorithm and, therefore, do not need to be set externally. The wave function $\Phi^{(i,k)}$ associated with a given quadrature node k and guess state i is obtained by solving the linear system

$$(z_k - \mathcal{H})\Phi^{(i,k)} = \Phi_{\text{guess}}^{(i)} \qquad (3.55)$$

that is defined in terms of the same projection operator $\Gamma_{z_k} = (z_k - \mathcal{H})$ as in vDMRG[IPI]. The Hamiltonian $\mathcal{H}$ can then be diagonalized within the S_M subspace to obtain the approximate eigenpairs within the energy interval I_E.

It should be noted that if Eq. (3.55) is solved exactly, the only approximation introduced by FEAST is the numerical integration in Eq. (3.54). Hence, a single FEAST iteration will return the eigenpairs to arbitrary accuracy for a dense enough quadrature grid. This is a major advantage compared to vDMRG[IPI], where the linear system

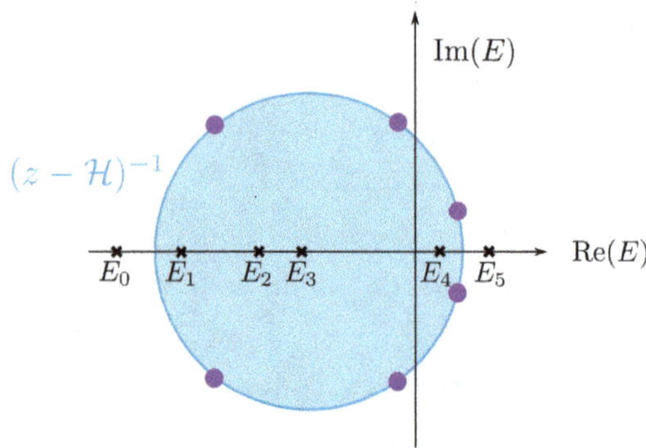

Figure 3.17.   Graphical representation of the FEAST algorithm. The blue circle represents the contour integral needed to calculate the projection operator $(z - \mathcal{H})^{-1}$ in the complex plane. The enclosed energy interval I_E contains four eigenvalues, namely $\{E_1, E_2, E_3, E_4\}$. In this example, the contour integral (blue circle) is approximated by a six-point Gauss-Hermite integration, as depicted by the purple quadrature points.

is solved N_{IPI} times in a sequential fashion. In vDMRG[FEAST], the $M \times N_{\mathrm{p}}$ linear systems of Eq. (3.55) are mutually independent and can be solved in parallel. By virtue of this trivial parallelization, FEAST is an efficient algorithm to compute multiple excited states simultaneously. vDMRG[FEAST] can also be applied to energy regions with a high density of states, as it does not require energy guesses for the shift parameter ω but the entire energy range can instead be calculated at once.

3.5.6. *Example — Vibrational transition energies of ethlyene*

We calculated the three lowest vibrational excitation energies of ethylene, *i.e.*, the same system that we studied in Section 3.4.3, with the excited state DMRG methods introduced above. As we show in Figure 3.18(a), the vDMRG[S&I] optimization converges within 2 sweeps for all ω values, therefore as fast as for ground state DMRG. The converged energy of all states matches the reference data obtained with vDMRG[ortho] (represented as dashed black lines in Figure 3.18). The same trend is observed for vDMRG[MaxO]. In this case, we constructed the reference MPS for the overlap

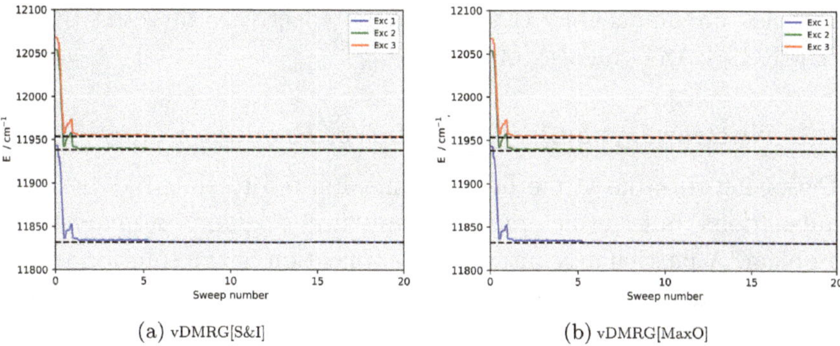

(a) vDMRG[S&I] (b) vDMRG[MaxO]

Figure 3.18. vDMRG energy convergence of the first (blue line), second (green line), and third (red line) vibrationally excited state of ethylene obtained with different excited state vDMRG variants. We set the shift parameter ω for vDMRG[S&I] to $11800\,\mathrm{cm}^{-1}$, $11900\,\mathrm{cm}^{-1}$, and $11950\,\mathrm{cm}^{-1}$ for the first, second, and third excited state, respectively. The reference MPS for vDMRG[MaxO] is constructed from the ONV associated with the target vibrational level. In both panels, we report the converged energies obtained with vDMRG[ortho] as dashed black lines. The bond dimension m was set to 20 in all calculations.

calculation from the ONV associated with the target state. As we show in Figure 3.18(b), the vDMRG[MaxO] convergence rate is as fast as for vDMRG[S&I]. Hence, both the maximum overlap criterion and the shift-and-invert transformation are sufficient to guide the optimization towards the targeted excited states in this example. Note that Figures 3.18(a) and (b) are nearly superimposable. In fact, vDMRG[MaxO] and vDMRG[S&I] solve the very same eigenvalue problem, and differ only in the criterion used to select the eigenpair which is propagated at each microiteration step. If the simulation parameters are chosen adequately such that the same state is targeted by both criteria, the convergence of the DMRG optimization will be identical for both methods. The optimized transition frequencies for the three vibrations are $820.16\ \mathrm{cm}^{-1}$, $926.46\ \mathrm{cm}^{-1}$, and $941.83\ \mathrm{cm}^{-1}$. As we showed in Ref. 78, both of these optimization algorithms become less stable when applied to high-energy states. The vDMRG[f] algorithm, possibly combined with vDMRG[MaxO], yields the highest numerical stability for these more challenging cases. Vibrationally excited states of large systems, for which the computational cost associated with vDMRG[f] becomes unmanageable, can instead be effectively targeted by the vDMRG[FEAST] algorithm.

3.6. Nuclear Dynamics with Matrix Product States

Tensor network-based methods are not limited to time-independent calculations, but can be applied to any kind of quantum many-body problem. A particularly interesting application is the solution of the time-dependent Schrödinger equation that reads, in Hartree atomic units,

$$\mathcal{H}|\Psi\rangle = \mathrm{i}\frac{\partial|\Psi\rangle}{\partial t}. \tag{3.56}$$

Methods aiming at solving Eq. (3.56) by encoding $|\Psi\rangle$ as an MPS are broadly referred to as time-dependent DMRG (TD-DMRG) algorithms.[96] The various TD-DMRG approaches which have been proposed in the literature differ in their computational strategy to solve Eq. (3.56), but they all share a common limitation, *i.e.*, they all represent a time-evolving many-body wave function as an MPS throughout the whole propagation. Whereas the accuracy of time-independent DMRG (TI-DMRG) is guaranteed by the area law under certain conditions,[36] there is no such time-dependent equivalent guaranteeing that a TD wave function can be encoded as a compact MPS. On the contrary, it has been observed that the wave function entanglement often increases with time under non-equilibrium conditions, a phenomenon known as the "entanglement barrier effect".[97] It should be noted that, as an additional complication, if the TD-DMRG wave function is not accurately represented as an MPS at a given time t, this inaccuracy may compromise the simulation accuracy for all subsequent times.

While these observations have questioned the reliability of TD-DMRG, it still remains a promising avenue for efficient quantum dynamics simulations for two reasons. First, even though the TD-DMRG efficiency cannot be guaranteed *a priori*, the area law does not apply to time-independent vibrational-structure problems either due to the presence of long-range many-body interaction terms. Nevertheless, DMRG was shown to produce sufficiently accurate results for large-scale full CI calculations. Hence, designing quantum chemical TD-DMRG algorithms is a key step to assess if and how the entanglement barrier effect impacts the simulation accuracy in practice. Second, techniques that are already exploited in modern quantum dynamics methods can be leveraged to tame the entanglement barrier effect. For instance, orbital optimization techniques enable adapting the one-particle basis to the non-equilibrium condition on-the-fly and can, therefore, enhance the TD-DMRG efficiency.[24, 98, 99]

In the following, we will first describe the most popular TD-DMRG methods, with particular emphasis on the tangent-space TD-DMRG theory,[91, 100] as it is the most promising approach for quantum chemical applications. We will follow the derivation that we reported in our previous works on vibrational[89] and electronic[101] dynamics with TD-DMRG. We will then discuss possible strategies that can be adopted to enhance the TD-DMRG efficiency and to tame the entanglement barrier effect.

3.6.1. *Entanglement barrier effect in TD-DMRG*

To illustrate the origin of the entanglement barrier effect, we approximate the solution of Eq. (3.56) for a time step Δt with a first-order Taylor expansion:

$$|\Psi(t + \Delta t)\rangle = |\Psi(t)\rangle - \mathrm{i}\Delta t \mathcal{H}|\Psi(t)\rangle. \tag{3.57}$$

If we now encode $|\Psi(t)\rangle$ as an MPS (see Eq. (3.4)) with time-dependent entries $M^{\sigma_i}_{a_{i-1},a_i}(t)$, and $\mathcal{H}$ as an MPO with entries $\mathcal{H}^{\sigma_i,\sigma'_i}_{b_{i-1},b_i}$ that we assume, for simplicity, to be time independent, then $\mathcal{H}|\Psi(t)\rangle$ can be encoded as follows:

$$\mathcal{H}|\Psi(t)\rangle = \sum_{b_1,\ldots,b_{L-1}} \sum_{a_1,\ldots,a_{L-1}} \sum_{\sigma,\sigma'} H^{\sigma_1,\sigma'_1}_{1,b_1}$$
$$\cdots H^{\sigma_L,\sigma'_L}_{b_{L-1},1} M^{\sigma'_1}_{1,a_1} \cdots M^{\sigma'_L}_{a_{L-1},1}|\sigma\rangle \tag{3.58}$$

$$= \sum_{\substack{a_1,\dots,a_{L-1} \\ b_1,\dots,b_{L-1}}} \sum_{\boldsymbol{\sigma}} \left(\underbrace{\sum_{\sigma_1'} H^{\sigma_1,\sigma_1'}_{1,b_1} M^{\sigma_1'}_{1,a_1}}_{N^{\sigma_1}_{1,b_1 a_1}} \right)$$

$$\cdots \left(\underbrace{\sum_{\sigma_L'} H^{\sigma_L,\sigma_L'}_{b_{L-1},1} M^{\sigma_L'}_{a_{L-1},1}}_{N^{\sigma_L}_{b_{L-1} a_{L-1},1}} \right) |\boldsymbol{\sigma}\rangle, \qquad (3.59)$$

where $N^{\sigma_i}_{b_{i-1}a_{i-1},b_i a_i}$ are the tensor entries of the resulting MPS. The MPS bond dimension between sites i and $(i+1)$ is b_i times larger than that of the original MPS, which gives rise to the entanglement barrier effect. We note, however, that the b_i factor is only an upper bound to the bond dimension increase. The MPS compression algorithms introduced above can be exploited to reduce, if possible, the bond dimension. We also note that the bond dimension growth increases with b_i, *i.e.*, with the MPO bond dimension, which in turn increases with the Hamiltonian complexity and, more specifically, with the extent of the long-range many-body correlations. Therefore, finding compact MPO representations is crucial to tame the entanglement barrier effect.

3.6.2. *State-of-the-Art TD-DMRG approaches*

TD-DMRG is maximally efficient for short-ranged Hamiltonians that can be expressed as

$$\mathcal{H} = \sum_{i=1}^{L-1} h_{i,i+1}, \qquad (3.60)$$

where $h_{i,i+1}$ is an operator coupling only neighboring sites i and $(i+1)$ of the DMRG lattice. In these cases, instead of approximating the solution to the time-dependent Schrödinger equation with a first-order Taylor series expansion, the propagator $e^{-i\mathcal{H}t}$ that determines the exact time evolution of the wave function for a time step t can be factorized based on the first-order Trotter approximation,

$$e^{-\mathrm{i}t\mathcal{H}} \approx \prod_{i\,\text{even}} e^{-\mathrm{i}h_{i,i+1}t} \prod_{i\,\text{odd}} e^{-\mathrm{i}h_{i,i+1}t}. \qquad (3.61)$$

Eq. (3.61) can be encoded as an MPO with bond dimension N_i for site i, with N_i being the dimension of the local basis for site i. The resulting propagation scheme, known as the time-evolving block decimation (TEBD) algorithm,[102] is the most efficient approach to propagate short-range Hamiltonians. TEBD variants that can be applied to long-range Hamiltonians have been proposed,[103,104] but they require applying so-called swap gates to bring non-neighboring interacting sites close to one another. This often leads to a drastic growth of the bond dimension m for complex Hamiltonians, with a consequent increase of the computational cost.

Methods that support arbitrarily complex MPOs are better suited for implementations supporting long-ranged Hamiltonians, and can be divided into two classes. The first one includes so-called "propagate-and-compress" approaches[105] that solve the TD Schrödinger equation by applying conventional integration algorithms, such as the Lanczos[106] or the Runge–Kutta[107,108] propagators, to MPS wave functions. Directly solving Eq. (3.57) belongs to this class of methods, as this corresponds to a straightforward Euler integration of the differential equation. The common limitation of all these methods is the need for calculating terms such as $\mathcal{H}^n|\Psi_{\mathrm{MPS}}\rangle$, with the accuracy of the integration algorithm growing with n. These terms are encoded by highly non-compact MPSs and, therefore, they must be compressed after each time step to keep their bond dimension fixed.

As an alternative, more appealing schemes based on the time-dependent variational principle have been developed. Instead of directly solving Eq. (3.56), these methods minimize the Dirac–Frenkel functional $F(t)$ at each time step, which is defined as[109]:

$$F(\Psi(t)) = \left\| \mathcal{H}|\Psi(t)\rangle - \mathrm{i}\frac{\partial \Psi(t)}{\partial t} \right\|^2. \qquad (3.62)$$

At each time t, $|\Psi(t)\rangle$ is expressed as an MPS with bond dimension m and the functional $F(t)$ is minimized with respect to the MPS entries to return the best wave function approximation. The first TD-DMRG algorithm relying on Eq. (3.62) was introduced by Haegeman *et al.*[110] and propagates all the MPS entries simultaneously for a given time step. Unfortunately, the equations of motion resulting from such an approach require the inversion of potentially ill-conditioned singular matrices and the algorithm is therefore prone to numerical instabilities.

3.6.3. *Tangent-space TD-DMRG*

The minimization problem of Eq. (3.62) is conveniently expressed as the following projected TD Schrödinger equation[91, 100]:

$$\mathcal{P}_{\mathrm{MPS(m)}}\mathcal{H}|\Psi(t)\rangle = \mathrm{i}\frac{\partial\Psi(t)}{\partial t}, \tag{3.63}$$

with $\mathcal{P}_{\mathrm{MPS(m)}}$ being the tangent-space projector onto the manifold composed by all MPSs with bond dimension m. In practice, the tangent-space projector implicitly compresses $\mathcal{H}|\Psi(t)\rangle$ so that it has the same bond dimension as the right-hand side MPS and, therefore, Eq. (3.63) can be solved exactly, without explicitly compressing the MPS at each time step. A key advantage of methods relying on Eq. (3.63) is that the closed-form expression for $\mathcal{P}_{\mathrm{MPS(m)}}$ reads as[91]

$$\mathcal{P}_{\mathrm{MPS(m)}} = \sum_{i=0}^{L-1} \sum_{a_i^{(\mathcal{L})}\sigma_{i+1}a_{i+1}^{(\mathcal{R})}} \underbrace{|a_i^{(\mathcal{L})}\sigma_{i+1}a_{i+1}^{(\mathcal{R})}\rangle\langle a_i^{(\mathcal{L})}\sigma_{i+1}a_{i+1}^{(\mathcal{R})}|}_{\mathcal{P}_i^{(1)}}$$

$$- \sum_{i=0}^{L-2} \sum_{a_{i+1}^{(\mathcal{L})}a_{i+1}^{(\mathcal{R})}} \underbrace{|a_{i+1}^{(\mathcal{L})}a_{i+1}^{(\mathcal{R})}\rangle\langle a_{i+1}^{(\mathcal{L})}a_{i+1}^{(\mathcal{R})}|}_{\mathcal{P}_i^{(2)}}. \tag{3.64}$$

Equation (3.64) defines $\mathcal{P}_{\mathrm{MPS(m)}}$ in terms of the so-called left- and right-renormalized bases. The left-renormalized basis for site i,

$|a_i^{(\mathcal{L})}\rangle$, is defined as

$$|a_i^{(\mathcal{L})}\rangle = \sum_{a_{i-1}\sigma_i} A^{\sigma_i}_{a_{i-1},a_i} |a_{i-1}^{(\mathcal{L})}\sigma_i\rangle. \tag{3.65}$$

The right-renormalized basis is defined analogously as

$$|a_{i-1}^{(\mathcal{R})}\rangle = \sum_{a_i\sigma_i} B^{\sigma_i}_{a_{i-1},a_i} |a_i^{(\mathcal{R})}\sigma_i\rangle, \tag{3.66}$$

where the $|a_0^{(\mathcal{L})}\rangle$ and $|a_L^{(\mathcal{R})}\rangle$ renormalized bases are two empty sets. In Eq. (3.64), the left- and right-renormalized bases for site i are defined with the MPS canonized with respect to the same site. Therefore, each term of Eq. (3.64) is defined based on a different MPS canonization, which makes the evaluation of the operator $\mathcal{P}_{\mathrm{MPS(m)}}$ not straightforward. This problem can be circumvented by solving Eq. (3.63) based on the Lie–Trotter splitting scheme, *i.e.*, by applying the terms appearing in Eq. (3.64) subsequently, for increasing i values, by adapting the MPS canonization consequently. Hence, the first differential equation to be solved is associated with $\mathcal{P}_0^{(1)}$ and determines the time evolution of the MPS tensor of the first site as

$$\sum_{a_0'\sigma_1'a_1'} H_{(a_0\sigma_1 a_1),(a_0'\sigma_1'a_1')} M^{\sigma_0'}_{a_0',a_1'} = \mathrm{i}\frac{\mathrm{d}}{\mathrm{d}t} M^{\sigma_0}_{a_0,a_1}, \tag{3.67}$$

which is solved for a time step Δt. We note that the $\boldsymbol{H}$ matrix introduced in Eq. (3.28) is obtained, for a given site l, by projecting the Hamiltonian operator onto the basis $|a_l^{(\mathcal{L})}\sigma_{l+1}a_{l+1}^{(\mathcal{R})}\rangle$. The MPS resulting from the solution of Eq. (3.67) defines the initial wave function for the second differential equation, which involves the $\mathcal{P}_0^{(2)}$ operator and determines the time evolution of the so-called zero-site tensor $N_{a_1,\tilde{a}_1}$ as[89]

$$\sum_{a_1'\tilde{a}_1'} H_{(a_1\tilde{a}_1),(a_1'\tilde{a}_1')} N_{a_1',\tilde{a}_1'} = -\mathrm{i}\frac{\mathrm{d}}{\mathrm{d}t} N_{a_1,\tilde{a}_1}. \tag{3.68}$$

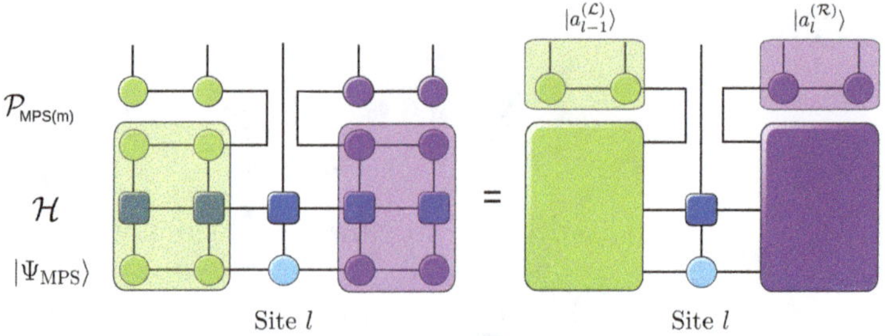

Figure 3.19. Graphical representation of the tensor network to be contracted in order to evaluate Eq. (3.67) for the site l of the DMRG lattice.

The matrix $H_{(a_1\tilde{a}_1),(a_1'\tilde{a}_1')}$ is, in this case, obtained by projecting the Hamiltonian $\mathcal{H}$ in the $|a_1^{(\mathcal{L})}a_1^{(\mathcal{R})}\rangle$ basis. Eqs. (3.67) and (3.68) are solved succesively for every site on the DMRG lattice to obtain the wave function at time t. Hence, for each time step the lattice is traversed in a sweeping fashion, as in TI-DMRG. A key advantage of tangent-space-based methods is that Eqs. (3.67) and (3.68) are linear differential equations with constant coefficients that can be solved to arbitrary high accuracy with Krylov-based methods.[111]

As we show in Figure 3.19, the left-hand side of Eq. (3.67) (and its generalization to any site l of the DMRG lattice) can be expressed in terms of the left and right boundaries that we already introduced for time-independent DMRG. Note that the graphical representation of Figure 3.19 holds for MPS canonized with respect to the same site at which the tangent-space projector is centered. This is, however, not a limitation, since the MPS normalization can be moved along the lattice while the terms of Eq. (3.64) are applied, as is done in TI-DMRG. The possibility of implementing the tangent-space TD-DMRG method with algorithms that are very similar to those of TI-DMRG is the key advantage of this method.

As already discussed above, propagating the MPS for a given time step Δt requires traversing the DMRG lattice to solve Eqs. (3.67) and (3.68) for all sites of the DMRG lattice. At the end of the propagation, the MPS is canonized with respect to the last site.

The canonization should, therefore, be shifted to the first site before applying the propagator for the next time step. As suggested in Ref. 100, the propagators can be applied in the reversed order instead, analogously to a backward sweep in conventional, time-independent DMRG. This approach corresponds to a second-order Trotter approximation of the solution to Eq. (3.63) and, therefore, enhances the method's accuracy and avoids unnecessary shifting of the MPS canonization.

As we already highlighted above, the main challenge associated with TD-DMRG simulations is the fact that the wave function entanglement evolves with time and the bond dimension m should be adapted correspondingly to yield a constant wave function accuracy over the whole propagation. However, the projection operator defined in Eq. (3.64) constrains the bond dimension m to be the same over the whole propagation. Such an approach does not allow for a dynamical adaptation of m. This means, in practice, that if the wave function entanglement is maximal at a given time t, then the bond dimension m yielding an accurate wave function representation at that time must be used for the whole propagation in order to obtain converged results. Moreover, the bond dimension m is constrained to be the same as that of the wave function at $t=0$, *i.e.*, at the beginning of the propagation. Therefore, the algorithm fails in describing the dynamical increase of the wave function entanglement.

All these limitations are addressed by the two-site (TS) TD-DMRG variant that generalizes the algorithm introduced above to propagate two neighbouring sites simultaneously, as is done in time-independent two-site DMRG. Without going into the algorithmic details, which can be found in Refs. 78 and 100, we note that TS-TD-DMRG allows for dynamically adapting the bond dimension along the propagation to the wave function entanglement. For this reason, this two-site variant is largely more accurate in the presence of rapidly varying time-dependent perturbations that induce a corresponding fast variation of the wave function entanglement.[101]

3.6.4. *Quantum chemical applications of real-time TD-DMRG*

As any quantum dynamics method, TD-DMRG has two primary application fields: the calculation of molecular spectra and the simulation of ultrafast molecular processes.

An absorption spectrum $I(\omega)$ can be expressed, within the dipole approximation and at $T = 0\,\mathrm{K}$ (the inclusion of thermal effects will be briefly discussed in the next section), as the integral over the dipole autocorrelation function $C_\mu(t)$ with

$$I(\omega) = \alpha \int_{-\infty}^{+\infty} C_\mu(t)\, e^{i\omega t}\, \mathrm{d}t = \alpha \int_{-\infty}^{+\infty} \langle \Psi | \mu e^{-i\mathcal{H}t} \mu | \Psi \rangle\, e^{i\omega t}\, \mathrm{d}t, \tag{3.69}$$

with $|\Psi\rangle$ being the ground state wave function and α a proportionality constant. The definition of the dipole operator μ depends on the simulation target. For vibrational spectroscopies, μ contains the nuclear contribution to the molecular dipole moment while, for electronic spectroscopies, it is the electric dipole operator.[101,112] Eq. (3.69) can be calculated with TD-DMRG by encoding $\mu|\Psi\rangle$ as an MPS and propagating the resulting wave function under the action of the Hamiltonian $\mathcal{H}$.[78] This procedure enables calculating spectra without resorting to complex sum-over-states expressions that are obtained with frequency-domain approaches. The time-domain route to molecular spectroscopy has been applied to TD-DMRG for calculating absorption spectra of molecular aggregates[105,108,113] based on excitonic Hamiltonians. The development of the tangent-space TD-DMRG method has paved the route towards extending this scheme to more complex *ab initio* vibronic Hamiltonians as well.[78,98]

Eq. (3.69) expresses a spectrum as the Fourier transform of the dipole autocorrelation function $C_\mu(t)$, which decays for $t \to +\infty$ due to dephasing and decoherence effects. The long-time dynamics, which are hard to simulate with TD-DMRG due to the entanglement barrier effect, have only a minor impact on the calculation of $I(\omega)$ and spectra converge remarkably fast with respect to the bond dimension.[78] However, spectra are not the only observables that can be extracted from a TD-DMRG propagation. The availability of the many-body wave function for all times t makes it possible to fully characterize the driving forces of non-equilibrium quantum dynamics. For instance, we showed that a TD-DMRG simulation can accurately identify the pathways that are followed by excitons to propagate along a molecular aggregate,[78, 113, 114] although the convergence of these observables with the bond dimension m is slower compared to the spectra.

3.6.5. *Example — Absorption spectrum of pyrazine*

We demonstrate the efficiency of the tangent-space TD-DMRG method on the simulation of the vibrationally resolved absorption spectrum of pyrazine. The presented data is taken from our work to the vibrational TD-DMRG theory,[89] where we applied TD-DMRG to the 4-mode vibronic Hamiltonian defined in Ref. 115, which is one of the most common benchmark systems for molecular quantum dynamics algorithms. This vibronic Hamiltonian describes the coupled S_1 and S_2 electronically excited states as

$$
\mathcal{H} = \begin{pmatrix} \mathcal{T}_{S_1}(\boldsymbol{Q}) & 0 \\ 0 & \mathcal{T}_{S_2}(\boldsymbol{Q}) \end{pmatrix} + \begin{pmatrix} \mathcal{V}_{S_1}(\boldsymbol{Q}) & \mathcal{V}_{12}(\boldsymbol{Q}) \\ \mathcal{V}_{12}(\boldsymbol{Q}) & \mathcal{V}_{S_2}(\boldsymbol{Q}) \end{pmatrix}, \tag{3.70}
$$

where $\mathcal{V}_{S_1}(\boldsymbol{Q})$ and $\mathcal{V}_{S_2}(\boldsymbol{Q})$ are the diabatic PES of the S_1 and S_2 states, respectively, and $\mathcal{V}_{12}(\boldsymbol{Q})$ is the non-adiabatic coupling between them. Both the PES and the non-adiabatic coupling are approximated as a second-order Taylor expansion in terms of the vibrational modes $\boldsymbol{Q}$.[115] By applying Eq. (3.69) to the vibronic Hamiltonian defined in Eq. (3.70), we simulated the vibronic spectrum of pyrazine for the $S_1 \leftarrow S_0$ and $S_2 \leftarrow S_0$ excitations.[78]

We report in Figure 3.20(a) the time-dependent autocorrelation function $C_\mu(t)$, taken from the data reported in our original TD-DMRG work.[89] $C_\mu(t)$ was qualitatively converged already with a bond dimension of $m = 10$ and quantitative differences compared to the $m = 20$ results were observed only after $100\,\mathrm{fs}$. No changes were observed by further increasing m to 40. As expected, the absorption spectrum, obtained from the Fourier transformation

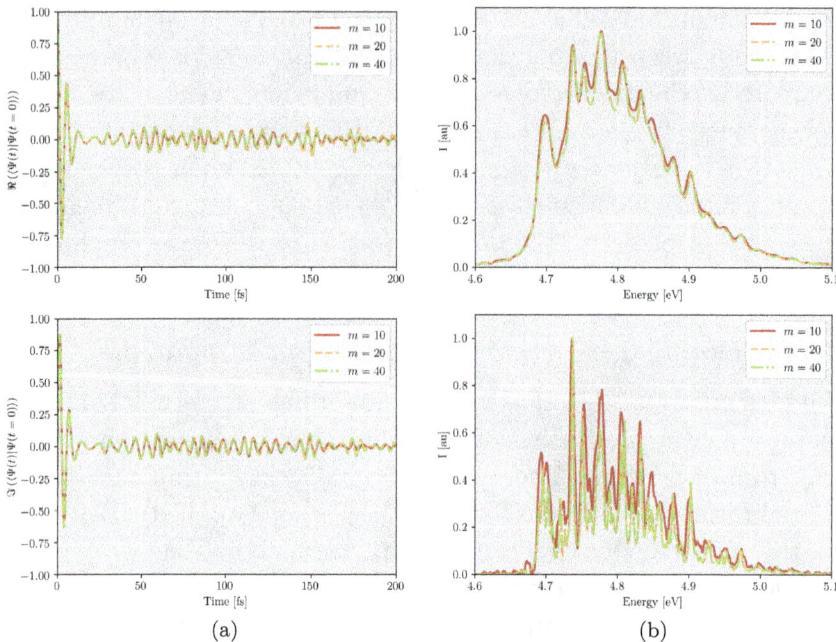

Figure 3.20.   Real (upper left panel) and imaginary (lower left panel) part of the time-dependent dipole autocorrelation function $C_\mu(t)$ calculated based on the 4-mode vibronic Hamiltonian of pyrazine as defined in Eq. (3.70). The corresponding spectrum obtained with Gaussian broadening functions with a half-width at half-maximum of $200\,\mathrm{cm}^{-1}$ (upper panel) and $50\,\mathrm{cm}^{-1}$ (lower panel) are reported in the right part of the figure. TD-DMRG calculations were carried out with $m = 10$ (solid red line), $m = 20$ (dashed orange line) and $m = 40$ (dashed green line). The time step Δt is $0.5\,\mathrm{fs}$ in all cases. This figure has been generated with data taken from Ref. 89.

of the autocorrelation function and reported in Figure 3.20(b), followed the same convergence pattern. For the higher-resolution spectrum, obtained with a half-width at half-maximum (HWHM) of $50\,\mathrm{cm}^{-1}$, small differences were observed between the intensity of the individual vibronic bands calculated with $m = 10$ and $m = 20$. Nevertheless, the band shape was fully converged with $m = 40$. By including broadening effects with Gaussian functions with an HWHM of $200\,\mathrm{cm}^{-1}$, the band shape was converged already with $m = 10$. As we noted,[89] the convergence of the spectrum with respect to m did not change substantially for larger vibronic

Hamiltonians. The absorption spectrum calculated based on the full-dimensional 24-mode Hamiltonian of pyrazine converged with a low m value of 100. This suggests that the entanglement barrier effect in practice only marginally impacts the simulation accuracy. Hence, TD-DMRG is a powerful method for simulating non-equilibrium chemical processes.

3.6.6. *Imaginary-time TD-DMRG: Ground state optimization and thermal ensembles*

The tangent-space TD-DMRG method can be extended to complex-valued times t and applied to two new classes of problems, namely to purely imaginary-time propagation and to time values that contain both a real and an imaginary part. The propagation algorithm obtained for purely imaginary time values t reads

$$\frac{\partial |\Psi_{\mathrm{MPS}}(t)\rangle}{\partial t} = -\mathcal{P}_{\mathrm{MPS(m)}}\mathcal{H}|\Psi_{\mathrm{MPS}}(t)\rangle, \qquad (3.71)$$

and converges to the ground state for $t \to +\infty$ independently of the initial wave function, provided that the guess MPS and the final, converged wave function are not orthogonal, a condition that is, in practice, almost always met in numerical simulations. Eq. (3.71) lays the foundation of many modern electronic-structure calculations, such as the full CI quantum Monte Carlo (FCIQMC)[116] and the auxiliary-field quantum Monte Carlo[117] methods. The imaginary-time variant of MCTDH is also routinely used to optimize anharmonic vibrational wave functions.[118–120] The imaginary-time TD-DMRG method (iTD-DMRG) is often less efficient than conventional TI-DMRG and requires more sweeps to converge. However, iTD-DMRG features the key advantage of allowing for a reliable optimization of the right eigenvalues of non-Hermitian Hamiltonians that are obtained through similarity transformation.[121] Consider, for instance, the CI equation associated with a Jastrow-like wave function *ansatz*:

$$\mathcal{H}e^{\mathcal{J}}|\Psi\rangle = Ee^{\mathcal{J}}|\Psi\rangle. \qquad (3.72)$$

The Jastrow factor can be revolved into the Hamiltonian operator as follows:

$$e^{-\mathcal{J}}\mathcal{H}e^{\mathcal{J}}|\Psi\rangle = E|\Psi\rangle, \tag{3.73}$$

and $|\Psi\rangle$ can be calculated as a right eigenvalue of the non-Hermitian operator $e^{-\mathcal{J}}\mathcal{H}e^{\mathcal{J}}$, usually referred to as the transcorrelated Hamiltonian.[121–123] For electronic-structure methods, the Jastrow factor can be tuned to include short-range correlation effects, so that $|\Psi\rangle$ can be encoded as a more compact CI wave function. Such an approach can drastically enhance the convergence of both FCIQMC[124] and DMRG.[121] Even though extensions of such schemes to vibrational-structure problems have been scarce, the generality of the transcorrelated approach makes it appealing also for applications to vibrational-structure calculation methods.

Further generalizing the TD-DMRG algorithm to time values that contain both a real and an imaginary part makes the method applicable to the simulation of thermal ensembles. Such a strategy has been applied to multiple TD-DMRG variants,[108, 125] including the tangent-space one,[126] to include temperature effects in the simulation of molecular spectra and of non-equilibrium excitonic dynamics. The generality of the MPS/MPO-based framework introduced above will make it possible to apply the very same theoretical framework also to more complex *ab initio* vibronic Hamiltonians.

3.6.7. *Enhancing the TD-DMRG efficiency*

The DMRG algorithm can be interpreted as an efficient algorithm to solve large-scale full CI problems in a given basis set. The accuracy of a full CI-based wave function is, however, not required for many chemical applications. For instance, MCTDH does not represent the wave function as a full CI expansion in the primitive basis (often constructed from distributed Gaussian[127] or discrete variable representation[128] functions). Instead, it contracts the primitive basis to obtain a pre-optimized modal basis and, then, applies full CI only to a subset of the resulting modal basis in the spirit of active space-based methods. The same idea can be exploited in tensor-based methods as well. Kurashige[98] combined the vibrational DMRG theory with the MCTDH algorithm to propagate both the single-particle functions and the MPS simultaneously. Therefore, the MPS is encoded as

$$|\Psi_{\mathrm{MPS}}(t)\rangle = \sum_{n_1,\ldots,n_L} \sum_{a_1,\ldots,a_{L-1}} M_{1,a_1}^{n_1}(t) M_{a_1,a_2}^{n_2}(t)$$

$$\cdots M_{a_{L-1},1}^{n_L}(t) \prod_{i=1}^{L} \chi_{n_i}^{i}(Q_i,t), \tag{3.74}$$

$\chi_{n_i}^{i}(Q_i,t)$ being the n_ith single-particle function of mode i. The main challenge in deriving the equation of motion associated with the *ansatz* of Eq. (3.74) is that the modal variation that is parallel to the modals that are included in Eq. (3.74) is already included in the propagation of the CI coefficients, expressed as MPSs in Eq. (3.74). To avoid double-counting, the equation-of-motion for the modals $\chi_{n_i}^{i}(Q_i,t)$ should only contain contributions arising from the modals that are not included in Eq. (3.74). Reference 98 circumvents this problem by applying an algorithm originally designed for MCTDH[129] that relies on the so-called single-hole function formalism.

The optimization scheme introduced above relies on the partition of the modal basis into an "active" and a "virtual" set, and DMRG is applied only to the former modals. However, Legeza and co-workers[99] introduced algorithms to further optimize the active modal basis to yield the most compact MPS wave function at all times. Originally introduced for time-independent problems,[130] this algorithm relies on the two-site TD-DMRG variant. After the two-site tensor is propagated, and before it is compressed to yield the target bond dimension m, the single-particle functions associated with the optimization sites are rotated based on a unitary transformation that is optimized to minimize the pair entropy. In this way, the wave function entanglement and, therefore, the bond dimension can be effectively reduced without compromising the accuracy. The success of this orbital optimization scheme as applied to electron dynamics suggests that it could also enhance the efficiency of nuclear dynamics simulations. We recall that vibrational modes are described in terms of distinguishable, bosonic particles, as discussed above. Therefore, in the n-mode second quantization framework the unitary transformation can only be applied between pairs of sites associated with the same mode, as otherwise basis functions associated with different particles would be mixed. As each site is associated with a different mode within the canonical quantization framework, the algorithm of Ref. 130 cannot be straightforwardly applied to the corresponding vDMRG variant. In this case, the pair entropy minimization principle cannot be applied, but the modals can still be optimized separately for each individual site.[131]

3.7. Conclusion and Outlook

We introduced the vDMRG method[48,78] to calculate vibrational ground and excited states of molecular systems. Even if the DMRG algorithm has originally been designed to target short-range spin Hamiltonians, we found[48] it to be, out-of-the-box, as efficient as alternative state-of-the-art methods when applied to vibrational-structure calculations. This shows that tensor network-based methods are a promising approach to overcome the current limits of large-scale vibrational-structure calculations.

However, many research routes remain to be explored to further enhance vDMRG. Following approaches originally devised for electronic-structure calculations,[132,133] vDMRG can be combined with modal optimization schemes to further augment the representation power of the vibrational tensor network states for both time-independent[134] and time-dependent[24] simulations. Moreover, vDMRG has, so far, only been applied to solve the vibrational full CI problem. The high accuracy of vDMRG is, however, only essential to describe strongly anharmonic molecular vibrations, while faster methods based on perturbation theory[135,136] are sufficient for an accurate description of weakly anharmonic modes. Hybrid variational-perturbative schemes are currently the method of choice to efficiently describe strongly-correlated electronic wave functions.[137,138] Extending this combined approach to vibrational problems would make it possible to efficiently target molecules with several dozen degrees of freedom.

Another, more radical improvement of the vDMRG algorithm could be achieved by generalizing the method beyond the matrix product state parameterization. As we highlighted in this chapter, the key component of vDMRG is the availability of an efficient optimization scheme that is based on the alternating least squares algorithm. This optimization procedure cannot be extended to multidimensional tensor network states, with few exceptions,[13, 58, 139] and this limitation has impeded the development of fully automated multidimensional tensor network-based methods so far. vDMRG has been applied mostly to moderately anharmonic potentials and, in these cases, an MPS yields a very compact representation of the CI wave function. However, for strongly anharmonic molecules,[140, 141] more complex tensor network states may provide a more flexible representation of the wave function than the MPS.

A key assumption underlying the theoretical framework described in this chapter is the Born–Oppenheimer (BO) approximation that simplifies the molecular Hamiltonian by disentangling the nuclear and electronic motions. Perturbation theory offers an effective way to account for the effects that are neglected by BO-based calculations,[142, 143] such as non-adiabatic couplings or quantum electrodynamics effects. The generality of the MPS/MPO-based framework makes it possible to straightforwardly include these terms in the definition of the molecular Hamiltonian. However, perturbative approaches rely on BO-based calculations that require calculating the PES, a task that may become challenging for strongly anharmonic molecules. Pre-Born–Oppenheimer (PreBO) methods overcome this problem by solving the full molecular Schrödinger equation by treating nuclear and electronic degrees of freedom on the same footing. Explicitly correlated algorithms are the method of choice for accurate PreBO calculations on small molecules,[144, 145] but their computational cost makes them applicable to few-particle molecules only. Orbital-based methods[146] have a much smaller computational cost, but they struggle with capturing strong nuclear-electron correlation effects and, therefore, do not yield spectroscopically accurate simulations. Tensor network states are inherently designed to capture strong correlation effects and are, therefore, the ideal candidate to fill the gap between explicitly correlated and orbital-based methods. Various PreBO DMRG variants are currently under development.[147,148] Although the development of these methods is still in its infancy, the possibility of efficiently encoding strong correlation effects in DMRG may make orbital-based PreBO methods amenable to accurate vibrational-structure calculations.

References

1. Shavitt, I. The history and evolution of configuration interaction. *Mol. Phys.* **1998**, *94*, 3–17.
2. Eriksen, J. J.; Anderson, T. A.; Deustua, J. E.; Ghanem, K.; Hait, D.; Hoffmann, M. R.; Lee, S.; Levine, D. S.; Magoulas, I.; Shen, J.; Tubman, N. M.; Whaley, K. B.; Xu, E.; Yao, Y.; Zhang, N.; Alavi, A.; Chan, G. K.-L.; Head-Gordon, M.; Liu, W.; Piecuch, P.; Sharma, S.; Ten-no, S. L.; Umrigar, C. J.; Gauss, J. The ground state electronic energy of benzene. *J. Phys. Chem. Lett.* **2020**, *11*, 8922–8929.
3. Williams, K. T.; Yao, Y.; Li, J.; Chen, L.; Shi, H.; Motta, M.; Niu, C.; Ray, U.; Guo, S.; Anderson, R. J.; Li, J.; Tran, L. N.; Yeh, C.-N.; Mussard, B.; Sharma, S.; Bruneval, F.; van Schilfgaarde, M.; Booth, G. H.; Chan, G. K.-L.; Zhang, S.; Gull, E.; Zgid, D.; Millis, A.; Umrigar, C. J.; Wagner, L. K. Direct comparison of many-body methods for realistic electronic hamiltonians. *Phys. Rev. X* **2020**, *10*, 011041.
4. White, S. R. Density matrix formulation for quantum renormalization groups. *Phys. Rev. Lett.* **1992**, *69*, 2863–2866.
5. White, S. R. Density-matrix algorithms for quantum renormalization groups. *Phys. Rev. B* **1993**, *48*, 10345–10356.
6. Chan, G. K.-L.; Dorando, J. J.; Ghosh, D.; Hachmann, J.; Neuscamman, E.; Wang, H.; Yanai, T. An Introduction to the Density Matrix Renormalization Group Ansatz in Quantum Chemistry. In *Frontiers in Quantum Systems in Chemistry and Physics*; Wilson, S., Grout, P. J., Maruani, J., Delgado-Barrio, G., Piecuch, P., Eds.; Springer Netherlands, 2008; pp. 49–65.
7. Chan, G. K.-L.; Zgid, D. The density matrix renormalization group in quantum chemistry. *Annu. Rep. Comput. Chem.* **2009**, *5*, 149–162.
8. Schollwöck, U. The density-matrix renormalization group in the age of matrix product states. *Ann. Phys.* **2011**, *326*, 96–192.

9. Chan, G. K.-L.; Sharma, S. The density matrix renormalization group in quantum chemistry. *Annu. Rev. Phys. Chem.* **2011**, *62*, 465–481.

10. Wouters, S.; Van Neck, D. The density matrix renormalization group for ab initio quantum chemistry. *Eur. Phys. J. D* **2013**, *31*, 395–402.

11. Keller, S. F.; Reiher, M. Determining factors for the accuracy of DMRG in chemistry. *Chimia* **2014**, *68*, 200–203.

12. Kurashige, Y. Multireference electron correlation methods with density matrix renormalisation group reference functions. *Mol. Phys.* **2014**, *112*, 1485–1494.

13. Olivares-Amaya, R.; Hu, W.; Nakatani, N.; Sharma, S.; Yang, J.; Chan, G. K.-L. The ab-initio density matrix renormalization group in practice. *J. Chem. Phys.* **2015**, *142*, 34102.

14. Szalay, S.; Pfeffer, M.; Murg, V.; Barcza, G.; Verstraete, F.; Schneider, R.; Legeza, Ö. Tensor product methods and entanglement optimization for ab initio quantum chemistry. *Int. J. Quantum Chem.* **2015**, *115*, 1342–1391.

15. Yanai, T.; Kurashige, Y.; Mizukami, W.; Chalupský, J.; Lan, T. N.; Saitow, M. Density matrix renormalization group for ab initio calculations and associated dynamic correlation methods: A review of theory and applications. *Int. J. Quantum Chem.* **2015**, *115*, 283–299.

16. Knecht, S.; Hedegård, E. D.; Keller, S.; Kovyrshin, A.; Ma, Y.; Muolo, A.; Stein, C. J.; Reiher, M. New approaches for ab initio calculations of molecules with strong electron correlation. *Chimia* **2016**, *70*, 244–251.

17. Baiardi, A.; Reiher, M. The density matrix renormalization group in chemistry and molecular physics: Recent developments and new challenges. *J. Chem. Phys.* **2020**, *152*, 040903.

18. Wang, H.; Thoss, M. Multilayer formulation of the multiconfiguration time-dependent Hartree theory. *J. Chem. Phys.* **2003**, *119*, 1289–1299.

19. Manthe, U. A multilayer multiconfigurational time-dependent Hartree approach for quantum dynamics on general potential energy surfaces. *J. Chem. Phys.* **2008**, *128*, 164116.

20. Vendrell, O.; Meyer, H.-D. Multilayer multiconfiguration time-dependent Hartree method: Implementation and applications to a Henon — Heiles Hamiltonian and to pyrazine. *J. Chem. Phys.* **2011**, *134*, 044135.

21. Römer, S.; Ruckenbauer, M.; Burghardt, I. Gaussian-based multiconfiguration time-dependent Hartree: A two-layer approach. I. Theory. *J. Chem. Phys.* **2013**, *138*, 064106.

22. Wang, H. Multilayer multiconfiguration time-dependent Hartree theory. *J. Phys. Chem. A* **2015**, *119*, 7951–7965.

23. Lubich, C.; Rohwedder, T.; Schneider, R.; Vandereycken, B.; Tensors, T.-t.; Vandereycken, B. Dynamical approximation by hierarchical tucker and tensor-train tensors. *SIAM J. Matrix Anal. Appl.* **2013**, *34*, 470–494.

24. Meyer, H.-D. Studying molecular quantum dynamics with the multiconfiguration time-dependent Hartree method. *WIREs Comput. Mol. Sci.* **2012**, *2*, 351–374.

25. Xie, Y.; Zheng, J.; Lan, Z. Full-dimensional multilayer multiconfigurational time-dependent Hartree study of electron transfer dynamics in the anthracene/C60 complex. *J. Chem. Phys.* **2015**, *142*, 084706.

26. Binder, R.; Lauvergnat, D.; Burghardt, I. Conformational dynamics guides coherent exciton migration in conjugated polymer materials: first-principles quantum dynamical study. *Phys. Rev. Lett.* **2018**, *120*, 227401.

27. Popp, W.; Polkehn, M.; Binder, R.; Burghardt, I. Coherent charge transfer exciton formation in regioregular P3HT: a quantum dynamical study. *J. Phys. Chem. Lett.* **2019**, *10*, 3326–3332.

28. Hammer, T.; Manthe, U. Iterative diagonalization in the state-averaged multi-configurational time-dependent Hartree approach: Excited state tunneling splittings in malonaldehyde. *J. Chem. Phys.* **2012**, *136*, 054105.

29. Leclerc, A.; Carrington, T. Calculating vibrational spectra with sum of product basis functions without storing full-dimensional vectors or matrices. *J. Chem. Phys.* **2014**, *140*, 174111.

30. Leclerc, A.; Thomas, P. S.; Carrington, T. Comparison of different eigensolvers for calculating vibrational spectra using low-rank, sum-of-product basis functions. *Mol. Phys.* **2017**, *115*, 1740–1749.

31. Thomas, P. S.; Carrington, T. Using nested contractions and a hierarchical tensor format to compute vibrational spectra of molecules with seven atoms. *J. Phys. Chem. A* **2015**, *119*, 13074–13091.

32. Godtliebsen, I. H.; Thomsen, B.; Christiansen, O. Tensor decomposition and vibrational coupled cluster theory. *J. Phys. Chem. A* **2013**, *117*, 7267–7279.

33. Godtliebsen, I. H.; Hansen, M. B.; Christiansen, O. Tensor decomposition techniques in the solution of vibrational coupled cluster response theory eigenvalue equations. *J. Chem. Phys.* **2015**, *142*, 024105.

34. Oseledets, I. V.; Dolgov, S. V. Solution of linear systems and matrix inversion in the TT-format. *SIAM J. Sci. Comput.* **2012**, *34*, A2718–A2739.

35. Muth, D.; McCulloch, I. P. From density-matrix renormalization group to matrix product states. *J. Stat. Mech. Theory Exp.* **2007**, 10014.

36. Hastings, M. B. An area law for one-dimensional quantum systems. *J. Stat. Mech. Theory Exp.* **2007**, P08024.

37. Ramasesha, S.; Pati, S. K.; Krishnamurthy, H. R.; Shuai, Z.; Brédas, J. L. Symmetrized density-matrix renormalization-group method for excited states of Hubbard models. *Phys. Rev. B* **1996**, *54*, 7598–7601.

38. Fano, G.; Ortolani, F.; Ziosi, L. The density matrix renormalization group method: Application to the PPP model of a cyclic polyene chain. *J. Chem. Phys.* **1998**, *108*, 9246–9252.

39. Anusooya, Y.; Pati, S. K.; Ramasesha, S. Symmetrized density matrix renormalization group studies of the properties of low-lying states of the poly-para-phenylene system. *J. Chem. Phys.* **1998**, *106*, 10230.

40. White, S. R.; Martin, R. L. Ab initio quantum chemistry using the density matrix renormalization group. *J. Chem. Phys.* **1999**, *110*, 4127–4130.

41. Chan, G. K.-L.; Head-Gordon, M. Highly correlated calculations with a polynomial cost algorithm: A study of the density matrix renormalization group. *J. Chem. Phys.* **2002**, *116*, 4462–4476.

42. Legeza, Ö.; Röder, J.; Hess, B. A. QC-DMRG study of the ionic-neutral curve crossing of LiF. *Mol. Phys.* **2003**, *101*, 2019–2028.

43. Marti, K. H.; Ondk, I. M.; Moritz, G.; Reiher, M. Density matrix renormalization group calculations on relative energies of transition metal complexes and clusters. *J. Chem. Phys.* **2008**, *128*, 014104.

44. Legeza, Ö.; Noack, R. M.; Sólyom, J.; Tincani, L. Applications of Quantum Information in the Density-Matrix Renormalization Group. In *Computational Many-Particle Physics*; Fehske, H., Schneider, R., Weiße, A., Eds.; Springer Berlin Heidelberg, 2008; pp. 653–664.

45. Marti, K. H.; Reiher, M. The density matrix renormalization group algorithm in quantum chemistry. *Z. Phys. Chem.* **2010**, *224*, 583–599.

46. Freitag, L.; Reiher, M. The Density Matrix Renormalization Group for Strong Correlation in Ground and Excited States. In *Quantum Chemistry and Dynamics of Excited States*; González, L., Lindh, R., Eds.; John Wiley & Sons, Inc., 2020; Chapter 7, pp. 205–245.

47. Rakhuba, M.; Oseledets, I. Calculating vibrational spectra of molecules using tensor train decomposition. *J. Chem. Phys.* **2016**, *145*, 124101.

48. Baiardi, A.; Stein, C. J.; Barone, V.; Reiher, M. Vibrational density matrix renormalization group. *J. Chem. Theory Comput.* **2017**, *13*, 3764–3777.

49. Larsson, H. R. Computing vibrational eigenstates with tree tensor network states (TTNS). *J. Chem. Phys.* **2019**, *151*, 204102.

50. Brorsen, K. R. Quantifying multireference character in multicomponent systems with heat-bath configuration interaction. *J. Chem. Theory Comput.* **2020**, *16*, 2379–2388.

51. Fetherolf, J. H.; Berkelbach, T. C. Vibrational heat-bath configuration interaction. *J. Chem. Phys.* **2021**, *154*, 074104.

52. Orús, R. Tensor networks for complex quantum systems. *Nat. Rev. Phys.* **2019**, *1*, 538550.

53. Penrose, R. Applications of Negative Dimensional Tensors. In *Combinatorial Mathematics and Its Applications*; Bose, R. C., Dowling, T. A., Eds.; UNC Press, 1969.

54. Kolda, T. G.; Bader, B. W. Tensor decompositions and applications. *SIAM Rev.* **2009**, *51*, 455–500.

55. Legeza, Ö.; Röder, J.; Hess, B. A. Controlling the accuracy of the density-matrix renormalization-group method: The dynamical block state selection approach. *Phys. Rev. B* **2003**, *67*, 125114.

56. Murg, V.; Verstraete, F.; Legeza, O.; Noack, R. M. Simulating strongly correlated quantum systems with tree tensor networks. *Phys. Rev. B* **2010**, *82*, 205105.

57. Nakatani, N.; Chan, G. K.-L. Efficient tree tensor network states (TTNS) for quantum chemistry: Generalizations of the density matrix renormalization group algorithm. *J. Chem. Phys.* **2013**, *138*, 134113.

58. Murg, V.; Verstraete, F.; Schneider, R.; Nagy, P. R.; Legeza, O. Tree tensor network state with variable tensor order: an efficient multireference method for strongly correlated systems. *J. Chem. Theory Comput.* **2015**, *11*, 1027–1036.

59. Gunst, K.; Verstraete, F.; Wouters, S.; Legeza, O.; Van Neck, D. T3NS: Three-legged tree tensor network states. *J. Chem. Theory Comput.* **2018**, *14*, 2026–2033.

60. Pižorn, I.; Verstraete, F. Fermionic implementation of projected entangled pair states algorithm. *Phys. Rev. B* **2010**, *81*, 245110.

61. Haghshenas, R.; O'Rourke, M. J.; Chan, G. K.-L. Conversion of projected entangled pair states into a canonical form. *Phys. Rev. B* **2019**, *100*, 054404.

62. Marti, K. H.; Bauer, B.; Reiher, M.; Troyer, M.; Verstraete, F. Complete-graph tensor network states: a new fermionic wave function ansatz for molecules. *New. J. Phys.* **2010**, *12*, 103008.

63. Kovyrshin, A.; Reiher, M. Tensor network states with three-site correlators. *New. J. Phys.* **2016**, *18*, 113001.

64. Kovyrshin, A.; Reiher, M. Self-adaptive tensor network states with multi-site correlators. *J. Chem. Phys.* **2017**, *147*, 214111.

65. Sandvik, A. W.; Vidal, G. Variational quantum Monte Carlo simulations with tensor-network states. *Phys. Rev. Lett.* **2007**, *99*, 220602.

66. Sabzevari, I.; Sharma, S. Improved speed and scaling in orbital space variational Monte Carlo. *J. Chem. Theory Comput.* **2018**, *14*, 6276–6286.

67. Kloss, B.; Reichman, D. R.; Lev, Y. B. Studying dynamics in two-dimensional quantum lattices using tree tensor network states. *Sci. Post Phys.* **2020**, *9*, 70.

68. Ceruti, G.; Lubich, C.; Walach, H. Time integration of tree tensor networks. *SIAM J. Num. Analysis* **2021**, *59*, 289–313.

69. Fisher, M. E. The renormalization group in the theory of critical behavior. *Rev. Mod. Phys.* **1974**, *46*, 597–616.

70. Östlund, S.; Rommer, S. Thermodynamic limit of density matrix renormalization. *Phys. Rev. Lett.* **1995**, *75*, 3537–3540.

71. Keller, S.; Dolfi, M.; Troyer, M.; Reiher, M. An efficient matrix product operator representation of the quantum chemical Hamiltonian. *J. Chem. Phys.* **2015**, *143*, 244118.

72. Fröwis, F.; Nebendahl, V.; Dür, W. Tensor operators: Constructions and applications for long-range interaction systems. *Phys. Rev. A* **2010**, *81*, 062337.

73. Hubig, C.; McCulloch, I. P.; Schollwöck, U. Generic construction of efficient matrix product operators. *Phys. Rev. B* **2017**, *95*, 035129.

74. Davidson, E. R. The iterative calculation of a few of the lowest eigenvalues and corresponding eigenvectors of large real-symmetric matrices. *J. Comput. Phys.* **1975**, *17*, 87–94.

75. Sleijpen, G. L. G.; Van der Vorst, H. A. A Jacobi–Davidson iteration method for linear eigenvalue problems. *SIAM Rev.* **2000**, *42*, 267–293.

76. Rissler, J.; Noack, R. M.; White, S. R. Measuring orbital interaction using quantum information theory. *Chem. Phys.* **2006**, *323*, 519–531.

77. Legeza, O.; Röder, J.; Hess, B. A. QC-DMRG study of the ionic-neutral curve crossing of LiF. *Mol. Phys.* **2003**, *101*, 2019–2028.

78. Baiardi, A.; Stein, C. J.; Barone, V.; Reiher, M. Optimization of highly excited matrix product states with an application to vibrational spectroscopy. *J. Chem. Phys.* **2019**, *150*, 094113.

79. Carter, S.; Culik, S. J.; Bowman, J. M. Vibrational self-consistent field method for many-mode systems: A new approach and application to the vibrations of CO adsorbed on Cu(100). *J. Chem. Phys.* **1997**, *107*, 10458–10469.

80. Bowman, J. M.; Carter, S.; Huang, X. MULTIMODE: A code to calculate rovibrational energies of polyatomic molecules. *Int. Rev. Phys. Chem.* **2003**, *22*, 533–549.

81. Kongsted, J.; Christiansen, O. Automatic generation of force fields and property surfaces for use in variational vibrational calculations of

anharmonic vibrational energies and zero-point vibrational averaged properties. *J. Chem. Phys.* **2006**, *125*, 124108.

82. Vendrell, O.; Gatti, F.; Lauvergnat, D.; Meyer, H.-D. Full-dimensional (15-dimensional) quantum-dynamical simulation of the protonated water dimer. I. Hamiltonian setup and analysis of the ground vibrational state. *J. Chem. Phys.* **2007**, *127*, 184302.

83. Hirata, S.; Hermes, M. R. Normal-ordered second-quantized Hamiltonian for molecular vibrations. *J. Chem. Phys.* **2014**, *141*, 184111.

84. Christiansen, O. A second quantization formulation of multimode dynamics. *J. Chem. Phys.* **2004**, *120*, 2140–2148.

85. Bowman, J. M. The self-consistent-field approach to polyatomic vibrations. *Acc. Chem. Res.* **1986**, *19*, 202–208.

86. Hansen, M. B.; Sparta, M.; Seidler, P.; Toffoli, D.; Christiansen, O. New formulation and implementation of vibrational self-consistent field theory. *J. Chem. Theory Comput.* **2010**, *6*, 235–248.

87. QCMaquis. 2021; https://github.com/qcscine/qcmaquis.

88. Hubig, C.; McCulloch, I. P.; Schollwöck, U.; Wolf, F. A. Strictly single-site DMRG algorithm with subspace expansion. *Phys. Rev. B* **2015**, *91*, 155115.

89. Baiardi, A.; Reiher, M. Large-scale quantum dynamics with matrix product states. *J. Chem. Theory Comput.* **2019**, *15*, 3481–3498.

90. Butscher, W.; Kammer, W. Modification of Davidson's method for the calculation of eigenvalues and eigenvectors of large real-symmetric matrices: "root homing procedure". *J. Comput. Chem.* **1976**, *20*, 313–325.

91. Lubich, C.; Oseledets, I.; Vandereycken, B. Time integration of tensor trains. *SIAM J. Numer. Anal.* **2015**, *53*, 917.

92. Dorando, J. J.; Hachmann, J.; Chan, G. K.-L. Targeted excited state algorithms. *J. Chem. Phys.* **2007**, *127*, 084109.

93. Saad, Y. *Numerical Methods for Large Eigenvalue Problems*. Society for Industrial and Applied Mathematics: Philapelphia, 2011.

94. Baiardi, A.; Kelemen, A. K.; Reiher, M. Excited-state DMRG made simple with FEAST. *J. Chem. Theory Comput.* **2022**, *18*(1), 415–430.

95. Polizzi, E. Density-matrix-based algorithm for solving eigenvalue problems. *Phys. Rev. B* **2009**, *79*, 115112.

96. Paeckel, S.; Köhler, T.; Swoboda, A.; Manmana, S. R.; Schollwöck, U.; Hubig, C. Time-evolution methods for matrix-product states. *Ann. Phys.* **2019**, *411*, 167998.

97. Schuch, N.; Wolf, M. M.; Vollbrecht, K. G. H.; Cirac, J. I. On entropy growth and the hardness of simulating time evolution. **2008**, *10*, 033032.

98. Kurashige, Y. Matrix product state formulation of the multiconfiguration time-dependent Hartree theory. *J. Chem. Phys.* **2018**, *149*, 194114.

99. Krumnow, C.; Eisert, J.; Legeza, Ö. Towards overcoming the entanglement barrier when simulating long-time evolution. 2019; arXiv:1904.11999.

100. Haegeman, J.; Lubich, C.; Oseledets, I.; Vandereycken, B.; Verstraete, F. Unifying time evolution and optimization with matrix product states. *Phys. Rev. B* **2016**, *94*, 165116.

101. Baiardi, A. Electron dynamics with the time-dependent density matrix renormalization group. *J. Chem. Theory Comput.* **2021**, *17*, 3320–3334.

102. Guifre, V. Efficient simulation of one-dimensional quantum many-body systems. *Phys. Rev. Lett.* **2004**, *93*, 40501–40502.

103. Stoudenmire, E. M.; White, S. R. Minimally entangled typical thermal state algorithms. *New. J. Phys.* **2010**, *12*, 055026.

104. Bauernfeind, D.; Aichhorn, M.; Evertz, H. G. Comparison of MPS based real time evolution algorithms for Anderson impurity models. 2019; arXiv:1906.09077.

105. Li, W.; Ren, J.; Shuai, Z. Numerical assessment for accuracy and GPU acceleration of TD-DMRG time evolution schemes. *J. Chem. Phys.* **2020**, *152*, 024127.

106. Frahm, L.-H.; Pfannkuche, D. Ultrafast ab-initio quantum chemistry using matrix product states. *J. Chem. Theory Comput.* **2019**, *15*, 2154–2165.

107. Ronca, E.; Li, Z.; Jimenez-Hoyos, C. A.; Chan, G. K.-L. Time-step targeting time-dependent and dynamical density matrix renormalization group algorithms with ab initio Hamiltonians. *J. Chem. Theory Comput.* **2017**, *13*, 5560–5571.

108. Ren, J.; Shuai, Z.; Chan, G. K.-L. Time-dependent density matrix renormalization group algorithms for nearly exact absorption and fluorescence spectra of molecular aggregates at both zero and finite temperature. *J. Chem. Theory Comput.* **2018**, *14*, 5027.

109. Moccia, R. Time-dependent variational principle. *Int. J. Quantum Chem.* **1973**, *7*, 779–783.

110. Haegeman, J.; Cirac, J. I.; Osborne, T. J.; Pižorn, I.; Verschelde, H.; Verstraete, F. Time-dependent variational principle for quantum lattices. *Phys. Rev. Lett.* **2011**, *107*, 070601.

111. Saad, Y. Analysis of some Krylov subspace approximations to the matrix exponential operator. *SIAM J. Numer. Anal.* **1992**, *29*, 209–228.

112. Neville, S. P.; Schuurman, M. S. A general approach for the calculation and characterization of x-ray absorption spectra. *J. Chem. Phys.* **2018**, *149*, 154111.

113. Kloss, B.; Reichman, D. R.; Tempelaar, R. Multiset matrix product state calculations reveal mobile Franck–Condon excitations under strong Holstein-type coupling. *Phys. Rev. Lett.* **2019**, *123*, 126601.

114. Borrelli, R.; Gelin, M. F. Simulation of quantum dynamics of excitonic systems at finite temperature: an efficient method based on thermo field dynamics. *Sci. Rep.* **2017**, *7*, 9127.

115. Raab, A.; Worth, G. A.; Meyer, H.-D.; Cederbaum, L. S. Molecular dynamics of pyrazine after excitation to the S2 electronic state using a realistic 24-mode model Hamiltonian. *J. Chem. Phys.* **1999**, *110*, 936–946.

116. Booth, G. H.; Thom, A. J. W.; Alavi, A. Fermion Monte Carlo without fixed nodes: A game of life, death, and annihilation in Slater determinant space. *J. Chem. Phys* **2009**, *131*, 054106.

117. Zhang, S.; Krakauer, H. Quantum Monte Carlo method using phase-free random walks with slater determinants. *Phys. Rev. Lett.* **2003**, *90*, 136401.

118. Meyer, H.-D.; Quéré, F. L.; Léonard, C.; Gatti, F. Calculation and selective population of vibrational levels with the multiconfiguration time-dependent Hartree (MCTDH) algorithm. *Chem. Phys.* **2006**, *329*, 179–192.

119. Wang, H. Iterative calculation of energy eigenstates employing the multilayer multiconfiguration time-dependent Hartree theory. *J. Phys. Chem. A* **2014**, *118*, 9253–9261.

120. Wodraszka, R.; Carrington, T. A new collocation-based multiconfiguration time-dependent Hartree (MCTDH) approach for solving the Schrdinger equation with a general potential energy surface. *J. Chem. Phys.* **2018**, *148*, 044115.

121. Baiardi, A.; Reiher, M. Transcorrelated density matrix renormalization group. *J. Chem. Phys.* **2020**, *153*, 164115.

122. Boys, S. F.; Handy, N. C.; Linnett, J. W. The determination of energies and wavefunctions with full electronic correlation. *Proc. R. Soc. Lond. A* **1969**, *310*, 43–61.

123. Handy, N. C. Energies and expectation values for be by the transcorrelated method. *J. Chem. Phys.* **1969**, *51*, 3205–3212.

124. Luo, H.; Alavi, A. Combining the transcorrelated method with full configuration interaction quantum monte carlo: Application to the homogeneous electron gas. *J. Chem. Theory Comput.* **2018**, *14*, 1403–1411.

125. Karrasch, C.; Bardarson, J. H.; Moore, J. E. Finite-temperature dynamical density matrix renormalization group and the drude weight of spin-1/2 chains. *Phys. Rev. Lett.* **2012**, *108*, 227206.

126. Li, W.; Ren, J.; Shuai, Z. Finite-temperature TD-DMRG for the carrier mobility of organic semiconductors. *J. Phys. Chem. Lett.* **2020**, *11*, 4930–4936.

127. Hamilton, I. P.; Light, J. C. On distributed Gaussian bases for simple model multidimensional vibrational problems. *J. Chem. Phys.* **1986**, *84*, 306–317.

128. Colbert, D. T.; Miller, W. H. A novel discrete variable representation for quantum mechanical reactive scattering via the Smatrix Kohn method. *J. Chem. Phys.* **1992**, *96*, 1982–1991.

129. Bonfanti, M.; Burghardt, I. Tangent space formulation of the Multi-Configuration Time-Dependent Hartree equations of motion: The projectorsplitting algorithm revisited. *Chem. Phys.* **2018**, *515*, 252–261.

130. Krumnow, C.; Veis, L.; Legeza, Ö.; Eisert, J. Fermionic orbital optimization in tensor network states. *Phys. Rev. Lett.* **2016**, *117*, 210402.

131. Brockt, C.; Dorfner, F.; Vidmar, L.; Heidrich-Meisner, F.; Jeckelmann, E. Matrix-product-state method with a dynamical local basis optimization for bosonic systems out of equilibrium. *Phys. Rev. B* **2015**, *92*, 241106.

132. Zgid, D.; Nooijen, M. The density matrix renormalization group self-consistent field method: Orbital optimization with the density matrix renormalization group method in the active space. *J. Chem. Phys.* **2008**, *128*, 144116.

133. Ma, Y.; Knecht, S.; Keller, S.; Reiher, M. Second-order self-consistent-field density-matrix renormalization group. *J. Chem. Theory Comput.* **2017**, *13*, 2533–2549.

134. Heislbetz, S.; Rauhut, G. Vibrational multiconfiguration self-consistent field theory: Implementation and test calculations. *J. Chem. Phys.* **2010**, *132*, 124102.

135. Barone, V. Anharmonic vibrational properties by a fully automated second-order perturbative approach. *J. Chem. Phys.* **2005**, *122*, 014108.

136. Heislbetz, S.; Rauhut, G. Vibrational multiconfiguration self-consistent field theory: Implementation and test calculations. *J. Chem. Phys.* **2010**, *132*, 124102.

Liu, F.; Kurashige, Y.; Yanai, T.; Morokuma, K. Multireference initio density matrix renormalization group (DMRG)-CASSCF and DMRG-CASPT2 study on the photochromic ring opening of spiropyran. *J. Chem. Theory Comput.* **2013**, *9*, 4462–4469.

138. Freitag, L.; Knecht, S.; Angeli, C.; Reiher, M. Multireference perturbation theory with Cholesky decomposition for the density matrix renormalization group. *J. Chem. Theory Comput.* **2017**, *13*, 451–459.

139. Verstraete, F.; Cirac, J. I. Renormalization algorithms for quantum-many body systems in two and higher dimensions. 2004; arXiv:cond-mat/0407066.

140. McCoy, A. B.; Braams, B. J.; Brown, A.; Huang, X.; Jin, Z.; Bowman, J. M. Ab initio diffusion Monte Carlo calculations of the quantum behavior of CH5+ in full dimensionality. *J. Phys. Chem. A* **2004**, *108*, 4991–4994.

141. Duong, C. H.; Gorlova, O.; Yang, N.; Kelleher, P. J.; Johnson, M. A.; McCoy, A. B.; Yu, Q.; Bowman, J. M. Disentangling the complex vibrational spectrum of the protonated water trimer, $H^+(H_2O)_3$, with two-color IR-IR photodissociation of the bare ion and anharmonic VSCF/VCI theory. *J. Phys. Chem. Lett.* **2017**, *8*, 3782–3789.

142. Pachucki, K.; Komasa, J. Nonadiabatic corrections to the wave function and energy. *J. Chem. Phys.* **2008**, *129*, 034102.

143. Pachucki, K.; Komasa, J. Nonadiabatic corrections to rovibrational levels of H2. *J. Chem. Phys.* **2009**, *130*, 164113.

144. Mátyus, E.; Reiher, M. Molecular structure calculations: A unified quantum mechanical description of electrons and nuclei using explicitly correlated Gaussian functions and the global vector representation. *J. Chem. Phys.* **2012**, *137*, 024104.

145. Bubin, S.; Pavanello, M.; Tung, W.-C.; Sharkey, K. L.; Adamowicz, L. BornOppenheimer and non-BornOppenheimer, atomic and molecular calculations with explicitly correlated Gaussians. *Chem. Rev.* **2013**, *113*, 36–79.

146. Pavošević, F.; Culpitt, T.; Hammes-Schiffer, S. Multicomponent quantum chemistry: integrating electronic and nuclear quantum effects via the nuclear–electronic orbital method. *Chem. Rev.* **2020**, *120*, 4222–4253.

147. Yang, M.; White, S. R. Density-matrix-renormalization-group study of a one-dimensional diatomic molecule beyond the Born–Oppenheimer approximation. *Phys. Rev. A* **2019**, *99*, 022509.

148. Muolo, A.; Baiardi, A.; Feldmann, R.; Reiher, M. Nuclear-electronic all-particle density matrix renormalization group. *J. Chem. Phys.* **2020**, *152*, 204103.

Chapter 4

Diffusion Monte Carlo Approaches for Studying Large Amplitude Vibrational Motions in Molecules and Clusters

Jacob M. Finney, Ryan J. DiRisio, and Anne B. McCoy*

*Department of Chemistry, University of Washington,
Seattle, WA 98195, USA*
abmccoy@uw.edu

4.1. Introduction

As we consider vibrational dynamics and how it is expressed through spectroscopy, we often rely on zero-order models that provide semi-quantitative descriptions of the systems and processes of interest. When successful, analysis of these models provides a set of quantum numbers that can be used to assign the spectra. Such models are also used to motivate the choice of basis sets that are used for more sophisticated treatments of the vibrational dynamics. For example, for molecules that remain localized near a single minimum in the potential, a zero-order description based on uncoupled oscillators can be extremely effective. When larger amplitude motions correspond to hindered rotations or internal rotations, such motions can be incorporated into the model. While such approaches are often very effective, when molecules sample multiple low-energy minima on the potential surface or contain strongly coupled, large amplitude vibrational motions, identifying an appropriate zero-order description can

145

become challenging. Additionally, the approaches that are taken to calculate the spectra can become system specific.

On the other hand, understanding properties of the ground state, and possibly low-energy excited states, of such molecular systems can often provide significant insights into how it samples its potential surface. Such insights can be used both for the interpretation of spectra and to motivate other approaches. A method that is well-suited for such studies is diffusion Monte Carlo (DMC).[1,2] This approach, which was credited to Fermi by Metropolis,[3] exploits the fact that when the time-dependent Schrödinger equation is expressed in imaginary time ($\tau = it/\hbar$),

$$\frac{d}{d\tau}\Psi(\mathbf{x},\tau) = \left[\frac{\hbar^2}{2}\sum_{k=1}^{n_{\text{atom}}}\frac{1}{m_k}\nabla_k^2 - V(\mathbf{x})\right]\Psi(\mathbf{x},\tau), \qquad (4.1)$$

has the same structure as the diffusion equation with a coordinate-dependent source-sink term

$$\frac{dC(\mathbf{x}(t))}{dt} = \left[D\nabla^2 - k(\mathbf{x})\right]C(\mathbf{x}(t)). \qquad (4.2)$$

Just as the diffusion equation can be solved by Monte Carlo approaches based on the evolution of coordinates of the particles that are diffusing, if we express the solution to the imaginary-time, time-dependent Schrödinger equation, $\Psi(\mathbf{x},\tau)$, as an ensemble of localized objects, called walkers, the computational machinery that has been used to obtain the equilibrium solution to the diffusion equation can also be applied to the generation of the equilibrium solution to the imaginary-time, time-dependent Schrödinger equation. While this may seem like a strange thing to calculate, $\lim_{\tau\to\infty}\Psi(\mathbf{x},\tau) = \Psi_0(\mathbf{x})$, which is the ground state solution to the corresponding time-independent Schrödinger equation. DMC has been used to study a broad range of chemical problems.[4–18] While much of the work has focused on solving the electronic Schrödinger equation, a significant challenge introduced in this application is the treatment of the nodes brought about by the requirement that the wave function must be antisymmetric with respect to permutation of the electrons. In some senses, though, DMC is better suited to vibrational problems, where the ground state is nodeless. Suhm and Watts,[2] and Buch,[6, 19, 20] pioneered work in this area. Their studies demonstrated the variety of insights that can be obtained from DMC calculations. Much of their work focused on hydrogen-bonded clusters, where the interacting molecules were constrained to be rigid.

Over the past 15 years, the increased interest in exotic molecules like H_5^{+}[4, 21–23] and CH_5^{+}[9, 24–27] have led to the need for methods that allow us to study molecules that have multiple strongly-coupled large amplitude vibrational motions. Critical to increased interest in these systems has been the availability of potential surfaces for a broad range of molecules that undergo large amplitude motions[4, 28–30] and the development of flexible general potentials,[31, 32] which can be used to explore nuclear quantum effects in assemblies of water molecules and protonated water clusters. Unlike the electronic structure problems, where the potential can be described by pairwise Coulombic interactions, once the Born–Oppenheimer approximation is invoked, the form for the vibrational potential loses its simplicity.

As we move from molecules with ten or fewer atoms to larger molecular clusters, we have found that the size of the ensemble that is needed to effectively sample the potential grows rapidly, and modifications to the DMC approach are needed. Over the past several years, we have explored a modified version of DMC, which employs guiding functions, to improve the sampling. In contrast to much of the previous work in using guiding functions in studies of vibrational problems, which focused on developing general functions of the intermolecular degrees of freedom,[4, 18, 33] this work focuses on the development of guiding functions that are functions of individual bond lengths and angles. We have applied this approach to water clusters with up to six water molecules,[12, 13] protonated water clusters, and molecules like CH_5^{+} that undergo large amplitude motions.[14] In the next section, we will describe this extension and explore some of the considerations that need to be be accounted for in applying the approach. We explore the method in the context of calculations on water, water hexamer, and CH_5^{+}.

4.2. Diffusion Monte Carlo

DMC provides a general approach for obtaining the ground state wave function and zero-point energy for an arbitrary molecular system.[1,2,34,35] To develop the working equations for DMC, the time-dependent Schrödinger equation is expressed as a propagator in imaginary time $\tau = it/\hbar$

$$|\Psi(\tau)\rangle = \exp[-\hat{H}\tau]|\Psi(0)\rangle. \tag{4.3}$$

At long times, $|\Psi(\tau)\rangle$ converges to the ground state solution to the corresponding time-independent Schrödinger equation, $|\Psi_0\rangle$, and the amplitude of the wave function decays as $\exp[-E_0\tau]$, where E_0 is the ground state energy. The simplest DMC algorithm can be developed by approximating the propagator in Eq. (4.3) by a Trotter expansion based on a split operator, and shift the potential energy by E_{ref}

$$|\Psi(\tau + \Delta\tau)\rangle = \exp[-(\hat{V} - E_{\text{ref}})\Delta\tau]\exp[-\hat{T}\Delta\tau]|\Psi(\tau)\rangle. \tag{4.4}$$

This shift is introduced as a parameter that is determined by the requirement that the amplitude of $|\Psi(\tau)\rangle$ remains constant. At the end of the simulation, the time-averaged value of E_{ref} provides the ground state energy for the system of interest. In contrast to other approaches discussed in this volume, where the wave function is expanded in a carefully chosen basis set, in DMC we represent the wave function by a weighted sum of localized functions, $\{g(\mathbf{x} - \mathbf{x}_i)\}$, commonly referred to as walkers

$$\Psi(\tau) = \sum_{i=1}^{N_{\text{w}}} w_i(\tau)g(\mathbf{x} - \mathbf{x}_i(\tau)) \tag{4.5}$$

where $\mathbf{x}$ represents the Cartesian coordinates of the atoms in the molecular system of interest, $\mathbf{x}_i$ is the geometry at which g is localized, and w_i provides the associated weight in the context of an ensemble of N_{w} walkers. As noted in Eq. (4.5), both the positions and weights of the walkers evolve according to Eq. (4.4).

Operationally, the simulation is divided into a series of short time steps, $\Delta\tau$. During each time step, the Cartesian coordinates of each of the atoms in each of the walkers are displaced based on a Gauss-random distribution with a width $\hbar(\Delta\tau/m_j)^{1/2}$, where m_j is the mass of the jth atom.[1] This provides an approximation to the action of the kinetic energy contribution to the propagator in Eq. (4.4) on one of the localized functions. Next, the potential energy is evaluated at the coordinates of each of the displaced walkers, and

$$P_i(\tau + \Delta\tau) = \exp[-(V(\mathbf{x}_i(\tau + \Delta\tau)) - E_{\mathrm{ref}}(\tau))\Delta\tau] \qquad (4.6)$$

is used to obtain $w_i(\tau + \Delta\tau)$.

Two approaches are used to update the weights. In the first, termed continuous weighting,

$$w_i(\tau + \Delta\tau) = P_i(\tau + \Delta\tau)w_i(\tau). \qquad (4.7)$$

This approach keeps the ensemble size constant, but can lead to numerical challenges as a small number of walkers can carry most of the weight. As a result, even for a large ensemble, as the simulation progresses the sampling can deteriorate. To circumvent this problem, we impose a range that the weights must remain within. If the weight of a walker becomes smaller than the minimum allowed weight, that walker is removed from the simulation and the walker with the largest weight is replicated. The weight of the replicated walker and its replica are each half the weight of the high-weight walker prior to replication. This ensures that the sum of the weights of the walkers and the number of walkers remain constant. A similar procedure is applied when the weight of a walker exceeds the maximum value. We have found that a minimum weight between 0.1 and 0.01 and a maximum weight of 20 provide stable simulations.[14] The second approach, referred to as discrete weighting, requires that all of the walkers have $w_i = 1$. In this case, after displacement $\mathrm{int}(P_i(\tau + \Delta\tau))$ is used to determine the number of walkers that should have the coordinates of the ith walker at the start of the next iteration of the propagation. The fractional part of $P_i(\tau + \Delta\tau)$ is compared to a random number between 0 and 1. If the fractional part of $P_i(\tau + \Delta\tau)$ is larger than the random number, an additional walker is placed at these coordinates. This approach has the advantage that all walkers contribute equally to the wave function, often leading to more efficient sampling, at the expense that the size of the ensemble fluctuates throughout the simulation.

The above procedure requires the evaluation of $E_{\text{ref}}(\tau)$. Following Anderson,[34]

$$E_{\text{ref}}(\tau) = \overline{V}(\tau) - \alpha \frac{W(\tau) - W(0)}{W(0)}. \tag{4.8}$$

Here $\overline{V}$ provides the ensemble average of the potential energy, while $W(\tau)$ provides the sum of the weights of the walkers. The second term in Eq. (4.8) introduces a penalty when the $W(\tau)$ deviates from $W(\tau = 0)$, where α is a simulation parameter, which controls the size of the fluctuations in E_{ref}. The effect of the size of α on the time evolution of E_{ref} has been discussed in a recent review (see Figure 6 therein).[36] We have found that a value of $\alpha = 0.5\hbar/\Delta\tau$ often provides a good choice for this parameter.[15,37] As noted above, the time-averaged value of E_{ref} provides the energy of the state of interest.

4.3. Introducing Guiding Functions

A common modification to the standard DMC algorithm is to introduce a guiding function, Φ_{T}, which are also referred to as trial wave functions.[38,39] The introduction of well-chosen guiding functions can improve the efficiency of DMC simulations, but Φ_{T} needs to be chosen with care.[14,18] As we will illustrate, one of the roles of the guiding function is to shift the walkers toward regions of configuration space where Φ_{T} has amplitude. Consequently, a poor choice of Φ_{T} can introduce bias and lead to poor results. At the same time, the potential energy is replaced by the local energy,

$$E_{\text{L}} = \frac{\hat{H}\Phi_{\text{T}}}{\Phi_{\text{T}}}. \tag{4.9}$$

If $\Phi_{\text{T}} = \Psi_0$, the local energy will be the ground state energy, and E_{ref} will be constant and equal to the ground state energy. This is not an interesting situation, as if we already know Ψ_0 there is no need to use DMC to evaluate it. More typically $\Phi_{\text{T}} \approx \Psi_0$. In this case E_{L} is not constant, but has a weaker functional dependence on the coordinates compared to the underlying potential.

To generate the working equations for this formulation of DMC,[38,39] we start with the imaginary-time time-dependent Schrödinger equation,

$$\frac{d\Psi}{d\tau} = \left[\sum_{j=1}^{n_{\text{atoms}}} \frac{\hbar^2}{2m_j} \nabla_j^2 - (V(\mathbf{x}) - E_{\text{ref}}) \right] \Psi(\mathbf{x}), \qquad (4.10)$$

and substitute f/Φ_{T} for Ψ, where f represents the product of the trial wave function and the wave function for the state of interest. If we define a drift term,

$$\vec{D}_j(\mathbf{x}) = \frac{\hbar^2}{m_j} \frac{1}{\Phi_{\text{T}}(\mathbf{x})} \vec{\nabla}_j \Phi_{\text{T}}(\mathbf{x}), \qquad (4.11)$$

upon substitution, Eq. (4.10) becomes

$$\frac{df}{d\tau} = \sum_{j=1}^{n_{\text{atoms}}} \left[\frac{\hbar^2}{2m_j} \nabla_j^2 - \vec{\nabla}_j \cdot \vec{D}_j(\mathbf{x}) \right] f(\mathbf{x})$$
$$-(E_L(\mathbf{x}) - E_{\text{ref}}) f(\mathbf{x}) \qquad (4.12)$$

The first and third terms in Eq. (4.12) are treated in the same manner as standard DMC simulations, described above. Two additional steps need to be introduced to account for the drift term. Following the diffusion step, each walker is displaced by $\mathbf{D}(\mathbf{x})\Delta\tau$, and the new coordinates of the ith walker are denoted as $\mathbf{x}_i'$. This move is accepted or rejected based on the the ratio of the probability of displacing from $\mathbf{x}_i' \to \mathbf{x}_i$ in a time step of $\Delta\tau$ to the probability of displacing from $\mathbf{x}_i \to \mathbf{x}_i'$ in the same time.[38,40,41] If the ratio is larger than 1, then the move is accepted. If not, it is only accepted if the ratio is larger than a random number selected in the range from 0 to 1. If the move is rejected, then $\mathbf{x}_i(\tau + \Delta\tau) = \mathbf{x}_i(\tau)$.[38,40]

To illustrate how this works, in Figure 4.1, we plot the quantities that comprise Eqs. (4.9–4.12) for a 1D model system based on a Morse oscillator that models the hydrogen-bonded OH bond in water dimer, with a frequency of $3704.5\,\text{cm}^{-1}$, an anharmonicity of $75.3\,\text{cm}^{-1}$, and an equilibrium OH bond length of $0.973\,\text{Å}$. The potential, V_{dim} is plotted with a solid black line in Figure 4.1(a).

     J.M. Finney, R.J. DiRisio, & A.B. McCoy

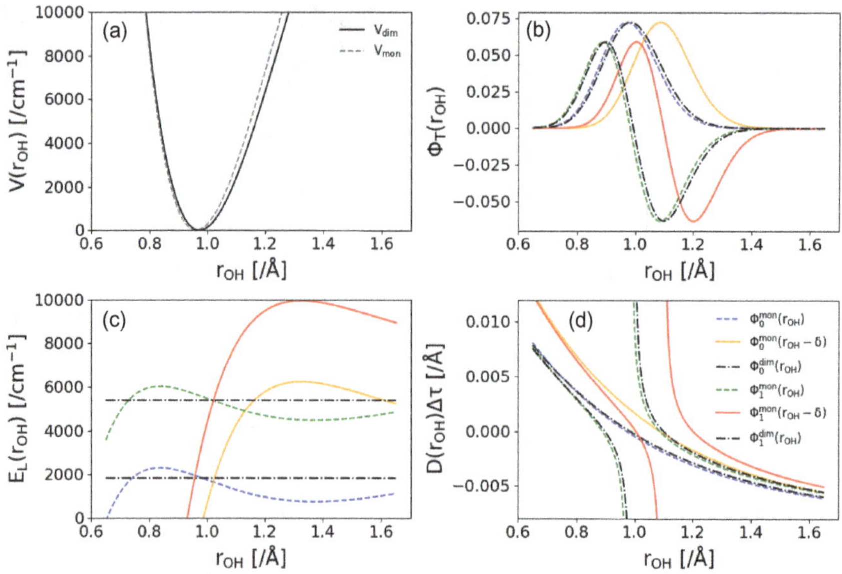

Figure 4.1.   Illustration of the relationship between the quantities described in Eqs. (4.9–4.12). (a) 1D Morse oscillator potentials model the hydrogen-bonded OH bond in water dimer, with a frequency of $3704.5\,\mathrm{cm}^{-1}$, an anharmonicity of $75.3\,\mathrm{cm}^{-1}$, and an equilibrium OH bond length of $0.973\,\text{Å}$ (V_dim black solid line), and a Morse oscillator potential that describes an OH bond in an isolated water molecule, with a frequency of $3884.8\,\mathrm{cm}^{-1}$, an anharmonicity of $86.9\,\mathrm{cm}^{-1}$, and an equilibrium OH bond length of $0.961\,\text{Å}$ (V_mon, grey dashed line),[42] plotted as a function of the OH bond length, r_OH. (b) Ground and first excited state wave functions for the Morse oscillators plotted in (a). (c) Local energy evaluated using Eq. (4.9) with V_dim, and the wave functions plotted in panel (b). (d) The value of the product of the drift term and $\Delta\tau$ with $\Delta\tau = 1$, where the drift term is evaluated using Eq. (4.11) based on the wave functions plotted in panel (b). In panels (b)-(d) the blue and green dashed curves provide results when we use $\Phi_n^\mathrm{mon}(r_\mathrm{OH})$, the gold and red solid lines provide the wave functions when we use $\Phi_n^\mathrm{mon}(r_\mathrm{OH} - \delta)$, where $\delta = 0.115\,\text{Å}$, which is 10 times the difference between $r_{\mathrm{OH},e}$ for the monomer and the dimer, and the black dotted lines provide the results when we use Φ_n^dim.

For Φ_T, we use the ground state solution to a Morse oscillator that describes an OH bond in an isolated water molecule, with a frequency of $3884.8\,\mathrm{cm}^{-1}$, an anharmonicity of $86.9\,\mathrm{cm}^{-1}$, and an equilibrium OH bond length of $0.961\,\text{Å}$.[42] This potential is plotted with a grey dashed line in Figure 4.1(a). The blue and green dashed curves in

Figure 4.1(b) show the $n = 0$ and 1 wave functions obtained using the Morse oscillator potential for water monomer, denoted Φ_n^{mon}. The gold and red solid lines show Φ_n^{mon} when the Δr_{OH} axis is shifted by 0.115 Å, which is 10 times the difference between $r_{\mathrm{OH},e}$ for the dimer and the monomer. This shift was chosen to explore what happens when a poor choice for Φ_T is made. Finally, the black dotted lines provide the results when we use Φ_n^{dim}.

Before discussing the results in panels (c) and (d) it is useful to recall that compared to the wave functions that describe the OH bond in water dimer, shown with dotted lines in Figure 4.1(b), the wave functions for the monomer have more amplitude at smaller values of r_{OH} and less amplitude at larger values of r_{OH}. Near $r_{\mathrm{OH}} = r_{\mathrm{OH},e}^{\mathrm{dim}}$, the local energy for the ground and excited state are roughly constant when the unshifted wave functions are used. Likewise, in this region, the drift term for the ground state vanishes, while it diverges for the $n = 1$ state. When the monomer wave function is shifted, the local energy becomes constant at larger values of r_{OH}. The shift in the wave function manifests only as a shift in the values of the drift term, as expected from Eq. (4.11).

If we only considered the local energy, we find that it decreases as r_{OH} is decreased, but remains roughly constant out to 1.6 Å. Based on this, we expect the ensemble of walkers to shift to smaller distances the longer the propagation is run. This will be more dramatic for the shifted wave functions, plotted in gold and red. This is exactly the behavior that is needed to obtain Ψ_n from the shifted version of Φ_n^{mon}. Considering only E_{L} one might expect the build up of amplitude at small values of r_{OH} to be problematic. The drift term, plotted in Figure 4.1(d), prevents this from happening by shifting the location of walkers away from regions where the amplitude of the trial wave function is small to regions where it is larger. This can be seen by the divergence of the drift term where there is a node in the wave function. As noted above, care needs to be taken in selecting Φ_T as a poor choice can lead to poor sampling. Characteristics of a poor choice for Φ_T include trial wave functions that do not have amplitude in regions of configuration space where the wave function that is

Table 4.1. Energies (in cm^{-1}) obtained from DMC calculations using V_{dim}, and the wave functions shown in Figure 4.1.

Φ_T	ΔE_0^a	$\Delta E_1^{left,b}$	$\Delta E_1^{right,b}$
$\Phi_n^{dim}(r_{OH})$	< 0.001	< 0.001	< 0.001
$\Phi_n^{mon}(r_{OH})$	$-0.43(1.39)$	$533.20(0.08)$	$-362.57(0.39)$
$\Phi_n^{mon}(r_{OH} - 0.011$ Å$)$	$0.10(0.12)$	$-25.20(0.38)$	$18.34(0.07)$
$\Phi_n^{mon}(r_{OH} - 0.115$ Å$)$	$-44.62(20.42)$	$-3246.76(34.19)$	$4092.42(0.39)$

[a] $\Delta E_0 = E_{DMC} - E_0^{dim}$, where $E_0^{dim} = 1833.42\,cm^{-1}$.
[b] $\Delta E_1^{left/right} = E_{DMC} - E_1^{dim}$, where $E_1^{dim} = 5387.37\,cm^{-1}$. These values were obtained from DMC calculations considering only one side of the node.

being modeled has amplitude, or if the node is not correctly placed for an excited state.

To illustrate the sensitivity of the quality of the results to the choice of Φ_T, we ran DMC simulations using the potential and wave functions shown in Figure 4.1 as well as $\Phi_n^{mon}(r_{OH} - 0.011$ Å$)$. The resulting energies are reported in Table 4.1. When we use Φ_n^{dim}, the resulting energies are in exact agreement with the expected energies. Ground state calculations using Φ_0^{mon} produce an accurate ground state energy, with an error of $0.4\,cm^{-1}$, and an uncertainty that is more than three times this error. Shifting Φ_0^{mon} by 0.011 Å, which is the difference between $r_{OH,e}$ for the dimer and monomer, improves the agreement, and the error is reduced to less than $0.1\,cm^{-1}$. When we shift Φ_0^{mon} by a much larger amount, the accuracy of the results deteriorates. In the case of excited states, the presence of the node introduces some complications, and separate calculations are performed for the regions where Φ_T has positive and negative amplitude, denoted as left and right, respectively, in Table 4.1. Overall, none of the results are as accurate as the ground state calculations, but the same general trends hold. The most accurate results are obtained when the trial wave function is shifted by 0.011 Å.

As we move from 1D model problems to real molecules and molecular clusters, we need to be careful in our choice of guiding functions. In DMC the high frequency vibrations often cause greater challenges than the low frequency ones. They are also easier ones to develop approximate wave functions for. Specifically, our approach to guided DMC is based on using trial wave functions that are products of 1D functions of the high frequency XH stretching, and HXH bending vibrations.[12–14] For example, for water clusters, we use a guiding function that is a product of 1D wave functions that are functions of OH bond lengths. These functions are obtained by solving for the ground state solution using a 1D cut through the potential for an isolated water molecule[43] along the OH stretch coordinate. For the bend, we use a 1D harmonic oscillator.[12] A similar approach was previously suggested by Suhm and Watts,[2] but it does not appear to have been applied to many studies. This is likely due to the lack of availability of potentials that accurately describe intra- and intermolecular motions in molecular clusters.

This approach is expected to work well for water clusters, as OH and HOH wave functions show only small changes when the isolated water molecule is placed in the context of a cluster. For example, in Figure 4.2(a) we compare the range of OH bond lengths in hydrogen-bonded OH bonds in water clusters with six or fewer water molecules to the ground state OH stretch wave function used in the development of Φ_T. As can be seen, the OH bond lengths generally deviate by less than $0.036\,\text{Å}$,[42] which is comparable to

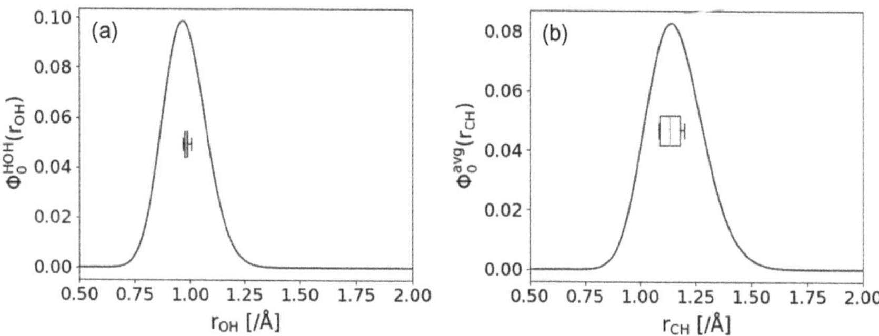

Figure 4.2. Comparison of the range of bond lengths in stationary point structures (box and whiskers plots) to the trial wave functions (plotted with blue solid lines) used in the development of Φ_T for DMC simulations. In panel (a) we focus on hydrogen-bonded OH bonds in water clusters with six or fewer water molecules[42] and in panel (b) we explore the stationary point structures of CH_5^+.[30]

the shift in the wave functions shown in Figure 4.1. This contrasts the situation for the CH bonds in CH_5^+, shown in Figure 4.2(b), which shows a much larger spread of CH bond lengths. In this case, if we used a single CH trial wave function, which is the average of the ground state wave functions for the five CH oscillators in the equilibrium geometry of the ion, to describe the five CH wave functions in this ion in our guided calculation, we obtain a zero-point energy of $10\,926(4)\,\mathrm{cm}^{-1}$,[14] which is roughly $10\,\mathrm{cm}^{-1}$ larger than the zero-point energy of this ion, $10\,917$.[9] A more sophisticated approach, which adjusts the form of the trial wave function depending on the local environment of the CH bond, will be needed to obtain an accurate zero-point energy for CH_5^+. The details of such an approach are discussed below.[14]

4.4. Evaluation of Excited States and Molecular Properties

The ability to evaluate ground state wave functions and energies allow us to obtain insights into systems that undergo large amplitude vibrational motions. For example, they allow us to explore the relative stability of various isomers of a molecular species of interest, and how this is affected by deuteration.[12, 13, 16, 17] On the other hand, much more can be learned from the ability to evaluate properties of the wave function and to calculate the energy and wave functions for excited states. In the discussion that follows, we will focus on DMC with guiding functions, in which the ensemble of walkers samples $f = \Psi \times \Phi_T$. One can consider the unguided situation as the special case when Φ_T is constant.

4.4.1. *Obtaining excited state wave functions*

Obtaining excited state wave functions with guiding functions requires little adjustment to the algorithm described above beyond using a trial wave function that describes the excited state of interest. If after the displacement of the walker, the phase of Φ_T has changed, the displacement is rejected, and the coordinates of the walkers are restored to their value prior to the displacement step. While this change appears straightforward, the sensitivity of the sampling to deficiencies in the trial wave function are more significant for excited states than for ground states. The source of the problem is illustrated by the divergence of the drift term at the position of the node, shown in 1D and Table 4.1. An improperly placed node can lead to poor sampling and errors in the energy obtained using DMC. This is why $\Delta E_1^{\text{left}} \neq \Delta E_1^{\text{right}}$ in Table 4.1.

To address this problem, prior to performing the DMC simulations, we identify the optimal position of the node or nodal surface. To accomplish this, we draw from approaches we developed for identifying nodes in unguided DMC calculations.[20,44,45] Specifically, we obtain an excited state trial wave function by replacing the 1D contribution to the trial wave function in the coordinate that is to be excited with the appropriate excited state based on the potential used to evaluate the ground state wave function. This procedure is likely to produce a trial wave function where the node is slightly shifted from the optimal position. To adjust the position of the node we take advantage of the fact that displacements of the walkers that change the phase of Φ_T are rejected. This allows us to run independent simulations for regions of configuration space where Φ_T is positive and negative. In one-dimension, if the node is correctly placed, the two simulations must provide identical values for the energy. In general, this becomes more complicated in higher dimensions. Since we are assuming a product form for Φ_T it is straightforward to obtain the nodal surface under this approximation of separability by requiring the energies obtained for the regions of configuration space where Φ_T is positive and negative be equal. Using this constraint, we run two DMC simulations in which at each time step the coordinate on which the wave function depends is shifted by a small amount. By making the displacement of the wave function a linear function of propagation time the approximate ground state energy will also change with τ. Plots of E_{ref} as a function of this displacement are made for both simulations, as illustrated in Figure 4.3. For this calculation, we are using Φ_1^{mon} in Figure 4.1(a) for the trial wave function to describe Ψ_1^{dim}. The point at which these two curves cross (0.993 Å and 5384 cm^{-1}) provides the optimal shift of the trial wave

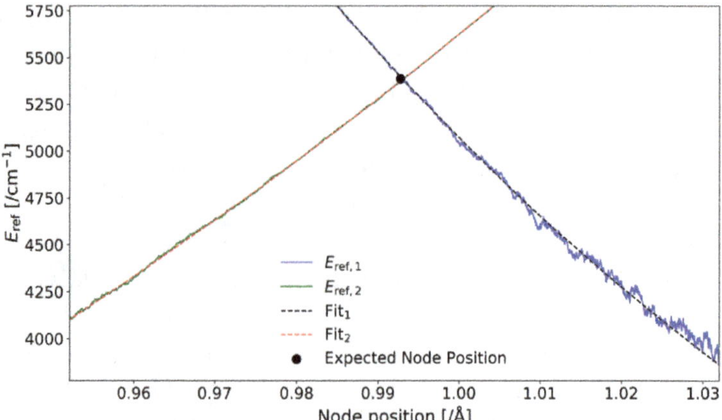

Figure 4.3. E_{ref} values plotted as functions of the position of the node, where the E_{ref} were obtained from a pair of simulations of the first excited state of a Morse oscillator that was constructed to approximate the hydrogen-bonded OH bond in water dimer, see Fig. 4.1, using Φ_1^{mon} as the guiding function. In the first simulation (blue) the walkers are initially placed on the left side of the node and Φ_1^{mon} is shifted to larger values of r_{OH} throughout the simulation. In the second simulation (green) the walkers are initially placed on the right side of the node and Φ_1^{mon} is shifted to smaller values of r_{OH} throughout the simulation. The two E_{ref} curves are plotted as functions of the position of the node and fit to cubic polynomials (red and black dashed lines) to identify the position of the node for this system. Finally, the black circle shows the expected location of the node and the corresponding energy.

function, and the corresponding energy is the energy of this excited state. To help illustrate the accuracy of the approach, the expected values of the node and energy are shown with a black circle, which is at 0.993 Å and 5387.37 cm^{-1}.

4.4.2. *Obtaining molecular properties*

Guided DMC provides a Monte Carlo sampling of f. As such, an integral over this distribution can be replaced by a sum over the walkers as

$$\int A(\mathbf{x})f(\mathbf{x})d\mathbf{x} = \sum_{i=1}^{N_W} w_i(\tau)A(\mathbf{x}_i(\tau)). \qquad (4.13)$$

In this way, the expectation value of a multiplicative operator $\hat{O} = O(\mathbf{x})$ can be evaluated using

$$\langle O \rangle = \int \Psi^2(\mathbf{x})O(\mathbf{x})d\mathbf{x} = \sum_{i=1}^{N_W} w_i(\tau)O(\mathbf{x}_i(\tau)) \left(\frac{\Psi(\mathbf{x}_i(\tau))}{\Phi_{\mathrm{T}}(\mathbf{x}_i(\tau))} \right) \quad (4.14)$$

while the projection of the probability amplitude onto an arbitrary coordinate, r, is evaluated using

$$P(r') = \int \Psi^2(\mathbf{x})\delta(r(\mathbf{x}) - r')d\mathbf{x} = \sum_{i=1}^{N_W} w_i(\tau)\delta(r(\mathbf{x}_i(\tau)) - r')$$

$$\times \left(\frac{\Psi(\mathbf{x}_i(\tau))}{\Phi_{\mathrm{T}}(\mathbf{x}_i(\tau))} \right).$$

$$(4.15)$$

Note operationally, $\delta(r(\mathbf{x}) - r')$ is replaced by a bin of width Δr.

Evaluation of the expressions in Eqs. (4.14) and (4.15) requires the ability to calculate Ψ/Φ_{T} at the coordinates of the walkers. Barnett *et al.* and Suhm and Watts[2,40] have shown that $(\Psi(\mathbf{x}_i)/\Phi_{\mathrm{T}}(\mathbf{x}_i))$ is proportional to the ratio of the weight of the ith walker at a propagation time of $\tau' = \tau + \delta\tau$ to the weight at propagation time τ.

$$\frac{\Psi(\mathbf{x}_i(\tau))}{\Phi_{\mathrm{T}}(\mathbf{x}_i(\tau))} \propto \frac{w_i(\tau')}{w_i(\tau)} \quad (4.16)$$

and, for example,

$$\langle O \rangle = \sum_{i=1}^{N_W} w_i(\tau)O(\mathbf{x}_i(\tau)) \left(\frac{\Psi(\mathbf{x}_i(\tau))}{\Phi_{\mathrm{T}}(\mathbf{x}_i(\tau))} \right) = \sum_{i=1}^{N_W} w_i(\tau')O(\mathbf{x}_i(\tau)).$$

$$(4.17)$$

Typically, $\delta\tau$ corresponds to roughly 200–300 a.u. of propagation time, and this approach is referred to as descendant weighting. For a discrete weighting simulation, this ratio is replaced by the number of walkers at time τ' that can be traced to the ith walker in the ensemble at time τ. An advantage of this approach is that it is straightforward to incorporate into the DMC simulation. The disadvantage is that the quality of the results can be sensitive to the value of $\delta\tau$, and the optimal value can depend on the system that is being studied. In addition, while the approach is effective for evaluating expectation values, its extension to the evaluation of matrix elements that involves the overlap of two different wave functions, while feasible, is challenging due to the fact that the ensembles of walkers used to sample the two states are not the same.[41]

An alternative approach takes advantage of the fact that Φ_T provides a good approximation to Ψ, at least for the coordinates that are included in the development of Φ_T. As such, $\Psi - \Phi_T = \delta$ can be assumed to be small. Following Singer and co-workers[8] and Barnett *et al.*,[40] based on this approximation for Ψ,

$$\Psi^2 = 2f - \Phi_T^2 + \delta^2 \approx 2f - \Phi_T^2. \tag{4.18}$$

With this relationship,

$$P(r') = 2\sum_{i=1}^{N_W} w_i(\tau)\delta(r(\mathbf{x}_i(\tau)) - r') - \phi_T^2(r') \tag{4.19}$$

where $\phi_T(r')$ is the 1D wave function that contributes to Φ_T, which is a function of r. Expectation values of operators can be obtained by a similar substitution into Eq. (4.13). By extension, if we are interested in evaluating $\langle\Psi'|O|\Psi\rangle$, we can also assume that $\Psi' = \Phi_T' + \delta'$, where δ' is also small. This leads to the approximation[20]

$$\langle\Psi'|O|\Psi\rangle = \sum_{i=1}^{N_W} w_i(\tau)O(\mathbf{x}_i(\tau))\left(\frac{\Phi_T'(\mathbf{x}_i(\tau))}{\Phi_T(\mathbf{x}_i(\tau))}\right) + \sum_{i'=1}^{N_W'} w_{i'}'(\tau)O(\mathbf{x}_{i'}(\tau))$$

$$\times \left(\frac{\Phi_T(\mathbf{x}_{i'}(\tau))}{\Phi_T'(\mathbf{x}_{i'}(\tau))}\right) - \int O(\mathbf{x})\Phi_T(\mathbf{x})\Phi_T'(\mathbf{x})d\mathbf{x}, \tag{4.20}$$

where the summations are over the distribution of walkers that represent f and f'.

To illustrate the relative accuracy of these treatments, we apply them to the Morse oscillators shown in Figure 4.1. In Figure 4.4(a), we plot the ground state probability amplitude. The results based on Φ_0^{dim} are shown in red, and are compared to the value of f and Ψ^2 obtained using descendant weighting and Eq. (4.19). As

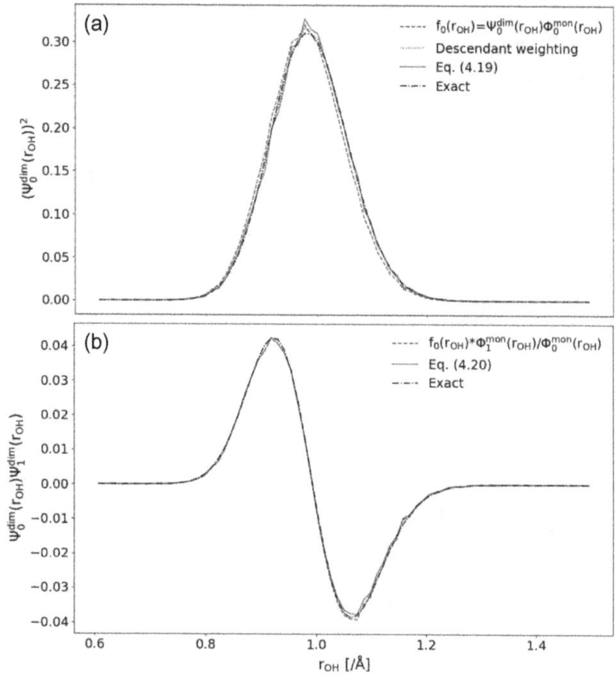

Figure 4.4. Several approximations to (a) $(\Psi_0^{dim})^2$ and (b) $\Psi_0^{dim}\Psi_1^{dim}$ are plotted as the function of the OH bond length. The black dotted lines provide the exact solutions to the one-dimensional potential shown in black dotted curves in Fig. 4.1(b). (a) The three approximations to the ground state probability amplitude include: using $f_0 = \Phi_0^{mon}$ (dashed blue line, identified as $f_0(r_{OH})$), using descendant weights as described in Eq. (4.15) (red dotted line), and using Eq. (4.19) (solid green line). (b) Two approximations to the overlap between the ground and first excited states. The first is obtained by multiplying $f_0(r_{OH})$ by the ratio of $\Phi_1^{mon}(r_{OH})/\Phi_0^{mon}(r_{OH})$ (dashed blue line). This corresponds to the first term in Eq. (4.20). The overlap that results from accounting for all three terms in Eq. (4.20) is plotted with a solid green line.

can be seen all three approaches yield accurate results. Likewise, the approximation to the overlap of Ψ_0^{dim} and Ψ_1^{dim} provided by Eq. (4.20) shows good agreement with the expected results. The distributions plotted in Figure 4.4 can also be used to evaluate expectation values or matrix elements of operators. For this discussion, we will focus on $x = r_{OH} - r_{OH,e}^{dim}$. The results of the various approximations are provided in Table 4.2. As can be seen, all of the approaches yield distributions that closely resemble the expected distribution (plotted with a black dash/dot line). The agreement

Table 4.2. Accuracy of approximate approaches for evaluating Ψ^2 and $\Psi_n \Psi_{n'}$ based on the Morse oscillator that describes the hydrogen-bonded OH oscillator in water dimer, see Figure 4.1.

Approach	$\langle \Psi_0 | x | \Psi_0 \rangle$[a]	$\langle \Psi_1 | x | \Psi_0 \rangle$[a]
Exact	0.0151	0.0700
Descendant Weighting	0.0150 (0.0007)	—
f	0.0095 (0.0006)	—
Eqs. (4.19) or (4.20)	0.0151 (0.0012)	0.0706 (0.0015)

[a] $x = r_{\mathrm{OH}} - r_{\mathrm{OH},e}^{\mathrm{dim}}$.

is poorest when we approximate Ψ^2 by f. This is more easily seen in the comparison of the values of $\langle x \rangle$, reported in Table 4.2, which have been evaluated either using descendant weights or the expressions provided in Eqs. (4.19) and (4.20). In the case of the overlaps, using Eq. (4.20) to obtain these quantities is much less computationally expensive than using descendant weighting.[41] These results are provided for 1D systems. When moving to larger systems the approximations provided in Eqs. (4.19) and (4.20) are only expected to provide accurate results when the guiding function is a function of the coordinate of interest.

4.5. Applications

With the formalism in place, it is helpful to consider molecular applications. We will discuss these results in the context of three systems. First we consider an isolated water molecule and the sensitivity of the results to the time step used for the simulations. We then discuss water hexamer, where the use of a guiding function can provide substantial savings in the computation time needed for the simulation. Finally we discuss the use of flexible trial wave functions for CH_5^+, where the distributions in Figure 4.2 indicate that a single trial wave function will cause challenge.

4.5.1. *Sensitivity of the ground state energy of water on the size of the time step*

Perhaps the simplest molecule that we can consider is water. With its three atoms, and relatively high vibrational frequencies, water

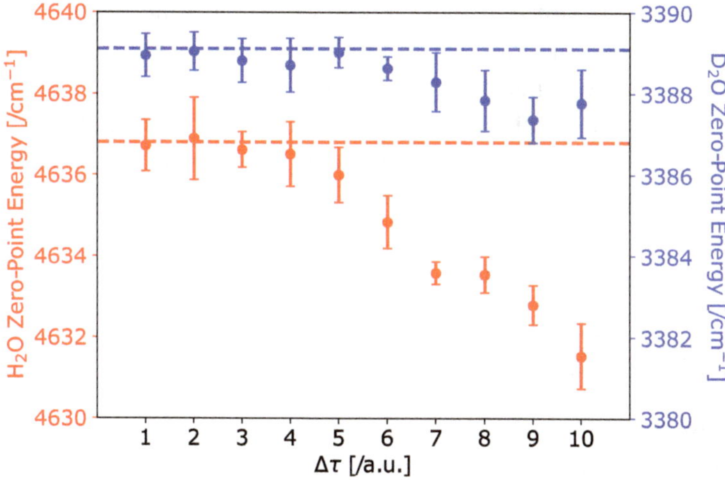

Figure 4.5. Calculated DMC zero-point energies for H_2O and D_2O shown as a function of the time step ($\Delta\tau$). The red dotted line indicates the zero-point energy for H_2O, and the blue dotted line indicates the zero-point energy for D_2O based on variational calculations (4636.8 and 3389.1 cm^{-1}).[46] These calculations are based on the standard DMC algorithm, *i.e.*, where Φ_T is constant.

provides a relatively straightforward molecule to use as an example for calculating the vibrational energies. We have run a series of unguided DMC simulations for both H_2O and D_2O, varying the time step used in the simulations. For all of these calculations, we use the potential developed by Partridge and Schwenke,[43] and employ an ensemble with 20 000 walkers, propagated for 200 000 a.u. The results are plotted in Figure 4.5, and the reported zero point energy is obtained by averaging E_{ref} from $\tau = 5000$ a.u. to the end of the simulation. As can be seen for time steps smaller than 4 a.u. for H_2O and below 6 a.u. for D_2O, the DMC simulations provide accurate values for the zero-point energy. When larger time steps are used, the quality of the results deteriorates. This is a consequence of the large frequency of the three vibrations in these molecules. Since the displacement of the atoms is proportional to $\sqrt{\Delta\tau}$, these relatively small time steps are required to obtain accurate results. This finding is important as we move to water clusters because similar values of $\Delta\tau$ will be needed to appropriately sample the intramolecular vibrational motions. When we use guided DMC, the calculations will all employ a 1 a.u. time step.

When we perform a calculation of the ground states of H_2O and D_2O, using a trial wave function that is a product of the local mode wave functions for the two OH or OD stretches and a harmonic treatment of the bend,[12] the calculated zero-point energies are 4637(1) and 3390(2) for H_2O and D_2O, respectively. In contrast to the unguided simulation, which required ensembles of at least 1000 walkers to obtain accurate results, the guided calculation produced accurate zero-point energies with as few as 100 walkers.

4.5.2. *Convergence of the zero-point energy of water hexamer with ensemble size*

As we move from the water monomer to water clusters, Mallory and Mandelshtam have shown that the ground state energies obtained from unguided simulations depend sensitively on the ensemble size.[16, 18] This is illustrated by the plot shown with dark blue diamonds in Figure 4.6. In this plot, we show the ground state energies as a function of $1/N_w$. The largest simulation corresponds to 10^6 walkers, with a time step of 10 a.u..[13] The red circles/solid line

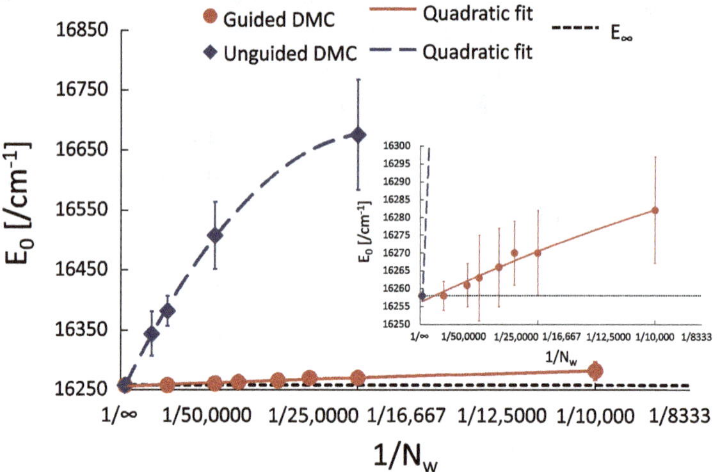

Figure 4.6.   Illustration of the convergence behavior of unguided (dark blue diamonds) and guided (red circles) DMC simulations for water hexamer with a time step of 1 a.u..[13] The ground state energies are plotted as a function of $1/N_w$. The results are fit to a quadratic function to aid the eye. The inset provides an expanded view (16 250 to 16 300 cm^{-1}) of the plot.

provide the results of guided DMC, for comparison. Even when 150 000 walkers are used with a time step of 1 a.u., the energy obtained using the unguided DMC simulation is showing a notable dependence on the ensemble size. In contrast, when we use guiding functions, which are identical to the one used we used for the water monomer, 50 000 walkers are required to obtain converged results.

To confirm that the use of guiding functions has not introduced bias into the simulations, we have also run several large simulations without guiding functions. When we use either 1 000 000 walkers with a time step of 10 a.u. or 1 a.u., we obtain ground state energies that are in good agreement with the ground state energies obtained from the guided simulation. The reason the guided approach works so well in this case is a reflection of the fact that the ground state wave function that describes an OH oscillator in the context of water hexamer is not that different from the ground state wave function for an OH oscillator for an isolated water molecule, plotted in Figure 4.1(b). In contrast, as noted above, this will not be the case for all systems.

4.5.3. *Flexible trial wave functions*

CH_5^+ provides an interesting system for illustrating a situation where the use of a product for the trial wave function is not appropriate. The equilibrium structure of CH_5^+ is shown in Figure 4.7(a) and 4.7(c). In Figure 4.7(b), we show the structures of two low-energy stationary points on the potential. Finally, the motions that connect the saddle point structures to the minima are shown with grey arrows. The associated normal mode frequencies are more than twice the associated barrier heights. Therefore, at the harmonic level the zero-point energy in these modes is sufficient for the ion to overcome these barriers. Analysis of the ground state probability amplitude obtained from DMC confirms this conclusion. If we take into account the symmetry of the ion, there are 120 equivalent minima, and these minima are connected via a series of 180 transition states, 60 of which correspond to the C_{2v} structure shown in the middle of the lower row of structures, while the remaining 120 correspond to the structure with C_s symmetry, shown in the middle of the upper

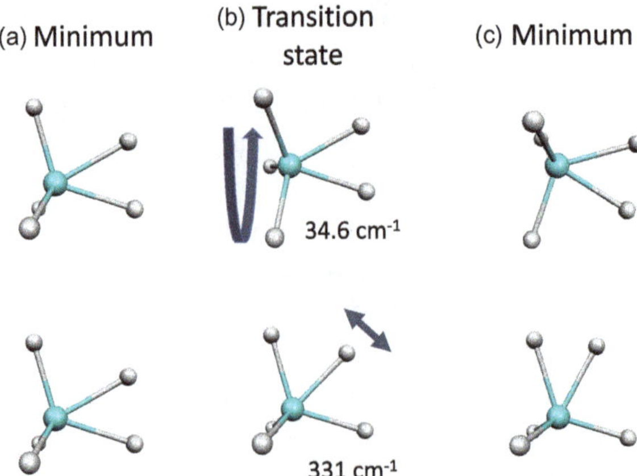

Figure 4.7. The three low-energy stationary point structures of CH_5^+. In (a) and (c) right columns, the structures correspond to minima on the potential. (b) Shows structures of transition states that connect two of these minima. The upper transition state corresponds to a 60° rotation of the CH_3 unit on the left side of the structure, while the lower transition state corresponds to the displacement of the CH bond in the upper right. Arrows are included to illustrate these motions, and the energies of the transition states relative to the minimum energy geometry are provided below the structures. These energies are based on the potential surface of Jin *et al.*[30]

row. More importantly for our discussion, the CH bond lengths in these structures vary from 1.08 to 1.20 Å.[30] Comparing this range of 0.12 Å to the ground state wave function for the lowest-frequency CH oscillator in the equilibrium structure (see Figure 4.2(b)), we find that the range spans roughly 40% of the full width at half maximum for that wave function. This is the reason the approach we used to develop trial wave functions for water clusters fails to provide accurate results for CH_5^+.

To address this challenge, we identified a correlation between the equilibrium CH bond length and the standard deviation among the HH′ distances involving the other four hydrogen atoms in the ion. This relationship is plotted in Figure 4.8, where the plotted points have been fit to a shifted exponential function. Additional details about this calculation can be found in Ref. 14. Using a single CH wave function as the trial wave function for each of the five CH oscillators in CH_5^+ gives a zero-point energy that is $10\,\mathrm{cm}^{-1}$ larger

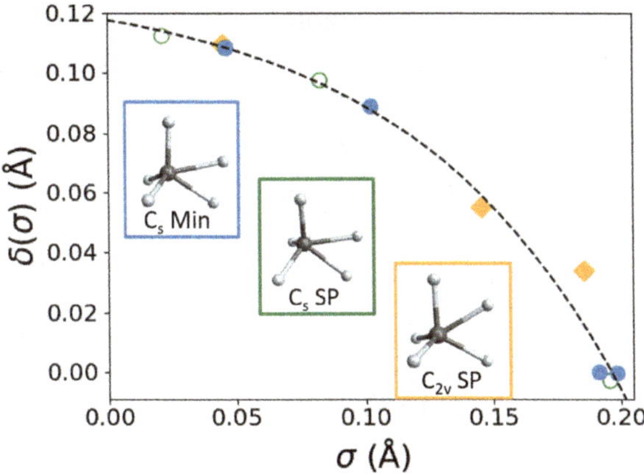

Figure 4.8.   Plot of the differences between each of the r_{CH} and the value of the longest CH distance in the C_s minimum-energy geometry of $\mathrm{CH_5^+}$ and the length of the CH bond of interest, δ, as a function of standard deviation of the H–H distances of this hydrogen atom, σ. These results are plotted for each of the three low-energy stationary point structures of $\mathrm{CH_5^+}$, shown in the insets. The values for the C_s minimum are plotted with blue filled circles, the values for the C_s saddle point are plotted with green open circles, while the values for the C_{2v} saddle point are shown with filled gold diamonds. The dotted line provides a fit of these values to an exponential function of the form $\delta(\sigma_i) = A\exp(B\sigma_i) + C$. A, B, and C are $-0.02389\,\text{Å}$, $6.29099\,\text{Å}^{-1}$, and $0.24620\,\text{Å}$, respectively. Reprinted with permission from Ref. 14. Copyright 2020 American Chemical Society.

than the energy obtained without guiding functions. When we allow the maximum in this wave function to be shifted based on the value of $\sigma_{\mathrm{HH'}}$, the calculated zero-point energy is 10 917(4) cm^{-1}, which is in excellent agreement with the value for the zero-point energy that is obtained without guiding functions, 10 918(7) cm^{-1}. As with water hexamer, the use of guided DMC allows us to employ significantly smaller ensembles. For unguided DMC, 20 000 walkers are needed to obtain an accurate zero-point energy, while 2000 walkers are sufficient for a guided DMC calculation.

4.6. Outlook

In this chapter, we have described approaches for studying molecules that undergo large-amplitude vibrational motions, which are based on DMC, through the introduction of carefully chosen trial wave functions. The introduction of guiding functions allows us to extend the size and types of molecular systems that can be studied using DMC. We have described some of the challenges introduced by using guiding functions as well as the opportunities they provide. Most importantly, these functions need to be selected with care as they can bias the results of DMC simulations. This is why our use of guiding functions has focused on intramolecular vibrations. Since the equilibrium bond lengths and vibrational frequencies of the molecular units that make up molecular clusters show relatively small variability with cluster size or the environment that the oscillator is in, developing trial wave functions from products of 1D functions of the high frequency intramolecular motions allows us to realize substantial savings in the size of the DMC calculations needed to achieve converged results without sacrificing their accuracy.[12,13] In some cases, like CH_5^+ or protonated water clusters, this approach requires minor modifications to account for changes in the equilibrium bond length or vibrational frequency of certain oscillators with low-frequency, large amplitude motions.[14]

We have also illustrated extensions to these ground state methods, by illustrating for 1D systems how guided DMC can be used to obtain excited state wave functions and energies. Work is currently underway to apply this approach to larger systems. With the ability to obtain ground and excited states, the next step is to obtain properties, like transition moments (*e.g.*, matrix elements of the dipole moment between two of these wave functions). This is complicated by the fact that in a typical implementation, each vibrational state is evaluated independently. As a result, these states are sampled at distinct sets of geometries, making the evaluation of matrix elements computationally demanding. We have extended the approach for obtaining probability amplitudes from $f = \Psi_0 \Phi_T$, described by Cho and Singer,[8] to allow us to obtain matrix elements using Eq. (4.20).

Finally, arguably the most time consuming part of the DMC simulation comes in the evaluation of the potential function. We have recently developed an extension of DMC to employ deep neural networks to fit the potential in the region that is relevant for a ground state DMC simulation.[46] Along with those described in this chapter, this advance is expected to greatly extend our capabilities for studying molecular systems that undergo large amplitude vibrational motions using DMC.

Acknowledgements

Support from the Chemistry Division of the National Science Foundation (CHE-1856125) is gratefully acknowledged. Parts of this work were performed using the Ilahie cluster, which was purchased using funds from a MRI grant from the National Science Foundation (CHE-1624430). This work was also facilitated through the use of advanced computational, storage, and networking infrastructure provided by the Hyak supercomputer system and funded by the STF at the University of Washington as well as resources of the National Energy Research Scientific Computing Center, a DOE Office of Science User Facility supported by the Office of Science of the U.S. Department of Energy under Contract No. DE-AC02-05CH11231 using NERSC award BES-ERCAP0020502. RJD was supported by a fellowship from The Molecular Sciences Software Institute under NSF grant OAC-1547580. Finally, the authors wish to thank Nicholas J. Vetterli and Victor G.-M. Lee for their contributions to work presented in this chapter.

References

1. Anderson, J. B. A random-walk simulation of the Schrödinger equation: H_3^+. *J. Chem. Phys.* **1975**, *63*, 1499–1503.
2. Suhm, M. A.; Watts, R. O. Quantum Monte Carlo studies of vibrational states in molecules and clusters. *Phys. Rep* **1991**, *204*, 293–329.
3. Metropolis, N.; Ulam, S. The Monte Carlo method. *J. Am. Stat. Assoc.* **1949**, *44*, 334–341.
4. Acioli, P. H.; Xie, Z.; Braams, B. J.; Bowman, J. M. Vibrational ground state properties of H_5^+ and its isotopomers from diffusion Monte Carlo calculations. *J. Chem. Phys.* **2008**, *128*, 104318.

5. Benoit, D.; Lauvergnat, D.; Scribano, Y. Does cage quantum delocalisation influence the translation-rotation states of molecules inside a clathrate? *Faraday Discuss.* **2018**, *212*, 533–546.

6. Sandler, P.; Buch, V.; Clary, D. C. Calculation of expectation values of molecular systems using diffusion Monte Carlo in conjunction with the finite field method. *J. Chem. Phys.* **1994**, *101*, 6353–6355.

7. Gregory, J. K.; Clary, D. C. Calculations of the tunneling splittings in water dimer and trimer using diffusion Monte Carlo. *J. Chem. Phys.* **1995**, *102*, 7817–7829.

8. Huang, X.; Cho, H. M.; Carter, S.; Ojamäe, L.; Bowman, J. M.; Singer, S. J. Full dimensional quantum calculations of vibrational energies of $H_5O_2^+$. *J. Phys. Chem. A* **2003**, *107*, 7142–7151.

9. Johnson, L. M.; McCoy, A. B. Evolution of structure in CH_5^+ and its deuterated analogs. *J. Phys. Chem. A* **2006**, *110*, 8213–8220.

10. Lin, Z.; McCoy, A. B. Signatures of large-amplitude vibrations in the spectra of H_5^+ and D_5^+. *J. Phys. Chem. Lett.* **2012**, *3*, 3690–3696.

11. Kelleher, P. J.; Johnson, C. J.; Fournier, J. A.; Johnson, M. A.; McCoy, A. B. Persistence of dual free internal rotation in $NH_4^+(H_2O)\cdot He_{n=0-3}$ ion–molecule complexes: Expanding the case for quantum delocalization in He tagging. *J. Phys. Chem. A* **2015**, *119*, 4170–4176.

12. Lee, V. G. M.; McCoy, A. B. An efficient approach for studies of water clusters using diffusion Monte Carlo. *J. Phys. Chem. A* **2019**, *123*, 8063–8070.

13. Lee, V. G. M.; Vetterli, N. J.; Boyer, M. A.; McCoy, A. B. Diffusion Monte Carlo studies on the detection of structural changes in the water hexamer upon isotopic substitution. *J. Phys. Chem. A* **2020**, *124*, 6903–6912.

14. Finney, J. M.; DiRisio, R. J.; McCoy, A. B. Guided diffusion Monte Carlo: A method for studying molecules and ions that display large amplitude vibrational motions. *J. Phys. Chem. A* **2020**, *124*, 9567–9577.

15. Petit, A.; McCoy, A. B. Diffusion Monte Carlo approaches for evaluating rotationally excited states of symmetric top molecules: Application to H_3O^+ and D_3O^+. *J. Phys. Chem. A* **2009**, *113*, 12706–12714.

16. Mallory, J. D.; Brown, S. E.; Mandelshtam, V. A. Assessing the performance of the diffusion Monte Carlo method as applied to the water monomer, dimer, and hexamer. *J. Phys. Chem. A* **2015**, *119*, 6504–6515.

17. Mallory, J. D.; Mandelshtam, V. A. Diffusion Monte Carlo studies of MB-pol $(H_2O)_{2-6}$ and $(D_2O)_{2-6}$ clusters: Structures and binding energies. *J. Chem. Phys.* **2016**, *145*, 064308.

18. Mallory, J. D.; Mandelshtam, V. A. Quantum melting and isotope effects from diffusion Monte Carlo studies of p-H_2 clusters. *J. Phys. Chem. A* **2017**, *121*, 6341–6348.

19. Buch, V. Treatment of rigid bodies by diffusion Monte Carlo: Application to the para-$H_2 \cdots H_2O$ and ortho-$H_2 \cdots H_2O$ clusters. *J. Chem. Phys.* **1992**, *97*, 726–729.

20. Severson, M. W.; Buch, V. Quantum Monte Carlo simulation of intermolecular excited vibrational states in the cage water hexamer. *J. Chem. Phys.* **1999**, *111*, 10866–10875.

21. Valdès, A.; Prosmiti, R.; Delgado-Barrio, G. Quantum-dynamics study of the H_5^+ cluster: Full dimensional benchmark results on its vibrational states. *J. Chem. Phys.* **2012**, *136*, 104302/1–6.

22. Cheng, T. C.; Jiang, L.; Asmis, K. R.; Wang, Y.; Bowman, J. M.; Ricks, A. M.; Duncan, M. A. Mid- and Far-IR spectra of H_5^+ and D_5^+ compared to the predictions of anharmonic theory. *J. Phys. Chem. Lett.* **2012**, *3*, 3160–3166.

23. Boyer, M. A.; Chiu, C. S.; McDonald, D. C.; Wagner, J. P.; Colley, J. E.; Orr, D. S.; Duncan, M. A.; McCoy, A. B. The role of tunneling in the spectra of H_5^+ and D_5^+ up to $7300\,cm^{-1}$. *J. Phys. Chem. A* **2020**, *124*, 4427–4439.

24. Asvany, O.; Kumar, P.; Redlich, B.; Hegeman, I.; Schlemmer, S.; Marx, D. Understanding the infrared spectrum of bare CH_5^+. *Science* **2005**, *309*, 1219–1222.

25. Asvany, O.; Yamada, K. M. T.; Brünken, S.; Potapov, A.; Schlemmer, S. Experimental ground-state combination differences of CH_5^+. *Science* **2015**, *347*, 1346–1349.

26. Thompson, K. C.; Crittenden, D. L.; Jordan, M. J. T. CH_5^+: Chemistry's chameleon unmasked. *J. Am. Chem. Soc.* **2005**, *127*, 4954–4958.

27. Kumar P, P.; Marx, D. Understanding hydrogen scramling and infrared spectrum of bare CH_5^+ based on *ab initio* siumulations. *Phys. Chem. Chem. Phys.* **2006**, *8*, 573–586.

28. McCoy, A. B.; Braams, B. J.; Brown, A.; Huang, X.; Jin, Z.; Bowman, J. M. Ab initio diffusion Monte Carlo calculations of the quantum behavior of CH_5^+ in full dimensionality. *J. Phys. Chem. A* **2004**, *108*, 4991–4994.

29. Braams, B. J.; Bowman, J. M. Permutationally invariant potential energy surfaces in high dimensionality. *Int. Rev. Phys. Chem.* **2009**, *28*, 577–606.

30. Jin, Z.; Braams, B. J.; Bowman, J. M. An ab Initio based global potential energy surface describing $CH_5^+ \rightarrow CH_3^+ + H_2$. *J. Phys. Chem. A* **2006**, *110*, 1569–1574.

31. Paesani, F. Getting the right answers for the right reasons: Toward predictive molecular simulations of water with many-body potential energy functions. *Acc. Chem. Res.* **2016**, *49*, 1844–1851.

32. Yu, Q.; Bowman, J. M. Communication: VSCF/VCI vibrational spectroscopy of $H_7O_3^+$ and $H_9O_4^+$ using high-level, many-body potential energy surface and dipole moment surfaces. *J. Chem. Phys.* **2017**, *146*, 121102.

33. Viel, A.; Whaley, K. B.; Wheatley, R. J. Blueshift and intramolecular tunneling of NH_3 umbrella mode in Hen4 clusters. *J. Chem. Phys.* **2007**, *127*, 194303.

34. Anderson, J. B. Quantum chemistry by random walk. H ^2P, H_3^+ D_{3h} $^1A_1'$, H_2 $^3\Sigma_u^+$, H_4 $^1\Sigma_g^+$, Be ^1S. *J. Chem. Phys.* **1976**, *65*, 4121–4127.

35. McCoy, A. B. Diffusion Monte Carlo approaches for investigating the structure and vibrational spectra of fluxional systems. *Int. Rev. Phys. Chem.* **2006**, *25*, 77–107.

36. Barone, V.; Alessandrini, S.; Biczysko, M.; Cheeseman, J. R.; Clary, D. C.; McCoy, A. B.; DiRisio, R. J.; Neese, F.; Melosso, M.; Puzzarini, C. Computational molecular spectroscopy. *Nat. Rev. Meth. Primers* **2021**, *1*, 38.

37. Petit, A. S.; Ford, J. E.; McCoy, A. B. Simultaneous evaluation of multiple rotationally excited states of H_3^+, H_3O^+, and CH_5^+ using diffusion Monte Carlo. *J. Phys. Chem. A* **2014**, *118*, 7206–7220.

38. Reynolds, P. J.; Ceperley, D. M.; Alder, B. J.; Lester Jr., W. A. Fixed-node quantum Monte Carlo for molecules. *J. Chem. Phys.* **1982**, *77*, 5593–5603.

39. Hammond, B. L.; Lester Jr., W. A.. Reynolds, P. J., Eds. *Monte Carlo Methods in Ab Initio Quantum Chemistry*. World Scientific Publishing Co.: Singapore, 1994.

40. Barnett, R.; Reynolds, P.; Lester Jr., W. A. Monte Carlo Algorithms for Expectation Values of Coordinate Operators. *J. Comput. Phys.* **1991**, *96*, 258–276.

41. Lee, V. G. M.; Madison, L. R.; McCoy, A. B. Evaluation of matrix elements using diffusion Monte Carlo wave functions. *J. Phys. Chem. A* **2019**, *123*, 4370–4378.

42. Boyer, M. A.; Marsalek, O.; Heindel, J. P.; Markland, T. E.; McCoy, A. B.; Xantheas, S. S. Beyond badger's rule: the origins and generality of the structure–spectra relationship of aqueous hydrogen bonds. *J. Phys. Chem. Lett.* **2019**, *10*, 918–924.

43. Partridge, H.; Schwenke, D. W. The determination of an accurate isotope dependent potential energy surface for water from extensive

ab initio calculations and experimental data. *J. Chem. Phys.* **1997**, *106*, 4618–4639.

44. Lee, H.-S.; Herbert, J. M.; McCoy, A. B. Adiabatic diffusion Monte Carlo approaches for studies of ground and excited state properties of van der Waals complexes. *J. Chem. Phys.* **1999**, *110*, 5481–5484.

45. McCoy, A. B. Diffusion Monte Carlo Approaches for Studying Systems that Undergo Large Amplitude Vibrational Motions. In *Recent Advances in Quantum Monte Carlo Methods*, Vol. 3; Anderson, J. B., Rothstein, S. M., Eds.; ACS Symposium Series 953, 2006; pp. 147–164.

46. DiRisio, R. J.; Lu, F.; McCoy, A. B. GPU-accelerated neural network potential energy surfaces for diffusion Monte Carlo. *J. Phys. Chem. A* **2021**, *125*, 5849–5859.

Chapter 5

Collocation Methods for Computing Vibrational Spectra

Tucker Carrington

*Department of Chemistry, Queen's University,
Kingston, Ontario K7L 3N6, Canada
Tucker.Carrington@queensu.ca*

5.1. Introduction

This chapter is about the advantages of collocation methods for solving the Schrödinger equation to compute vibrational spectra of molecules with between 4 and 6 atoms. The key advantage of collocation is that it obviates the need for quadrature. Throughout the chapter, I shall assume that it is sufficient to consider vibration on a single Born–Oppenheimer potential energy surface(s) (PES). The focus is on methods for solving the time-independent Schrödinger equation

$$\hat{H}\psi_n(\mathbf{x}) = E_n\psi_n(\mathbf{x}), \tag{5.1}$$

but many of the ideas can also be applied to solve the time-dependent Schrödinger equation. Differences between the energies E_n are transition energies. The Hamiltonian, $\hat{H}$, is the sum of a kinetic energy operator (KEO) and a PES. In this paper, the PES is assumed to be known as a function of some coordinates. To solve the Schrödinger equation one must choose coordinates in which to write the KEO. Often one chooses coordinates so that the corresponding KEO is known. The KEO is always a sum of terms each of which is a coefficient multiplied by a second (or possibly a first) derivative with respect to the coordinates. In general, the coefficients are coordinate dependent. In some cases, it may be better to use coordinates in terms of which the KEO cannot be written down and to numerically evaluate the coefficients at values of the coordinates.[1-4]

5.2. The Variational Method

A conceptually simple and general way to solve Eq. (5.1) is to replace $\psi_n(\mathbf{x})$ with

$$\psi_n(\mathbf{x}) = \sum_{i=1} u_{in} b_i(\mathbf{x}), \tag{5.2}$$

where $b_i(\mathbf{x})$ is one of a set of known basis functions, multiply on the left by $b_{i'}(\mathbf{x})$ and integrate to obtain a matrix eigenvalue problem.[5] The matrix eigenvalue problem is

$$\mathbf{HU} = \mathbf{SUE}, \tag{5.3}$$

where $\mathbf{S}$ is an overlap matrix, $\mathbf{H}$ is a Hamiltonian matrix, and $\mathbf{U}$ is a matrix of eigenvectors. Each element of $\mathbf{S}$ and $\mathbf{H}$ is an integral. The dimensionality of the integrals is the number of vibrational coordinates. In simple cases, it is possible to choose orthonormal basis functions so that elements of the matrix representation of the KEO are known in equation form. If integrals are done exactly, $\mathbf{S} = \mathbf{I}$. In general, at least some of the integrals must be done by multi-dimensional quadrature. Eq. (5.3) is often known as the variational method. Like the variational method in introductory quantum chemistry books, if all the integrals are done exactly the smallest eigenvalue is an upper limit for the exact ground state energy. In addition, the other eigenvalues are upper bounds for excited state energies.

Two things make using Eq. (5.3) computationally difficult: (1) for molecules with more than about 4 atoms the matrices are large and computing eigenvalues and eigenvectors is costly; (2) computing matrix elements themselves requires high-dimensional quadrature. The size of the KEO, overlap, and the PES matrices is equal to the number of basis functions required to obtain converged solutions of the Schrödinger equation. The most obvious way to do a variational calculation is to compute and store all of the matrix elements of $\mathbf{S}$ and $\mathbf{H}$. Ideas have been developed that make it possible to evaluate Hamiltonian matrix–vector products and therefore use iterative eigensolvers without storing $\mathbf{S}$ and $\mathbf{H}$. They are simple when the basis is a direct product (also called a tensor product).[6] In this chapter, I focus not on reducing the cost of computing eigenvalues and eigenvectors, but on computing the matrices that are fed into an eigensolver.

5.3. Sum of Product Potential Energy Surfaces

A popular way to eradicate the problem posed by the need to compute matrix elements of $\mathbf{S}$ and $\mathbf{H}$ is to use an orthogonal direct product basis and to insist that the PES must be a sum of products (SOP). In a direct product basis each function is a product of univariate functions,

$$b_{i_1,i_2,\ldots,i_D}(x_1, x_2, \ldots, x_D) = b_{i_1}^{(1)}(x_1)b_{i_2}^{(2)}(x_2)\cdots b_{i_D}^{(D)}(x_D);$$

$$i_k = 1, 2, \ldots, b+1; \quad k = 1, 2, \ldots, D. \qquad (5.4)$$

When the PES is a SOP and the basis is a direct product every multi-dimensional integral is a product of 1D integrals which are trivial to evaluate by quadrature. In one fell swoop all the problems introduced by the need to do multi-dimensional quadrature are eliminated. A SOP also has the important advantage that it reduces the cost of the matrix–vector products one must evaluate to use an iterative eigensolver.[7,8]

Direct product basis SOP PES methods are extremely useful and several chapters in this book are about such methods. The most popular direct product basis SOP PES program is the Heidelberg multiconfiguration time-dependent Hartree (MCTDH) package (see Chapter 7). It makes calculations on large molecules and molecular systems possible for people who are not familiar with basis sets, quadratures, *etc.* The program requires that it be possible to store vectors whose size is n^D, where n is a representative number of univariate basis functions (called single particle functions in the MCTDH literature) and D is the number of coordinates. There is no need to store **S** and **H**. There are also direct product SOP methods that do not even require storing vectors with n^D elements, but impose a tensor form on computed wavefunctions.[9–11] They are related to DMRG which is reviewed in Chapter 3. DMRG is well established as a method for computing ground states. Chapter 1 also reviews methods that exploit a SOP PES. Coupled cluster methods have been applied to vibrational problems and are reviewed in Chapter 2; they also require that the PES be in SOP form.

There are problems for which direct product SOP PES methods are poorly suited. For example, methods that require storing vectors with as many elements as there are direct product basis functions cannot be used if it is not possible to store n^D numbers in memory. There are many methods that exploit advantages of an SOP PES but mitigate the memory cost by using only a subset of the functions in a direct product basis. This idea of pruning a direct product basis can also be applied to MCTDH. There are also multi-layer MCTDH methods that do not use a basis of products of univariate functions.[14–16] All of these ideas reduce the memory required.

A different problem for direct product SOP PES methods is not related to the product basis, but to the requirement that the PES be a SOP. For molecules with more than about 5 atoms it is common to compute PES derivatives (often up to quartic derivatives) at the equilibrium structure and represent the PES as a "force field". Chapter 7 describes the calculation of force fields. For some molecules the best known PESs are force fields. For many, and especially smaller, molecules it is possible to make very accurate PESs by fitting parameters of some function so that it nearly reproduces *ab initio* energies computed at a large number of points. Machine learning methods have made this almost routine in many cases.[17] Fitted PESs are rarely in SOP form. A PES that is fitted to a SOP in one set of coordinates is not a SOP in another set of coordinates. There are established tools for converting a non-SOP PES into SOP form.[18–22] Once the PES has been converted, all **S** and **H** elements are products of 1D integrals, if a product basis is used. Clearly, it would be nice to avoid the need to convert an accurate PES to SOP form. Moreover, when a PES is converted to SOP some accuracy is lost. Often scientists who construct a PES work hard to achieve great accuracy and it is a shame to throw away even some of that accuracy. One way to increase the accuracy of a SOP PES is to increase the number of terms. However, the cost of a variational method using a SOP PES depends on the number of terms in the PES.

5.4. Using Quadrature with a General PES

The SOP-PES direct product basis method is advantageous, but if one wishes to compute vibrational energy levels on a general PES there is no compelling reason to use a direct product basis. Nonetheless, using a direct product basis does make calculations simpler. This is particularly the case if one uses a discrete variable representation (DVR).[6] When using a DVR, the matrix representing the PES is diagonal and one only needs to evaluate the PES at the DVR points. The only commonly used DVRs are direct products and this unfortunately means that the size of the basis scales exponentially with the number of atoms in the molecule. A DVR is a disguised direct product quadrature. When using a general PES one is free to use a better basis, *i.e.*, a basis with fewer functions, but with which one can nonetheless compute converged energies and wavefunctions. Such better basis functions incorporate coupling between coordinates and are not products of univariate functions. Regrettably, it is difficult to use both a DVR and multivariate basis functions. They can be combined only when using a simultaneous diagonalization DVR.[23, 24]

If the PES is not a SOP and a direct product basis is too large then one confronts the need to somehow do multi-dimensional quadrature. There is no need to use quadrature to calculate elements of the matrix representing the PES. It is only necessary to be able to evaluate matrix–vector products. A simple and useful basis has functions that are products of functions of subsets of the coordinates. The basis functions are often called contracted.[25–28] For example, for a 6D problem with coordinates $q_1, q_2, q_3, q_4, q_5, q_6$, one might use basis functions that are products of two functions depending on three coordinates,

$$b_{i_A}^A(q_1, q_2, q_3) b_{i_B}^B(q_4, q_5, q_6). \tag{5.5}$$

In the MCTDH community, this idea is called "mode combination".[29,30] A contracted basis function can be written $b_{i_A}^A(\mathbf{Q_A})$ $b_{i_B}^B(\mathbf{Q_B})$, where $\mathbf{Q_A}$ and $\mathbf{Q_B}$ each include three coordinates and are sometimes called "logical" coordinates. An obvious way to use a general PES and a $b_{i_A}^A(\mathbf{Q_A}) b_{i_B}^B(\mathbf{Q_B})$ basis is to compute matrix vector products by evaluating the basis functions and the PES on a direct product quadrature grid. This requires storing a vector with as many components as there are quadrature points. For a 12D problem there are about 10^{12} quadrature points and the vector is too large. Paradoxically, although it is fairly simple to reduce the basis size by using contracted functions, it is impossible to store the potential values required to use a general PES. It is, however, possible to use two direct product quadrature grids, one for $\mathbf{Q_A}$ and one for $\mathbf{Q_B}$, but not store the potential on the full product grid. This is done by storing an intermediate matrix labeled by basis functions for $\mathbf{Q_A}$ and points for $\mathbf{Q_B}$.[31–33]

Quadrature methods are still the workhorses of computational quantum dynamics. They can be used with a non-direct product basis. However, unless both the KEO and the PES are SOPs they do have disadvantages. (1) It is hard to use quadrature if it is not easy to apply the KEO to the chosen basis functions and it is costly to deal with a complicated KEO having many terms with coordinate-dependent coefficients; (2) it is necessary to choose quadrature points and weights so that all integrals are computed accurately.

5.5. The Collocation Method

Collocation is a simple alternative to quadrature. To use collocation, wavefunctions are represented as linear combinations of basis functions, as in Eq. (5.2), but the coefficients are obtained differently. One way of deriving the equation for the coefficients is to substitute Eq. (5.3) into the Schrödinger equation and to multiply on the left by a Dirac delta function before integrating. The matrix equation for the coefficients is[34]

$$\mathbf{HU} = \mathbf{BUE}. \tag{5.6}$$

When the number of collocation points equals the number of basis function, all of the matrices are square. Note that $\mathbf{H}, \mathbf{U}$, and $\mathbf{E}$ in Eqs. (5.6) and (5.3) are not the same. For a 1D problem, $\mathbf{H}$ in Eq. (5.6) is,

$$\mathbf{H} = \mathbf{T} + \mathbf{VB}, \tag{5.7}$$

where

$$\mathbf{T}_{ai} = \hat{K} b_i(\boldsymbol{x}) \Big|_{x=x_a} \quad ; \quad \mathbf{B}_{ai} = b_i(\boldsymbol{x}) \Big|_{x=x_a} \quad ; \quad \mathbf{V}_{aa'} = V(\boldsymbol{x_a})\delta_{aa'}. \tag{5.8}$$

$\hat{K}$ is the appropriate KEO and the index a labels a collocation point. There is no need to compute integrals, but one must still choose basis functions and now also collocation points.

Collocation has two significant disadvantages. First, $\boldsymbol{H}$ is not symmetric. Second, Eq. (5.6) is a generalized eigenvalue problem. A consequence of the first disadvantage is that it is difficult to use iterative eigensolvers that require storing only a few vectors (*e.g.*, Lanczos methods which require storing only two vectors). A consequence of the second disadvantage is that it is necessary to deal with the inverse of $\boldsymbol{B}$.

Collocation also has important advantages. In most variational calculations, coordinates and basis functions are chosen so that matrix elements of terms in the KEO or factors in terms of the KEO are exact and available in closed form. For example, one often uses normal coordinates and harmonic oscillator basis functions or polyspherical coordinates[35–38] and Legendre and Chebyshev or plane-wave basis functions. When using collocation there is no basis representation of the KEO, which opens the door to optimizing coordinates and basis functions. It is straightforward to use collocation with a complicated KEO. In curvilinear coordinates, the general

KEO can be written,

$$\hat{K} = \sum_{j,l}^{3N_{at}-6} \left(G_{jl}(\boldsymbol{x}) \frac{\partial}{\partial x_j} \frac{\partial}{\partial x_l} \right) + \sum_{j}^{3N_{at}-6} \left(H_j(\boldsymbol{x}) \frac{\partial}{\partial x_j} \right) + V'(\boldsymbol{x}).$$

When using a variational method, it is necessary to compute matrix elements of the functions $H_j(\mathbf{x})$ and $G_{jl}(\mathbf{x})$; when doing a collocation calculation, it is only necessary to evaluate H_j and G_{jl} at points. In fact, for a collocation calculation, there is no need to know H_j and G_{jl} everywhere, they only need to be known at the collocation points.[39] This facilitates using a "numerical" KEO.[1-4] It is also possible to use a space-fixed KEO and basis functions that are functions of internal coordinates.[40] The most obvious advantage of collocation is that it completely eliminates the need to evaluate integrals of the PES. It is only necessary to evaluate the PES at the collocation points. In a variational calculation, one must choose quadrature points so that integrals are accurate.

When the basis is small one can store $\mathbf{H}$ and $\mathbf{B}$ and use a direct method (the QZ algorithm[41] is probably the best) to solve

$$(T + VB)U = BUE, \tag{5.9}$$

however, if the matrices are too large to store then it is necessary to use an iterative eigensolver and this means dealing with $\mathbf{B}^{-1}$. One can solve either $(G = BU)$

$$(TB^{-1} + V)G = GE, \tag{5.10}$$

or

$$(B^{-1}T + B^{-1}VB)U = UE. \tag{5.11}$$

In both cases, one must do matrix-vector products with $\boldsymbol{B}^{-1}$. If the matrices are large it is better to avoid inverting $\boldsymbol{B}$.

5.5.1. *Collocation with a direct product basis and a direct product grid*

It may seem that there is no need for collocation with a direct product basis and a direct product grid. It has been known for decades how to evaluate sums sequentially when using a direct product basis and direct product quadrature so that spectra may be calculated with an iterative eigensolver.[7] As I stress in Section 5.4, it is when the basis is *not* a direct product of 1D bases that quadrature is difficult to use. Nonetheless, it is useful to think about collocation with a direct product basis and a direct product grid. There are several reasons: (1) In some cases it is possible to use fewer collocation points than quadrature points when using a direct product basis. (2) Standard methods use univariate basis functions that are classical orthogonal polynomials (or linear combinations of them) and Gauss quadratures with which one can compute accurate potential and overlap matrix elements. It may be advantageous to use other univariate (maybe non-orthogonal) functions and to use points that are not Gauss points. In this case quadrature is usually less accurate. (3) Thinking about the direct product calculation helps to understand ideas required to use collocation with pruned (and much smaller) product bases and grids. In the pruned case, using quadrature is harder because the quadrature grids for a single coordinate must be nested.

When both the collocation grid and the basis are direct products, then $\boldsymbol{B}$ is a Kronecker product,

$$\boldsymbol{B} = \boldsymbol{B_1} \otimes \boldsymbol{B_2} \cdots \boldsymbol{B_D}, \tag{5.12}$$

and

$$\boldsymbol{B}^{-1} = \boldsymbol{B_1}^{-1} \otimes \boldsymbol{B_2}^{-1} \cdots \boldsymbol{B_D}^{-1}. \tag{5.13}$$

The size of the matrices $\boldsymbol{B_k}$, $k = 1, \ldots, D$ is the number of basis functions (or number of points) for a *single* coordinate. $(\boldsymbol{B_k})_{a_k, i_k} = b_{i_k}^{(k)}(x_{a_k}^{(k)})$. The size of $\boldsymbol{B}$ is about 10^D and the size of $\boldsymbol{B_k}$ is about 10. It is straightforward to apply a Kronecker product of inverses (Eq. (5.13)) to the vector that represents a function in a direct product basis.

5.5.1.1. *Using collocation with the MCTDH method*

If one is going to use a direct product basis then for many problems the best choice is the MCTDH basis. This makes the combination of collocation and MCTDH particularly useful. MCTDH is a successful method for solving the Schrödinger equation.[42,43] The MCTDH univariate functions are called single-particle functions (SPFs). MCTDH is usually used with a SOP PES. Manthe's correlation DVR (CDVR) is a quadrature-type idea that makes it possible to use a non-SOP PES.[44]

An MCTDH wavefunction is an expansion in a direct product basis

$$\Psi(x_1,\ldots,x_D,t) = \sum_{i_1=1}^{n_1} \cdots \sum_{i_D=1}^{n_D} A_{i_1,\ldots,i_D}(t)\, \varphi_{i_1}^{(1)}(x_1,t)\, \varphi_{i_2}^{(2)}(x_2,t)$$
$$\cdots \varphi_{i_D}^{(D)}(x_D,t). \tag{5.14}$$

In an MCTDH calculation, one solves two coupled equations to compute $A_{i_1,\ldots,i_D}(t)$ and $\varphi_{i_k}^{(k)}(x_k,t)$. The equations are derived from a variational principle. To use MCTDH in conjunction with collocation one must derive new equations.[45] With the choice of the right gauge, the collocation-MCTDH (C-MCTDH) equations have exactly the same form as the standard MCTDH equations. The equation for $A_{i_1,i_2,\ldots,i_D}(t)$ is

$$\sum_{i_1,\ldots,i_D} \mathrm{i}\dot{A}_{i_1,\ldots,i_D}(t) \prod_{k=1}^{D} \varphi_{i_k}^{(k)}\left(r_{a_k}^{(k)}(t),t\right)$$
$$= \sum_{i_1,\ldots,i_D} \langle r_{a_1}^{(1)}(t)| \cdots \langle r_{a_D}^{(D)}(t)|\,\hat{H}\,|\varphi_{i_1}^{(1)}(t)\rangle \cdots |\varphi_{i_D}^{(D)}(t)\rangle\, A_{i_1,\ldots,i_D}.$$
$$\tag{5.15}$$

Using the fact that the inverse of a Kronecker product is a Kronecker product of inverses one obtains

$$\mathrm{i}\dot{A}_{i_1,\ldots,i_D}(t) = \sum_{a_1,\ldots,a_D}\sum_{j_1,\ldots,j_D} \left[\mathbf{B}^{(1)^{-1}}\right]_{i_1,a_1}$$
$$\cdots \left[\mathbf{B}^{(D)^{-1}}\right]_{i_D,a_D} H_{\substack{a_1,\ldots,a_D \\ j_1,\ldots,j_D}} A_{j_1,\ldots,j_D}(t). \tag{5.16}$$

For the SPFs, one finds the equation,

$$\sum_{i_k} \rho_{i'_k,i_k}^{(k)} \, \mathrm{i} \, |\dot{\varphi}_{i_k}^{(k)}(t)\rangle$$

$$= \sum_{i_k} \hat{\mathcal{H}}_{i'_k,i_k}^{(k)} \, |\varphi_{i_k}^{(k)}(t)\rangle$$

$$- \sum_{j_k} \underbrace{\sum_{i_k,a_k} |\varphi_{i_k}^{(k)}(t)\rangle \left[\mathbf{B}^{(k)^{-1}}\right]_{i_k,a_k} \langle r_{a_k}^{(k)}(t)| \, \hat{\mathcal{H}}_{i'_k,j_k}^{(k)} \, |\varphi_{j_k}^{(k)}(t)\rangle}_{=\hat{P}^{(k)}} ,$$

$$(5.17)$$

with the projector $\hat{P}^{(k)}$, the density matrix $\rho_{i'_k,i_k}^{(k)} = \sum_{j_1,\ldots,j_D}^{(\neg k)} A_{j_1,\ldots,i'_k,\ldots,j_D}^* A_{j_1,\ldots,i_k,\ldots,j_D}$ and the (new) mean-field matrix

$$\hat{\mathcal{H}}_{i'_k,i_k}^{(k)} = \sum_{j_1,\ldots,j_D}^{(\neg k)} \sum_{a_1,\ldots,a_D}^{(\neg k)} \sum_{i_1,\ldots,i_D}^{(\neg k)} A_{j_1,\ldots,i'_k,\ldots,j_D}^* \prod_{l \neq k} \left[\mathbf{B}^{(l)^{-1}}\right]_{j_l,a_l}$$

$$\times \left(\bigotimes_{l \neq k} \langle r_{a_l}^{(l)}(t)|\right) \hat{H} \left(\bigotimes_{l \neq k} |\varphi_{i_l}^{(l)}(t)\rangle\right) A_{i_1,\ldots,i_D}. \quad (5.18)$$

The upper limit $\neg k$ on the sum in the definition of the density matrix means that the sum over j_k is omitted.

Eqs. (5.16) and (5.17) have the same form as the standard MCTDH equations and can therefore be solved using the machinery[46,47] developed to solve the standard MCTDH equations (with small changes). To solve Eq. (5.17), we use collocation and a set of points denser than the grid with which Eq. (5.16) is solved.

5.5.2. *Using collocation with a pruned product basis*

A pruned product basis is one obtained from a direct product basis by removing functions thought to be superfluous. If $D > 6$, pruning a basis can reduce the size of the basis by many orders of magnitude. The simplest pruning scheme is to remove functions for which $i_1 + i_2 + \cdots + i_D > b + D$, where b is a convergence parameter. By choosing a point set with the same number of points as there are basis functions, it is possible to use collocation. However, one must confront a major problem: it is difficult to compute the product of $\boldsymbol{B}^{-1}$ and a vector. It is only easy when both the grid and the basis are direct products. $\boldsymbol{B}$ transforms a column of $\boldsymbol{U}$, *i.e.*, a vector of basis coefficients, to a vector whose elements are values of a function at collocation points. $\boldsymbol{B}^{-1}$ transforms a vector of values at points to a vector of basis coefficients. In many ways $\boldsymbol{B}$ and $\boldsymbol{B}^{-1}$ are like the matrices used to transform between a DVR and the corresponding finite basis representation (FBR). With few exceptions, the FBR–DVR transformation is only done when both the quadrature grid and the finite basis are direct products.[6] A pruned product basis is not a direct product basis. In this section, I outline a scheme for computing the product of $\boldsymbol{B}^{-1}$ and a vector that works when both the basis and the grid are non-direct products.[48–51]

A general way to prune a direct product grid is to divide the univariate functions, from which the direct product is built, into levels and to use a basis of products of functions of different coordinates, subject to a constraint on the levels. For example, for coordinates x_1 and x_2, one might choose to put functions labeled by $i = 1$ and $i = 2$ into level $\ell = 1$, functions labeled by $i = 1$, $i = 2$, $i = 3$, and $i = 4$ into level $\ell = 2$, *etc.*, and to stipulate that the pruned basis includes products of functions for which $\ell_1 + \ell_2 \leq b$. The levels are nested in the sense that a function in level $\ell_k - 1$ is also in level ℓ_k. A fairly general pruning condition can be obtained by using the condition $G_1(\ell_1) + G_2(\ell_2) + \cdots \leq b + D$. The functions $G_k(\ell_k)$ must be non-decreasing.[52] The number of functions in a level is m_ℓ. In this chapter, I shall explain how the ideas work for the simple case with $G_k(\ell_k) = \ell_k$ and $m_\ell = \ell$. In that case $G_1(\ell_1) + G_2(\ell_2) + \cdots \leq b + D$ is equivalent to $i_1 + i_2 + \cdots i_D \leq b + D$ mentioned in the previous paragraph. Equations for the general case are more complicated.[48, 51–53]

The next problem is to make a set of collocation points that is the same size as the pruned basis. The pruned basis has structure conveyed by the definition of the levels and the pruning condition, and in order to be able to evaluate matrix–vector products with $\boldsymbol{B}^{-1}$ it is imperative that the grid have the same structure. Smolyak[54] proposed ideas for making such grids for multi-dimensional quadrature and they are now referred to as sparse grids. For each coordinate x_k, a set of points $a_k = 1, 2, \ldots$ is divided into levels. Each level has as many points as there are functions in the corresponding level used to make the pruned basis. The points are nested in the sense that a point in level $\ell_k - 1$ is also in level ℓ_k. The points of the collocation grid are those for which $a_1 + a_2 + \cdots i_D \leq b + D$.

Having defined the pruned basis and the collocation grid, we can now consider the evaluation of the product of $\boldsymbol{B}^{-1}$ and a vector labeled by basis indices. In a direct product basis, it is simple to both invert $\boldsymbol{B}$ and to apply the resulting matrix to the vector. In the non-direct product case $\boldsymbol{B}$ is not a Kronecker product and new ideas are required. It is conceptually possible to write

$$\mathbf{B} = \mathbf{C}^T \mathbf{B}^{(\mathrm{kron})} \mathbf{C}, \tag{5.19}$$

where $\mathbf{C} \in \mathbb{R}^{N_{\mathrm{full}} \times N_{\mathrm{sparse}}}$ is a chopping matrix[49, 55] which is an $N_{\mathrm{full}} \times N_{\mathrm{full}}$ identity matrix from which the columns for which $\|\mathbf{i} - \mathbf{1}\|_1 > b$ have been removed and $\mathbf{B}^{(\mathrm{kron})}$ is a Kronecker product with the form of Eq. (5.12). Because matrix–vector products with $\mathbf{C}$, $\mathbf{B}^{(\mathrm{kron})}$, and $\mathbf{C}^T$ are all straightforward, it is easy and inexpensive to evaluate matrix–vector products with $\mathbf{B}$. Matrix–vector products with $\boldsymbol{B}^{-1}$ appear to be costly, owing to the fact that it is not obtained from a Kronecker product by removing rows and columns.

If it were somehow possible to chop the Kronecker product of the inverse of $\mathbf{B}^{(\mathrm{kron})}$ rather than inverting the chopped Kronecker product $\mathbf{B}^{(\mathrm{kron})}$ everything would be easy. By carefully choosing univariate basis functions one can in fact ensure that

$$\left[\mathbf{C}^T \mathbf{B}^{(\mathrm{kron})} \mathbf{C} \right]^{-1} = \mathbf{C}^T \left[\mathbf{B}^{(\mathrm{kron})} \right]^{-1} \mathbf{C}. \tag{5.20}$$

However, one wishes to choose univariate basis functions to reduce the size of the (multi-d) basis required to converge wavefunctions and not in order to simplify the evaluation of matrix–vector products. Fortunately, it is possible to find univariate basis functions that do both. We call the new basis functions ZAPPL functions. ZAPPL functions of x_k are linear combinations of the original basis functions $b_{i_k}^{(k)}(x_k)$, chosen to reduce the size of the basis. They are defined so that a ZAPPL function is Zero At Points in Previous Levels. When we first introduced ZAPPL functions we called them hierarchical functions because the hierarchical functions commonly used by mathematicians are zero at points in previous levels.[56] Commonly used hierarchical functions are piecewise linear functions. They are not good basis functions for representing smooth (wave)functions. Using the recipe of Refs. 48, 50, 51 and 57, it is straightforward to make ZAPPL functions from *any* $b_{i_k}^{(c)}(x_k)$. No compromise is necessary between working with basis functions to reduce the size of the multi-d basis and working with basis functions that facilitate evaluating matrix–vector products. When using a ZAPPL basis, the individual $\mathbf{B}_k$ matrices are lower triangular. Eq. (5.20) then follows from the standard block Gaussian elimination formula for the inverse of a matrix consisting of four blocks, if basis functions excluded from the pruned basis are all zero at the collocation points. Computing matrix–vector products with $\boldsymbol{B}^{-1}$ is also crucial for making a multi-d interpolant that enables one to write a function, known at the points of the Smolyak grid, as a linear combination of the pruned basis functions.[50] The same ideas can therefore be used to convert values of a PES known at collocation points to pruned basis representation of the PES.[58]

As explained above, the collocation point set is made from sets of points for each coordinate and a rule, such as $a_1 + a_2 + \cdots i_D \leq b + D$, for combining them. It remains to explain how the points for a single coordinate are chosen. There are several options.[59] Recently, we have been using Pseudo-Gauss and Leja points.[51–53, 60] In our pruned-basis collocation calculations we have used univariate functions, $b_{i_k}^{(k)}(x_k)$ that are eigenfunctions of 1D cut Hamiltonians, harmonic oscillator functions, and MCTDH single particle functions.

The effectiveness of pruned bases has been demonstrated in several calculations. Using MCTDH SPFs, energy levels of CH_2NH were computed with Leja points. The pruning made it possible to reduce the size of the basis by more than two orders of magnitude.[49] In this calculation, the KEO is approximated by neglecting contributions involving π_α, the so-called vibrational angular momentum. An advantage of collocation is the ability to deal with complicated KEOs. In calculations on methane with a harmonic basis, we used the full KEO which has momentum cross-terms and coordinate-dependent coefficients.[52]

5.6. Conclusion

In this chapter, I have briefly reviewed several ways of applying collocation to compute vibrational states. Applications are given in the papers I have cited. The main advantage of collocation is that it obviates the need for quadrature. When the PES is a SOP and the basis functions are products of univariate functions, there is no need to avoid quadrature because all the integrals required to use a variational method can be factorized. It is only if one wishes to use a general PES that collocation is a useful tool. Even in that case, collocation is not the only option. There are excellent variational methods. To use these methods, it is necessary to choose a quadrature good enough to accurately calculate elements of the variational Hamiltonian matrix. Most of these methods require, when used with a general PES, storing values of the PES at the quadrature points. Because not much is known about constructing non-direct product quadrature grids, it is common to use direct product quadratures. Direct product grids are unmanageably large when $D > 9$. One solution is to use contracted basis functions and store an intermediate matrix.[31–33] Collocation is another solution. It works because it is not necessary to use collocation points with which one could accurately evaluate integrals. This makes it possible to use pruned grids with basis functions that are not classical orthogonal polynomials.

Some possible advantages of collocation have not yet been fully explored or exploited. An important advantage of collocation is that it facilitates dealing with a complicated KEO. In this chapter, I have focused on the fact that it is nice not to have to evaluate matrix elements of the PES. A nasty KEO would have (at least) D^2 terms each being a product of two momentum operators and a coordinate-dependent coefficient. To use this KEO with a variational method, it would be necessary to calculate matrix elements for each of the D^2 terms. The number of KEO integrals is a factor of D^2 larger than the number of potential integrals! For most coordinate systems in common use the KEO is not this nasty (it is in normal coordinates, in general), but in optimized coordinates it might be. When using collocation, it is only necessary to evaluate the coordinate-dependent coefficients at points. Another advantage of collocation is that there is no need to choose basis functions to simplify the calculation of integrals. In principle, one could use very complicated basis functions that depend on a subset or even on all of the coordinates. We know that the choice of the collocation points becomes less important as the quality of the basis improves. It is possible to use more points than basis functions.[40] Recently, collocation has been used with 3D and 4D MCTDH single particle functions.[61] Little is known about how to choose collocation points when basis functions are not products of univariate functions, however, collocation opens the door to using much more general basis functions. For example one could use semi-classical wavefunctions and use collocation to correct the semi-classical approximation.

Acknowledgments

The financial support of the Natural Sciences and Engineering Research Council is gratefully acknowledged. The research described in this perspective was done with Gustavo Avila-Blanco, Sergei Manzhos, Robert Wodraszka, and Emil Zak and I am grateful for their collaboration.

References

1. Meyer, R. Flexible models for intramolecular motion, a versatile treatment and its application to glyoxal. *J. Mol. Spectrosc.* **1979**, *76*, 266–300.
2. Lauvergnat, D.; Nauts, A. Exact numerical computation of a kinetic energy operator in curvilinear coordinates. *J. Chem. Phys.* **2002**, *116*, 8560.
3. Mátyus, E.; Czakó, G.; Császár, A. G. Toward black-box-type full- and reduced-dimensional variational (ro) vibrational computations. *J. Chem. Phys.* **2009**, *130*, 134112.
4. Laane, J.; Harthcock, M. A.; Killough, P. M.; Bauman, L. E.; Cooke, J. M. Vector representation of large-amplitude vibrations for the determination of kinetic energy functions. *J. Mol. Spectrosc.* **1982**, *91*, 286.

5. Carney, G. D.; Sprandel, L. L.; Kern, C. W. Variational approaches to vibration-rotation spectroscopy for polyatomic molecules. *Adv. Chem. Phys.* **1978**, *37*, 305.

6. Light, J. C.; Carrington, T. Jr. Discrete-variable representations and their utilization. *Adv. Chem. Phys.* **2000**, *114*, 263–310.

7. Bramley, M. J.; Carrington, T. A general discrete variable method to calculate vibrational energy levels of threeand fouratom molecules. *J. Chem. Phys.* **1993**, *99*, 8519–8541.

8. Manthe, U.; Koeppel, H. Dynamics on potential energy surfaces with a conical intersection: Adiabatic, intermediate, and diabatic behavior. *J. Chem. Phys.* **1990**, *93*, 345.

9. Leclerc, A.; Carrington, T. Calculating vibrational spectra with sum of product basis functions without storing fulldimensional vectors or matrices. *J. Chem. Phys.* **2014**, *140*, 174111.

10. Thomas, P. S.; Carrington, T. Jr. Using nested contractions and a hierarchical tensor format to compute vibrational spectra of molecules with seven atoms. *J. Phys. Chem. A* **2015**, *119*, 13074–13091.

11. Rakhuba, M.; Oseledets, I. Calculating vibrational spectra of molecules using tensor train decomposition. *J. Chem. Phys.* **2016**, *145*, 124101.

12. Bowman, J. M.; Carter, S.; Huang, X. MULTIMODE: A code to calculate rovibrational energies of polyatomic molecules. *Int. Rev. Phys. Chem.* **2003**, *22*, 533–549.

13. Rauhut, G. Configuration selection as a route towards efficient vibrational configuration interaction calculations. *J. Chem. Phys.* **2007**, *127*, 184109.

14. Wang, H.; Thoss, M. Multilayer formulation of the multiconfiguration time-dependent Hartree theory. *J. Chem. Phys.* **2003**, *119*, 1289–1299.

15. Manthe, U. A multilayer multiconfigurational time-dependent Hartree approach for quantum dynamics on general potential energy surfaces. *J. Chem. Phys.* **2008**, *128*, 164116.

16. Vendrell, O.; Meyer, H.-D. Multilayer multiconfiguration time-dependent Hartree method: Implementation and applications to a Henon-Heiles Hamiltonian and to pyrazine. *J. Chem. Phys.* **2011**, *134*, 044135.

17. Manzhos, S.; Carrington, T. Neural network potential energy surfaces for small molecules and reactions. *Chem. Rev.* **2021**, *121*, 10187–10217.

18. Jaeckle, A.; Meyer, H.-D. Product representation of potential energy surfaces. *J. Chem. Phys.* **1996**, *104*, 7974.

19. Schrder. M.; Meyer, H.-D. Transforming high-dimensional potential energy surfaces into sum-of-products form using Monte Carlo methods. *J. Chem. Phys.* **2017**, *147*, 064105.

20. Schrder, M. Transforming high-dimensional potential energy surfaces into a canonical polyadic decomposition using Monte Carlo methods. *J. Chem. Phys.* **2020**, *152*, 024108.

21. Pelaez, D.; Meyer, H.-D. The multigrid POTFIT (MGPF) method: Grid representations of potentials for quantum dynamics of large systems. *J. Chem. Phys.* **2013**, *138*, 014108.

22. Manzhos, S.; Carrington, T. Using neural networks to represent potential surfaces as sums of products. *J. Chem. Phys.* **2006**, *125*, 194105-1–194105-5.

23. Dawes, R.; Carrington, T. A multidimensional discrete variable representation obtained by simultaneous diagonalization. *J. Chem. Phys.* **2004**, *121*, 726–736.

24. van Harrevelt, R.; Manthe, U. Multidimensional time-dependent discrete variable representations in multiconfiguration Hartree calculations. *J. Chem. Phys.* **2005**, *123*, 064106.

25. Carter, S.; Handy, N. C. A variational method for the determination of the vibrational ($J = 0$) energy levels of acetylene, using a Hamiltonian in internal coordinates. *Comput. Phys. Commun.* **1988**, *51*, 49.

26. Bowman, J. M.; Gazdy, B. A truncation/recoupling method for basis set calculations of eigenvalues and eigenvectors. *J. Chem. Phys.* **1991**, *94*, 454.

27. Henderson, J. R.; Tennyson, J. All the vibrational bound states of H_3^+. *Chem. Phys. Lett.* **1990**, *173*, 133.

28. Bačić, Z.; Light, J. C. Theoretical methods for rovibrational states of floppy molecules. *Annu. Rev. Phys. Chem.* **1989**, *40*, 469.

29. Worth, G. A.; Meyer, H.-D.; Cederbaum, L. S. Relaxation of a system with a conical intersection coupled to a bath: A benchmark 24-dimensional wave packet study treating the environment explicitly. *J. Chem. Phys.* **1998**, *109*, 3518.

30. Beck, M. H.; Jaeckle, A.; Worth, G. A.; Meyer, H.-D. The multiconfiguration time-dependent Hartree (MCTDH) method: A highly efficient algorithm for propagating wave packets. *Phys. Rep.* **2000**, *324*, 1–105.

31. Bramley, M. J.; Carrington, T. Calculation of triatomic vibrational eigenstates: Product or contracted basis sets, Lanczos or conventional eigensolvers? What is the most efficient combination? *J. Chem. Phys.* **1994**, *101*, 8494–8507.

32. Wang, X.-G.; Carrington, T. New ideas for using contracted basis functions with a Lanczos eigensolver for computing vibrational spectra of molecules with four or more atoms. *J. Chem. Phys.* **2002**, *117*, 6923–6934.

33. Felker, P. M.; Bačić, Z. H_2O–CO and D_2O–CO complexes: Intra- and intermolecular rovibrational states from full-dimensional and fully coupled quantum calculations. *J. Chem. Phys.* **2020**, *153*, 074107.

34. Yang, W.; Peet, A. C. The collocation method for bound solutions of the Schrdinger equation. *Chem. Phys. Lett.* **1988**, *153*, 98–104.

35. Gatti, F.; Iung, C.; Menou, M.; Justumm, Y.; Nauts, A.; Chapuisat, X. Vector parametrization of the N-atom problem in quantum mechanics. *J. Chem. Phys.* **1998**, *108*, 8804.

36. Mladenović, M. Rovibrational Hamiltonians for general polyatomic molecules in spherical polar parametrization. I. Orthogonal representations. *J. Chem. Phys.* **2000**, *112*, 1070–1081.

37. Chapuisat, X.; Belfhal, A.; Nauts, A. N-body quantum-mechanical Hamiltonians: Extrapotential terms. *J. Chem. Phys.* **1991**, *149*, 274–304.

38. Gatti, F.; Iung, C. Exact and constrained kinetic energy operators for polyatomic molecules: The polyspherical approach. *Phys. Rep.* **2009**, *484*, 1.

39. Avila, G.; Carrington, T. Jr. Solving the Schroedinger equation using Smolyak interpolants. *J. Chem. Phys.* **2013**, *139*, 134114.

40. Manzhos, S.; Carrington, T. Using an internal coordinate basis and a space-fixed Cartesian coordinate kinetic energy operator to compute a vibrational spectrum with Gaussian basis functions and rectangular collocation. *J. Chem. Phys.* **2016**, *145*, 224110.

41. Moler, C. B.; Stewart, G. W. An algorithm for generalized matrix eigenvalue problems. *SIAM J. Numer. Anal.* **1973**, *10*, 241–256.

42. Manthe, U.; Meyer, H.-D.; Cederbaum, L. S. Wave-packet dynamics within the multiconfiguration Hartree framework: General aspects and application to NOCl. *J. Chem. Phys.* **1992**, *97*, 3199–3213.

43. Meyer, H.-D., Gatti, F., Worth, G. A., Eds. *Multidimensional Quantum Dynamics: MCTDH Theory and Applications.* Wiley-VCH: Weinheim, 2009.

44. Manthe, U. A time-dependent discrete variable representation for (multiconfiguration) Hartree methods. *J. Chem. Phys.* **1996**, *105*, 6989.

45. Wodraszka, R.; Carrington, T. A new collocation-based multi-configuration time-dependent Hartree (MCTDH) approach for solving the Schrdinger equation with a general potential energy surface. *J. Chem. Phys.* **2018**, *148*, 044115.

46. Beck, M. H.; Meyer, H.-D. An efficient and robust integration scheme for the equations of motion of the multiconfiguration time-dependent Hartree (MCTDH) method. *Z. Phys. D* **1997**, *42*, 113–129.

47. Manthe, U. On the integration of the multi-configurational time-dependent Hartree (MCTDH) equations of motion. *Chem. Phys.* **2006**, *329*, 168–178.

48. Avila, G.; Carrington, T. Using collocation and a hierarchical basis to solve the vibrational Schrdinger equation. *J. Chem. Phys.* **2015**, *143*, 214108.

49. Wodraszka, R.; Carrington, T. A pruned collocation-based multiconfiguration time-dependent Hartree approach using a Smolyak grid for solving the Schrdinger equation with a general potential energy surface. *J. Chem. Phys.* **2019**, *150*, 154108.

50. Wodraszka, R.; Carrington, T. Jr. Efficiently Transforming from Values of a Function on a Sparse Grid to Basis Coefficients. In *Sparse Grids and Applications — Munich 2018*; Bungartz, H.-J., Garcke, J., Pflüger, D., Eds.; Springer International, 2018.

51. Zak, E. J.; Carrington, T. Using collocation and a hierarchical basis to solve the vibrational Schrdinger equation. *J. Chem. Phys.* **2019**, *150*, 204108.

52. Avila, G.; Carrington, T. Computing vibrational energy levels of CH4 with a Smolyak collocation method. *J. Chem. Phys.* **2017**, *147*, 144102.

53. Avila, G.; Carrington, T. Reducing the cost of using collocation to compute vibrational energy levels: Results for CH2NH. *J. Chem. Phys.* **2017**, *147*, 064103.

54. Smolyak, S. A. Quadrature and interpolation formulas for tensor products of certain classes of functions. *Sov. Math. Dokl.* **1963**, *4*, 240.

55. Brown, J.; Carrington, T. Assessing the utility of phase-space-localized basis functions: Exploiting direct product structure and a new basis function selection procedure. *J. Chem. Phys.* **2016**, *144*, 244115.

56. Bungartz, H.-J.; Griebel, M. Sparse grids. *Acta Numer.* **2004**, *13*, 1–123.

57. Valentin, J.; Pflueger, D. Fundamental Splines on Sparse Grids and their Application to Gradient-Based Optimization. *Sparse Grids and Applications — Miami 2016*; Garcke, J., Pflüger, D., Webster, C. G., Zhang, G., Eds.; Springer International, 2016; pp. 229–251.

58. Avila, G.; Carrington, T. Using multi-dimensional Smolyak interpolation to make a sum-of-products potential. *J. Chem. Phys.* **2015**, *143*, 044106-1–044106-9.

59. Avila, G.; Carrington, T. Solving the Schroedinger equation using Smolyak interpolants. *J. Chem. Phys.* **2013**, *139*, 134114.

60. Avila, G.; Carrington, T. Non-product quadrature grids for solving the vibrational Schrödinger equation. *J. Chem. Phys.* **2009**, *131*, 174103–174115.

61. Wodraszka, R.; Carrington, T. A collocation-based multi-configuration time-dependent Hartree method using mode combination and improved relaxation. *J. Chem. Phys.* **2009**, *131*, 174103–174115.

Chapter 6

Vibration-Rotation-Tunneling Levels and Spectra of Van der Waals Molecules

Ad van der Avoird

Theoretical Chemistry, Institute for Molecules and Materials
Radboud University, 135 Heyendaalseweg, 6525 AJ Nijmegen, Netherlands
A.vanderAvoird@theochem.ru.nl

6.1. Introduction

Non-covalent interactions play an important role in nature, in particular in biological systems. Ideal objects to study these interactions are Van der Waals (VdW) molecules: complexes of molecules (or molecular ions or radicals) held together by non-covalent forces. These forces are weaker than the chemical bonds in stable molecules, but they can vary in strength by orders of magnitude: from weak VdW interactions between rare gas atoms or non-polar molecules such as benzene, to strong hydrogen bonding between water molecules, for example. VdW molecules can be formed in cold supersonic molecular beams and spectroscopically studied in high resolution. Microwave and millimeter spectra probe rotational transitions which yield information about their structure. When they contain quadrupolar nuclei, such as nitrogen or deuterium, and the corresponding hyperfine splittings can be resolved, the spectra provide additional information about the orientations of the individual molecules in the complex. Far-infrared or terahertz spectra directly probe the intermolecular vibrations, and thereby the intermolecular forces. Infrared spectra show the frequency shifts of the molecular vibrations caused by the interactions with other molecules in the complex, as well as combination bands that originate from the intermolecular vibrations. To extract all of this information from the spectra and, more specifically, to critically test intermolecular interaction potentials that can nowadays be obtained from *ab initio* electronic structure calculations, it is essential that one can accurately compute the bound rovibrational energy levels of a VdW molecule from a given intermolecular potential energy surface. The energy differences between the levels are the transition frequencies; the line intensities can be obtained from the calculated wave functions when a dipole function is also available. The calculated spectra can then be directly compared with measured high-resolution spectra and the accuracy of the underlying intermolecular potential can be verified. Furthermore, the calculated spectra are very useful in assigning the lines in the measured spectra. The present chapter describes the methods by which the vibration-rotation-tunneling (VRT) states of VdW molecules can be calculated and presents some illustrative examples.

The (an)harmonic vibrations in molecules (treated in most chapters of this book) start from a given equilibrium structure, corresponding to the global minimum in the potential surface. The coordinates in rovibrational level calculations obey Eckart conditions, which minimize the coupling between vibrations and rotations. Actually, there may be more than one equilibrium geometry, *i.e.*, multiple equivalent global minima in the potential surface, but this is physically unimportant as long as the barriers between these minima are so high that there is no tunneling between them. Special methods have been devised for floppy molecules with relatively low barriers between equivalent minima. Typical examples are the internal rotation in ethane or methanol, and umbrella-inversion tunneling in ammonia. In VdW molecules only the constituent molecules are usually assumed to be nearly rigid, while all internal motions of the molecules in the complex are nearly free, in principle, and are strongly coupled to the overall rotation of the complex. Not only the force constants and higher derivatives of the potential at the equilibrium geometry, but a global potential surface is needed to calculate the bound states of such complexes. Our description of the methods to calculate the VRT states of VdW molecules begins with the definition of an appropriate set of coordinates. Next, we discuss the form of the Hamiltonian and the choice of a suitable basis set to compute its eigenstates, which are delocalized over multiple minima with tunneling between them. Finally, we outline various methods applied to obtain these eigenstates.

Another important aspect is symmetry. When calculating the vibrational states of a molecule, one commonly uses the point group of its equilibrium structure. Also the equilibrium geometry of a VdW molecule has a certain point group symmetry, but the symmetry group actually used is the permutation-inversion (PI) group, sometimes called the molecular symmetry group.[1] This PI group contains all "feasible" permutations of the nuclear coordinates in a VdW molecule that relate its different equivalent equilibrium structures (global minima in the potential). A permutation is called feasible when tunneling between the minima related by it leads to an observable splitting of the energy levels, a definition that depends on the resolution at which the spectra are measured. The point group of the equilibrium geometry is a subgroup of the PI group, but the latter usually contains many more operations. The use of PI symmetry considerably simplifies the calculation of the VRT levels of a VdW molecule and, moreover, is helpful in the interpretation of the corresponding states. Permutations of the nuclei also interchange their spin coordinates and the nuclear spin states of a VdW complex can also be adapted to the irreducible representations (irreps) of the PI group. Since the nuclei are either bosons or fermions and the total wave function must be symmetric or antisymmetric under permutations of identical nuclei, the PI symmetry of the VRT states is directly related to the symmetry of the nuclear spin states. The conversion of different nuclear spin states into each other is very slow, and the states of a VdW complex with different PI symmetries can often be considered as distinct species. Correspondingly, the measured spectra can be decomposed as overlapping spectra originating from the different symmetry species, weighted with their nuclear spin multiplicities (the nuclear spin statistical weights). Examples will be given to illustrate these symmetry aspects.

6.2. Theory

6.2.1. *Coordinates and Hamiltonian*

Most of the basic theory to obtain the VRT states of VdW molecules is given in the review by van der Avoird *et al.*[2] It begins with the choice of the coordinates in which the Hamiltonian and its eigenfunctions are expressed. The most convenient set of coordinates for a VdW complex consisting of two arbitrary molecules A and B is defined with respect to a body-fixed (BF) frame. In general, such a BF frame is obtained from the laboratory or space-fixed (SF) frame by a rotation over three Euler angles. For VdW molecules the simplest form of the kinetic operator in the Hamiltonian is obtained if one chooses a two-angle embedded BF frame with its z-axis along the intermolecular vector $\boldsymbol{R} \equiv \boldsymbol{R}_{AB}$ that points from the center of mass of molecule A to that of molecule B. The orientation of this frame is defined by two Euler angles (Θ, Φ), the polar angles of the vector $\boldsymbol{R}$ in the SF frame. The orientations of arbitrary (non-linear) molecules A and B with respect to this BF frame are given by six Euler angles $\boldsymbol{\omega}_X = (\alpha_X, \beta_X, \gamma_X)$ with X = A and B. The same Euler angles appear in the intermolecular potential, which depends only on the difference angle $\alpha = \alpha_B - \alpha_A$, the dihedral angle of the complex. The Hamiltonian in this two-angle embedded BF frame was first derived by Brocks *et al.*,[3] and rederived in a different manner by van der Avoird *et al.*[2] It reads

$$\hat{H} = -\frac{\hbar^2}{2\mu R}\frac{\partial^2}{\partial R^2}R + \frac{\hat{J}^2 + \hat{j}_{AB}^2 - 2\hat{j}_{AB}\cdot\hat{J}}{2\mu R^2}$$

$$+ \hat{H}_A + \hat{H}_B + \hat{V}(R, \beta_A, \gamma_A, \alpha, \beta_B, \gamma_B), \tag{6.1}$$

where R is the length of the vector $\boldsymbol{R}$, $\mu = m_A m_B/(m_A + m_B)$ is the reduced mass of the complex, $\hat{\boldsymbol{J}}$ is the total angular momentum operator, $\hat{\boldsymbol{j}}_{AB} = \hat{\boldsymbol{j}}_A + \hat{\boldsymbol{j}}_B$ is the sum of the monomer angular momentum operators, and $\hat{H}_A, \hat{H}_B$ are the monomer Hamiltonians. As derived by van der Avoird *et al.*,[2] the operator $\hat{J}^2 + \hat{j}_{AB}^2 - 2\hat{\boldsymbol{j}}_{AB}\cdot\hat{\boldsymbol{J}}$ is the BF equivalent of the end-over-end rotation operator $\hat{L}^2$ in the SF frame. It includes the Coriolis coupling between the internal rotations $\boldsymbol{j}_A$ and $\boldsymbol{j}_B$ and the overall rotation $\boldsymbol{J}$.

The potential $\hat{V}(R, \beta_A, \gamma_A, \alpha_B, \beta_B, \gamma_B)$ can be obtained from *ab initio* electronic structure calculations with α_A set to zero, preferentially by a method that is size-consistent.[4] Currently, the golden standard is the coupled-cluster method with single and double excitations and the perturbative inclusion of triples [CCSD(T)], which is implemented in most program packages for electronic structure calculations and is known to yield rather accurate results. In some cases,[5,6] it turned out that a higher level of electron correlation, obtained by the full inclusion of triple excitations and the perturbative inclusion of quadruples [CCSDT(Q)], was needed to explain the observed spectrum of a VdW molecule. Electronic structure methods to compute accurate intermolecular potentials are not the topic of this chapter, however.

In most calculations on VdW molecules it is assumed that the monomers A and B are rigid, and that the monomer Hamiltonians are simply the rigid rotor operators

$$\hat{H}_X(\boldsymbol{\omega}_X) = A_X \hat{j}_{Xa}^2 + B_X \hat{j}_{Xb}^2 + C_X \hat{j}_{Xc}, \tag{6.2}$$

where A_X, B_X, C_X are the rotational constants of monomer $X = A$ or B, and $\hat{j}_{Xa}, \hat{j}_{Xb}, \hat{j}_{Xc}$ are the components of the monomer rotational angular momentum $\hat{\boldsymbol{j}}_X$ on the principal axes a, b, c. The rigid monomer assumption is justified, because the intramolecular vibrations usually have much higher frequencies than the intermolecular vibrations against the relatively weak non-covalent forces, and one can make a Born–Oppenheimer (BO)-like adiabatic separation between these different types of motions. Strictly, such an adiabatic treatment implies that one starts from a potential surface $V(R, \beta_A, \gamma_A, \alpha, \beta_B, \gamma_B, \boldsymbol{q}_A, \boldsymbol{q}_B)$ that depends on the internal coordinates $\boldsymbol{q}_A$ and $\boldsymbol{q}_B$ of the constituent molecules, with monomer Hamiltonians $\hat{H}_X(\boldsymbol{q}_X, \boldsymbol{\omega}_X)$ that also depend on $\boldsymbol{q}_X$. In the first step, one solves the monomer vibrational problems with fixed intermolecular coordinates in the potential. When this is done for many dimer geometries, the intermolecular potential $V_{v_A, v_B}(R, \beta_A \gamma_A, \alpha_B, \beta_B, \gamma_B)$ thus obtained from the energies of a specific vibrational state v_A, v_B of molecules A and B can be used in the second step to compute the VRT states of the VdW molecule. This is rarely done, however, in most cases one simply calculates the intermolecular potential $V(R, \beta_A, \gamma_A, \alpha_B, \beta_B, \gamma_B)$ with the monomer coordinates fixed, either at the monomer equilibrium geometry or at its vibrationally averaged geometry. It has been demonstrated[7] on the example of Ar-HF that the latter choice gives more accurate results, in comparison with the energy levels from a full-dimensional treatment of the internal motions.

Better than to fix the molecular geometries is to calculate the vibrational wave functions of the free molecules A and B depending on q_A and q_B and to average the full potential $V(R, \beta_A, \gamma_A, \alpha, \beta_B, \gamma_B, q_A, q_B)$ over these wave functions. This procedure has been applied to the ground and vibrationally excited states of CO–H$_2$[5,6] and H$_2$O–H$_2$.[8–10] Also the full adiabatic treatment described above has been applied,[11–13] to the example of the water dimer with six intermolecular degrees of freedom and six (3+3) intramolecular ones. This provides not only the intermolecular VRT levels, but also the shifts of the vibrational frequencies of the donor and acceptor molecules in the hydrogen-bonded water dimer. This (6+6)D adiabatic model of the water dimer has later been extended[14] to a crude-adiabatic 12D treatment that uses a product basis of two 3D monomer vibrational wave functions calculated for isolated monomers and the same 6D intermolecular basis as in the rigid-monomer treatments (see Eq. (6.3)). The set of monomer basis products was first limited to the five vibrational states of each H$_2$O monomer with the lowest energy, and then further reduced to 10 functions in total for which the sum of the monomer energies is less than $3,800\,\mathrm{cm}^{-1}$. The intermolecular VRT levels from this 12D treatment agree well with those of the (6+6)D adiabatic treatment, the shifts of the monomer vibrational frequencies show larger differences. An alternative method that also includes both the intra- and intermolecular degrees of freedom has been developed by Felker and Bačić[15] and applied in 9D calculations on various H$_2$O-containing complexes.[16–18] Also this method is based on the calculation of the vibrational states of the monomers, here combined with the calculation of the intermolecular VRT states in a rigid-monomer model. These preparatory calculations produce a set of wave functions, both for the monomer vibrations and for the intermolecular VRT states. A limited set of these wave functions is then used in a direct product basis to compute the eigenstates of the full flexible-monomer Hamiltonian. An important conclusion is that only the wave functions of the lower intermolecular VRT states need to be included to obtain realistic shifts of the monomer vibrational frequencies, although these frequencies are much higher than the energy of the highest included intermolecular VRT state.

6.2.2. *Basis set*

Just as one needs a global potential — with multiple equivalent global minima in most cases — to compute the VRT states of VdW molecules, one needs a global basis. The basis in the radial coordinate R must cover the range from the repulsive wall in the potential at small R to the long range region where the potential decays as R^{-n}. For non-polar molecules the smallest n equals 6, from the attractive dispersion forces; for neutral polar molecules the smallest n equals 3, from dipole–dipole interactions. When either A or B is an ion, the smallest n equals 4, from the dipole induced-dipole attraction. When A and B are both ions, with opposite charges, the resulting complex is very strongly bound by the Coulomb attraction and can probably not be considered as a VdW molecule. Especially VRT states close to the dissociation limit are very sensitive to the long range interactions, so one must carefully choose the upper limit of R in order to converge such states. The angular basis must cover the complete range of the angular coordinates of both molecules A and B.

An angular basis that is complete for a non-linear molecule consists of the symmetric rotor functions, or Wigner D-functions, $D_{mk}^{(j)}(\boldsymbol{\omega})^*$.[19] This basis can be used to describe the rotations of the molecules A and B in the complex, which are hindered by the anisotropic intermolecular potential. A similar basis can be used for the overall rotation of the complex, and the full basis in BF coordinates is given by

$$| n, j_A, k_A, j_B, k_B, j_{AB}, K; J, M \rangle$$

$$= \chi_n(R) \left[\frac{(2j_A + 1)(2j_B + 1)(2J + 1)}{256\pi^5} \right]^{1/2}$$

$$\times \sum_{m_A m_B} D^{(j_A)}_{m_A k_A}(\boldsymbol{\omega}_A)^* D^{(j_B)}_{m_B k_B}(\boldsymbol{\omega}_B)^*$$

$$\times \langle j_A m_A; j_B m_B \, | \, j_{AB} K \rangle D^{(J)}_{MK}(\Phi, \Theta, 0)^*. \qquad (6.3)$$

The symmetric rotor basis functions of molecules A and B are coupled with Clebsch–Gordan coefficients $\langle j_A m_A; j_B m_B \, | \, j_{AB} K \rangle$.[19] The quantum numbers J and M are exact quantum numbers.

The monomer kinetic energy operators in Eq. (6.2) are diagonal in this basis, except when the monomers are asymmetric rotors with $A_X > B_X > C_X$. The angular kinetic energy operator in Eq. (6.1) is diagonal except for the off-diagonal Coriolis coupling terms in the operator $2j_{AB} \cdot J$, which couple basis functions with different K. This implies that K, the projection of the total angular momentum J on the dimer axis $\boldsymbol{R}$, is not an exact quantum number. Still, K may be used to distinguish different states: states with $K = 0$ are called Σ states, states with $|K| = 1$ are called Π states.

How the matrix elements of the potential over the basis are calculated depends on the method, see Sec. 6.2.3. In Ref. 2, it is explained how the potential can be expanded in angular basis functions, similar to those in Eq. (6.3), but with $J = K = 0$ and no overall rotation functions. An analytic expression for the matrix elements of the potential is given in Eq. (18) of Ref. 2. It is simpler than the corresponding expression for a SF basis in Eq. (17) of that reference. A further, more important, advantage of the BF basis over the SF basis is that all matrix elements of the potential with $K' \neq K$ are zero, so that the number of matrix elements that must be computed is considerably reduced. If one uses a pseudo-spectral or collocation method, the analytic basis in Eq. (6.3) is accompanied by a numerical quadrature grid basis. The matrix of the potential is diagonal in this grid basis; the diagonal elements are simply the values of the potential at the grid points.

Various strategies can be chosen for the radial basis functions $\chi_n(R)$. They can computed by a sinc function discrete variable representation (DVR)[20,21] on an equidistant grid in R and contracted by the procedure described in Ref. 8. The pseudo-spectral method explained in Sec. 6.2.3 uses sine functions in the analytic radial basis and a Fourier-type DVR for the corresponding grid basis. Other alternatives are the use of a potential-optimized DVR[22] or a Morse-type DVR basis.[23]

6.2.3. *Methods to compute eigenstates*

When the Hamiltonian matrix is not too large and can be stored in computer memory, the eigenvalues and eigenvectors can be computed efficiently by transforming the matrix to tri-diagonal form with Householder transformations. Subsequently, the eigenvalues are computed with the QR algorithm. For larger dimensions, iterative methods such as the Lanczos[24,25] or Davidson[26] algorithms are preferred. In the Lanczos algorithm, a three-term recursion relation is used to transform the Hamiltonian to tri-diagonal form and only a few vectors need to be kept in memory. Every iteration requires a Hamiltonian-matrix vector multiplication, which can be implemented without storing the entire matrix. Once the eigenvalues are found the eigenvectors must be computed separately with inverse iteration. Mathematically, the Lanczos recursion generates an orthogonal basis, but numerically orthogonality is lost and more iterations are needed than would otherwise be the case. This problem is solved by the Davidson algorithm. This is a subspace method in which in every iteration the subspace is extended and the memory requirement scales with the number of iterations. The Davidson algorithm does not suffer from loss of orthogonality, and the number of iterations is reduced by using the diagonal of the Hamiltonian matrix as a preconditioner.

These methods can be applied in different ways. One can expand the potential as explained above, and calculate the non-zero matrix elements of the Hamiltonian over the basis of Eq. (6.3) analytically. This procedure was used in calculations of the VRT states of the water dimer with the Davidson algorithm.[27] An efficient alternative is the use of a pseudo-spectral method as implemented by Leforestier *et al.*[28,29] with the angular basis in Eq. (6.3), combined with a grid basis. The angular grid is a Gauss–Legendre quadrature grid in the angles β_A and β_B, multiplied with Fourier grids in each of the remaining internal angles γ_A, γ_B, and α. The radial analytic basis in this approach consists of sine functions, defined over a finite range of equidistant R values that form the radial grid. The kinetic energy operators in Eqs. (6.1) and (6.2) act on the analytic basis, which yields rather simple expressions, while the potential is calculated in the grid basis in which it is diagonal. The diagonal elements of the potential matrix are simply the values of the potential at the grid points. Leforestier's program[28,29] is based on the Lanczos method and includes very efficient transformations of the Lanczos vectors from the grid basis to the analytic basis, and vice versa. Also this method has been applied to the water dimer, not only in rigid-monomer calculations,[29] but also in $(6 + 6)$D adiabatic calculations with flexible monomers.[11–13] Another approach of this type is the use of collocation methods, see Chapter 5.

A totally different approach to calculate VRT states is by the application of scattering methods. This approach is implemented in the program BOUND,[30] which is part of the scattering program package MOLSCAT.[31] In scattering calculations with the coupled-channel (or close-coupling) method, one mostly uses SF coordinates, not BF coordinates, and the angular channel basis is the SF equivalent of the BF basis in Eq. (6.3). Instead of using a radial basis $\chi_n(R)$ one solves a set of coupled second-order differential equations, the coupled-channel equations. In calculations for molecular collisions, the scattering states are continuum states and the energy can be freely chosen. The second-order differential equations are solved by propagation on a grid of R values, starting at some small R in the repulsive wall where the radial wave functions are still zero, and generating the solution at larger R values with a log-derivative or Airy propagator. Bound states occur only at specific discrete energies, which have to be determined. The program BOUND approaches this problem by solving the coupled-channel equations by propagation both from small R and from large R, with an energy lower than the potential at these extreme R values. In general, this leads to a mismatch of the wave functions at intermediate R. Only when the energy chosen coincides with a correct bound state energy, a proper wave function is obtained. An algorithm to determine this energy is part of the program BOUND. The method is rather time-consuming, but produces very accurate VRT levels, also for states close to the dissociation limit.

6.2.4. *Symmetry aspects*

It is well known in quantum chemistry that the electronic states of molecules obey the point group symmetry of the nuclear framework, and that their vibrations obey the point group symmetry of the equilibrium structure. The symmetry of the electronic states is based on the BO approximation, saying that in electronic structure calculations the nuclei may be clamped. The point group symmetry of the vibrational states is related with the basic postulate of quantum mechanics that the nuclear wave functions should be symmetric under permutations of identical nuclei when they are bosons with integer nuclear spin quantum number I, and antisymmetric when they are fermions with half-integer I. Actually, the point group used in calculations of the vibrational states of a semi-rigid molecule is isomorphic with the group of "feasible" permutation-inversion (PI) operations. The idea of "feasibility" in this context, defined in the Introduction, is due to Hougen[32] and Longuet-Higgins.[33] There is a one-to-one correspondence between the point group operations and the PI operations: the rotations in the point group correspond to nuclear permutations, the inversion and the reflections correspond to inversion in the PI group or to a combined permutation-inversion operation. The PI operations are more general than the point group operations: they describe not only the symmetry of the vibrational wave functions in the BF frame fixed to the molecule, but also determine "equivalent rotations" of this BF frame with respect to the SF frame.

When molecules are flexible, in the sense that they exhibit internal motions with low-energy barriers, additional PI operations are feasible. Examples are molecules with internal rotations, such as methanol (CH_3OH) of which the equilibrium geometry has C_s symmetry, while the PI group also contains the cyclic permutations P_{123} and P_{132}[a] that correspond to the internal rotation of the CH_3 group, relative to the OH fragment. The PI group in this case is G_6, isomorphic with C_{3v}. Another example is NH_3, which has a pyramidal equilibrium geometry with C_{3v} symmetry, but has a low energy barrier for inversion to an equivalent pyramidal structure. Inversion tunneling splittings are indeed observed, additional permutations P_{12}, P_{13}, and P_{23} become feasible, and the PI group is G_{12}, isomorphic with D_{3h}.

In VdW molecules it is mostly assumed that the constituent molecules are (semi-)rigid. The PI group contains the PI operations that reflect the symmetry of the equilibrium structure, as well as all PI operations that transform this structure into equivalent equilibrium structures, while keeping the monomers rigid. If we take for example the VdW complex $H_2O–H_2$, the PI group G_8 is generated by the permutation P_{12} that interchanges the H nuclei in H_2O, the permutation P_{34} that interchanges the H nuclei in H_2, and inversion E^* of the $H_2O–H_2$ complex as a whole. Inversion of the individual monomers is not a symmetry operation anymore, because of the interaction with the other monomer. For a dimer with identical monomers A and B, such as $H_2O–H_2O$ with PI group G_{16}, also the interchange P_{AB} of the monomers is a symmetry operation. The ammonia dimer, $NH_3–NH_3$, has PI group G_{36} when the monomers are assumed to be rigid. The generators are P_{123} and P_{456} that rotate the NH_3 fragments about their threefold symmetry axes, $P_{23} P_{56} E^*$ that corresponds to reflection σ_v of both monomers, and the interchange P_{AB}. When we include the inversion of each of the NH_3 monomers, additional generators are P_{23} and P_{56} and the symmetry group is G_{144}. The adaptation of the basis functions

---

[a] A permutation P_{ij} interchanges indices i and j, a permutation P_{ijk} replaces i by j, j by k, and k by i.

in Eq. (6.3) to the irreducible representations (irreps) of the PI group considerably simplifies the calculation of the VRT states. This is illustrated in Ref. 8 for H_2O–H_2, in Refs. 27 and 29 for H_2O–H_2O, and in Refs. 34 and 35 for NH_3–NH_3 with rigid and flexible monomers, respectively. In addition, the PI symmetry yields strict selection rules for transitions between the VRT states, see Sec. 6.2.5.

The postulate that the nuclear wave functions should be symmetric or antisymmetric under permutations of identical nuclei holds for simultaneous permutations of the spatial and spin coordinates of the nuclei. The adaptation of the VRT states to the irreps of the PI group discussed above concerns the spatial coordinates. Also the nuclear spin functions can be adapted to the irreps of the PI group. Wave functions that are symmetric under nuclear permutations (for bosons) require that the spatial and spin wave functions belong to the same irrep, for antisymmetric (fermion) wave functions they must belong to associated irreps.[36] This requirement leads to the concept of nuclear spin multiplicity, also called nuclear spin statistical weight.[1] The simplest example is the H_2 molecule containing protons with $I = 1/2$: spatial states with even rotational quantum number j are even under P_{12}, and must have odd nuclear spin functions with $I = 0$, spatial states with odd j are odd under P_{12}, and must have even nuclear spin functions with $I = 1$. The corresponding species are called para- and ortho-hydrogen (pH_2 and oH_2), with nuclear spin multiplicities 1 and 3, respectively.

In H_2O, the permutation P_{12} corresponds to rotation about the twofold symmetry axis, which is the principal b-axis: pH_2O with nuclear spin weight 1 must have even k_b, oH_2O with nuclear spin weight 3 must have odd k_b. The water molecule is an asymmetric rotor and k_b is not a good quantum number; the rotational states of such molecules are usually labeled with $j_{k_a k_c}$, where k_a and k_c are approximate quantum numbers. It follows from the correlation of the $j_{k_a k_c}$ asymmetric rotor states with prolate and oblate symmetric rotor states j_{k_a} and j_{k_c} and the symmetry of the latter states under the equivalent rotations of P_{12} that states with even $k_a + k_c$ belong to pH_2O, states with odd $k_a + k_c$ to oH_2O. Hence the ground state of pH_2O is 0_{00}, the ground state of oH_2O is 1_{01}. The PI group $G_{12} \simeq D_{3h}$ of flexible NH_3 has the irreps $A_1^{\pm}$, $A_2^{\pm}$, and $E^{\pm}$. States with $A_1^{\pm}$ symmetry are "Pauli-forbidden", they do not occur because there are no nuclear spin functions with the appropriate symmetry. States with $A_2^{\pm}$ symmetry have symmetric rotor quantum numbers $k = 0(\text{modulo } 3)$ and belong to oNH_3 with nuclear spin weight 12. States with $E^{\pm}$ symmetry have symmetric rotor quantum numbers $k = \pm1(\text{modulo } 3)$ and belong to pNH_3 with nuclear spin weight 6. A factor of 3 in these weights originates from the multiplicity of the $I = 1$ spin functions of ^{14}N.

For VdW molecules with different constituents the nuclear spin weights are simply the products of the nuclear spin weights of the monomers. So in H_2O–H_2 there are four different species: pH_2O–pH_2 with spin weight 1, oH_2O–pH_2 with spin weight 3, pH_2O–oH_2 with spin weight 3, and oH_2O–oH_2 with spin weight 9. Also in NH_3–H_2 there are four species: oNH_3–pH_2 with weight 12, pNH_3–pH_2 with weight 6, oNH_3–oH_2 with weight 36, and pNH_3–oH_2 with weight 18. When a VdW molecule consists of identical monomers, the situation is more complicated, because the spatial and nuclear spin wave functions must also be adapted to P_{AB}. This implies that the para–ortho and ortho–para species are indistinguishable; there is just a single mixed para–ortho species with spatial wave functions that are interconverted by P_{AB}. The wave functions of the para–para and ortho–ortho species are symmetric or antisymmetric under P_{AB}. For the examples of H_2O–H_2O and NH_3–NH_3 we refer to the papers cited above.

The conversion between different nuclear spin species is very slow, because nuclear spins interact only weakly with other quantities. This implies that different nuclear spin species of the same molecule act as distinct molecules in many processes. In spectroscopy this implies that the spectra are composed of overlapping spectra originating from different nuclear spin species, with alternating line intensities, because of different nuclear spin weights. This applies also to the spectra of VdW molecules.

6.2.5. *Line intensities and spectra*

Not only the VRT levels are calculated by most methods, but also the corresponding wave functions. So, when a dipole moment surface is available, one can calculate dipole transition moments. Accurate dipole moment surfaces can be obtained from *ab initio* electronic structure calculations. In high-resolution spectroscopy the transition frequencies are very accurate, but the line intensities are usually much less accurate. Therefore, it is often sufficient to use an approximate dipole function. For VdW molecules with at least one polar constituent such an approximate dipole function can be obtained by taking the permanent dipole of that constituent and expressing it, first with respect to the BF frame by using the orientation of the constituent with respect to this frame, and then with respect to the SF frame with

$$\mu_m^{\mathrm{SF}}(R, \boldsymbol{\omega}_A^{\mathrm{SF}}, \boldsymbol{\omega}_B^{\mathrm{SF}}, \Phi, \Theta, 0) = \sum_k \mu_k^{\mathrm{BF}}(R, \boldsymbol{\omega}_A^{\mathrm{BF}}, \boldsymbol{\omega}_B^{\mathrm{BF}}) D_{mk}^{(1)}(\Phi, \Theta, 0)^*.$$

$$(6.4)$$

Also the dipole moment induced in the other monomer has sometimes been included in such an approximate dipole function.

Transitions between the intermolecular VRT levels of VdW complexes are usually observed in far-infrared or terahertz spectra, while transitions involving a simultaneous vibrational excitation of one of the monomers in the complex are observed in infrared spectra. Such intramonomer vibrations can be taken into account, even when the monomer vibrational wave functions are not explicitly computed. It is easily demonstrated by formally including the lower and upper state of the monomer transition in the wave functions, that the dipole function for the intermolecular states is just the monomer dipole transition moment. This transition dipole must then be expressed in the BF and SF coordinates of the complex, just as described above for the permanent dipole.

After the discussion in Sec. 6.2.4 of the PI symmetry and of different nuclear spin species, let us now consider the selection rules.

The dipole moment is a multiplicative vector operator with three components that are invariant under all permutations of identical nuclei and change sign under inversion. Let us say that it belongs to the irrep A_1^- of the PI group. This yields a set of strict selection rules: transitions are allowed only when the irrep product of the upper and lower states in a transition contains irrep A_1^-. Since the distinction between different nuclear spin species is based on the symmetry under pure permutations, this implies that only transitions within the same nuclear spin species are allowed. When also vibrational transitions within one of the monomers are involved and when the lower and upper state in this monomer transition have different symmetries, one must be especially careful. We illustrate this point with the example of H_2O–H_2.[9] The spectra discussed in that paper involve a transition within the H_2O monomer, with the transition dipole moment perpendicular to the twofold symmetric axis. The ground state of H_2O is symmetric under P_{12}, but the monomer excited state in this transition is antisymmetric. To make combined intra- and intermolecular transitions allowed, the different permutation symmetry of the intramolecular states must be "compensated" by the intermolecular VRT wave functions of the ground and excited state, since the total dipole operator is symmetric under all permutations. So, if one considers only the intermolecular VRT states involved, it seems that transitions between pH_2O–H_2 and oH_2O–H_2 occur, but in reality this is not the case.

A rotational selection rule follows from Eq. (6.4), which describes the BF to SF frame transformation of a vector property. It implies that only transitions with $\Delta J = 0$ (Q-band) and $\Delta J = \pm 1$ (P- and R-bands) are allowed, as usual.

The method used to calculate dipole transition matrix elements depends on the method chosen to calculate the VRT states, discussed in Sec. 6.2.3. If the potential is expanded in angular functions and its matrix elements are calculated analytically, as explained in Ref. 2, also the dipole moment can be expanded as in Eqs. (35) and (36) of that reference and its matrix elements can also be calculated analytically. A general formula is given in the Appendix of Ref. 9, which also contains a simplified formula that can be used when the approximate dipole function is given by the permanent dipole of a polar monomer.

When a pseudo-spectral method is used with a combined analytic and grid basis, or a method that is solely based on a grid basis, one can exploit the fact that the dipole moment is a multiplicative operator, just as the potential. Multiplicative operators are diagonal in a grid basis, so when also the calculated eigenstates are expressed in that basis, the calculation of the dipole transition matrix simply involves the computation of the dipole moment components on the grid points, multiplied with the values of the lower and upper eigenstate at those points.

When the VRT states are calculated for sufficiently high values of J to include all states occupied at a given temperature T, and for all irreps of the PI group, and the dipole transition moments are calculated as described above, one can generate complete theoretical spectra. The dipole transition moments can be substituted into the formula

$$\frac{\pi N_A g_i}{3\hbar^2 \epsilon_0 c Z}(E_{i',J'} - E_{i,J})\left[\exp(-E_{i,J}/kT) - \exp(-E_{i',J'}/kT)\right]$$

$$\times \sum_{M',M,m} \left|\langle i'J'M'|\mu_m^{\mathrm{SF}}|iJM\rangle\right|^2, \tag{6.5}$$

for the infrared absorption coefficient for a transition from an initial state i, J with energy $E_{i,J}$ to a final state i', J' with energy $E_{i',J'}$. The quantum numbers M and M' are the SF projections of the rotational angular momenta J and J', g_i are the nuclear spin weights, and Z is the partition function

$$Z = \sum_{i,J} g_i(2J+1)\exp(-E_{i,J}/kT). \tag{6.6}$$

The symbol N_A in the prefactor stands for Avogadro's number, ϵ_0 for the vacuum dielectric constant, c for the speed of light, and $\hbar$ for the reduced Planck constant. The transition frequencies are the energy differences between the VRT levels. The theoretical spectra may be compared with measured spectra and are often very helpful in their assignment. When the VRT levels are well converged by using a sufficiently large basis, the comparison of the *ab initio* calculated spectra with measured high-resolution spectra provides a very critical check of the quality of *ab initio* intermolecular potentials.

In some cases, such as for various deuterated versions of NH_3–H_2,[37,38] the high-resolution spectra reveal hyperfine splittings due to the nuclear quadrupole coupling of the ^{14}N and D nuclei with $I = 1$. When it is assumed that the weak non-covalent interactions between the constituent molecules do not change the electric field gradients at the nuclei, the observed splittings are directly proportional to the expectation values of the second order Legendre polynomials $\langle P_2(\cos \beta_X) \rangle$. This provides direct information of the average orientations of the monomers $X = A$ and B in the complex.

6.2.6. *Open-shell systems*

The idea that one can solve the nuclear motion problem with a potential energy surface obtained from electronic structure calculations is based on the BO or adiabatic separation between the electronic and nuclear motions. This idea is generally applied for the calculation of the rovibrational states of a molecule, as well as for the calculation of the VRT states of VdW molecules. The basic condition for the validity of the BO approximation is that the energy separation between different electronic states is considerably larger than the nuclear kinetic energy. The electronic states can then be calculated with the neglect of the effect of the nuclear kinetic energy operator, *i.e.*, with clamped nuclei. Variation of the positions of the clamped nuclei then yields the electronic energies as a function of the nuclear positions, which is the potential energy surface that can be used in the nuclear motion problem.

The BO approximation breaks down for molecules with degenerate electronic states. This leads to non-adiabatic effects, such as Renner–Teller and Jahn–Teller coupling. Also when at least one of the molecules in a VdW complex has a degenerate electronic state, the BO approximation is not valid. Let us consider the example of OH–HCl.[39,40] The OH radical has a $^2\Pi$ electronic ground state, which is spatially twofold degenerate. Adiabatic OH–HCl potential energy surfaces calculated with clamped nuclei are also degenerate for linear geometries of the OH–HCl complex and at large OH–HCl separations. But also for other geometries the energy difference between the two electronic states that correlate asymptotically with the Π state of OH remains small and the BO approximation is not valid. Another example is the Br–HCN complex,[41,42] in which the Br atom has a threefold spatially degenerate ground state (^2P). The complex has three adiabatic potential energy surfaces which are degenerate at large Br–HCN distances, while two of them are also degenerate for all distances at linear geometries. Also spin–orbit coupling is important in these systems, as well as smaller spin–spin couplings, spin-rotation couplings, *etc.*

In multidimensional systems the non-adiabatic coupling between nearby electronic states cannot be eliminated, but it can be strongly reduced and the singularities that occur for example at linear geometries of OH–HCl can be avoided by a transformation of the adiabatic states to a set of diabatic states. The coupling that originates from the nuclear kinetic energy operator can then be neglected, but instead one has to deal with $n \times n$ matrices of diabatic potential surfaces, where n is the number of nearly degenerate states (2 for OH–HCl, 3 for Br–HCN). The transformation from adiabatic to diabatic states is relatively simple when the complex is planar, such as OH–Ar or Br–HCN, because the states are distinguished by even/odd reflection symmetry (A'/A'') with respect to this plane. When $n = 2$, as in OH–Ar, the diabatic potentials are simply the sum and difference of the adiabatic potentials $V_{A'}$ and $V_{A''}$, when $n = 3$ the transformation is rather simple also.[43] When two molecules are involved, as in OH–HCl, the geometry of the complex is non-planar, in general, and one has to derive the adiabatic to diabatic transformation by using the wave functions and properties of the complex. A general and powerful method is the multi-property-based algorithm developed by Karman *et al.*[44]

VdW complexes with open-shell constituents that do not have spatially degenerate ground states cause no special problem, except that spin-spin and spin-rotation couplings have to be taken into account. Examples are $O_2(^3\Sigma_g^-)$-$O_2(^3\Sigma_g^-)$,[45] $NH(^3\Sigma^-)$-He, and $NH(^3\Sigma^-)$-Ne.[46,47]

Most of the electronically excited states of molecules are open-shell systems. So also when at least one of the molecules in a VdW complex is excited, one has to deal with the problems outlined above.

6.2.7. *Additional comments*

A special treatment is necessary when the monomers in VdW molecules have degenerate vibrational states. In such cases, one must take into account vibrational angular momentum, which couples in first order to the angular momenta of the internal monomer rotations and the overall rotation of the complex. An example is the He-H_3^+ complex in the vibrationally excited ν_2 state of H_3^+. Both theoretical and experimental high-resolution spectra of this complex have been produced.[48,49]

Special attention is also needed when calculating VRT states close to the dissociation limit of the VdW complex with a radial basis set. As illustrated by work on He-H_2^+,[50] the convergence of such states requires a large number of radial basis functions, over a range that extends to very large values of R.

Another important issue is the calculation of resonances at energies (just) above the dissociation limit. Such resonances are sometimes observed in high-resolution spectra as broadened lines at higher frequencies. The line width gives information on the lifetime of a resonance. It would lead too far to treat this subject in the present chapter. Instead, we refer to a recent review on the calculation of resonances by various methods.[51]

6.3. Illustrative Results

In this section we illustrate the above theory by discussing the work on some VdW complexes, both closed-shell and open-shell.

6.3.1. *H_2O-H_2*

The H_2O-H_2 complex was already mentioned several times, here we discuss the work on this system[8-10,52] in more detail. Infrared spectra were measured for this complex in which either the bend mode near $1600\,\mathrm{cm}^{-1}$ (see Ref. 53) or the symmetric-asymmetric stretch combination near $7{,}200\,\mathrm{cm}^{-1}$ (see Ref. 9) of H_2O was excited. Various side bands originating from intermolecular vibrations were observed in these spectra. Calculations on the VRT states of the complex[8, 52] and its H_2O–D_2 isotopologue[10] were performed with the methods outlined above. They were based on a 9D H_2O–H_2 *ab initio* potential surface[54] that depends on the five intermolecular coordinates: the three intramolecular coordinates of H_2O and the H_2 bond length. This potential was first averaged over the vibrational ground state wave functions of H_2O and H_2 and the resulting 5D intermolecular potential was used in calculations of the VRT states of the H_2O–H_2 complex. The agreement between the experimental spectrum of the complex[53] for the bend mode of H_2O and the calculated transition frequencies[8] is very good.

More extensive calculations were made[9] for the combination band near $7{,}200\,\mathrm{cm}^{-1}$. The 9D potential was averaged both over the ground and excited state monomer wave functions and used in calculations of the VRT levels of the ground and excited state complex. Also the ground and excited state rotational constants of H_2O were used in these calculations, together with the ground state rotational constant of H_2. A transition dipole function was obtained with the assumption that it can be represented by the H_2O monomer transition dipole moment, see Sec. 6.2.5, expressed in the coordinates of this monomer in the complex. The symmetric–asymmetric stretch combination band involves the dipole component perpendicular to the twofold symmetry axis of H_2O, and antisymmetric under the permutation P_{12} that interchanges the two H nuclei. Also this aspect was discussed in Sec. 6.2.5. Finally, theoretical infrared spectra were generated with this transition dipole function and the wave functions of the ground and excited VRT states.

Figure 6.1 shows the equilibrium geometry of the H_2O–H_2 complex, which might be regarded as a hydrogen-bonded structure with H_2 as the donor and H_2O as the acceptor in the hydrogen bond. It also shows the structure at a local minimum corresponding to a hydrogen-bonded structure with H_2O as the donor and H_2 as the

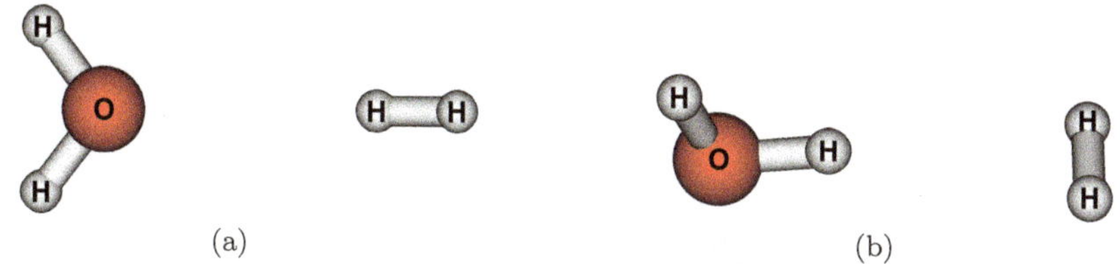

(a)   (b)

Figure 6.1. (a) Planar equilibrium geometry of H_2O–H_2 with C_{2v} symmetry. Distance between the centers of mass $R_e = 5.82\,a_0$, binding energy $D_e = 235.14\,\mathrm{cm}^{-1}$. (b) Metastable structure with $R_e = 6.07\,a_0$ and $D_e = 199.40\,\mathrm{cm}^{-1}$.

acceptor. The complex has binding energies $D_e = 235.14\,\mathrm{cm}^{-1}$ in the global minimum, and $D_e = 199.40\,\mathrm{cm}^{-1}$ in the local minimum. The planar equilibrium geometry of H_2O–H_2 has C_{2v} symmetry, which corresponds to the PI group G_4 with the permutation P_{12} and the inversion E^* as generators. The PI group used in the calculation of the VRT states is $G_8 \simeq D_{2h}$, with the additional permutation P_{34} that interchanges the H nuclei of H_2. A 2D cut of the ground state intermolecular potential is displayed in Figure 6.2.

Cuts of the wave functions of the lowest states with total angular momentum $J = 0$ are shown in Figure 6.3. All wave functions in the four panels in this figure are strongly delocalized, which implies that the internal rotations of the H_2O and H_2 monomers in the H_2O–H_2 complex are only weakly hindered. Still, the ground state of pH_2O–pH_2 is clearly centered at the global minimum of the potential, *cf.* Figure 6.2. The wave functions of the mixed ortho–para and para–ortho dimers also shown in Figure 6.3, and also that of the ortho–ortho dimer, are pushed away from the global minimum by the nodal planes imposed by the nature of the rotational wave functions of oH_2O (with odd $k_a + k_c$) and oH_2 (with odd j). Since the internal monomer rotations in H_2O–H_2 are only weakly hindered, such nodal planes have a strong effect on the overall shape of the wave function.

Figure 6.4 shows the first excited Σ ($K = 0$) state of pH_2O–pH_2, the lowest Π ($|K| = 1$) states of oH_2O–pH_2 and pH_2O–oH_2, and the ground state of oH_2O–oH_2, which is a Π state. Also all of these states are strongly delocalized, with additional nodal planes in the Π states.

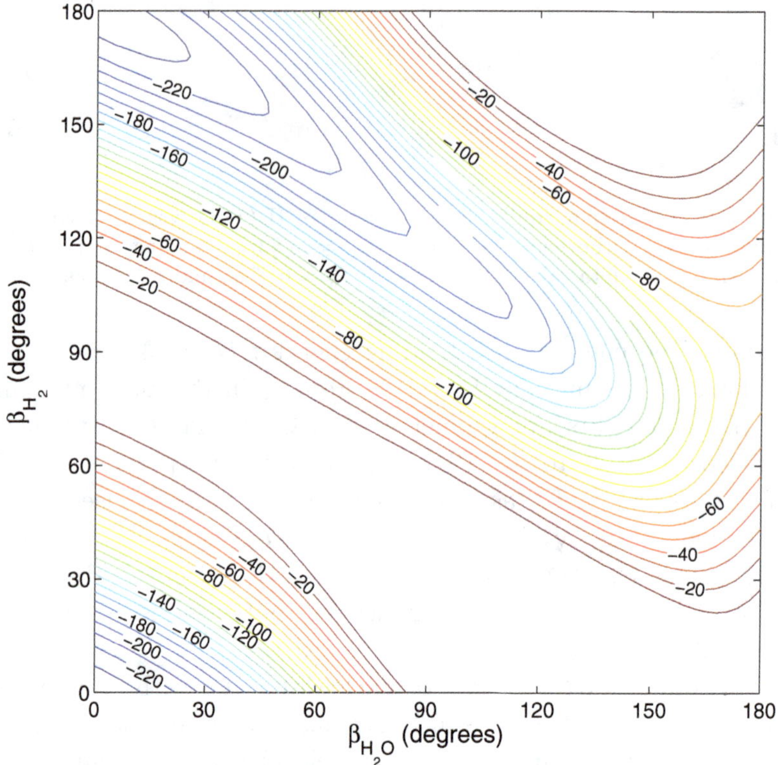

Figure 6.2. Cut of the 5D potential surface of H_2O–H_2 (in cm^{-1}) for planar geometries with R optimized to find the energy minimum for all angles.

The dissociation energy D_0 of the H_2O–H_2 complex follows directly from the calculated energies of the ground VRT state. It is interesting that the different para–ortho mixtures of the complex have very different ground state energies: $-33.57\,cm^{-1}$ for pH_2O–pH_2, $-12.83\,cm^{-1}$ for oH_2O–pH_2, $65.08\,cm^{-1}$ for pH_2O–oH_2, and $83.43\,cm^{-1}$ for oH_2O–oH_2. Also the dissociation energies D_0 of the different para–ortho combinations of the complex are quite different: $33.57\,cm^{-1}$ for pH_2O–pH_2, $36.63\,cm^{-1}$ for oH_2O–pH_2, $53.60\,cm^{-1}$ for pH_2O–oH_2, and $59.04\,cm^{-1}$ for oH_2O–oH_2. The D_0 values take into account that the complex dissociates into the corresponding para and ortho species. The $j_{k_a k_c} = 1_{01}$ ground state of oH_2O is higher in energy by $23.80\,cm^{-1}$ than the 0_{00} ground

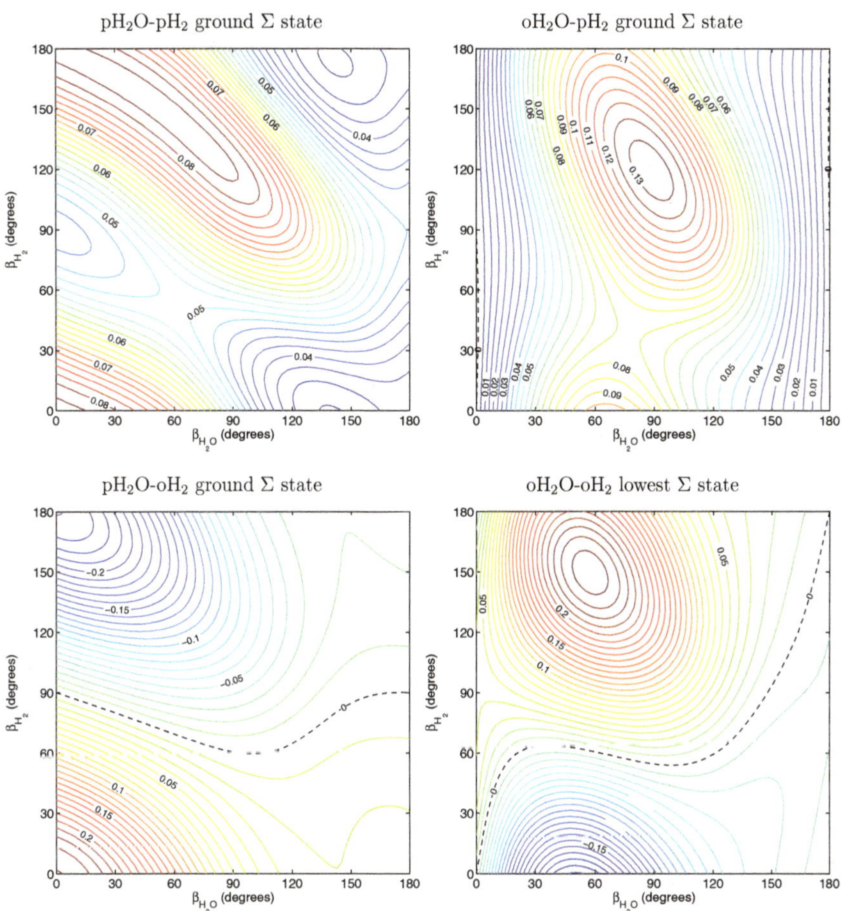

Figure 6.3. Wave functions for $J = 0$, planar geometries, $R = 6.30\,a_0$.

state of pH$_2$O. And the $j = 1$ ground state of oH$_2$ is higher in energy by $118.68\,\mathrm{cm}^{-1}$ than the $j = 0$ ground state of pH$_2$. Such strongly different D_0 values are another sign of the fact that the internal rotations in H$_2$O–H$_2$ are only weakly hindered. A further remarkable observation is that all of these D_0 values are much smaller than the binding energy $D_e = 235.14\,\mathrm{cm}^{-1}$. Clearly, the intermolecular vibrations and hindered internal rotations in H$_2$O–H$_2$ have a large zero-point energy, due to the relatively low monomer masses and large rotational constants. This is in strong contrast

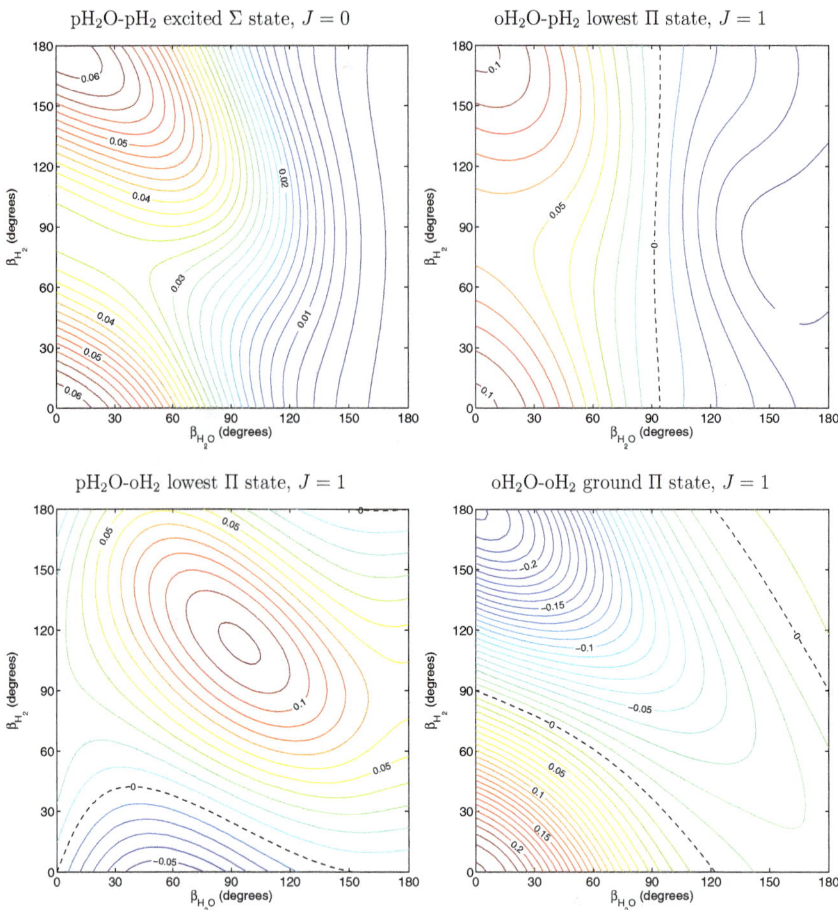

Figure 6.4. Wave functions for $J = 0$ and 1, planar geometries, $R = 6.30\,a_0$.

with O_3–N_2, for which the VRT states will be discussed in the next section.

Figure 6.5 shows the measured infrared spectrum of H_2O–H_2 for excitation to the overtone state of H_2O at about $7200\,\mathrm{cm}^{-1}$, in comparison with the *ab initio* calculated spectrum. The side bands in this spectrum next to the H_2O monomer transition band are due to excitations of the intermolecular VRT states. It is clear from this figure that the agreement between experiment and theory is good. The assignment of the side bands follows from the calculated spectra.

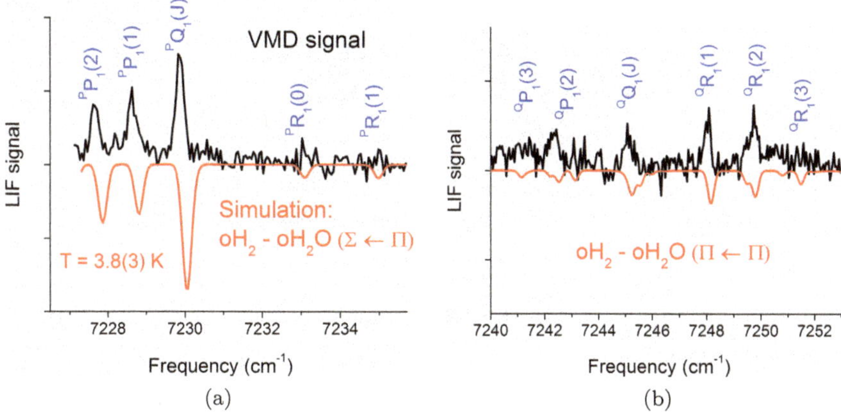

Figure 6.5. Comparison between first principles *ab initio*/dynamical theory calculations (colored lines, downward) and experimental spectra (black lines, upward). Excellent agreement between experiment and theory (rotational temperature 3.8 K) allows assignment of the observed structure to (a) $\Sigma \leftarrow \Pi$ and (b) $\Pi \leftarrow \Pi$ internal rotor bands of oH$_2$O–oH$_2$ (red). Lines are labeled in symmetric top notation according to ${}^{\Delta K}\Delta J_{K''}(J'')$.

6.3.2. O_3–N_2

The VRT states of the O_3–N_2 complex were recently calculated[55] with the same method as used for H$_2$O–H$_2$ and also the G_8 symmetry is the same, but the results are very different. Here, the generators of G_8 are P_{12}, interchanging the equivalent outer O atoms of O$_3$, P_{34}, interchanging the N atoms, and inversion, E^*. The nuclear spin weights are different from those in H$_2$O–H$_2$, because both ^{16}O and ^{14}N are bosons, with nuclear spin $I = 0$ and $I = 1$, respectively. This implies that only states even under P_{12} are physically allowed, states that are odd under P_{12} are Pauli-forbidden. In other words, they have nuclear spin weight zero. Among the allowed species one can distinguish O$_3$–oN$_2$ with even j_{N_2} functions and nuclear spin weight 6, and O$_3$–pN$_2$ with odd j_{N_2} functions and nuclear spin weight 3. The calculations were based on a 5D rigid-monomer potential.[55] The calculation of the VRT states required a larger basis of free internal rotor functions than for H$_2$O–H$_2$, because, as shown below, the orientations of the monomers are much more localized. The dipole function used to generate theoretical far-infrared

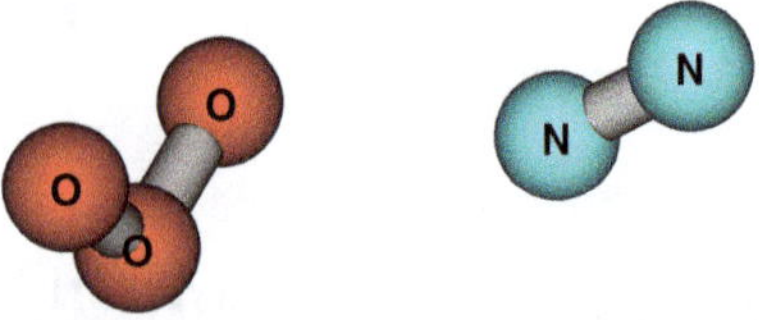

Figure 6.6. Equilibrium geometry of O_3–N_2 with C_s symmetry. Distance between the centers of mass $R_e = 6.69\,a_0$, binding energy $D_e = 348.88\,\mathrm{cm}^{-1}$.

spectra was an approximate one, it just consisted of the permanent dipole of O_3 expressed in the coordinates of this monomer in the complex.

Figure 6.6 shows the equilibrium geometry of the O_3–N_2 complex, with the N_2 axis nearly perpendicular to the O_3 plane and symmetry C_s, corresponding to the PI operation $P_{12}E^*$. The additional PI operations that generate G_8 are E^* and P_{34}. Some 2D cuts of the *ab initio* O_3–N_2 potential are shown in Figure 6.7.

An important difference with H_2O–H_2 is that the VRT states of different ortho–para modifications, O_3–oN_2 and O_3–pN_2 in this case, have almost the same energies. The minute difference between the ground state energies of -220.8517 and $-220.8515\,\mathrm{cm}^{-1}$ is the tunneling splitting between equivalent minima in the potential with the two N atoms interchanged. The O_3–oN_2 ground state with even j_{N_2} is the lower tunneling state, the O_3–pN_2 with odd j_{N_2} is the upper tunneling state. This tunneling splitting cannot be observed, even in high-resolution spectra, because transitions between O_3–oN_2 and O_3–pN_2 do not occur. Still, the fact that this tunneling splitting is so small indicates that O_3–N_2 is a relatively rigid VdW molecule. This is confirmed by the observation that also the excited states of O_3–oN_2 and O_3–pN_2 have nearly the same energies.

The dissociation energies D_0 of O_3–oN_2 and O_3–pN_2 are 220.85 and $224.83\,\mathrm{cm}^{-1}$, respectively. Since the ground state energies of the two species are practically the same, the difference is caused by the $j = 1$ ground state energy of pN_2 being $3.98\,\mathrm{cm}^{-1}$ higher than the $j = 0$ ground state energy of oN_2. The difference between $D_e = 348.88\,\mathrm{cm}^{-1}$ and these D_0 values, *i.e.*, the zero-point energy associated with the intermolecular vibrations, is relatively much

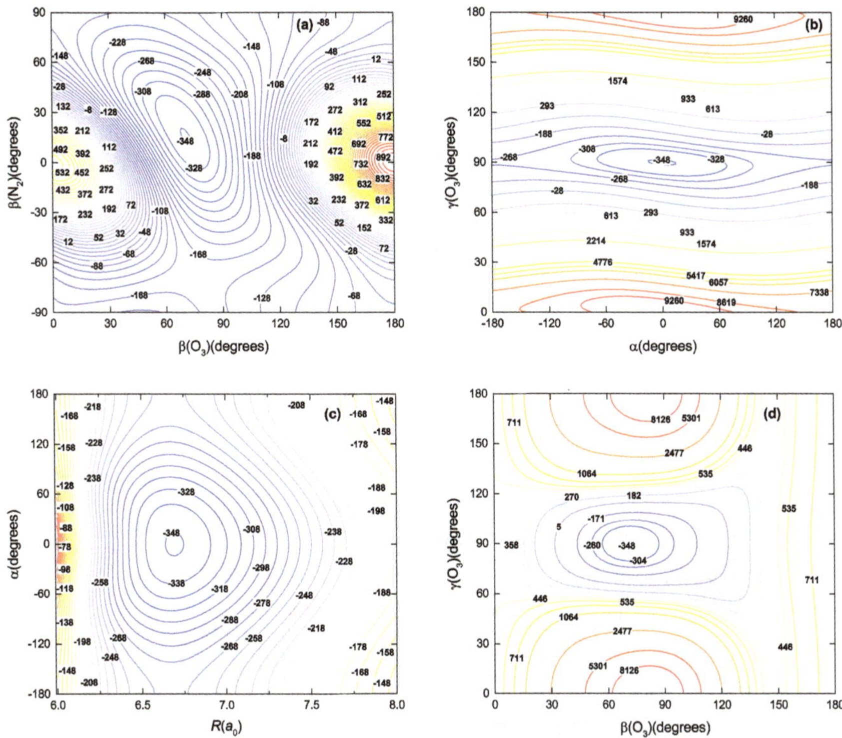

Figure 6.7. Some 2D cuts of the 5D potential surface of O_3–N_2 (in cm^{-1}) with the coordinates not shown at their equilibrium values.

smaller than for H_2O–H_2. This is due to the larger mass and smaller rotational constants of the monomers.

Another confirmation of the near-rigidity of O_3–N_2 follows from the pattern of the energy levels calculated for $J = 0$ and $J = 1$. Each $J = 0$ level is accompanied by three $J = 1$ levels, one Σ $(K = 0)$ and two Π $(K = \pm 1)$ levels. The $J = 1$ levels that "belong to" the ground state $J = 0$ level may be considered as rotationally excited states of the complex. A strong indication of the near-rigidity of O_3–N_2 is that the energies of these rotationally excited states, 0.132, 0.457, and 0.466 cm^{-1}, are nearly the same as the rigid-rotor levels at 0.135, 0.459, and 0.468 cm^{-1} calculated with the inertia tensor and the corresponding rotational constants of the O_3–N_2 complex at its equilibrium geometry. The higher $J = 0$ levels calculated are vibrationally excited states with energies 23.25, 23.81, 42.98, 45.04, and 47.39 cm^{-1} relative to the ground state, and the accompanying $J = 1$ levels are the corresponding rovibrationally excited states.

Let us now look at the wave functions of the ground and excited $J = 0$ vibrational states, shown for O_3–oN_2 in Figure 6.8 and very similar to the wave functions of O_3–pN_2. A further confirmation of the near-rigidity of O_3–N_2 is that the ground state wave function in Figure 6.8(a) is well localized around the global minimum shown in Figure 6.7, much more so than the ground state of pH_2O–pH_2 in Figure 6.3. The other panels in Figure 6.8 show vibrationally excited states, also rather well localized, with nodal planes indicating the nature of the excitations: torsion, O_3 wag and N_2 bend, torsion overtone, O_3 twist, and stretch excited. These terms seem to indicate that the vibrations in this VdW molecule are similar to those in "normal" molecules. But the comparison of the vibrational excitation energies with harmonic frequencies calculated from the force constants at the equilibrium geometry shows that even a nearly rigid VdW molecule is still quite different from a normal molecule held together by covalent bonds.

Finally, Figure 6.9 shows stick spectra for O_3–oN_2, very similar to those of O_3–pN_2, with the line intensities calculated from the VRT wave functions and the approximate dipole function mentioned above. The infrared or far-infrared spectrum of O_3–N_2 has not been measured yet.

6.3.3. *OH–HCl*

The procedure to calculate the VRT states of open-shell VdW molecules for which the BO approximation breaks down is illustrated on OH–HCl. The OH radical has a $^2\Pi$ ground state with electronic angular momenta $\Lambda = \pm 1$ around the O–H axis. It has one unpaired electron and total spin $S = 1/2$. The axial components of the spin with $\sigma = \pm 1/2$, combined with the spatial $\Lambda = \pm 1$ components, yield the approximate quantum number Ω that adopts the values of $\pm 3/2$ and $\pm 1/2$. States with $\Omega = \pm 3/2$ and $\pm 1/2$ are split by 139.2 cm^{-1} by spin–orbit coupling. The state with (approximately) $\Omega = \pm 3/2$ is the lower spin–orbit state (conventionally called F_1), the state with

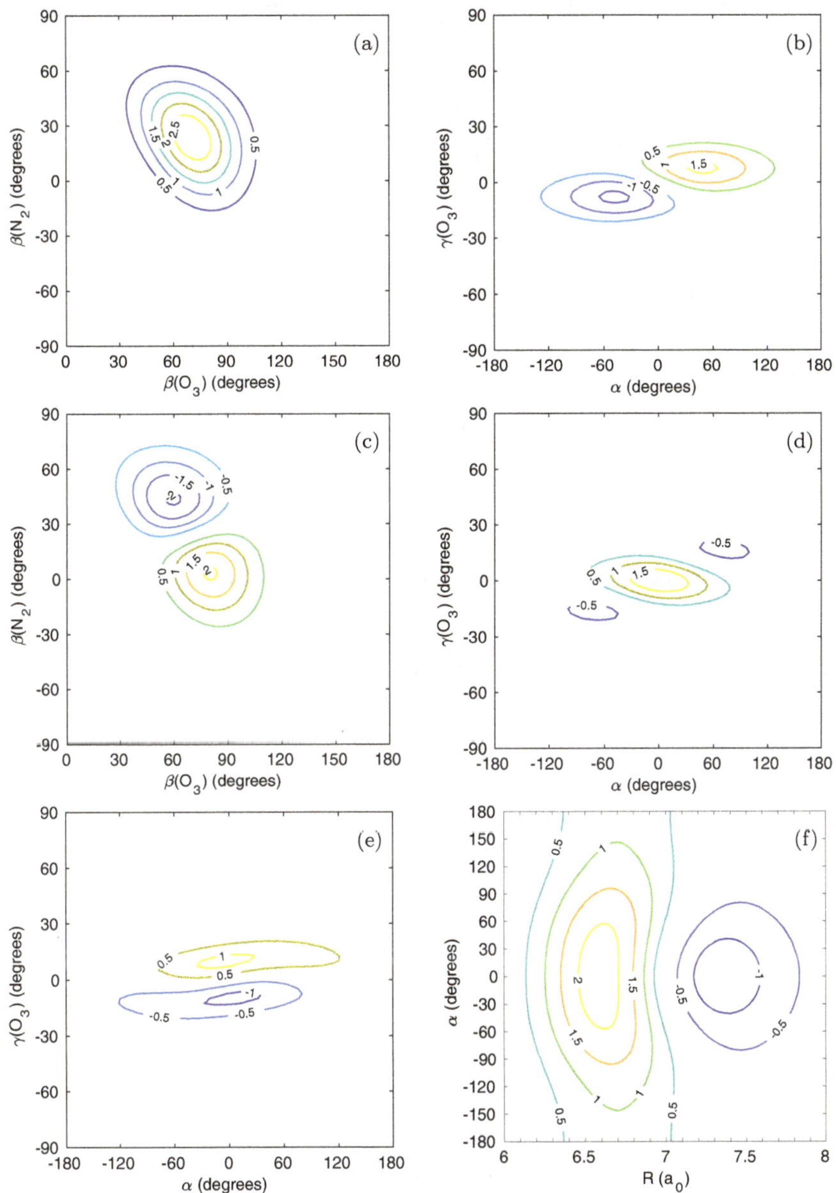

Figure 6.8. Wave functions for $J = 0$: (a) Ground state, (b) torsionally excited state at $23.25\,\mathrm{cm}^{-1}$, (c) O_3 wag and N_2 bend excited state at $23.81\,\mathrm{cm}^{-1}$, (d) torsion overtone at $42.98\,\mathrm{cm}^{-1}$, (e) O_3 twist excited state at $45.04\,\mathrm{cm}^{-1}$, (f) stretch excited state at $47.39\,\mathrm{cm}^{-1}$. The distance R is fixed at $6.75\,a_0$, except in panel (f), and the angles not shown are fixed at their equilibrium values.

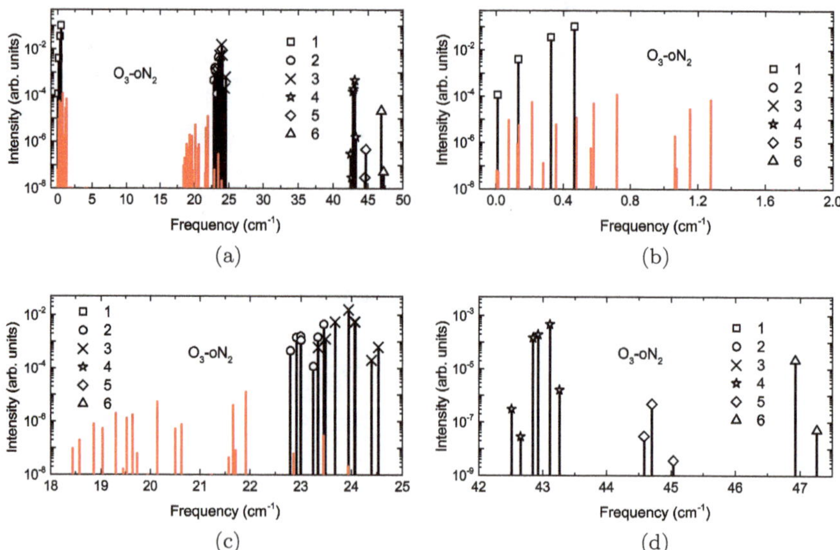

Figure 6.9. Stick spectra representing transition frequencies and intensities at $T = 5\,$K calculated with the bound states of O_3–oN_2: (a) global view; (b), (c), and (d) detailed figures for the regions 0–2 cm^{-1}, 18–25 cm^{-1}, and 42–48 cm^{-1}. The symbols with the sticks denote the following transitions: (1) pure rotational, (2) torsion, (3) O_3 wag, N_2 bend, (4) torsion overtone, (5) O_3 twist, (6) intermolecular stretch. The red sticks represent hot bands. The spectrum of O_3–pN_2 looks very similar.

$\Omega = \pm 1/2$ is the upper spin-orbit state (called F_2). Both of these states are further split into parity doublets, but these splittings are small, about 0.05 cm^{-1} for the ground state.

OH–HCl is a hydrogen-bonded system with two equilibrium structures, see Figure 6.10. The most stable structure with HCl as proton donor and OH as acceptor has binding energy $D_e = 1123\,$cm^{-1}, the structure with OH as donor and HCl as acceptor corresponds to a local minimum in the OH–HCl potential with well depth $D_e = 655\,$cm^{-1}. The two adiabatic intermolecular potentials that asymptotically correlate with the Π ground state of OH were calculated in Ref. 39, and are displayed in Figure 6.11. The two hydrogen-bonded structures shown in Figure 6.10 were found on the lowest adiabatic potential. The transformation of these adiabatic potentials to a 2×2 matrix of diabatic potentials $V_{\Lambda',\Lambda}$ is unitary

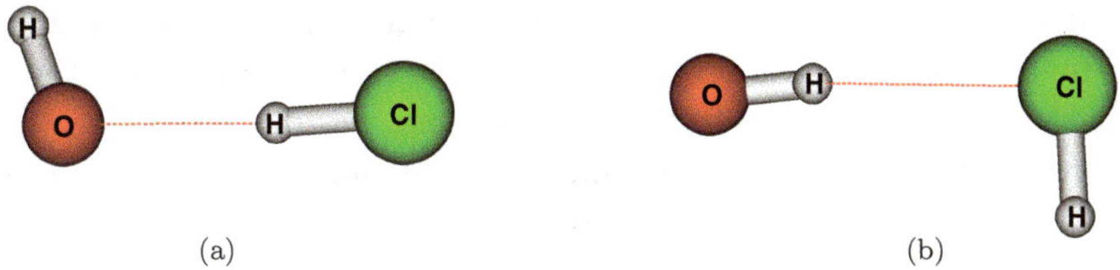

Figure 6.10. (a) Planar equilibrium geometry of OH-Cl with C_s symmetry. Distance between the centers of mass $R_e = 6.361\,a_0$, binding energy $D_e = 1123\,\mathrm{cm}^{-1}$. (b) Metastable structure with $R_e = 6.646\,a_0$, and $D_e = 655\,\mathrm{cm}^{-1}$.

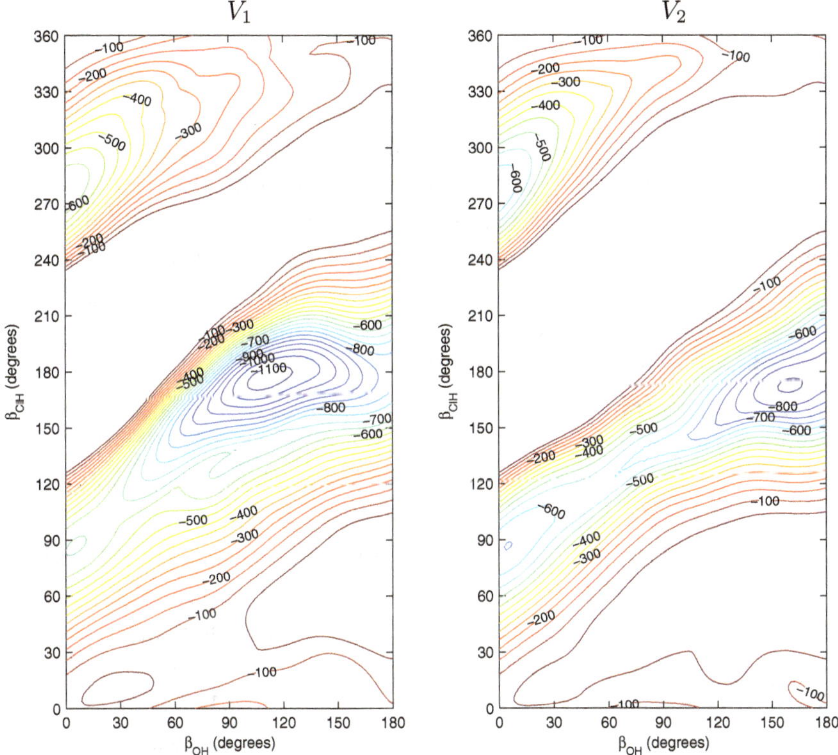

Figure 6.11. Adiabatic potentials V_1 and V_2 in cm^{-1} for optimized R. In order to cover all planar geometries, the range of β_{ClH} was extended to $360°$. The lower half of the figure for $0° \leq \beta_{ClH} \leq 180°$ corresponds to $\phi = 180°$, the upper half for $180° \leq \beta_{ClH} \leq 360°$ to $\phi = 0°$.

and can be parameterized by a mixing angle. This mixing angle is a function of the geometry of the complex, just as the potentials, and can be extracted from the electronic wave function of the complex calculated at each geometry.[39] The diabatic potentials $V_{1,1}$ and $V_{-1,-1}$ on the diagonal of the 2×2 matrix are identical, the off-diagonal diabatic potentials $V_{1,-1}$ and $V_{-1,1}$ are each other's complex conjugate. Also the expansion of the diabatic potentials in angular basis functions is explained in Ref. 39.

The calculation of the VRT states of the OH–HCl complex is described in Ref. 40. They were expanded in the basis of Eq. (6.3), formally multiplied with the electronic wave functions $|\Omega\rangle = |\Lambda\rangle|\sigma\rangle$ of the open-shell OH radical. Since HCl is a linear molecule, the basis label k_B equals zero. Also OH is linear, so as far as the rotational basis is concerned, k_A would also be zero. But since the OH basis also contains the electronic wave functions $|\Omega\rangle$, k_A is equal to Ω and ranges over the values $\pm 1/2, \pm 3/2$. The diagonal diabatic potentials couple basis functions with the same Λ and the same Ω, the off-diagonal diabatic potentials couple basis functions with different $\Lambda = \pm 1$, *i.e.*, they couple the upper spin-orbit state with $\Omega = \pm 1/2$ to the lower state with $\Omega = \mp 3/2$.

The symmetry of OH–HCl is very simple. The planar equilibrium geometry has C_s symmetry, corresponding to the PI operation E^*. There are no equivalent nuclei, hence there is no additional permutation symmetry and the PI group is $G_2 \equiv \{E, E^*\}$

The calculated dissociation energy is $D_0 = 685\,\mathrm{cm}^{-1}$, which might be compared with the binding energy $D_e = 1123\,\mathrm{cm}^{-1}$ at the global minimum in the lowest adiabatic potential for the hydrogen-bonded structure with HCl as the donor. The density distribution in Figure 6.12(a) shows that the ground state is rather well localized around this global minimum. We note here that the wave functions contain four electronic components with $\Omega = \pm 3/2$ and $\pm 1/2$ and they are complex-valued, so we plot density distributions, instead of wave functions. Low-lying rovibronic states that correlate with the $OH(^2\Pi_{3/2})$ ground state were found at 14 and 26 cm^{-1} above the ground state. The OH–HCl stretch fundamental frequency equals 93.6 cm^{-1}, the lowest bend excited states (involving a coupled bend

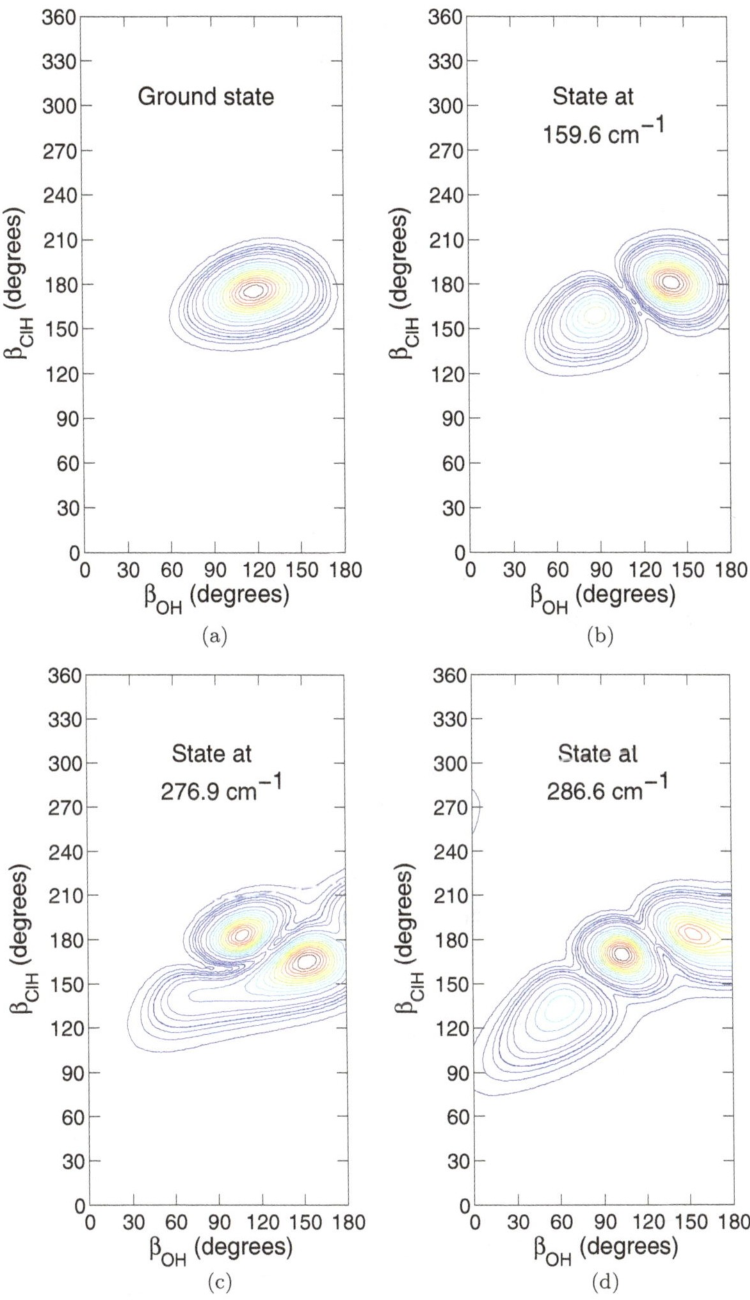

Figure 6.12. Density distributions for planar geometries of the ground state and some bend excited states.

motion of both fragments) were found in the region of $150\text{--}160\,\mathrm{cm}^{-1}$ above the ground state. Density distributions for some of these bend excited states are shown in Figure 6.12. The lowest excited state for which the density distribution contours approach the local minimum in the V_1 potential that corresponds to the hydrogen-bonded structure with OH as the donor lies at $286.6\,\mathrm{cm}^{-1}$, see Figure 6.12(d). In some of the excited states the vibrationally averaged geometry is non-planar. Important non-adiabatic effects that involve both adiabatic electronic states were observed especially in the excited states. The results indicate that neither the diabatic nor the adiabatic picture holds, and that the bound states of OH–HCl could not have been calculated reliably on a single potential energy surface.

It is interesting also to look at the effects of spin-orbit coupling. It followed from the calculations in Ref. 40 that the splitting of $139.2\,\mathrm{cm}^{-1}$ between the $|\Omega| = 3/2$ and $|\Omega| = 1/2$ spin-orbit states of free OH is largely quenched by the anisotropic interaction with HCl. This was understood with a "computer experiment" in which the spin-orbit coupling constant A_0 is first set to zero, and then switched back to its physical value. The effect of the A_0 value on the lower levels of OH–HCl is shown in Figure 6.13. When A_0 is set to zero, the level splittings are caused by the anisotropic potential and, in particular, by the off-diagonal coupling potential $V_{1,-1}$. The ground state has a planar geometry, on average, with a rovibronic wave function that is rather well localized around the minimum in the lower adiabatic potential V_1. The minimum in the higher adiabatic potential V_2 is higher by $266\,\mathrm{cm}^{-1}$ and it is expected that the corresponding state will also be higher than the ground state by about this amount. The large splitting between these two electronic states is caused by the diabatic coupling potential $V_{1,-1}$. Since this splitting is so large, the lower rovibronic states of OH–HCl can be considered in first approximation as if the complex has only a single electronic (adiabatic) state. It has a bent geometry with OH making an angle of about $110°$ with the intermolecular axis and HCl nearly aligned with this axis. Such a geometry corresponds to a prolate rotor and the rotational states can be labeled with the

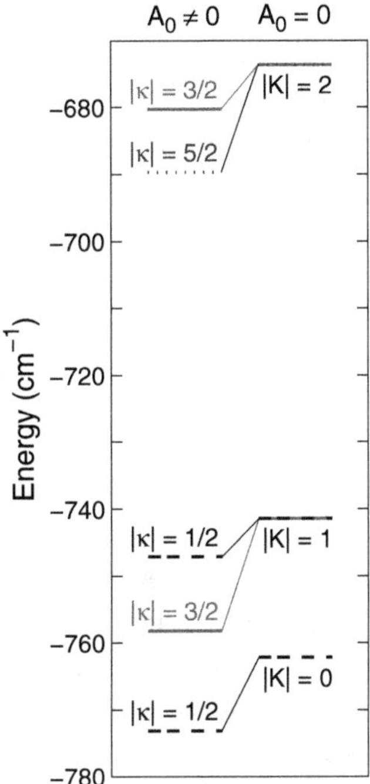

Figure 6.13.   Comparison of energy levels with and without spin–orbit coupling.

quantum number K, the projection of the total angular momentum J (without spin) on the dimer z-axis. The energy levels calculated for $A_0 = 0$ in Figure 6.13 indeed show the energy pattern $E_K = AK^2$ of a prolate near-symmetric rotor with K values 0, ± 1, ± 2, and a rotational constant A of about $21\,\mathrm{cm}^{-1}$. This value is close to the value calculated for OH–HCl in its equilibrium geometry. Spin has not been considered so far, but since for $A_0 = 0$ the spin is completely free, one can simply include the spin without affecting the energies. The values $\sigma = \pm 1/2$ of the $S = 1/2$ spin projection on the dimer axis must be added to the values of K to obtain the projection κ of total J. For $K = 0$ one obtains $\kappa = \pm 1/2$, for $K = \pm 1$ one gets $\kappa = \pm 1/2$ and $\pm 3/2$, for $K = \pm 2$ one gets $\kappa = \pm 3/2$ and $\pm 5/2$. Such a splitting pattern and the corresponding quantum numbers are in perfect agreement with the energy levels shown at the right-hand side of Figure 6.13.

When the spin–orbit coupling is switched back to its physical value of $A_0 = -139.2\,\mathrm{cm}^{-1}$, one obtains the energy levels on the left-hand side of Figure 6.13. The spin–orbit coupling operator mixes the ground electronic state, which is an antisymmetric combination of the OH substates with $\Lambda = \pm 1$, with the excited state that corresponds to the symmetric combination and, thereby, unquenches the angular momentum Λ. Since these electronic states are split by nearly $300\,\mathrm{cm}^{-1}$ through the effect of $V_{1,-1}$, this mixing will only occur to a limited extent. The rather small splittings of the levels with $K = \pm 1$ and $K = \pm 2$ which were degenerate when $A_0 = 0$ and the small energy lowerings of the levels show that indeed the orbital angular momentum Λ stays nearly quenched.

In summary, it could be concluded from this "computer experiment" that the energy level pattern for the lower states of the OH–HCl complex is completely understood. It is in first instance dominated by the large off-diagonal coupling potential $V_{1,-1}$ that quenches the orbital angular momentum Λ of the $\mathrm{OH}(^2\Pi)$ state and gives rise to a large splitting of the electronic states. Spin–orbit coupling partly unquenches the electronic orbital angular momentum and gives further (smaller) splittings and an overall energy lowering of the levels.

Acknowledgment

I thank Gerrit Groenenboom for a longtime collaboration, from which I learned a lot.

References

1. Bunker, P. R.; Jensen, P. *Molecular Symmetry and Spectroscopy*, 2nd ed. NRC Research Press: Ottawa, 1998.
2. van der Avoird, A.; Wormer, P. E. S.; Moszynski, R. From intermolecular potentials to the spectra of van der waals molecules, and vice versa. *Chem. Rev.* **1994**, *94*, 1931–1974.
3. Brocks, G.; van der Avoird, A.; Sutcliffe, B. T.; Tennyson, J. Quantum dynamics of non-rigid systems comprising two polyatomic fragments. *Mol. Phys.* **1983**, *50*, 1025.

4. Szabo, A.; Ostlund, N. S. *Modern Quantum Chemistry.* Dover: New York, 1996.

5. Jankowski, P.; McKellar, A. R. W.; Szalewicz, K. Theory untangles the high-resolution infrared spectrum of the ortho-H_2-CO van der waals complex. *Science.* **2012**, *336*, 1147.

6. Jankowski, P.; Surin, L. A.; Potapov, A.; Schlemmer, S.; McKellar, A. R. W.; Szalewicz, K. A comprehensive experimental and theoretical study of H_2-CO spectra. *J. Chem. Phys.* **2013**, *138*, 084307.

7. Jeziorska, M.; Jankowski, P.; Szalewicz, K.; Jeziorski, B. On the optimal choice of monomer geometry in calculations of intermolecular interaction energies: Rovibrational spectrum of Ar-HF from two- and three-dimensional potentials. *J. Chem. Phys.* **2000**, *113*, 2957.

8. van der Avoird, A.; Nesbitt, D. J. Rovibrational states of the H_2O-H_2 complex; an *ab initio* calculation. *J. Chem. Phys.* **2011**, *134*, 044314.

9. Ziemkiewicz, M. P.; Pluetzer, C.; Nesbitt, D. J.; Scribano, Y.; Faure, A.; van der Avoird, A. Overtone vibrational spectroscopy in H_2-H_2O complexes: a combined high level theoretical *ab initio*, dynamical and experimental study. *J. Chem. Phys.* **2012**, *137*, 084301.

10. van der Avoird, A.; Scribano, Y.; Faure, A.; Weida, M. J.; Fair, J. R.; Nesbitt, D. J. Intermolecular potential and rovibrational states of the H_2O-D_2 complex. *Chem. Phys.* **2012**, *399*, 28–38.

11. Leforestier, C.; van Harrevelt, R.; van der Avoird, A. Vibration-rotation-tunneling levels of the water dimer from an ab initio potential surface with flexible monomers. *J. Phys. Chem. A.* **2009**, *113*, 12285–12294.

12. Szalewicz, K.; Leforestier, C.; van der Avoird, A. Towards the complete understanding of water by a first-principles computational approach. *Chem. Phys. Lett.* **2009**, *482*, 1–14.

13. Leforestier, C.; Szalewicz, K.; van der Avoird, A. Spectra of water dimer from a new *ab initio* potential with flexible monomers. *J. Chem. Phys.* **2012**, *137*, 014305.

14. Wang, X.-G.; Carrington, T. Using monomer vibrational wavefunctions to compute numerically exact (12D) rovibrational levels of water dimer. *J. Chem. Phys.* **2018**, *148*, 074108.

15. Felker, P. M.; Bačić, Z. Weakly bound molecular dimers: Intramolecular vibrational fundamentals, overtones, and tunneling splittings from full-dimensional quantum calculations using compact contracted bases of intramolecular and low-energy rigid-monomer intermolecular eigenstates. *J. Chem. Phys.* **2019**, *151*, 024305.

16. Felker, P. M.; Bačić, Z. Benzeneh$_2$o and benzenehdo: fully coupled nine-dimensional quantum calculations of flexible H_2O/HDO intramolecular vibrational excitations and intermolecular states of the dimers, and

their infrared and raman spectra using compact bases. *J. Chem. Phys.* **2020**, *152*, 124103.

17. Felker, P. M.; Bačić, Z. H_2OCO and D_2OCO complexes: Intra- and intermolecular rovibrational states from full-dimensional and fully coupled quantum calculations. *J. Chem. Phys.* **2020**, *153*, 074107.

18. Liu, Y.; Li, J.; Felker, P. M.; Bačić, Z. $HClH_2O$ dimer: an accurate full-dimensional potential energy surface and fully coupled quantum calculations of intra- and intermolecular vibrational states and frequency shifts. *Phys. Chem. Chem. Phys.* **2021**, *23*, 7101–7114.

19. Brink, D. M.; Satchler, G. R. *Angular Momentum*, 3rd ed. Clarendon: Oxford, 1993.

20. Colbert, D. T.; Miller, W. H. A novel discrete variable representation for quantum mechanical reactive scattering via the s-matrix kohn method. *J. Chem. Phys.* **1992**, *96*, 1982.

21. Groenenboom, G. C.; Colbert, D. T. Combining the discrete variable representation with the s-matrix kohn method for quantum reactive scattering. *J. Chem. Phys.* **1993**, *99*, 9681–9696.

22. Harris, D. O.; Engerholm, G. G.; Gwinn, W. D. Calculation of matrix elements for one-dimensional quantum-mechanical problems and the application to anharmonic oscillators. *J. Chem. Phys.* **1965**, *43*, 1515.

23. Wei, H.; Carrington, T. The discrete variable representation of a triatomic hamiltonian in bond length-bond angle coordinates. *J. Chem. Phys.* **1992**, *97*, 3029.

24. Lanczos, C. An iteration method for the solution of the eigenvalue problem of linear differential and integral operators. *J. Res. Natl. Bur. Stand.* **1950**, *45*, 255.

25. Cullum, J. K.; Willoughby, R. A. *Lanczos Algorithms for Large Symmetric Eigenvalue Computations*. Birkhäuser: Boston, 1985.

26. Davidson, E. R. The iterative calculation of a few of the lowest eigenvalues and corresponding eigenvectors of large real-symmetric matrices. *J. Comput. Phys.* **1975**, *17*, 87.

27. Groenenboom, G. C.; Wormer, P. E. S.; van der Avoird, A.; Mas, E. M.; Bukowski, R.; Szalewicz, K. Water pair potential of near spectroscopic accuracy: II. vibration-rotation-tunneling levels of the water dimer. *J. Chem. Phys.* **2000**, *113*, 6702–6715.

28. Leforestier, C. Grid method for the wigner functions. Application to the van der waals system $Ar-H_2O$. *J. Chem. Phys.* **1994**, *101*, 7357.

29. Leforestier, C.; Braly, L. B.; Liu, K.; Elrod, M. J.; Saykally, R. J. Fully coupled six-dimensional calculations of the water dimer vibration-rotation-tunneling states with a split wigner pseudo spectral approach. *J. Chem. Phys.* **1997**, *106*, 8527.

30. Hutson, J. M.; Le Sueur, C. R. Bound: a program for bound states of interacting pairs of atoms and molecules, version 2020.0, 2020a; https://github.com/molscat/molscat.

31. Hutson, J. M.; Le Sueur, C. R. Molscat: a program for non-reactive quantumscattering calculation on atomic and molecular collisions, version 2020.0, 2020b; https://github.com/molscat/molscat.

32. Hougen, J. T. Classification of rotational energy levels for symmetric-top molecules. *J. Chem. Phys.* **1962**, *37*, 1433.

33. Longuet-Higgins, H. C. The symmetry groups of non-rigid molecules. *Mol. Phys.* **1963**, *6*, 445–460.

34. Olthof, E. H. T.; van der Avoird, A.; Wormer, P. E. S.; Loeser, J. G.; Saykally, R. J. The nature of monomer inversion in the ammonia dimer. *J. Chem. Phys.* **1994**, *101*, 8443–8454.

35. Olthof, E. H. T.; van der Avoird, A.; Wormer, P. E. S. Is the NH_3-dimer hydrogen bonded? *J. Mol. Structure (Theochem)* **1994**, *307*, 201–215.

36. Johnston, D. F. Group theory in solid state physics. *Rept. Progr. Phys.* **1960**, *23*, 66.

37. Surin, L. A.; Tarabukin, I. V.; Hermann, M.; Heyne, B.; Schlemmer, S.; Kalugina, Y. N.; van der Avoird, A. Ab initio potential energy surface and microwave spectrum of the NH_3N_2 van der waals complex. *J. Chem. Phys.* **2020**, *152*, 234304.

38. Tarabukin, I.; Surin, L.; Hermanns, M.; Heyne, B.; Schlemmer, S.; Lee, K.; McCarthy, M.; van der Avoird, A. Rotational spectroscopy and bound state calculations of deuterated NH_3H_2 van der waals complexes. *J. Mol. Spectr.* **2021**, *377*, 111442.

39. Wormer, P. E. S.; Kłos, J. A.; Groenenboom, G. C.; van der Avoird, A. *Ab initio* computed diabatic potential energy surfaces of OH-HCl. *J. Chem. Phys.* **2005**, *122*, 244325.

40. Groenenboom, G. C.; Fishchuk, A. V.; van der Avoird, A. Bound states of the $OH(^2\Pi)$-HCl complex on *ab initio* diabatic potentials. *J. Chem. Phys.* **2009**, *131*, 124307.

41. Fishchuk, A. V.; Merritt, J. M.; Groenenboom, G. C.; van der Avoird, A. *Ab initio* treatment of the chemical reaction precursor complex $Br(^2P)$–HCN; 2. bound state calculations and infrared spectra. *J. Phys. Chem. A* **2007**, *111*, 7270–7281.

42. Fishchuk, A. V.; Merritt, J. M.; van der Avoird, A. *Ab initio* treatment of the chemical reaction precursor complex $Br(^2P)$–HCN; 1. adiabatic and diabatic potential surfaces, *J. Phys. Chem. A* **2007**, *111*, 7262–7269.

43. Kłos, J.; Chałasiński, G.; Szczęśniak, M. M.; Werner, H.-J. Ab initio calculations of adiabatic and diabatic potential energy surfaces of

$Cl(^2P)\ldots HCl(\Sigma^+)$ van der waals complex. *J. Chem. Phys.* **2001**, *115*, 3085.

44. Karman, T.; van der Avoird, A.; Groenenboom, G. C. Communication: Multiple-property-based diabatization for open-shell van der waals molecules. *J. Chem. Phys.* **2016**, *144*, 121101.

45. Karman, T.; Koenis, M. A. J.; Banerjee, A.; Parker, D. H.; Gordon, I. E.; van der Avoird, A.; van der Zande, W. J.; Groenenboom, G. C. O_2-O_2 and O_2-N_2 collision-induced absorption mechanisms unravelled. *Nature Chem.* **2018**, *10*, 549.

46. Kerenskaya, G.; Schnupf, U.; Heaven, M. C.; van der Avoird, A. Bound state spectroscopy of NH-He. *J. Chem. Phys.* **2004**, *121*, 7549.

47. Kerenskaya, G.; Schnupf, U.; Heaven, M. C.; van der Avoird, A.; Groenenboom, G. C. Experimental and theoretical investigation of the $A^3\Pi - X^3\Sigma^-$ transition of NH/D-Ne. *Phys. Chem. Chem. Phys.* **2005**, *7*, 846–854.

48. Harding, M. E.; Lipparini, F.; Gauss, J.; Gerlich, D.; Schlemmer, S.; van der Avoird, A. The He-H_3^+ complex: I. Vibration-rotation-tunneling states and transition probabilities. *J. Chem. Phys.* **2022**, *156*, 144307.

49. Salomon, T.; Brackertz, S.; Asvany, O.; Savić, I.; Gerlich, D.; Harding, M. E.; Lipparini, F.; Gauss, J.; van der Avoird, A.; Schlemmer, S. The He-H_3^+ complex: II. Infrared predissociation spectrum and energy term diagram. *J. Chem. Phys.* **2022**, *156*, 144308.

50. Asvany, O.; Schlemmer, S.; van der Avoird, A.; Szidarovszky, T.; Csaszar, A. G. Vibrational spectroscopy of H_2He^+ and D_2He^+. *J. Mol. Spectr.* **2021**, *377*, 111423.

51. Császár, A. G.; Simkó, I.; Szidarovszky, T.; Groenenboom, G. C.; Karman, T.; van der Avoird, A. Rotational-vibrational resonance states. *Phys. Chem. Chem. Phys.* **2020**, *22*, 15081–15104.

52. Wang, X.-G.; Carrington, T. Theoretical study of the rovibrational spectrum of H_2O-H_2. *J. Chem. Phys.* **2011**, *134*, 044313.

53. Weida, M. J.; Nesbitt, D. J. High-resolution diode laser study of H_2-H_2O van der waals complexes: H_2O as proton acceptor and the role of large amplitude motion. *J. Chem. Phys.* **1999**, *110*, 156.

54. Valiron, P.; Wernli, M.; Faure, A.; Wiesenfeld, L.; Rist, C.; Kedžuch, S.; Noga, J. R12-calibrated H_2O-H_2 interaction: Full dimensional and vibrationally averaged potential energy surfaces. *J. Chem. Phys.* **2008**, *129*, 134306.

55. Kalugina, Y. N.; Egorov, O.; van der Avoird, A. Ab initio study of the O_3-N_2 complex: Potential energy surface and rovibrational states. *J. Chem. Phys.* **2021**, *155*, 054308.

Chapter 7

Vibrational and Rovibrational Spectroscopy Applied to Astrochemistry

Ryan C. Fortenberry[*‡] and Timothy J. Lee[†,§]

*Department of Chemistry & Biochemistry, University of Mississippi,
University, MS 38677-1848, USA
†MS245-3, Planetary Systems Branch,
Space Science and Astrobiology Division,
NASA Ames Research Center, Moffett Field, CA 94035, USA
‡r410@olemiss.edu
§timothy.j.lee@nasa.gov

7.1. Introduction

Quantum chemistry and electronic structure theory have played a crucial role in the detection of molecules in space largely since molecules have been detected in space. In the early- and mid-1970s radioastronomical observations were increasing the census of extraterrestrial molecules nearly from zero. Observation of the protonated nitrogen cation (N_2H^+) was among the first detection studies and also among the first that utilized the self-consistent field (SCF) Hartree–Fock (HF) approach to compute an equilibrium geometry and provide corresponding rotational constants for comparison to astronomically derived spectral constants.[1,2] Similarly, detections of C_2H, C_4H, and C_3N also employed a similar approach utilizing quantum chemical *ab initio* theoretical data for corroborating evidence of detection.[3–5] During this era, the famed "X-ogen" lines were reported,[6] but these lines were conclusively linked to HCO^+ by Herbst and Klemperer[7] once more through the use of corroborating rotational constants derived from quantum chemically computed equilibrium molecular geometries.

In the decades since, quantum chemistry has spread its application in astrochemistry to more than just providing optimized geometries and the associated, complementary rotational constants, although this is still a vital service from quantum chemistry. With the growth in telescopic power and technology, notably for air- and space-based observatories, the infrared region of the electromagnetic spectrum is one of the, if not the, most important wavelength ranges for modern observations. IR photons can penetrate dust and also provide emissions from optically thick regions where other wavelengths become opaque or convoluted. The ongoing usage of the *Stratospheric Observatory for Infrared Astronomy* (SOFIA) and the upgraded CRyogenic high-resolution InfraRed Echelle Spectrograph (CRIRES+) instrument at the European Southern Observatory (ESO) as well as the recently launched *James Webb Space Telescope* will usher in a new era of unprecedented data for the near-, mid-, and far-IR regions. Hence, only the throughput and bottom-up approach of quantum chemistry and electronic structure theory can hope to provide the necessary volume and labeling of reference data for such large amounts of information to be gleaned from these observatories.

Recent work has already shown direct application of quantum chemistry to observations from SOFIA. Rovibrational lines determined, in part, from high-level, modern electronic structure computations have led to the first mid-IR detection of HNC and $H^{13}CN$ toward the Orion Hot Core.[8] Additionally, quantum chemical studies have provided the full set of fundamental anharmonic vibrational frequencies for protonated cyclopropenylidene $(c\text{-}C_3H_3{}^+)$[9] leading to its IR detection in the laboratory[10] as well as providing the target frequency range for ongoing SOFIA and CRIRES+ searches for this postulated precursor to larger interstellar hydrocarbon chemistry. Furthermore, the determination of dense molecular line lists of rotational and rovibrational states is also a vital service provided by quantum chemical computations where thousands of lines can be provided for a single molecule without any gaps in the data. These are often empirically refined for a few known lines providing exceptional accuracies for the vast majority of unknown lines.[11–15] Such studies help to eliminate the so-called spectral "weeds" by pruning the unknown transitions of known molecules allowing for the potential observation of unknown transitions of unknown molecules (the "flowers"). Hence, the role of quantum chemistry in astrochemistry continues to grow especially with regard to molecular vibrations and their associated physics.

This present work will highlight the role of quantum chemistry in the determination of vibrational and rovibrational properties and spectral data for numerous applications to astrochemistry. While this review is intended to be broad, the applications of electronic structure theory and quantum chemistry to astrochemistry are even broader and seemingly endless. The single molecule nature of traditional quantum chemistry and the flexibility of study for terrestrially unstable molecules makes it a natural fit for astrochemistry. Consequently, the applications of quantum chemistry, even computations of molecular vibrations, would stretch beyond any single review. Herein, we will focus on the emerging work where molecular vibrations utilizing electronic structure computations are vital including the aforementioned molecular line lists as well as the observation and constraining of polycyclic aromatic hydrocarbons (PAHs), small molecules of astrobiological significance, and even more exotic behaviors including molecular vibrations of electronically excited states. These areas highlight the ways in which the intramolecular motions of atoms within molecules hold vital significance for pushing the boundaries of knowledge for astrophysical implications. In so many ways, molecular vibrations are the fingerprints of astrophysics, and quantum chemistry provides the needed reference data for comparison.

7.2. Computational Aspects

7.2.1. *Theoretical framework*

Most quantum chemical computations addressing molecular vibrations in some way rely upon the harmonic approximation. Such is sensible when molecular vibrations are the smallest, meaningful contributor to the total energy, as is the case for computing reaction energies. However, when the molecular vibrations are the focus, more accurate representations are required. The venerable quartic force field (QFF), or fourth-order Taylor series approximation to the internuclear Hamiltonian's potential, has largely become the most-often utilized method for producing the anharmonic potential. Mathematically, the QFF is represented as:

$$V = \frac{1}{2}\sum_{ij} F_{ij}\Delta_i\Delta_j + \frac{1}{6}\sum_{ijk} F_{ikj}\Delta_i\Delta_j\Delta_k + \frac{1}{24}\sum_{ijkl} F_{ijkl}\Delta_i\Delta_j\Delta_k\Delta_l,$$

$$(7.1)$$

where the $F_{ij...}$ terms are the force constants while the $\Delta_i\Delta_j \ldots$ terms are the displaced distances for coordinates i, j, *etc.*, from a given reference geometry,[16,17] typically on the order of 0.005 Å and 0.005 radians. The reader should note that the reference point and the first derivative needed for a more complete Taylor series expansion appear to be omitted from Eq. (7.1). In truth, the reference point is set to zero by construction, and the first derivative should be zero by definition if the reference geometry is, in fact, a minimum.

The coordinates (i terms in Eq. (7.1)) can take on many forms from Cartesians to Morse-cosine coordinates,[18] but most QFF computations discussed herein utilize symmetry-internal coordinates, where applicable, or the simple-internal coordinates if necessary. The symmetry-internals are made up of simple-internals so the two often work hand-in-hand. The INTDER program[19] developed by Wesley Allen and coworkers at the University of Georgia has proven to be a valuable tool in the construction and translation of coordinates especially between simple-internals, symmetry-internals, and Cartesians.

Regardless, such a representation as Eq. (7.1) has been in use in quantum chemical electronic structure theory for many decades, and the evolution of such an approach for astrochemistry has been recently reviewed by numerous groups.[20–23] In truth, this is the lowest level anharmonic potential that can be constructed for consistently physically meaningful representation of molecular vibrations. However, for most molecules, even those with seemingly bizarre electronic structure (bizarre in the terrestrial laboratory sense), the QFF is sufficient to describe the intramolecular atomic motions' potential energy.

However, the QFF is not the total computation. Once the potential has been constructed, it must be conjoined to the kinetic and kinetic-potential interaction portions of a larger vibrational model. The Watson Hamiltonian is the usual vehicle for such with rovibrational perturbation theory likely among the most straightforward means of using the QFF:

$$H = \frac{1}{2}\sum_{\alpha\beta}(J_\alpha - \pi_\alpha)\mu_{\alpha\beta}(J_\beta - \pi_\beta) - \frac{1}{2}\sum_k \frac{\partial^2}{\partial Q_k^2} - \frac{1}{8}\sum_\alpha \mu_{\alpha\alpha} + V(\mathbf{Q}).$$

$$(7.2)$$

In the above Eq. (7.2), J_α represents the total angular momentum of a given cardinal direction ($x, y,$ or z) denoted by α or β. The π_α term is the total vibrational angular momentum of the same direction; $\mu_{\alpha\beta}$ is the inverse of the moment of inertia tensor for the given geometric coordinates; Q_k is a single normal coordinate; $\mathbf{Q}$ is the set of all normal coordinates; and $V(\mathbf{Q})$ is the potential portion, the QFF in this case.[9,24]

Like the QFF being the lowest-level anharmonic potential, even just second-order vibrational perturbation theory (VPT2) is enough to generate representations of molecular vibrations that are directly applicable to astrochemical problems. This has been extensively utilized through various VPT2 computer programs including the SPECTRO program[25] developed by Jeff Gaw, Nick Handy, and co-workers at Cambridge University with additions coming from the NASA Ames group. In short, VPT2 utilizes the harmonic approximation as the unperturbed portion of the computation. The cubic and quartic terms (third- and fourth-derivatives) in the potential represent the perturbation. Specifically, the Hamiltonian is split into the $\hat{H}_0$ (harmonic oscillator or quadratic terms from Eq. (7.1)) zeroth-order Hamiltonian as well as the two perturbation pieces: the $\hat{H}_1$ operator leading to the first-order correction to the energy comprised of cubic terms from the QFF and the $\hat{H}_2$ operator leading to the second-order correction to the energy comprised of the quartic terms from the QFF.[24,26,27] Both the first-order and second-order corrections to the energy come from the first-order wave function as is standard in Rayleigh–Schrödinger perturbation theory.[28] As a result, the energy of a state can written as

$$E_i = \langle \Psi^{(i)} | \hat{H}_0 | \Psi^{(i)} \rangle + \sum_{n \neq i} \frac{|\langle \Psi^{(i)} | \hat{H}_1 | \Psi^{(n)} \rangle|^2}{\langle \Psi^{(i)} | \hat{H}_0 | \Psi^{(i)} \rangle - \langle \Psi^{(n)} | \hat{H}_0 | \Psi^{(n)} \rangle}$$
$$+ \langle \Psi^{(i)} | \hat{H}_2 | \Psi^{(i)} \rangle, \tag{7.3}$$

or, equivalently,

$$E_i = \epsilon_i + \sum_{n \neq i} \frac{|\langle \Psi^{(i)} | \hat{H}_1 | \Psi^{(n)} \rangle|^2}{\epsilon_i - \epsilon_n} + \langle \Psi^{(i)} | \hat{H}_2 | \Psi^{(i)} \rangle. \tag{7.4}$$

As with most Rayleigh–Schrödinger perturbation theory extensions such as the electronic structure MP2 theory,[29] VPT2 constructs a portion of the resulting perturbation energy to be solved in terms of the normalized square, or inner-product space, of the integral divided by the difference between the unperturbed (harmonic oscillator in this case) eigenvalues of two separate states i and n.[28] The $\hat{H}_2$ term is constructed in much the same way as the cubic, inner-product sum, but only the given $\langle \Psi^{(i)} | \hat{H}_2 | \Psi^{(i)} \rangle$ piece survives by rules of matrix elements producing a simplified form of the anharmonic vibrational state energy in Eq. (7.4).[27] There are other forms of this energy (with one given later in Eq. (7.10) for asymmetric-top molecules), but the above is a more generic form of the VPT2 energy/frequency formulation.

Differently, vibrational configuration interaction (VCI) theory often utilized within the MULTIMODE program[30,31] developed by Stuart Carter and Joel Bowman along with their students and collaborators (and discussed in Chapter 8 of this book) is another means of utilizing the QFF for computing molecular vibrational frequencies as well as other variational approaches. VCI has much more stringent requirements on the potential than VPT2 largely because variational methods access larger portions of the PES and, thus, require that it be positive definite or, in other words, not have any holes and not turn over at large coordinate values, largely since an actual wave function is created. These types of computations often require a global or semi-global potential energy surface (PES). However, if proper limiting behavior can be built into the QFF, even the smaller PES contained within a QFF can allow for QFFs to be utilized within VCI or other variational approaches.[18,32,33] Both VPT2 and VCI usage of QFFs have benefits and drawbacks, and if the potential cannot be accurately represented, neither will perform well. Hence, the electronic structure computations of the displaced molecular geometries in any PES/QFF have really driven how meaningful quantum chemical computations of molecular vibrations are for application to astrochemistry.

7.2.2. *QFFs in practice*

Agnostic to the level of theory, QFFs are implemented beginning with a tightly converged, optimized, minimum geometry. This is essential since the relative energies for the Δ_i step sizes (from Eq. (7.1)) can often be on the order of sub-milliHartrees even if the total electronic energy computed can be hundreds or thousands of Hartrees. The tight geometry convergence (10^{-6} Å or rad.) also requires that the computed correlation energies also be tightly converged (10^{-14} Hartrees) which also requires that the reference energy be tightly converged (10^{-16} Hartrees) which requires that the integrals must be tightly converged (10^{-20} Hartrees or probability for S integrals).[9]

From this optimized, reference geometry, a set of coordinates must be constructed in order to define the QFF properly. These coordinates are the i, j, k, l indices given in Eq. (7.1). The coordinates can take on multiple forms,[18] but the most successful are typically the symmetry-internal coordinates. These are simply the symmetry-conserving coordinates made up of linear combinations of the simple bond stretches, angle bends, and torsional bends. Just like and closely related to the number of internal degrees of freedom, the number of coordinates for non-linear molecules is $3N - 6$, where N is the number of atoms in the molecule being examined. Typically and from practice, the heurestic for non-cyclic molecules is that if there are N atoms, the symmetry-internal coordinate system will require $N - 1$ stretches, $N - 2$ bends, and $N - 3$ torsions or out-of-plane bends. A simple coordinate set example is for triatomic ($N = 3$) water: the symmetric and antisymmetric bond stretches (2) as well as the bend therein (1) with no torsions. Such coordinates are akin to normal coordinates but are not mass-weighted allowing them to be utilized for different isotopic substitutions without having to recompute the QFF.

Displacements of these coordinates define the Δ_n terms in Eq. (7.1). Due to the vanishing integral rule, only those coordinates or products of coordinates that transform as the totally symmetric irreducible representation of the molecular point group are non-zero. For the water example, the quadratic or harmonic terms that survive are the second derivatives of each symmetry coordinate and the combined first derivatives of the bend and symmetric stretch. Hence, symmetry-internal coordinates reduce the total number of geometry points required for the QFF. The number of points grows geometrically with the number of atoms (N), and higher symmetry can alleviate such growth to some degree.

Once the coordinates are defined and the displaced geometries (often referred to simply as points) are constructed, the single-point electronic structure energies are computed at the desired level of theory. These are then gathered and turned into relative energies in order to reduce the numerical noise[34] for a least squares fitting. The function fit is defined from the displacements of the coordinates and the resulting energies. The output of the fitting is the $F_{ij...}$ in Eq. (7.1). The function is then refit with the new stationary point (minimum in this case) included so that the gradients are truly zeroed and the remaining force constants are as precise as possible. These force constants can then be transformed as desired, and our experience is that construction of Cartesian coordinate force constants from simple-internal coordinate force constants allows for the most flexibility in the actual VPT2 comptuations. At this point, the VPT2 procedure then produces the anharmonic frequencies and spectroscopic constants for the molecule being examined. While QFFs are independent of the energy type defining them, the choice of electronic structure method often is the principle dictation in the accuracy of the QFF conjoined to VPT2.

7.3. Predicting Vibrational and Rovibrational Spectra and Rovibrational Spectroscopic Constants to Identify Molecules in Astronomical Observations

The growth in accuracy of electronic structure methods, especially during the 1990s and 2000s, greatly enhanced the applicability of QFF-based anharmonic frequency computations to molecular vibrations with an eye towards astrochemistry. The "gold standard"[35] of coupled cluster theory at the singles, doubles, and perturbative triples level [CCSD(T)][36] especially with correlation consistent basis sets[37,38] has revolutionized *ab initio* quantum chemistry across nearly all fields ushering in notable accuracy for acceptable computational cost. However, the difference in chemical accuracy (typically thought of as 1 kcal/mol) and spectroscopic accuracy (typically $1\,\mathrm{cm}^{-1}$) is orders of magnitude and many more significant digits. Higher convergence criteria are therefore required of both optimized, reference geometries as well as the energies themselves in QFFs that are tightly fit via a least squares procedure to construct the force constants. While CCSD(T)/cc-pVQZ was able to produce spectroscopic accuracy on a few occasions,[32,39,40] QFFs employing this and similar levels of theory were still in error compared to experiment by a variable amount on the order of tens of cm^{-1}.

Further enhancements to the actual energies used to produce the QFF have been able to reduce this error in vibrational spectroscopy to single wavenumbers routinely.[41] This consistency began with the use of a composite approach based on CCSD(T). Taking the CCSD(T) energy out to the complete basis set (CBS) limit and including perturbative effects for core electron correlation, relativity, and even higher-order electron correlation emerged in the mid-to-late 2000s.[42,43] Over the past decade, such an approach has consistently been employed to produce experimentally comparable results for vibrational, rotational, and rovibrational spectroscopic data playing part in the aforementioned laboratory detection of c-$C_3H_3^{+}$[9,10] as well as the detection of HSS (or S_2H) toward the Horsehead nebula among other successes.[44,45]

7.3.1. *Small molecules in their ground electronic states*

7.3.1.1. *CCSD(T)-based methods*

While water is often the physical chemists' favorite molecule for testing various chemical models, astrochemically, employing H_2O as a test subject actually makes physical sense. After the uber-abundant diatomics in H_2 and CO, the most common molecule in various astrochemical environments is water.[46] Water was first identified in an astronomical observation in 1969 towards Sagittarius B2 (the nebula surrounding the center of the Milky Way Galaxy) as well as the Orion nebula via radioastronomical observation.[47] Water is also present in every body observed within our own Solar System and is the pervasive molecule in both gas- and condensed-phase astrochemistry. In 2008, it along with HO_2^+, were the first molecules to be examined with high-level composite energy QFF VPT2 computations.[42]

Water and HO_2^+ both show that the most accurate fundamental anharmonic vibrational frequencies of the methods explored include an initial CBS extrapolation involving the aug-cc-pVTZ/QZ/5Z basis sets using a three-point formula[48]:

$$E(l) = A + B\left(l + \frac{1}{2}\right)^{-4} + C\left(l + \frac{1}{2}\right)^{-6}. \tag{7.5}$$

While two-point extrapolations with the aug-cc-pVQZ/5Z basis sets behave in a similar fashion, the computational cost of the triple-ζ level is so small in comparison that its utilization is an addition that provides more certainty to the result with little cost. A two-point extrapolation with only the triple- and quadruple-ζ bases, however, is not as accurate. Inclusion of Douglas-Kroll scalar relativity[49] as a difference between the CCSD(T)/cc-pVTZ-DK energies with and without relativity included is added to the CBS energy. This provides a $\sim 3\,\mathrm{cm}^{-1}$ gain in accuracy for all three modes. Finally, explicit and additive inclusion of core electron correlation is shown to have a non-negligible contribution to the final anharmonic fundamental vibrational frequencies, as well.

The higher-order electron correlation is computed in this first approach via the averaged coupled-pair functional (ACPF) modification to multireference configuration interaction theory.[50] The difference in this energy at the cc-pVTZ level from the CCSD(T)/cc-pVTZ is also then added to the CBS+rel. energy for the QFF points giving a further shift of $\sim 5\,\mathrm{cm}^{-1}$ and producing better agreement with experiment. The ACPF/cc-pVQZ computations are much more costly than the triple-ζ and most often have a less than $1\,\mathrm{cm}^{-1}$ shift in the fundamental frequencies. The so-called CBS+rel.+ACPF/TZ QFF coupled to VPT2 is within $4\,\mathrm{cm}^{-1}$ of experimental for all three fundamentals of water. While the same QFF with VCI is similarly accurate, both VPT2 and VCI produce stretching frequencies nearly indistinguishable from one another but differ by $4\,\mathrm{cm}^{-1}$ for the bend. Additionally, step sizes of $0.005\,\text{Å}$ and 0.005 radians (the Δ_i values in Eq. (7.1)) are shown to be optimal in balancing energy shifts with known problems in minimizing higher-order numerical contamination of the force constants. $HO_2{}^+$ performs similarly giving indication that such an approach for computing anharmonic vibrational frequencies is a viable means for computing these values.[42]

This approach was verified shortly thereafter on a pair of anions: $NH_2{}^-$ and CCH^-.[43] The C_2H radical was one of the first molecules observed in astrophysical media,[3] and anion forms of its longer family members (C_4H^-, C_6H^-, and C_8H^-) have also been observed.[51–55] NH_2 has also been observed towards Sgr B2[56] even though the anion form has been famously not detected in the same source despite ongoing searches.[57] In either case, the CBS+core+rel.+ACPF/TZ QFF behaves similarly compared to experiment for these two molecules of potential astrophysical significance further verifying the method. The major addition in this study is the inclusion of a core electron correlation composite term through the use of the Martin–Taylor (MT) core correlating basis set which actually pushes the CBS+core QFF frequencies away from experiment. This is brought back into agreement with experiment by including the relativity and ACPF terms.[43] Inclusion of the core correlation explicitly in the terms used for CBS extrapolations or as a composite energy term does not affect the overall results by more than a single cm^{-1} or so, but it slows down the computation of the CBS terms by significantly more than the cost of the two composite energies. The implication is that the composite approach is the more effective execution for this piece of the QFF.

The method, however, was modified, once more a few years later in examination of c-/l-$C_3H_3{}^+$,[9] again of note for larger PAH synthesis.[58] Herein, the reference geometry is shown to require only CCSD(T)/aug-cc-pV5Z corrected compositely for core electron correlation for some geometrical parameter R:

$$R = R_{5Z} + (R_{MT(core)} - R_{MT(valence)}). \tag{7.6}$$

Addition of the other relativistic or higher-order correlation terms does not affect the reference geometry enough to warrant their inclusion. The MT core electron correlating basis set[59] is used at this point. From this geometry, the CBS+rel.+core+ACPF QFF is coupled to both VPT2 and VCI agreeing well with each other (within $10\,cm^{-1}$) and with the available experimental data. Again, the theoretically computed anharmonic frequencies have informed experimental analysis of the ν_4 stretch of the cyclic isomer which was found to be $0.6\,cm^{-1}$ less than the computed value at $3131.7\,cm^{-1}$.[10]

Such accuracies greatly informed the computation of the molecular vibrations for the *trans*-HOCO radical.[60] HOCO is believed to be a necessary intermediate in the synthesis of CO_2 in the Earth and Martian atmospheres and likely informs the carbon budget of similar gaseous regions.[61–63] At this point, the nomenclature for the QFF procedure is streamlined from CBS+rel.+core+ACPF to CcCRE with the "C", "R", "cC", and "E" terms coming from their respective counterparts in the form:

$$E(CcCRE) = E_{CBS(TQ5)} + (E_{MT(core)} - E_{MT(valence)}) + (E_{DK(rel)}$$
$$- E_{DK(norel)}) + (E_{CCSDT/TZ} - E_{CCSD(T)/TZ}). \tag{7.7}$$

Additionally, the ACPF term is dropped and replaced with a CCSDT/TZ computation. The ACPF computations are the most costly, as are the CCSDT, but the latter is a reduction in cost over the former. Even so, the "E" term in HOCO shifts the frequencies upon its inclusion both positively and negatively from the CcCR values upon its inclusion and only shifts the frequencies by typically $\sim 1\,\mathrm{cm}^{-1}$, $4\,\mathrm{cm}^{-1}$ at the most. In fact, the ν_2 C=O stretch at $1852.567\,\mathrm{cm}^{-1}$[64] is actually more accurate with the CcCR QFF VPT2 approach than CcCRE QFF VPT2. The CcCRE QFF VCI ν_1 O$-$H stretch at $3634.4\,\mathrm{cm}^{-1}$ is exceptionally close to the gas-phase experimental value at $3635.702\,\mathrm{cm}^{-1}$,[65] on the other hand. Even so, the computational cost and inconsistency of the E term in either form has led to its omission from most of the subsequent QFF computations.

With the maturity of the CcCR approach, the anharmonic fundamental vibrational frequencies for a large number of molecules with astrochemical significance have been able to be computed. Some had existing experimental data, some did not, but none had a complete set of fundamental vibrational frequencies provided to the level that the CcCR approach could accomplish. These are given in Table 7.1.

While this list appears to be long, it represents over eight years of work. This highlights the fact that the CcCR method, even without the E term, is still exceptionally time consuming. While computer hardware has progressed over the preceding decade, the original *trans*-HOCO QFF energy point computations for 743 points and seven energies at each point required over 20,000 CPU hours. While more contemporary hardware could likely reduce this by several factors as can distributed computations, high-performance computer clusters, and even cloud computing,[109, 127] the fact remains that a simple tetratomic molecule still requires an inordinately large amount of computer time to produce spectroscopic accuracy, which is nearly achieved for the CcCR QFF.[41] However, larger molecules, potentially with tens of thousands or even millions of points, are the next frontier in computing molecular vibrations of astronomically relevant species.

Table 7.1.   Molecules analyzed via CcCR QFFs to date.

cis-HOCO$^{./-}$	Ref. 66	*cis*- & *trans*-HNNS	Ref. 67
HOCO^{+a}	Ref. 68	ArHAr$^+$, NeHNe$^+$, ArHNe$^+$,	
		HeHHe$^+$, HeHNe$^+$, & HeHAr^{+d}	Refs. 69, 70
HOCS$^+$	Ref. 71	HPSi & HSiP	Ref. 72
CH$_2$CN$^-$	Ref. 73	NS$_2^{./-}$	Ref. 74
l-C$_3$H^{+75b}	Ref. 76	c-CNN, c-CNC$^-$, c-HCNN$^+$, and c-N$_3^+$	Ref. 77
C$_3$H$^-$	Ref. 78	HOSO	Ref. 79
NNOH$^+$	Ref. 80	CH$_2$NH$_2^{+f}$	Ref. 81
cis- and *trans*-HOCS and HSCO	Ref. 82	MgO, Mg$_2$O$_2$, MgCH$_2$, MgH$_2$, & MgF$_2$	Refs. 83–85
protonated acetylene (C$_2$H$_3^+$)	Ref. 86	*cis*- & *trans*-HSSH$^+$ as well as SSH$_2^+$	Refs. 87, 88
c-C$_3$H$^-$	Ref. 89	c-SiC$_2$H$_2$	Ref. 90
NeH$_2^+$ & ArH$_2^{+c}$	Ref. 91	H$_2$S$^+$	Ref. 92
ArH$_3^+$, Ar$_2$H$_3^+$, & Ar$_3$H$_3^+$	Refs. 93, 94	ArCH$_2^+$ & ArNH$_2^+$	Refs. 95, 96
CCOH$^-$	Ref. 97	:CNH$_2^+$ & :CCH$_2$	Ref. 98
OCHCO$^+$, NNHNN$^+$, NN-HCO$^+$, & CO-HNN^{+d}	Refs. 99–102	oxywater cation (H$_2$OO$^+$)	Ref. 103
SNO$^{./-}$ & OSN$^{./-}$	Refs. 104, 105	NCNCN$^-$ (C$_2$N$_3^-$)	Ref. 106
C$_3$P$^-$	Ref. 107	Isomers of [Al,N,C,O]	Ref. 108
SiCH$^-$	Ref. 109	c-C$_2$NH$_2^+$	Ref. 110
ArOH$^+$, NeOH$^+$, ArNH$^+$, NeCCH$^+$, ArCCH$^+$, ArCN$^+$, & NeON^{+e}	Refs. 111–114	OAlOH, AlOH, & HAlNP	Refs. 115, 116
HOX and HXO, X = Si$^+$, P, S$^+$, and Cl	Ref. 117	HCCOH	Ref. 118
SPSi	Ref. 119	HSO$_2$	Ref. 120

[a]Known in the ISM.[121, 122]

[b]Claimed[123] and later confirmed as an interstellar molecule in the Horsehead nebula.

[c]Following from detection of ArH$^+$.[124]

[d]These are collectively called proton-bound complexes.

[e]The ArOH$^+$ computations informed later experimental observation of this molecule.[125]

[f]Later observed experimentally.[126]

7.3.1.2. *Explicitly correlated QFFs*

The advent of explicitly correlated wave functions[128] and their modern incarnations like the F12b formalism in CCSD(T) is a promising avenue in moving QFF technology forward.[129, 130] This approach effectively reduces the size of the basis set necessary to reach the effective CBS limit and, in turn, likely the number of energies needed to be computed at each QFF energy point. The earliest exploration into explicit correlation for computing CCSD(T)-based QFFs of molecules with astrochemical significance is largely coincident with the earliest versions of the CcCR QFF.[131] A usual suspect list of H_2O, N_2H^+, NO_2^+, and C_2H_2 were explored with emerging R12 flavors of explicit correlation over a decade ago. In the case of the first three molecules, CCSD(T)-R12 performed as well as or slightly worse than the CCSD(T)/CBS results. However, the previously documented out-of-plane bending issues in $C-C$ multiply bonded systems[40, 132–134] was seemingly minimized with the R12 treatment. Such behavior produces erroneous and non-physical imaginary harmonic frequencies, among other issues, for certain levels of theory (MP2/aug-cc-pVDZ for example), but this is, again, reduced for explicitly correlated methods. However, at the time, the explicitly correlated methods alone were not robust enough on their own to provide accuracies akin to those that CcCRE had begun to provide. A major reason why is likely that the R12 approach being used was not sufficient for explicitly including core correlation. Even so, such approaches have crept into usage recently due to the huge cost savings in total computational time.

Specifically, CCSD(T)-F12b/cc-pVTZ-F12 (called F12-TZ from hereon) QFFs have been explored more recently in order to see if the computational cost can be reduced. In fact, they can. In a study comparing more than 20 molecules' worth of fundamental vibrational frequencies between CcCR and F12-TZ, the explicitly correlated method differs by less than $5\,cm^{-1}$ typically.[135, 136] The rotational constants do not fair as well, but the small shifts between F12-TZ and CcCR frequencies are often hovering around the available experimental data. Additionally, the reduction in computational time is on the scale of multiple orders of magnitude! As a result, this has

Table 7.2.   Molecules analyzed via F12-TZ QFFs to date.

Mg_2F_4	Ref. 85
c-(C)C_3H_2	Ref. 137
CH_2ClH^+	Ref. 138
CO_3 & C_2O_3	Ref. 139
$MgSiO_3$	Ref. 140
c-(CH)$C_3H_2{}^+$	Ref. 141
cis- & *trans*-HCSH as well as H_2SC	Ref. 142
SiO_2, SiO_3, Si_2O_3, & Si_2O_4	Ref. 143
ammonia borane (BH_3NH_3)	Ref. 144
AlH_2OH, $HMgOH$, AlH_2NH_2, & $HMgNH_2$	Ref. 145
AlH_3OH_2, SiH_3OH, & SiH_3NH_2	Ref. 146

opened the use of QFFs to even larger molecules beyond six atoms in some cases. Notable results are listed in Table 7.2.

In addition, core electron correlation can now be included in explicitly correlated computations. Work in this area has shown that computing CCSD(T)-F12b/cc-pCVTZ-F12 QFFs slows down the computations and may or may not increase the accuracy.[147] However, this initial examination was only on inorganic oxide dimers of the form $(MO)_2$ where M = Mg, Al, Si, P, S, Ca, and Ti, and these may have properties unique from molecules comprised of typical upper-*p*-block atoms and hydrogen. Regardless, in this case the core correlation is treated within the single energy (and not as a composite correction term). The vibrational frequencies are nearly always less than valence F12-TZ by an average of $3.5\,\mathrm{cm}^{-1}$. These are typically closer to the corresponding CcCR values but slow down the computations by a factor of roughly 13 in this case. However, the fewer core electrons in C, N, and O compared to S, Ca, and Ti will likely reduce this cost in lighter elements. Work is currently ongoing in this area, but promising results are on the horizon.

7.3.2. *Small molecules in excited vibrational states*

Rotational spectra of vibrationally excited states have been observed in circumstellar and interstellar media, most notably for the C_6H radical[148] and SiS.[149] Similarly, acetylene,[150] carbon dioxide,[151] the C_{60} and C_{70} buckyballs,[152] and a few other molecules lacking dipole moments have been directly observed vibrationally and/or rovibrationally, as well.[153] Consequently, the rotational constants of vibrationally excited states are also useful reference data from quantum chemical computations as well as the combination bands and overtones, especially for common astrochemical molecular species.

In every instance mentioned above in Tables 7.1 and 7.2, the CcCR QFF provides the zero-point averaged rotational constants, but these only require cubic terms. This is also true for being able to compute the vibrationally averaged rotational constants for all of the fundamentals, as well as other excited vibrational states.[154,155] Hence, these values fall out of the QFF VPT2 computations within SPECTRO. These have been reported for nearly all of the molecules listed in the previous section.

However, other notable examples still remain. Foremost is SiC_2. This molecule has been observed toward the carbon-rich star IRC+10 216[156] as has the triatomic inverse of CSi_2 (SiCSi)[157] implying a diversity of carbon-silicon chemistry. The ν_3 antisymmetric Si$-$C stretching frequency in SiC_2 is computed[158] and experimentally known[159] to be exceptionally low in frequency at $196.37\,cm^{-1}$. Hence, this mode can be thermally excited implying that overtones and combination bands would serve as a possible thermometer in regions where SiC_2 has been previously detected.[158] The first overtone lies at $507.7\,cm^{-1}$ with $3\nu_2$ at $701.0\,cm^{-1}$. Even though VPT2 slightly underreports the associated vibrational frequencies, empirically scaling the known rotational constants with those computed for the excited vibrational states allows for what should be accurate semi-empirical rovibrational data provided for this potential molecular thermometer.[158]

7.3.3. *Small molecules in excited electronic states*

While most of the molecules uniquely detected in various astrophysical media have come from radioastronomical observation with the next-most originating with infrared observation, some have come from detection of electronic spectral signatures.[153] Most notable is $C_{60}{}^{+}$[160,161] and its correlation with several of the diffuse interstellar bands, a series of unattributed UV-to-near-IR absorption features whose provenance is still nearly a complete mystery after more than a century.[162–168] However, the first three molecules observed beyond our Solar System were also detected from their UV-visible spectra towards the star ζ Ophiuchi in the 1930s and 40s: CH, CN, and CH$^+$.[165,169–171] Additionally, the electronic spectral features of other small molecules are of notable importance for planetary and cometary observations within our solar system[172–176] making fully rovibronic spectral characterizations necessary for such molecules.[177]

The easiest way to treat molecular vibrations of electronically excited states with QFFs is not to treat them as electronically excited at all. Variationally accessible and/or the lowest energy spin states different from the ground electronic state require no additional treatment beyond either standard CcCR or F12-TZ. This has been first implemented for CcCR QFFs with 1 $^3A'$ HCN, HNC, HCO$^+$, and HOC$^+$.[178,179] Additionally, the lowest quartet state of H$_2$SS^{+}[88] has been examined via QFFs and VPT2. The bent/linear electronic isomers of NS$_2$[74] along with the cyclic and bent electronic isomers of HC$_2$N[180] and HPSi/HSiP[72] have been computed in this manner, as well. The dipole bound excited states of CP$^-$ and C$_2$P^{-}[181] have also been computed in a variationally accessible ground state approach using specially constructed basis sets.[182] Even so, this is an exceedingly limited approach that requires an extremely narrow set of circumstance to be present in order to be useful.

A more generic approach would require moving beyond CCSD(T), though, in spite of its balance between accuracy and computational cost. Since CCSD(T) is a combination of an iterative CCSD method and a perturbational (T) correction, any electronic excitation method will be complicated. Within coupled-cluster theory, equation-of-motion (EOM) CCSD[183,184] is the standard excited state approach, but CCSD lacks the necessary correlation to produce highly accurate results for electronically excited states.[185] Other, approximate triples methods like CC3[186–188] are more accurate and produce wave functions open to electronic occupation shifts, but they are often exceedingly costly. There are more, exotic, triples-including, excited state coupled cluster methods, but none have thus far been able to garner a consensus among the community like CCSD(T) or even EOM-CCSD.

Hence, attempts have been made to compute QFFs based on EOM-CCSD with CC3 corrections. A CcCRE-like QFF has been constructed for $1\ ^2A''$ HOC where this state is actually variationally accessible allowing for direct comparison to a trusted method in CcCR.[189] The so-called EOM-CcCE QFF energy is defined as

$$E(\text{EOM} - \text{CcCE}) = E_{\text{CBS(TQ5)}} + \left(E_{\text{MT(core)}} - E_{\text{MT(valence)}}\right)$$
$$+ \left(E_{\text{CC3/TZ}} - E_{\text{CCSD/TZ}}\right), \tag{7.8}$$

with all terms implied to be for the same electronic state computed via EOM. This QFF has mixed results with the EOM-CcCE computations varying by as much as $25\,\text{cm}^{-1}$ compared to the ground state CcC results.[189] While this use of explicitly computed anharmonic frequencies in electronically excited states is better than simple harmonics or even scaled harmonics, this method is not as accurate as would be desired compared to the ground state methods or even that which is necessary for comparison to laboratory astrophysical simulation. A refinement to this approach utilizing a scaling term from the ground electronic state in the difference in the CCSD-based CcCE QFF results and those from true CCSD(T)-based CcC (or CcCR) produces unobserved fundamentals for the excited $1\ ^2\Pi$ state of the C_2H radical.[190] However, the errors are still on the order of tens of cm^{-1} for other known fundamentals and vibronic excitations. Again, even this approach is not as accurate as would be desired based on the performance of ground electronic state methods. Even so, a similar approach has been utilized in computing the $2\ ^1A_1$ c-C_3H^- dipole bound state,[191] but problems in the electronic structure of the $C-C$ antisymmetric stretching coordinate have led to only the a_1 and b_1 modes being reliably produced.

A recent breakthrough in this area has combined both CCSD(T) and EOM-CCSD in a novel way. This so-called (T)+EOM approach computes both the "ground state" (T) energy and the "excited state"

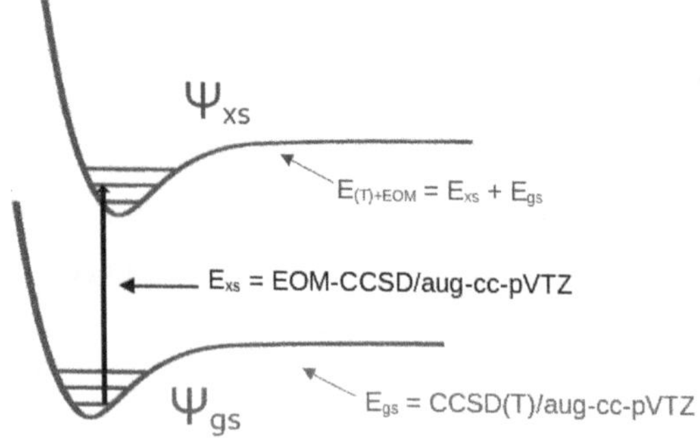

Figure 7.1. A visual depiction of how the excited state (Ψ_{XS}) is accessed for computing QFFs via the (T)+EOM approach.

EOM-CCSD energy and adds both to the CCSD energy[192] and is visually depicted in Figure 7.1. More precisely:

$$E_{(T)+EOM} = E_{CCSD} + E_{(T)} + E_{EOM-CCSD}, \qquad (7.9)$$

where E_{CCSD} is the total reference and correlation energy up to the CCSD level, $E_{(T)}$ is just the (T) contribution to the energy, and $E_{EOM-CCSD}$ is only the excitation energy. In so doing, (T)+EOM/CcCR QFFs can be created that are equivalent in energy to those determined in Eq. (7.7) with the appropriate terms included. For the sample set of HOO, HNF, HNO, and HCF species with variationally accessible excited states, (T)+EOM/CcCR and standard CcCR often agree to better than $1\,\mathrm{cm}^{-1}$ and (with one exception) as far as $12\,\mathrm{cm}^{-1}$ for an average difference of less than $7\,\mathrm{cm}^{-1}$. The one exception is the ν_2 bend in HOO, but the F12-TZ and (T)+EOM/CcCR QFF frequencies agree to better than $2\,\mathrm{cm}^{-1}$ at 1195.1 cm^{-1} implying that CcCR itself may be in error for this frequency.[192] In any case, the (T)+EOM/CcCR method is currently being pushed and tested for new molecules in order to ascertain its robustness, and other excited state methods for computing QFFs are under development in order to provide fully rovibronic spectral data for molecules of astrochemical interest.

7.3.4. *Large molecules in their ground electronic states*

PAHs, buckyballs, and large carbonaceous molecules are believed to contain most of the carbon in the Universe not already tied up in CO or CO_2. Some estimates put this as high as 20% of all carbon atoms.[58] These also likely serve as reservoirs for the carbon observed in other small molecules potentially even those with astrobiological significance. With the confirmed detections of benzonitrile, cyanonaphthalene, and even indene[193–196] in addition to earlier reports of benzene from IR observations,[197] there is now no doubt of PAHs existing in astrophysical environments. Even though some PAHs have small dipole moments, or functionalizing them clearly increases them, most PAHs will have negligible dipole moments if any at all. Hence, IR detections of PAHs are the most viable means of observing this entire class of molecules. The molecular vibrational and anharmonic fundamental frequencies are now needed more than ever if methods can be brought to bear to compute these values for comparison.

The largest CCSD(T)-based QFF produced to date has been on a mere eight atoms for ammonia borane.[144] The issue with going to larger molecules is that the number of points scales geometrically with the addition of each atom even for the sparsity of the PES provided by the QFF. While, again, explicitly correlated methods can reduce the computational cost of each point and higher symmetry molecules like those with a D_{2h} point group can in some ways reduce the number of points and the cost of each point, moving beyond 10 atoms requires moving beyond the safety of CCSD(T) and *wave function*-based methods. While benzene could be computed with F12-TZ due to its high symmetry requiring "only" somewhere around 22,000 points for the QFF, most astronomers do not consider this molecule to be a PAH (since it's mono- and not polycyclic). That moniker for some does not appear until the seven rings of circum-benzene, or coronene, are present in the molecule. Coronene requires more than one million QFF points. For larger systems like PAHs, other methods must be employed.

Density functional theory (DFT) is really the most viable means of computing energies for placement into QFFs for subsequent VPT2 analysis. While various flavors of DFT fail inexplicably at times, this approach is the only one that steps beyond the Hartree–Fock approximation with enough accuracy to give hope for providing somewhat meaningful descriptions of molecular vibrations. One approach is to take DFT harmonic computations and scale them.[198–200] Such a methodology has even provided molecular vibrational data for molecules with up to 384 atoms.[201] However, such methods cannot compensate for inconsistent anharmonicities or, most notably, treatment of vibrational resonances like QFFs and VPT2 can.[202] As a result, recent work has shown that DFT-computed QFFs can provide this, if the proper functional and basis set can be chosen.

The first PAHs examined via QFFs and VPT2 include the linear acene family: naphthalene, anthracene, and tetracene.[203, 204] The B971 functional[205] coupled to the TZ2P basis set gives accurate results for decent computational time cost even with the added tight convergence requirements. The QFFs are displaced via normal mode coordinates, which necessitates an initial computation of the harmonic vibrational frequencies. The double-harmonic intensities are also computed at this same level of theory. VPT2 from SPECTRO differs from that available in Gaussian16[206] mostly in that the resonance treatment, specifically coupling of symmetry-blocked Fermi resonance polyads for a limited energy range, are available. The utilization of resonance treatment produces fundamentals, overtones, combination bands, and their intensities — including the ability to describe intensity borrowing more completely within the polyad — all corrected for the resulting frequency shifts giving exceptional accuracy compared to experiment from the same work.[203, 204] Some QFF VPT2 anharmonic fundamental vibrational frequencies agree to experimental precision. Others vary by $\sim$10 cm^{-1}. The visual spectra, though, are nearly indistinguishable between theory and experiment in this case as shown in Figure 7.2 for anthracene, taken from Ref. 203. Such is not true for the harmonic spectra, scaled or otherwise, largely due to the lack of resonance treatment.

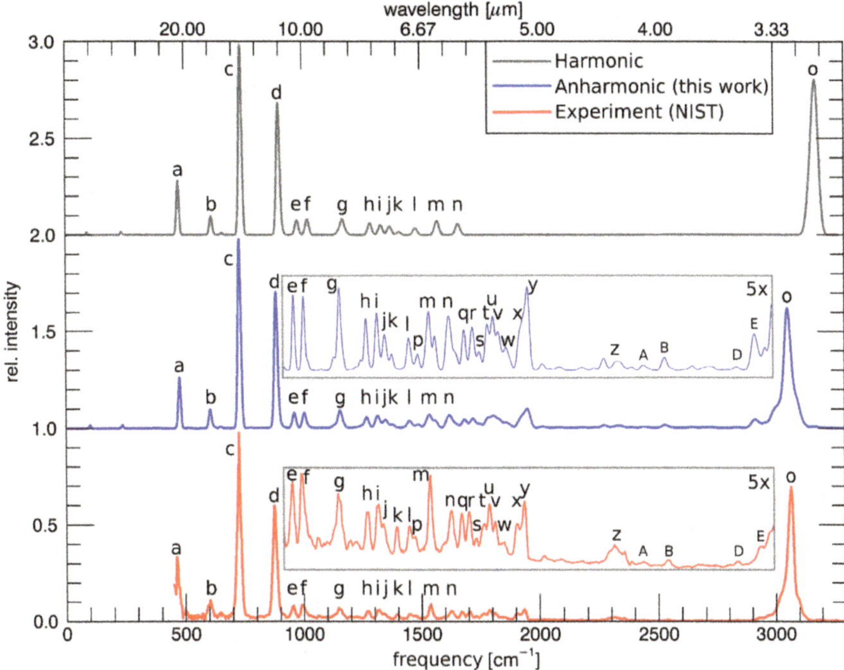

Figure 7.2. Full IR spectrum of anthracene calculated using two methods, harmonic (top) and anharmonic (this work) (middle), compared with gas-phase spectra at 300 K (bottom). A selected region is shown with the relative intensities increased by a factor of five; from Ref. 203.

With so many vibrational frequencies in such a high density, the resonances are essential for proper treatment of the vibrational/ IR spectrum.[203]

This work has been followed by similar analysis of larger PAHs including benz[a]anthracene, chrysene, phenanthrene, pyrene, and triphenylene[207] as well as the methylated PAHs 9-methylanthracene and 9,10-dimethylanthracene in addition to the hydrogenated PAHs 9,10-dihydroanthracene, 9,10-dihydrophenanthrene, 1,2,3,4-tetrahydronaphthalene, and 1,2,3,6,7,8-hexahydropyrene.[208] The inclusion of the aliphatic groups in the latter study further informs correlation between theory and experiment where isolation of pure PAHs is nearly impossible. The visual spectra are, once again, nearly indistinguishable for the substituted PAHs, and the accuracies are good. However, some errors creep into the results for a few of the C–H stretches and in some of the intensities.[208] While such results are still wildly successful, the aliphatic contributions are more in error than the aromatic frequencies. This result implies that one DFT functional cannot be applied to all molecules for a reduction of computational cost. B971/TZ2P works exceptionally well for PAHs, but, while still good, is not as predictive for aliphatics leading to use of B3LYP[209–211] with the double-ζ N07D basis set[212] in such cases.[208] Hence, more refinement in the choice of methods will be required for moving into functionalized PAHs. Even so, the use of B971/TZ2P QFFs with resonance-treated VPT2 is promising for advancing the study of molecular vibration spectra for PAHs.

A major remaining issue for the QFF computations of large molecules is that anharmonic DFT still is not tractable for the computation of coronene-level PAHs or larger. The density of IR bands likely arising from PAHs and related species is virtually optically opaque in some regions,[213,214] and the diversity of possible PAHs, including those with aliphatic portions, becomes an exercise in statistical probability and stochastic geometry. Consequently, computational methods are really needed to create trustworthy, but high-throughput spectral characterization into the astrochemical literature. Of course, work is ongoing in this area in order to produce meaningful, anharmonic vibrational spectra for larger PAHs.

7.4. Computing Rovibrational Line Lists for Eliminating "Weeds" and Providing Data for Modeling the Opacity of Exoplanet Atmospheres (Absorption and Emission)

QFFs provide sparse PESs that are relatively quick to compute. However, the accuracies begin to break down in moving beyond the fundamental vibrational frequencies and their corresponding rotational constants and spectroscopic data. A larger PES is required in order to provide the most accurate descriptions of the full rovibrational spectral values, also called rovibrational line lists. This usually involves a global PES. Even so, the two approaches have some similarities. Most notably, the principle energy computed at each point on the surface can be computed through any flavor of quantum chemical method desired. Some of the most accurate line lines are still based on the CcCRE energy with the ACPF "E" term included as defined above in Eq. (7.7). Even for small molecules, such energies are costly on the scale of hundreds of thousands of points on the PES. Hence, lower levels of theory are often brought to bear initially in order to sample the PES geometries and select a subset of those points that are required to represent a chosen energy range. Once the higher-level PESs are computed as informed by the lower level guidance, these PESs are then fit to very high tolerances. Refinement of the PES using known empirical parameters to very high accuracy serves to enhance significantly the accuracy of the line list, usually by $\sim$3 orders of magnitude. This has led to the Best Theory + Reliable High-Resolution Experiment (BTRHE) strategy and has produced line lists of notable accuracy for several molecules discussed below.[215] Other groups have provided similar line lists for similar astrochemical applications,[11,216,217] but the present discussion will largely focus on work from the NASA Ames group.

The selection of molecules for such line lists must be done with care as it may take years to produce a quality PES, empirical fit, variational computation, and even dipole moment surface (for the intensities) at a low enough noise level. As mentioned above, water is an obvious candidate.[218] Additionally, atmospherically relevant species also are sensible for both terrestrial applications, exoplanetary considerations, and planetary modeling. This has led to the selection of CO_2, NH_3, and SO_2[15,215] for the other molecules thus far examined via such high-level analysis. Hence, these molecules will be present across all spectral ranges not just the ones in which they have previously been detected. This implies that they will exhibit unknown transitions even though they are known molecules. These may be overtones and combination bands but are much more often rotational lines of various vibrational quantum levels. Such spectral lines are called "weeds" since they do not add to the chemical inventory of a certain celestial body and often add little to the understanding of the underlying physical conditions. The only way to trim these "weeds" in order to find the "flowers" (the unknown transitions of unknown molecules) is to provide highly accurate line lists for common molecules and remove them from the observed astronomical spectra.

Another important application for these highly accurate rovibrational line lists is modeling the opacity of planetary atmospheres, especially exoplanets.[219] The study of exoplanets, planets outside our solar system, has exploded in recent years and the next phase of these studies will be to characterize the atmospheres of exoplanets to determine their nature, and ultimately whether they may be able to support life. In the near term, JWST will most likely be used to characterize the atmospheres of the closest known exoplanets, and many of these are very hot, approaching $2000\,\mathrm{K}$ or more. Hence the rovibrational line lists for the molecules discussed here need to be highly accurate, but also extend to very high rovibrational energies in order to cover the appropriate temperature range.

7.4.1. *Small, stable, abundant molecules: CO_2*

The first PES produced for CO_2 is isotopic-independent with a root-mean-squared (RMS) error compared to known experimental lines of $0.0156\,\mathrm{cm}^{-1}$ for 6873 $J = 0 - 117$ transitions for the principle isotopologue.[220] The isotopologues increase the RMS by an order of magnitude, but this is still a small error. The initial dipole moment surface (DMS) is able to produce intensities to within 20% of the experimental values providing a useful high-resolution benchmark for subsequent analysis at or below $296\,\mathrm{K}$.[220] A CCSD(T)/aug-cc-pVQZ DMS is able to refine the accuracy of the intensities and further increasing the J values up to 150 and the temperature up to $1000\,\mathrm{K}$.[221]

Additional refinements allow the PES for all 13 major (and some minor) isotopologues of CO_2 to stretch up to $18,000\,\mathrm{cm}^{-1}$ and $1500\,\mathrm{K}$ while also providing line shapes for planetary bodies including the Earth but also Mars and Venus with hotter conditions possible.[222] Line accuracies at such higher energies are better than 0.05 cm^{-1} implying a well-suited description of the line lists for

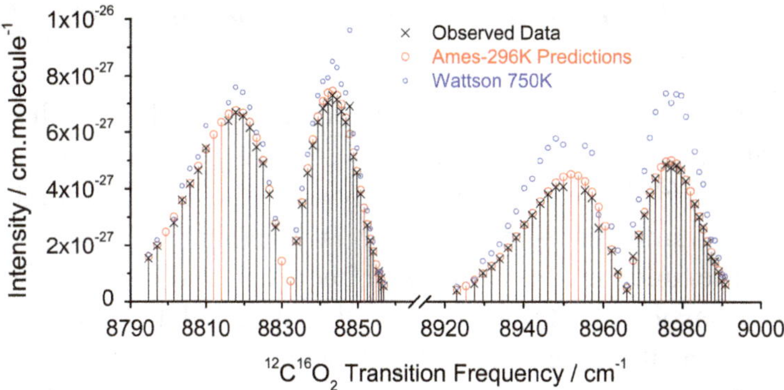

Figure 7.3. Measured 5001r−00001 (r = 3, 4) line positions and intensities, compare to Ames-296 K line list and Wattson 750 K (also called High-T or HOT-CO$_2$) line list; from Ref. 221.

unknown transitions. In addition, problems in existing experimental data have been identified as demonstrated in Figure 7.3 (Figure 2 from Ref. 221). In this figure, excellent agreement between the Ames line list and recent experiments for two specific rovibrational bands is found for both line positions and intensities, while the HITRAN2008 intensities are two orders of magnitude too low. While studies of CO$_2$ reach into the 1500 K range, planetary conditions of Venus, hot Jupiter exoplanets, and circumstellar envelopes will extend beyond this temperature range requiring higher energy line lists. Hence, this line list is performing exceptionally well at these higher temperatures and can be usefully extended even further up the temperature scale in order to provide even more rovibrational lines for CO$_2$ and its isotopologues.[223,224]

In the latest DMS computed for CO$_2$, intensities are now predicted to within a few percent for some strong bands but at least to within 90% of experiment.[224] In fact, new experiments are actually needed at this point because the possibility remains that the computations are more accurate than the current experiments. Additionally, computing accurate rovibrational line intensities requires accurate total internal partition sums (TIPS), also called internal partition functions, and the Ames CO$_2$ and isotopologue line lists have been used to yield very accurate TIPS.[225]

7.4.2. *Small, stable molecules with large amplitude motions: NH$_3$*

Not all molecules lend themselves to straightforward study, however. The barrier to linearity in water is relatively high at more than 11,000 cm^{-1},[226, 227] but the inversion barrier, the so-called "umbrella" motion, of ammonia is much lower at roughly 2,000 cm^{-1}, below many of the fundamental vibrational frequencies. While such features can play havoc on a QFF at times,[41] the global PES required for line lists can, if treated properly, alleviate these concerns in many ways.[15] Granted, the proper choice of coordinates for a QFF can also circumvent these issues, specifically including NH$_3$,[18] but not in all cases.[94, 99] Regardless, ammonia is known to exist in interstellar regions,[228] protoplanetary disks,[229] planetary atmospheres,[230] and other astrophysical regions. While not as common as water, this simple hydride plays a vital role in many astrochemical pathways requiring the most accurate of computations for comparable results.

The first highly accurate line list for NH$_3$, again, utilizes the CcCRE-type (Eq. (7.7)) energy with ACPF sometimes and sometimes not included as well as diagonal Born–Oppenheimer corrections.[232] The initial fitting for the $J = 0-2$ bands gives a RMS compared to 13 database benchmarks of 0.023 cm^{-1}, and these do not reduce for theoretical benchmarks past 10,000 cm^{-1}. The inversion is computed to be 1784.19 cm^{-1} within 3 cm^{-1} of previously computed values.[232] The ammonia line list is further corrected utilizing a second-order correction for the Born–Oppenheimer approximation and additional experimental refinements.[231] Figure 7.4 (Figure 3 from Ref. 231) shows how the NH$_3$ PES improves by two orders of magnitude from purely *ab initio* (HSL-0), to initial refinement (HSL-1), to refinement and inclusion of non-adiabatic Born–Oppenheimer breakdown terms (HSL-2). Additionally, the HSL-2 PES is used to compute the purely rotational transitions for ^{15}NH$_3$ and compare to the Cologne database,[233] and the RMS error is a mere 0.00034 cm^{-1}. Further, a deficiency of the ^{15}NH$_3$ rovibrational energy levels is identified and tied to a problem with an effective Hamiltonian model used to fit the experimental data.[231] As a result, the accuracy

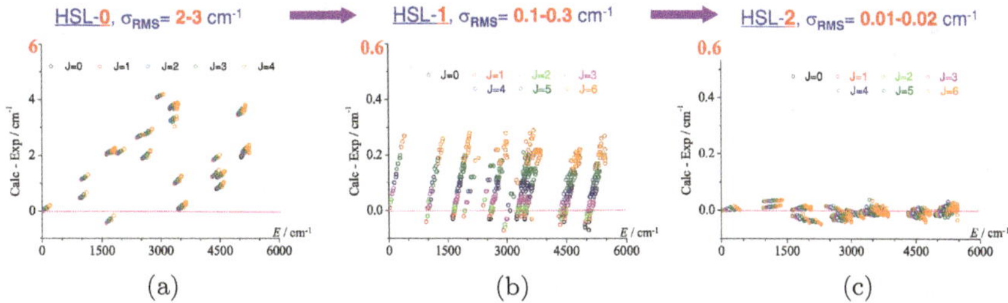

Figure 7.4. 2 orders of magnitude rms error reduction from HSL-0 to HSL-1 to the latest HSL-2. From left to right, (a), (b), and (c) are for HSL-0/1/2, respectively; from Ref. 231.

of the PES opens the possibility for exploring existing, unassigned HITRAN data. Strong mixing between the $2\nu_4$ overtone and the $2\nu_2 + \nu_4$ combination bands shows to be the source of deficiencies in the experimental modeling of $2\nu_4$, and new predictions of the $4\nu_2$ band and its associate lines result from this analysis.[234] Building upon this notable strength of the ammonia line lists, comparison to experiment has been able to elucidate the origins for several hundred spectral features including nearly 4800 NH_3 positions and intensities between $6300\,cm^{-1}$ and $7000\,cm^{-1}$ providing unprecedented clarity for such higher-order transition lines.[235]

7.4.3. *Small, stable molecules containing a heavy atom: SO_2*

The lessons of both ammonia and carbon dioxide also serve in the line list creation for SO_2. Sulfur dioxide is of notable significance for the Venusian atmosphere[237] especially in the sulfuric acid cycle[238] and the interstellar medium[239] among other sources. Its study also builds off of the approaches refined from the other two molecules from the NASA Ames group discussed above and can be found online (http://huang.seti.org/). The BTRHE approach for this molecule begins with a CcCR PES, a CCSD(T)/aug-cc-pV(Q+d)Z DMS, and an empirical fitting to HITRAN data.[12] The lines fit for SO_2 at $296\,K$ cover up to $5,500\ cm^{-1}$, have uncertainties of less than $0.03\,cm^{-1}$ per each line, and produce intensities to within 15% or better of experiment, though the line list is likely to be reliable up to about $8000\,\mathrm{cm}^{-1}$, within 15% of experiment. Extension of the PES to the $^{32/33/34/36}\mathrm{S}$ and $^{18}\mathrm{O}$ isotopologues perform similarly well as the standard isotopologue.[240] Additional work in conjunction with the ExoMol group from the University College London has been able to produce a whopping 1.3 billion lines for temperatures up to 2,000 K giving incredibly fine grained results for a more complete line list of SO_2.[241] The refined $^{32}\mathrm{S}^{16}\mathrm{O}^{18}\mathrm{O}$ line list also provides alternatives for questionable empirical assignments and experimentally missing bands.[236] As shown in Figure 7.5 (Figure 6 from Ref. 236), assigning experimental rovibrational lines for $^{32}\mathrm{S}^{16}\mathrm{O}^{18}\mathrm{O}$ can be difficult because there will also be lines from the pure $^{32}\mathrm{S}^{16}\mathrm{O}_2$ and

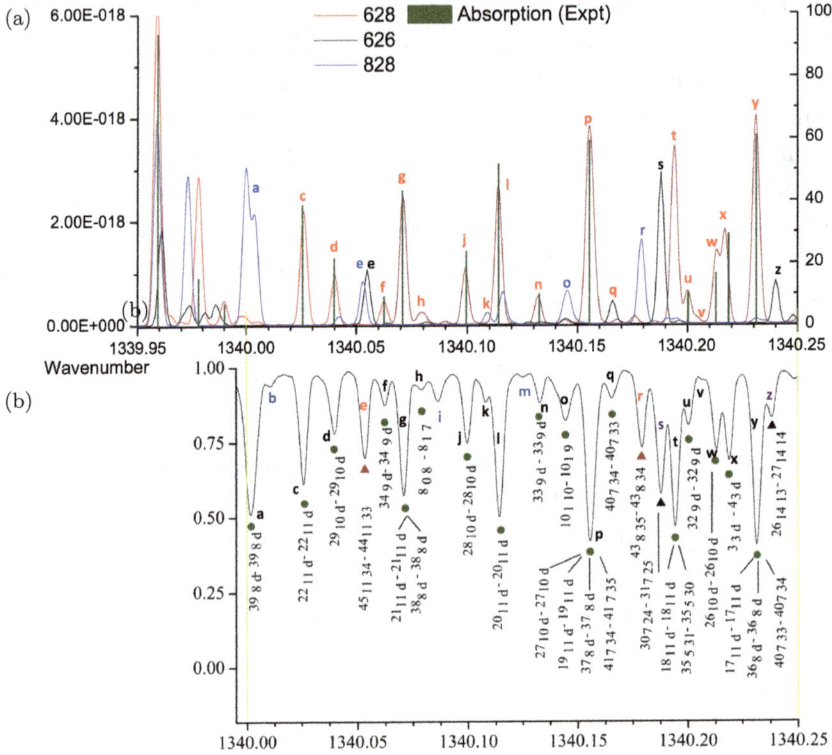

Figure 7.5. (a) Ames-296K list-based IR simulations vs. (b) experimental spectrum analysis. Discrepancies between the Ames analysis and the reported experimental assignments are as obvious as the very good agreement. $\sigma = 0.001\,\mathrm{cm}^{-1}$ in the Ames Guassian convolution; from Ref. 236.

$^{32}S^{18}O_2$ isotopologues. Consequently, those need to be assigned first, whereas such problems do not exist for the computed rovibrational line lists. Extension to the $^{32/33/24}S^{18}O_2$, $^{32}S^{18}O_2$ and $^{16}O^{32}S^{18}O$ isotopologues largely completes the 296 K lines for all notable versions of sulfur dioxide.[13,224] Further refinement for SO_2 will require additional high-resolution rovibrational experiments that go above $8,000\,cm^{-1}$, as well as high-resolution rotational experiments that include "hot" microwave transitions, *i.e.*, transitions from excited rotational levels.[242] Even so, the line lists in these regions based on PESs fit to the lower temperature and lower energy data can serve as an initial guide for future experimental analysis, as has already been the case.

The goal for such line lists still remains two-fold: (a) to provide a means of removing the unknown transitions of known molecules, and (b) to provide data for modeling the opacity of planetary atmospheres, especially exoplanets. While CO_2, NH_3, and SO_2 (in addition to water)[216,218,227] represent a small sample of notable species, they have a huge potential impact on the regions where they are present. Hence, removal of their spectral features is key in finding new molecular species hidden among their lines. Additionally, other molecules' line lists have been computed by the ExoMol group (https://www.exomol.com), the Reims-Tomsk group (http://theorets.tsu.ru; http://theorets.univ-reims.fr), and even more rovibronic spectral data are currently being proposed to the community.[177] Future directions will include potential bio- and even technosignature molecules where the density of lines can serve as definitive evidence for such molecules beyond a few features (or even just a single feature as has been claimed recently for phosphine[243]). Consequently, the BTRHE approach as well as CcCRE PES data will continue to inform future work in producing line lists for astrochemical "gardening" in pruning "weeds" in search of "flowers".

7.5. Simulating Cascade Emission Spectra of Large PAH Molecules (Emission)

Most of the astrochemistry vibrational and rovibrational spectroscopy discussed earlier involves predicting the absorption spectroscopy of small or even large molecules for comparison to either astronomical observations or laboratory experiments. The one exception is modeling the opacities of (exo)planetary atmospheres as the astronomical observations will be of the emission coming from these (exo)planetary atmospheres. In that case, however, the line lists are so complete that computing the density of states is reliable and, subsequently, the opacity computations are relatively straightforward, even if computationally intensive. Trying to simulate the IR emission spectra from astronomical observations of the interstellar medium (ISM), circumstellar shells, or other low-density astrophysical environments is wholly different, though. In such environments, the emission spectrum is dominated by bands that were originally referred to as the unidentified infrared bands (UIBs or, equivalently, UIRs), but are now commonly referred to as the aromatic infrared bands (AIBs).[244–246] The AIBs are widely believed to emanate from emission of relatively large PAH molecules (*ca.* 50 carbon atoms).[244]

The astrophysical environments from which the AIBs originate are harsh and generally have a large flux of ultraviolet (UV) photons due to the presence of nearby stars. Hence, small molecules are readily photodissociated and do not survive. Large PAH molecules, on the other hand, can absorb even a 12 eV photon and survive provided the excess energy is radiated away before absorbing another such photon.[244] The mechanisms behind this are straightforward: (a) a UV photon is absorbed, exciting the large PAH molecule into an excited electronic state; (b) the large PAH molecule reverts to the ground electronic state due to non-adiabatic coupling with the excess energy then deposited into molecular vibrations; (c) because the large PAH has so many vibrational degrees of freedom and due to the density of states, most individual vibrational degrees of freedom will not have very much energy deposited into them; (d) this gives the molecule time to start emitting IR photons; (e) after each photon is emitted, the excess energy in the molecule is redistributed due to intermolecular vibrational relaxation (IVR); and, finally, (f) steps (d) and (e) continue until the PAH molecule is once again in its ground vibrational state, which is referred to as a cascade emission spectrum.[244, 247] The PAH molecule is then ready to absorb another UV photon and repeat the process. The time scale for steps (b) and (e) is very rapid $\sim 10^{-8}$ s or less, whereas the time scale for step (d) is $\sim 10^{-6}$ s, and the time scale for the entire cascade spectrum is $\sim 10^{-3}$ s.[244] The challenge for us, then, is to compute a simulated cascade emission spectrum of a large PAH molecule, and this cannot be done with, for example, a line list as the theoretical methods applied there are much too expensive to be used on a large PAH molecule. In the sections below, the most commonly used approach thus far is discussed, which makes use of scaled harmonic vibrational frequencies. However, a fully anharmonic vibrational frequency approach was developed only a few years ago and work on refining it continues presently.

7.5.1. *Cascade emission spectra from scaled harmonic frequencies*

For this section, the model used in the NASA Ames PAH IR spectroscopic database (denoted PAHdb) is examined.[200, 248–250] Several factors are considered including band position, band shape, line width, and relative band intensities. For the band position, since most anharmonic corrections are negative (*i.e.*, lower than a given transition energy), the highest energy photon that could be emitted for a given vibrational mode is known to correspond to the molecule relaxing from the first excited state for that mode to the ground vibrational state (the zero-point energy level). To see this, examination of the VPT2 vibrational energy level formulae for asymmetric tops[24, 26, 154, 155, 251, 252] is useful:

$$E(\nu) = \Sigma_{k=1}^{N}\omega_k \left(\nu_k + \frac{1}{2}\right) + \Sigma_{k\leq l}^{N}X_{kl}\left(\left(\nu_k + \frac{1}{2}\right)\left(\nu_l + \frac{1}{2}\right)\right), \quad (7.10)$$

where ω_k is the harmonic frequency of vibrational mode k, ν_k is the vibrational quantum number for mode k, and X_{kl} is the anharmonic constant connecting modes k and l. This is in many ways a reformulation of Eq. (7.4) from earlier. Since the anharmonic constants are generally negative, as one moves to higher quantum numbers ν_k, the vibrational energy levels will become closer together. In a cascade emission spectrum, many of the emitted photons will not originate from the first excited vibrational state, and, therefore, the band position will be somewhat lower in energy than the fundamental. In order to approximate this effect using only scaled harmonic vibrational frequencies, the band positions can be shifted. This is probably most important for the C−H stretches because they have the largest anharmonic correction generally, and they will tend to emit more when the molecule is fairly energetic. That is, in the cascade emission process, the C−H stretching modes will be statistically less populated as the molecule emits more photons and its internal vibrational energy is reduced. In fact, near the end of the cascade emission process, the photons being emitted will almost all be the low energy vibrational frequencies that occur in the far-IR.[248, 253]

The band shapes are generally taken to be either Lorentzian emission profiles or Gaussian emission profiles, and there are arguments in favor of both.[254] The NASA Ames PAHdb has adopted a Lorentzian emission profile for individual PAH molecules as given by

$$\mathcal{P}(\nu) = \frac{1}{\pi} \frac{\frac{1}{2}\Gamma}{(\nu - \nu_i)^2 + (\frac{1}{2}\Gamma)^2}, \tag{7.11}$$

where $\frac{1}{2}\Gamma$ is the full width half max (FWHM) line width and ν_i is the scaled harmonic vibrational frequency of mode k.[248] The FWHM line width is allowed to be frequency dependent where the mid-IR range is given a FWHM of 10–30 cm^{-1}, while vibrational modes in the far-IR (in the 15–20 μm or 667–500 cm^{-1} range) are given FWHM values of between 4–8 cm^{-1}. When using the NASA Ames PAHdb to fit astronomical spectra using many PAH molecules, Gaussian profiles have been found to perform better in this situation.

Relative band intensities are obtained by multiplying the intrinsic band intensity by a blackbody at an average emission temperature or by using a somewhat more sophisticated model involving the specific PAH molecule's heat capacity. In Ref. 248, there are some details of the cascade emission process that are not described well enough to know for sure how the NASA Ames PAHdb handles them. For example, the cascade emission process is seemingly not followed explicitly, but, rather, it adopts a canonical description of the ensemble. Thus, the emission cascade is evaluated through a statistical mechanics approach, but details of this are not given, though some additional details are given in Appendix A of Ref. 255. Another consequence of using scaled harmonic frequencies is that the only vibrational bands with a non-zero IR intensity are the fundamental vibrational frequencies, and, thus, the density of states just ends up being the total number of fundamental vibrational frequencies. This winds up being a significant undercount for a large PAH molecule.

7.5.2. *Fully anharmonic cascade emission IR spectra*

In 2018, Mackie *et al.*[247] showed how to use the data from an anharmonic VPT2 analysis (including harmonic frequencies, anharmonic constants, and coupling terms used in the polyad matrices) to create a fully anharmonic cascade emission IR spectrum of a PAH molecule, including the incorporation of polyad resonance matrices for fundamental vibrational frequencies as well as bands originating from excited vibrational states. This publication built upon their earlier work where VPT2 was applied to the IR absorption spectrum of a PAH molecule at low temperatures and then compared with high-resolution experiments in the C–H stretch region or low-resolution spectra covering most of the IR region (not much below $\sim$500 cm^{-1}).[203, 207, 208, 256, 257] In the earlier works, the importance of including resonance polyads is demonstrated especially in comparing to the high-resolution experiments in the C–H stretch region, where very good agreement was found. In these studies, Mackie *et al.* use a DFT functional basis set approach to compute the QFF, wherein the combined DFT functional and basis set had been shown previously to yield reasonable results for larger molecules.[258, 259] The incorporation of the polyads into the VPT2 analysis follows earlier work, as well,[202, 260] but the method to distribute the IR intensities across the bands included in the polyads according to the eigenfunctions of the polyad matrix is novel.[203]

Modifying the VPT2 analysis for low-temperature IR absorption spectra to compute a cascade emissions spectrum, the exact type of spectra that astronomers observe, poses several new challenges. First, since the PAH molecules would have very high internal energies, *i.e.*, be very hot, temperature-dependent IR spectra for PAH molecules must be computed. This had previously been done for several PAH molecules in the context of IR absorption spectra, and in particular Mackie *et al.* adopt an approach using a Wang–Landau style walk, a Monte Carlo sampling method, that had previously been used in conjunction with QFFs for PAH molecules.[261–263] The first step involves using a Wang–Landau walk in order to estimate the vibrational density of states (DOS) required for a temperature-dependent spectrum. Following the calculation of the DOS, a second Wang–Landau walk is used to compute an energy dependent spectrum. The full details will not be expanded here, but the interested reader is referred to Mackie *et al.*[247] for the details. One point to note, however, is that Mackie *et al.* have to modify the procedure slightly in order to compute the temperature-dependent emission spectra as opposed to the temperature-dependent absorption spectrum. A second point to note is that Mackie *et al.* explicitly follow the cascade emission spectrum, reducing the internal energy in the PAH molecule once a photon is emitted, and allow for a redistribution of this internal energy via IVR using the DOS.

An important innovation introduced by Mackie *et al.* is to diagonalize the polyad matrices not just for the fundamental vibrational frequencies but also for excited vibrational frequencies. They point out that for excited vibrational frequencies, say for example the overtone of a C−H stretch, the states involved in the polyad are similarly excited by one quantum in the C−H stretch of interest. Since the underlying coordinate system used in VPT2 is the normal coordinates, they note that the polyad matrices are closely related. The off-diagonal elements only depend on the difference between the states so these matrix elements do not change, and only the diagonal elements change. Thus, the polyad matrices are re-diagonalized for all necessary states, and, again, the IR intensity for the bands resulting from the polyad are distributed according to the eigenfunctions of the polyad diagonalization. To show the importance of including the polyads in temperature-dependent spectra, Figure 1 from Mackie *et al.*[247] is reproduced here (Figure 7.6) where the blue spectrum has

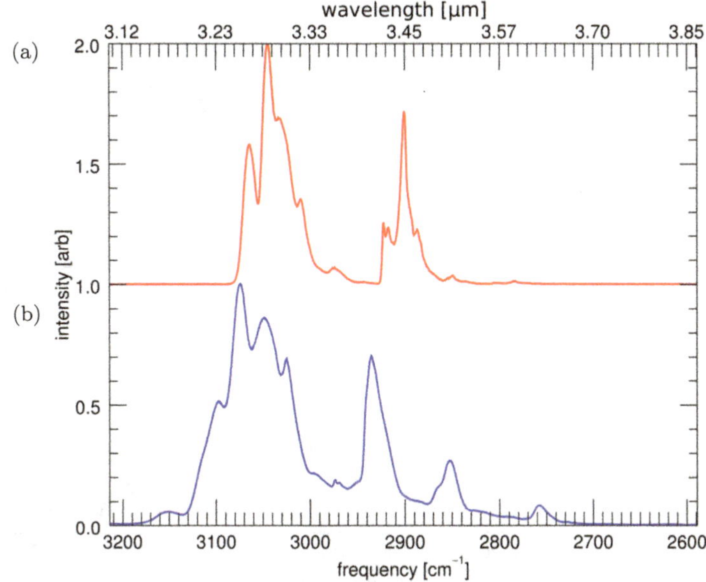

Figure 7.6. The effect of including polyads (blue, (b)) and excluding polyads (red, (a)) from the anharmonic temperature-dependent calculations of 9-methylanthracene using the Wang–Landau approach; from Ref. 247.

included polyads, and the red spectrum is where polyads are excluded for 9-methylanthracene. There is clearly a difference in these spectra, highlighting the importance of including polyads in the temperature-dependent spectra. Further, Figure 2 of Mackie *et al.* (Figure 7.7 here) is reproduced as well, which demonstrates the importance of re-diagonalizing the polyad matrices for excited vibrational states. Spectra from several temperatures are given, and the importance of a particular combination band is shown and becomes more pronounced at higher temperatures.

Several other important phenomena have been investigated in computing the cascade emission spectra of a PAH molecule, such as how a band profile changes with temperature when there are two bands close to each other and how the cascade emission spectrum changes depending on the starting temperature (or energy of the absorbed UV photon). Mackie *et al.* also examine "snap-shot"

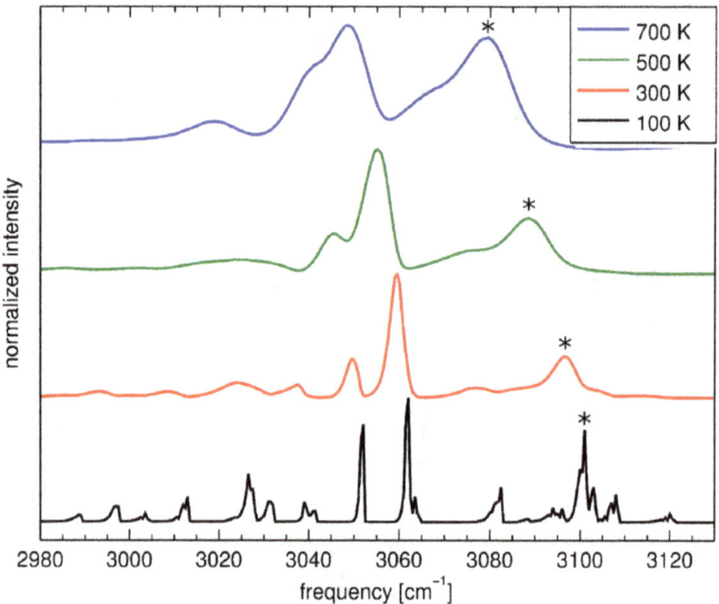

Figure 7.7.   The theoretical infrared spectrum of tetracene showing an increase in the relative intensity of a combination band (marked with an "*") as the internal temperature is increased; from Ref. 247.

emission spectra as opposed to a cascade emission spectrum, and, in a separate study, Chen *et al.*[264] show that an experiment conducted by Wagner *et al.* in 2000[265] is in fact a snap-shot emission spectrum of pyrene as opposed to a cascade emission spectrum. In the 2018 study, Mackie *et al.*[247] do not investigate the use of line shapes, but more recently Mackie *et al.*[266] examine using Gaussian emission profiles together with the inclusion of rotational broadening for the 11.2 μm feature of anthracene, tetracene, and pentacene and how this affects the overall band profile. There is still much work to be done in order to determine the best line shapes to use as well as the development of techniques which will allow QFFs to be computed for larger PAH molecules, and this work continues.

In this section, there are some important aspects that have not yet been addressed. Firstly, it is reasonable to ask whether VPT2 is applicable to highly excited PAH molecules. To see why this is the case, consider that a 12 eV photon will contain about $100{,}000\,\text{cm}^{-1}$ in excess energy. However, the PAH molecules thought to exist in the harsh environments where they are observed by astronomers should contain 40–50 carbon atoms (various estimates exist, but this is on the low end),[247] which means that with the necessary hydrogen atoms to complete the PAH molecule, there will easily be more than 100 vibrational degrees of freedom. This implies that even with $100{,}000\,\text{cm}^{-1}$ in excess vibrational energy, all of the energy could be accommodated with the average ν quantum number less than 1. Considering the density of states, clearly, in most instances, there will not be too much energy (*i.e.*, large ν quantum numbers) in any one vibrational mode or degree of freedom. Hence VPT2 should perform well in these instances provided that the molecules are tightly bound and do not contain large-amplitude motions. Since PAH molecules, often referred to as "chicken-wire" molecules, form a rigid molecular framework, there should not be large-amplitude motions. One exception to this is examining a PAH molecule with an aliphatic chain replacing a hydrogen atom as in 9-methylanthracene. In these cases, Mackie *et al.*[208] just treat the vibrational degrees of freedom which exhibit a large-amplitude motion, such as the hindered rotation of the methyl group, at the harmonic level of theory. In general, however, PAH molecules are exceedingly well suited to be treated with VPT2 provided the large number of resonances are treated properly through polyads.

7.6. Conclusions

Quantum chemistry is uniquely suited to provide novel insights for molecular vibrations of molecules relevant to astronomical studies. Various astrophysical regions exhibit conditions too harsh or otherwise too difficult to replicate in the laboratory. Computational simulation is well-suited to provide data for molecules found in such regions if only the methods employed are accurate enough. This work shows that advanced electronic structure theory computations (like the CcCR composite approach) tied to rather straightforward vibrational methods (such as VPT2) can produce exceptional comparison to experiment. Such methods can then move beyond replication and on to prediction. The unique and novel vibrational, rotational, and rovibrational spectra data for more than 50 molecules have been produced within the past decade with benchmarks for such methods implying errors on the average of $7\,\text{cm}^{-1}$ or less, and often much less. Additionally, development is currently moving beyond simply providing rovibrational spectral data for small molecules. Modern work is stretching electronic structure theory to compute relevant molecular vibration data for electronically excited states, ever larger molecules, highly accurate molecular line lists, and even full infrared cascade emission spectra. All of these studies are geared towards expanding the inventory of attributed astrophysical spectral features in numerous astronomical objects ranging from circumstellar envelopes to photodissociation regions to exoplanetary atmospheres and even to the diffuse interstellar medium. The computational results are now, in many ways, going hand-in-hand with experiment and, in some circumstances, beginning to show signs of superseding experiment. Regardless, the volume of the throughput, granular nature of the results, and definitive descriptions of the simulated spectral features makes quantum chemistry well-placed to be a principle partner in the elucidation of molecular vibrations for molecules of astrochemical interest.

Acknowledgments

RCF is supported by NASA Grant NNX17AH15G, NSF Grant OIA-1757220, and startup funds provided by the University of Mississippi. TJL gratefully acknowledges financial support from the 17-APRA17-0051, 18-APRA18-0013, 18-2XRP18 2-0046, and NASA 20-EW20_2-0144 grants. Megan Davis of the University of Mississippi is acknowledged for providing Figure 7.1. Additionally, the authors would like to thank Dr. David W. Schwenke of the NASA Ames Research Center and Prof. Alexander G. G. M. Tielens of the Leiden Observatory for assistance in editing the manuscript.

References

1. Turner, B. E. U93.174 — A new interstellar line with quadrupole hyperfine splitting. *Astrophys. J.* **1974**, *193*, L83–L84.

2. Green, S.; Montgomery, J. A.; Thaddeus, P. Tentative identification of U93.174 as the molecular Ion N_2H^+. *Astrophys. J.* **1974**, *193*, L89–L91.

3. Tucker, K. D.; Kutner, M. L.; Thaddeus, P. *Astrophys. J.* **1974**, *193*, L115–L119.

4. Guèlin, M.; Green, S.; Thaddeus, P. *Astrophys. J.* **1978**, *224*, L27–L30.

5. Wilson, S.; Green, S. Theoretical study of the butadiynyl and cyanoethynyl radicals — support for the identification of C3N in IRC + 10216. *Astrophys. J.* **1977**, *212*, L87–L90.

6. Buhl, D.; Snyder, L. E. Unidentified interstellar microwave line. *Nature* **1970**, *228*, 267.

7. Herbst, E.; Klemperer, W. Is X-Ogen HCO^+? *Astrophys. J.* **1974**, *188*, 255–256.

8. Nickerson, S.; Rangwala, N.; Colgan, S. W. J.; DeWitt, C.; Huang, X.; Acharyya, K.; Drozdovskaya, M.; Fortenberry, R. C.; Herbst, E.; Lee, T. J. The first mid-infrared detection of HNC in the interstellar medium: Probing the extreme environment toward the orion hot core. *Astrophys. J.* **2021**, *907*, 51.

9. Huang, X.; Taylor, P. R.; Lee, T. J. Highly accurate quartic force field, vibrational frequencies, and spectroscopic constants for cyclic and linear $C_3H_3^+$. *J. Phys. Chem. A* **2011**, *115*, 5005–5016.

10. Zhao, D.; Doney, K. D.; Linnartz, H. Laboratory gas-phase detection of the cyclopropenyl cation (c-$C_3H_3^+$). *Astrophys. J. Lett.* **2014**, *791*, L28.

11. Tennyson, J. Accurate variational calculations for line lists to model the vibration-rotation spectra of hot astrophysical atmospheres. *Wiley Interdis. Rev.: Comput. Mol. Sci.* **2012**, *2*, 698–715.

12. Huang, X.; Schwenke, D. W.; Lee, T. J. A highly accurate potential energy surface and initial IR line list of $^{32}S^{16}O_2$ up to $8000\,cm^{-1}$. *J. Chem. Phys.* **2014**, *140*, 114311.

13. Huang, X.; Schwenke, D. W.; Lee, T. J. Quantitative validation of ames IR intensity and new line lists for $^{32/33/24}S^{18}O_2$, $^{32}S^{18}O_2$ and $^{16}O^{32}S^{18}O$. *J. Quant. Spectrosc. Radiat. Transf.* **2019**, *225*, 327–336.

14. Li, H. Y.; Tennyson, J.; Yurchenko, S. N. ExoMol line lists — XXXII. The rovibronic spectrum of MgO. *Mon. Not. R. Astron. Soc.* **2019**, *486*, 2351–2365.

15. Fortenberry, R. C.; Huang, X.; Schwenke, D. W.; Lee, T. J. Limited rotational and rovibrational line lists computed with highly accurate

quartic force fields and *Ab Initio* dipole surfaces. *Spectrochim. Acta A* **2014**, *119*, 76–83.

16. Cramer, C. J. *Essentials of Computational Chemistry: Theories and Models*, 2nd ed. Wiley: West Sussex, England, 2004.

17. Lee, T. J.; Martin, J. M. L.; Taylor, P. R. An accurate *ab initio* quartic force field and vibrational frequencies for CH_4 and its isotopomers. *J. Chem. Phys.* **1995**, *102*, 254–261.

18. Fortenberry, R. C.; Huang, X.; Yachmenev, A.; Thiel, W.; Lee, T. J. On the use of quartic force fields in variational calculations. *Chem. Phys. Lett.* **2013**, *574*, 1–12.

19. Allen, W. D. *et al. INTDER* 2005 is a general program written by W. D. Allen and coworkers, which performs vibrational analysis and higher-order non-linear transformations, 2005.

20. Barone, V.; Biczysko, M.; Puzzarini, C. Quantum chemistry meets spectroscopy for astrochemistry: Increasing complexity toward prebiotic molecules. *Acc. Chem. Res.* **2015**, *48*, 1413–1422.

21. Puzzarini, C.; Barone, V. Diving for accurate structures in the ocean of molecular systems with the help of spectroscopy and quantum chemistry. *Acc. Chem. Res.* **2018**, *51*, 548–556.

22. Fortenberry, R. C.; Lee, T. J. Computational vibrational spectroscopy for the detection of molecules in space. *Annu. Rep. Comput. Chem.* **2019**, *15*, 173–202.

23. Puzzarini, C.; Barone, V. The challenging playground of astrochemistry: An integrated rotational spectroscopy–quantum chemistry strategy. *Phys. Chem. Chem. Phys.* **2020**, *22*, 6507–6523.

24. Watson, J. K. G. Aspects of Quartic and Sextic Centrifugal Effects on Rotational Energy Levels. In *Vibrational Spectra and Structure*; Durig, J. R., Ed.; Elsevier: Amsterdam, 1977; pp 1–89.

25. Gaw, J. F.; Willets, A.; Green, W. H.; Handy, N. C. SPECTRO: A Program for the Derivation of Spectroscopic Constants from Provided Quartic Force Fields and Cubic Dipole Fields. In *Advances in Molecular Vibrations and Collision Dynamics*; Bowman, J. M., Ratner, M. A., Eds.; JAI Press, Inc.: Greenwich, Connecticut, 1991; pp. 170–185.

26. Watson, J. K. G. Simplification of the molecular vibration-rotation Hamiltonian. *Mol. Phys.* **1968**, *15*, 479–490.

27. Franke, P. R.; Stanton, J. F.; Douberly, G. E. How to VPT2: Accurate and intuitive simulations of CH stretching infrared spectra using VPT2+K with large effective Hamiltonian resonance treatments. *J. Phys. Chem. A* **2021**, *125*, 1301–1324.

28. Szabo, A.; Ostlund, N. S. *Modern Quantum Chemistry: Introduction to Advanced Electronic Structure Theory*. Dover: Mineola, NY, 1996.

29. Møller, C.; Plesset, M. S. Note on an approximation treatment for many-electron systems. *Phys. Rev.* **1934**, *46*, 618–622.

30. Carter, S.; Bowman, J. M.; Handy, N. C. Extensions and tests of "multimodes": A code to obtain accurate vibration/rotation energies of many-mode molecules. *Theor. Chem. Acc.* **1998**, *100*, 191–198.

31. Bowman, J. M.; Carter, S.; Huang, X. MULTIMODE: A code to calculate rovibrational energies of polyatomic molecules. *Int. Rev. Phys. Chem.* **2003**, *22*, 533–549.

32. Dateo, C. E.; Lee, T. J.; Schwenke, D. W. An accurate quartic force-field and vibrational frequencies for HNO and DNO. *J. Chem. Phys.* **1994**, *101*, 5853–5859.

33. Carter, S.; Bowman, J. M. Variational calculations of rotational-vibrational energies of CH_4 and isotopomers using an adjusted ab initio potential. *J. Phys. Chem. A* **2000**, *104*, 2355–2361.

34. Morgan, W. J.; Fortenberry, R. C.; Schaefer III, H. F.; Lee, T. J. Vibrational analysis of the ubiquitous interstellar molecule cyclopropenylidene (c-C_3H_2): The importance of numerical stability. *Mol. Phys.* **2019**, *118*, e1589007.

35. Helgaker, T.; Ruden, T. A.; Jørgensen, P.; Olsen, J.; Klopper, W. *A Priori* calculation of molecular properties to chemical accuracy. *J. Phys. Org. Chem.* **2004**, *17*, 913–933.

36. Raghavachari, K.; Trucks, G. W.; Pople, J. A.; Head-Gordon, M. A fifth-order perturbation comparison of electron correlation theories. *Chem. Phys. Lett.* **1989**, *157*, 479–483.

37. Dunning, T. H. Gaussian basis sets for use in correlated molecular calculations. I. The atoms boron through neon and hydrogen. *J. Chem. Phys.* **1989**, *90*, 1007–1023.

38. Kendall, R. A.; Dunning, T. H.; Harrison, R. J. Electron affinities of the first-row atoms revisited. Systematic basis sets and wave functions. *J. Chem. Phys.* **1992**, *96*, 6796–6806.

39. Lee, T. J.; Scuseria, G. E. The vibrational frequencies of ozone. *J. Chem. Phys.* **1990**, *93*, 489–494.

40. Lee, T. J.; Huang, X.; Dateo, C. E. The effect of approximating some molecular integrals in coupled-cluster calculations: Fundamental frequencies and rovibrational spectroscopic constants for isotopologues of cyclopropenylidene. *Mol. Phys.* **2009**, *107*, 1139–1152.

41. Gardner, M. B.; Westbrook, B. R.; Fortenberry, R. C.; Lee, T. J. Highly-accurate quartic force fields for the prediction of anharmonic rotational constants and fundamental vibrational frequencies. *Spectrochim. Acta A* **2021**, *248*, 119184.

42. Huang, X.; Lee, T. J. A procedure for computing accurate *ab initio* quartic force fields: Application to HO_2^+ and H_2O. *J. Chem. Phys.* **2008**, *129*, 044312.

43. Huang, X.; Lee, T. J. Accurate *ab initio* quartic force fields for NH_2^- and CCH^- and rovibrational spectroscopic constants for their isotopologs. *J. Chem. Phys.* **2009**, *131*, 104301.

44. Fortenberry, R. C.; Francisco, J. S. On the detectability of the $\tilde{X}^2 A''$ HSS, HSO, and HOS radicals in the interstellar medium. *Astrophys. J.* **2017**, *835*, 243.

45. Fuente, A.; Goicoechea, J. R.; Pety, J.; Gal, R. L.; Martín-Doménech, R.; Gratier, P.; Guzmán, V.; Roueff, E.; Loison, J. C.; Caro, G. M. M. *et al.* First detection of interstellar S_2H. *Astrophys. J.* **2017**, *851*, L49.

46. Heikkilä, A.; Johansson, L. E. B.; Olofsson, H. Molecular abundance variations in the magellanic clouds. *Astron. Astrophys.* **1999**, *344*, 817–847.

47. Cheung, A. C.; Rank, D. M.; Townes, C. H.; Thornton, D. D.; Welch, W. J. Detection of water in interstellar regions by its microwave radiation. *Nature* **1969**, *221*, 626.

48. Martin, J. M. L.; Lee, T. J. The atomization energy and proton affinity of NH_3. An *ab initio* calibration study. *Chem. Phys. Lett.* **1996**, *258*, 136–143.

49. Douglas, M.; Kroll, N. M. Quantum electrodynamical corrections to the fine structure of helium. *Ann. Phys.* **1974**, *82*, 89–155.

50. Gdanitz, R. J.; Ahlrichs, R. The averaged coupled-pair functional (ACPF): A size-extensive modification of MR CI(SD). *Chem. Phys. Lett.* **1988**, *143*, 413–420.

51. McCarthy, M. C.; Gottlieb, C. A.; Gupta, H.; Thaddeus, P. Laboratory and astronomical identification of the negative molecular Ion C_6H^-. *Astrophys. J.* **2006**, *652*, L141–L144.

52. Cernicharo, J.; Guèlin, M.; Agùndez, M.; Kawaguchi, K.; McCarthy, M.; Thaddeus, P. Astronomical detection of C_4H^-, the second interstellar anion. *Astron. Astrophys.* **2007**, *467*, L37–L40.

53. Brünken, S.; Gupta, H.; Gottlieb, C. A.; McCarthy, M. C.; Thaddeus, P. Detection of the carbon chain negative ion C_8H^- in TMC-1. *Astrophys. J.* **2007**, *664*, L43–L46.

54. Remijan, A. J.; Hollis, J. M.; Lovas, F. J.; Cordiner, M. A.; Millar, T. J.; Markwick-Kemper, A. J.; Jewell, P. R. Detection of C_8H^- and comparison with C_8H toward IRC+10216. *Astrophys. J.* **2007**, *664*, L47–L50.

55. Fortenberry, R. C. Interstellar anions: The role of quantum chemistry. *J. Phys. Chem. A* **2015**, *119*, 9941–9953.

56. van Dishoeck, E. F.; Jansen, D. J.; Schilke, P.; Phillips, T. G. Detection of the interstellar NH_2 radical. *Astrophys. J.* **1993**, *416*, L83–L86.

57. Persson, C. M.; Hajigholi, M.; Hassel, G. E.; Olofsson, A. O. H.; Black, J. H.; Herbst, E.; Müller, H. S. P.; Cernicharo, J.; Wirström, E. S.; Olberg, M. *et al.* Upper limits to interstellar NH^+ and *para*-NH_2 abundances. Herschel-HIFI observations towards Sgr B2 (M) and G10.6-0.4 (W31C). *Astron. Astrophys.* **2014**, *567*, A130.

58. Allamandola, L. J. PAHs and Astrobiology. In *PAHs and the Universe: A Symposium to Celebrate the 25th Anniversary of the PAH Hypothesis*; Joblin, C., Tielens, A. G. G. M., Eds.; EAS Publication Series: Cambridge, UK, 2011; pp. 305–318.

59. Martin, J. M. L.; Taylor, P. R. Basis set convergence for geometry and harmonic frequencies. Are h functions enough? *Chem. Phys. Lett.* **1994**, *225*, 473–479.

60. Fortenberry, R. C.; Huang, X.; Francisco, J. S.; Crawford, T. D.; Lee, T. J. The *trans*-HOCO radical: Fundamental vibrational frequencies, quartic force fields, and spectroscopic constants. *J. Chem. Phys.* **2011**, *135*, 134301.

61. Francisco, J. S.; Muckerman, J. T.; Yu, H.-G. HOCO radical chemistry. *Acc. Chem. Res.* **2010**, *43*, 1519–1526.

62. Yu, H.; Muckerman, J.; Francisco, J. Direct *ab initio* dynamics study of the OH plus HOCO reaction. *J. Phys. Chem. A* **2005**, *109*, 5230–5236.

63. Yu, H.-G.; Muckerman, J. T.; Francisco, J. S. Quantum force molecular dynamics study of the reaction of O atoms with HOCO. *J. Chem. Phys.* **2007**, *127*, 094302.

64. Sears, T. J.; Fawzy, W. M.; Johnson, P. M. Transient diode-laser absorption-spectroscopy of the ν_2 fundamental of *trans*-HOCO and DOCO. *J. Chem. Phys.* **1992**, *97*, 3996–4007.

65. Petty, J. T.; Moore, C. B. Transient infrared-absorption spectrum of the ν_1 fundamental of *trans*-HOCO. *J. Mol. Spectrosc.* **1993**, *161*, 149–156.

66. Fortenberry, R. C.; Huang, X.; Francisco, J. S.; Crawford, T. D.; Lee, T. J. Vibrational frequencies and spectroscopic constants from quartic force fields for *cis*-HOCO: The radical and the anion. *J. Chem. Phys.* **2011**, *135*, 214303.

67. Fortenberry, R. C. The rovibrational nature of *cis*- and *trans*-HNNS: A possible nitrogen molecule progenitor. *J. Chem. Phys.* **2016**, *145*, 204302.

68. Fortenberry, R. C.; Huang, X.; Francisco, J. S.; Crawford, T. D.; Lee, T. J. Quartic force field predictions of the fundamental vibrational frequencies and spectroscopic constants of the cations $HOCO^+$ and $DOCO^+$. *J. Chem. Phys.* **2012**, *136*, 234309.

69. Fortenberry, R. C. Rovibrational characterization of the proton-bound, noble gas complexes: $ArHNe^+$, $ArHAr^+$, and $NeHNe^+$. *ACS Earth Space Chem.* **2017**, *1*, 60–69.

70. Stephan, C. J.; Fortenberry, R. C. The interstellar formation and spectra of the noble gas, proton-bound $HeHHe^+$, $HeHNe^+$ and $HeHAr^+$ complexes. *Mon. Not. R. Astron. Soc.* **2017**, *469*, 339–346.

71. Fortenberry, R. C.; Huang, X.; Francisco, J. S.; Crawford, T. D.; Lee, T. J. Fundamental vibrational frequencies and spectroscopic constants of $HOCS^+$, $HSCO^+$, and isotopologues via quartic force fields. *J. Phys. Chem. A* **2012**, *116*, 9582–9590.

72. Fortenberry, R. C.; Francisco, J. S. Factors affecting the spectroscopic observation of bridged HPSi and linear HSiP in astrophysical/interstellar media. *Astrophys. J.* **2017**, *843*, 124.

73. Fortenberry, R. C.; Crawford, T. D.; Lee, T. J. The potential interstellar anion CH_2CN^-: Spectroscopic constants, vibrational frequencies, and other considerations. *Astrophys. J.* **2013**, *762*, 121.

74. Fortenberry, R. C.; Thackston, R.; Francisco, J. S.; Lee, T. J. Toward the laboratory identification of the not-so-simple NS_2 neutral and anion isomers. *J. Chem. Phys.* **2017**, *147*, 074303.

75. Huang, X.; Fortenberry, R. C.; Lee, T. J. Spectroscopic constants and vibrational frequencies for l-C_3H^+ and isotopologues from highly-accurate quartic force fields: The detection of l-C_3H^+ in the horsehead nebula PDR questioned. *Astrophys. J. Lett.* **2013**, *768*, 25.

76. Brünken, S.; Kluge, L.; Stoffels, A.; Asvany, O.; Schlemmer, S. Laboratory rotational spectrum of l-C_3H^+ and confirmation of its astronomical detection. *Astrophys. J.* **2014**, *783*, L4.

77. Fortenberry, R. C.; Lee, T. J.; Huang, X. Towards completing the cyclopropanylidene cycle: Rovibrational analysis of cyclic N_3^+, CNN, $HCNN^+$, and CNC^-. *Phys. Chem. Chem. Phys.* **2017**, *19*, 22860–22869.

78. Fortenberry, R. C.; Huang, X.; Crawford, T. D.; Lee, T. J. High-accuracy quartic force field calculations for the spectroscopic constants and vibrational frequencies of $1^1A'$ l-C_3H^-: A possible link to lines observed in the horsehead nebula PDR. *Astrophys. J.* **2013**, *772*, 39.

79. Fortenberry, R. C.; Francisco, J. S.; Lee, T. J. Quantum chemical rovibrational analysis of the HOSO radical. *J. Phys. Chem. A* **2017**, *121*, 8108–8114.

80. Huang, X.; Fortenberry, R. C.; Lee, T. J. Protonated nitrous oxide, $NNOH^+$: Fundamental vibrational frequencies and spectroscopic constants from quartic force fields. *J. Chem. Phys.* **2013**, *139*, 084313.

81. Thackston, R.; Fortenberry, R. C. Quantum chemical spectral characterization of $CH_2NH_2^+$ for remote sensing of Titan's atmosphere. *Icarus* **2018**, *299*, 187–193.

82. Fortenberry, R. C.; Huang, X.; McCarthy, M. C.; Crawford, T. D.; Lee, T. J. Fundamental vibrational frequencies and spectroscopic constants of *cis*- and *trans*-HOCS, HSCO, and isotopologues via quartic force fields. *J. Phys. Chem. B* **2014**, *118*, 6498–6510.

83. Kloska, K. A.; Fortenberry, R. C. Gas-phase spectra of MgO molecules: A possible connection from gas-phase molecules to planet formation. *Mon. Not. R. Astron. Soc.* **2018**, *474*, 2055–2063.

84. Bassett, M. K.; Fortenberry, R. C. Magnesium in the formaldehyde: The theoretical rovibrational analysis of $\tilde{X}\ ^3B_1$ $MgCH_2$. *J. Mol. Spectrosc.* **2018**, *344*, 61–64.

85. Palmer, C. Z.; Fortenberry, R. C. Rovibrational considerations for the monomers and dimers of magnesium hydride (MgH_2) and magnesium fluoride (MgF_2). *J. Phys. Chem. A* **2018**, *122*, 7079–7088.

86. Fortenberry, R. C.; Huang, X.; Crawford, T. D.; Lee, T. J. Quartic force field rovibrational analysis of protonated acetylene, $C_2H_3^+$, and its isotopologues. *J. Phys. Chem. A* **2014**, *118*, 7034–7043.

87. Fortenberry, R. C.; Francisco, J. S. A possible progenitor of the interstellar sulfide bond: Rovibrational characterization of the hydrogen disulfide cation $HSSH^+$. *Astrophys. J.* **2018**, *856*, 30.

88. Fortenberry, R. C.; Trabelsi, T.; Francisco, J. S. Hydrogen sulfide as a scavenger of sulfur atomic cation. *J. Phys. Chem. A* **2018**, *122*, 4983–4987.

89. Fortenberry, R. C.; Huang, X.; Crawford, T. D.; Lee, T. J. Quantum chemical rovibrational data for the interstellar detection of c-C_3H^-. *Astrophys. J.* **2014**, *796*, 139.

90. Fortenberry, R. C.; Novak, C. M.; Lee, T. J. Rovibrational analysis of c-SiC_2H_2: Further evidence for out-of-plane bending issues in correlated methods. *J. Chem. Phys.* **2018**, *149*, 024303.

91. Theis, R. A.; Morgan, W. J.; Fortenberry, R. C. ArH_2^+ and NeH_2^+ as global minima in the $Ar^+/Ne^+ + H_2$ reactions: Energetic, spectroscopic, and structural data. *Mon. Not. R. Astron. Soc.* **2015**, *446*, 195–204.

92. Morgan, W. J.; Huang, X.; Schaefer III, H. F.; Lee, T. J. Astrophysical sulfur in diffuse and dark clouds: The fundamental vibrational frequencies and spectroscopic constants of hydrogen sulfide cation (H_2S^+). *Mon. Not. R. Astron. Soc.* **2018**, *480*, 3483–3490.

93. Theis, R. A.; Fortenberry, R. C. Trihydrogen cation with neon and argon: Structural, energetic, and spectroscopic data from quartic force fields. *J. Phys. Chem. A* **2015**, *119*, 4915–4922.

94. McDonald II, D. C.; Rittgers, B.; Theis, R. A.; Fortenberry, R. C.; Marks, J. H.; Leicht, D.; Duncan, M. A. Infrared spectroscopy and anharmonic theory of $H_3^+Ar_{2,3}$ complexes: The role of symmetry in solvation. *J. Chem. Phys.* **2020**, *153*, 134305.

95. Fortenberry, R. C.; Ascenzi, D. $ArCH_2^+$: A detectable noble gas molecule. *Chem. Phys. Chem.* **2018**, *19*, 1–6.

96. Fortenberry, R. C. The $ArNH_2^+$ noble gas molecule: Stability, vibrational frequencies, and spectroscopic constants. *J. Mol. Spectrosc.* **2019**, *357*, 4–8.

97. Fortenberry, R. C.; Lee, T. J. Rovibrational and energetic analysis of the hydroxyethynyl anion ($CCOH^-$). *Mol. Phys.* **2015**, *113*, 2012–2017.

98. Fortenberry, R. C.; Lee, T. J.; Inostroza-Pino, N. The possibility of $:CNH_2^+$ within Titan's atmosphere: Rovibrational analysis T of $:CNH_2^+$ and $:CCH_2$. *Icarus* **2019**, *321*, 260–265.

99. Fortenberry, R. C.; Yu, Q.; Mancini, J. S.; Bowman, J. M.; Lee, T. J.; Crawford, T. D.; Klemperer, W. F.; Francisco, J. S. Communication: Spectroscopic consequences of proton delocalization in $OCHCO^+$. *J. Chem. Phys.* **2015**, *143*, 071102.

100. Yu, Q.; Bowman, J. M.; Fortenberry, R. C.; Mancini, J. S.; Lee, T. J.; Crawford, T. D.; Klemperer, W.; Francisco, J. S. The structure, anharmonic vibrational frequencies, and intensities of $NNHNN^+$. *J. Phys. Chem. A* **2015**, *119*, 11623–11631.

101. Fortenberry, R. C.; Lee, T. J.; Francisco, J. S. Towards the astronomical detection of the proton-bound complex $NN-HCO^+$: Implications for the spectra of protoplanetary disks. *Astrophys. J.* **2016**, *819*, 141.

102. Fortenberry, R. C.; Lee, T. J.; Francisco, J. S. Quantum chemical analysis of the $CO-HNN^+$ proton-bound complex. *J. Phys. Chem. A* **2016**, *120*, 7745–7752.

103. Del Rio, W. A.; Fortenberry, R. C. Rotational and vibrational fingerprints of the oxywater cation (H_2OO^+), a possible precursor to abiotic O_2. *J. Mol. Spectrosc.* **2019**, *364*, 111183.

104. Fortenberry, R. C.; Francisco, J. S. Quartic force field-derived vibrational frequencies and spectroscopic constants for the isomeric pair SNO and OSN and isotopologues. *J. Chem. Phys.* **2015**, *143*, 084308.

105. Fortenberry, R. C.; Francisco, J. S. Energetics, structure, and rovibrational spectroscopic properties of the sulfurous anions SNO^- and OSN^-. *J. Chem. Phys.* **2015**, *143*, 184301.

106. Dubois, D.; Sciamma-O'Brien, E.; Fortenberry, R. C. The fundamental vibrational frequencies and spectroscopic constants of the dicyanoamine anion, $NCNCN^-$ ($C_2N_3^-$): Quantum chemical analysis for astrophysical and planetary environments. *Astrophys. J.* **2019**, *883*, 109.

107. Fortenberry, R. C.; Lukemire, J. A. Electronic and rovibrational quantum chemical analysis of C_3P^-: The next interstellar anion? *Mon. Not. R. Astron. Soc.* **2015**, *453*, 2824–2829.

108. Trabelsi, T.; Davis, M. C.; Fortenberry, R. C.; Francisco, J. S. Spectroscopic investigation of [Al,N,C,O] refractory molecules. *J. Chem. Phys.* **2019**, *151*, 244303.

109. Fortenberry, R. C.; Thackston, R. Optimal cloud use of quartic force fields: The first purely commercial cloud computing based study for rovibrational analysis of $SiCH^-$. *Int. J. Quant. Chem.* **2015**, *115*, 1650–1657.

110. Bera, P. P.; Huang, X.; Lee, T. J. Highly accurate quartic force field and rovibrational spectroscopic constants for the azirinyl cation (c-$C_2NH_2^+$) and its isomers. *J. Phys. Chem. A* **2020**, *124*, 362–370.

111. Theis, R. A.; Fortenberry, R. C. Potential interstellar noble gas molecules: $ArOH^+$ and $NeOH^+$ rovibrational analysis from quantum chemical quartic force fields. *Mol. Astrophys.* **2016**, *2*, 18–24.

112. Novak, C. M.; Fortenberry, R. C. Theoretical rovibrational analysis of the covalent noble gas compound $ArNH^+$. *J. Mol. Spectrosc.* **2016**, *322*, 29–32.

113. Novak, C. M.; Fortenberry, R. C. The rovibrational spectra of three, stable noble gas molecules: $NeCCH^+$, $ArCCH^+$, and $ArCN^+$. *Phys. Chem. Chem. Phys.* **2017**, *19*, 5230–5238.

114. Fortenberry, R. C.; Gwaltney, S. R. $NeON^+$: An atom *AND* a molecule. *ACS Earth Space Chem.* **2018**, *2*, 491–495.

115. Fortenberry, R. C.; Trabelsi, T.; Francisco, J. S. Anharmonic frequencies and spectroscopic constants of OAlOH and AlOH: Strong bonding but unhindered motion. *J. Phys. Chem. A* **2020**, *124*, 8834–8841.

116. Fortenberry, R. C.; Trabelsi, T.; Francisco, J. S. Theoretical rovibrational characterization of HAlNP: Weak bonding but strong intensities. *J. Mol. Spectrosc.* **2020**, *377*, 111422.

117. Kitchens, M. J. R.; Fortenberry, R. C. The rovibrational nature of closed-shell third-row triatomics: HOX and HXO, X = Si^+, P, S^+, and Cl. *Chem. Phys.* **2016**, *472*, 119–127.

118. Dallas, J. D.; Westbrook, B. R.; Fortenberry, R. C. Anharmonic vibrational frequencies and spectroscopic constants for the detection of ethynol in space. *Frontiers Astron. Space Sci.* **2021**, *7*, 626407.

119. Finney, B.; Fortenberry, R. C.; Francisco, J. S.; Peterson, K. A. A Spectroscopic case for SPSi detection: The third-row in a single molecule. *J. Chem. Phys.* **2016**, *145*, 124311.

120. Fortenberry, R. C.; Francisco, J. S. Anharmonic fundamental vibrational frequencies and spectroscopic constants of the potential HSO_2 radical astromolecule. *J. Chem. Phys.* **2021**, *155*, 114301.

121. Minh, Y. C.; Irvine, W. M.; Ziurys, L. M. Observations of interstellar HOCO$^+$: Abundance enhancements towards the galactic center. *Astrophys. J.* **1988**, *334*, 175–181.

122. Minh, Y. C.; Brewer, M. K.; Irvine, W. M.; Friberg, P.; Johansson, L. E. B. Abundance and chemistry of interstellar HOCO$^+$. *Astron. Astrophys.* **1991**, *244*, 470–476.

123. Pety, J.; Gratier, P.; Guzmán, V.; Roueff, E.; Gerin, M.; Goicoechea, J. R.; Bardeau, S.; Sievers, A.; Petit, F. L.; Bourlot, J. L. et al. The IRAM-30 m line survey of the horsehead PDR II. First detection of the *l*-C_3H^+ hydrocarbon cation. *Astron. Astrophys.* **2012**, *548*, A68.

124. Barlow, M. J.; Swinyard, B. M.; Owen, P. J.; Cernicharo, J.; Gomez, H. L.; Ivison, R. J.; Krause, O.; Lim, T. L.; Matsuura, M.; Miller, S. et al. Detection of a noble gas molecular Ion, $^{36}ArH^+$, in the crab nebula. *Science* **2013**, *342*, 1343–1345.

125. Wagner, J. P.; McDonald II, D. C.; Duncan, M. A. An argon–oxygen covalent bond in the $ArOH^+$ molecular Ion. *Angew. Chem. Int. Ed.* **2018**, *57*, 5081–5085.

126. Wagner, J. P.; Giles, S. M.; Duncan, M. A. Gas phase infrared spectroscopy of the $H_2C{=}NH_2^+$ methaniminium cation. *Chem. Phys. Lett.* **2019**, *726*, 53–56.

127. Thackston, R.; Fortenberry, R. C. The performance of low-cost commercial cloud computing as an alternative in computational chemistry. *J. Comput. Chem.* **2015**, *36*, 926–933.

128. Lee, T. J.; Schaefer III, H. F. The classical and nonclassical forms of protonated acetylene, $C_2H_3^+$: Structures, vibrational frequencies, and infrared intensities from explicitly correlated wave functions. *J. Chem. Phys.* **1986**, *85*, 3437–3443.

129. Adler, T. B.; Knizia, G.; Werner, H.-J. A simple and efficient CCSD(T)-F12 approximation. *J. Chem. Phys.* **2007**, *127*, 221106.

130. Knizia, G.; Adler, T. B.; Werner, H.-J. Simplified CCSD(T)-F12 Methods: Theory and benchmarks. *J. Chem. Phys.* **2009**, *130*, 054104.

131. Huang, X.; Valeev, E. F.; Lee, T. J. Comparison of one-particle basis set extrapolation to explicitly correlated methods for the calculation of accurate quartic force fields, vibrational frequencies, and spectroscopic constants: Application to H_2O, N_2H^+, NO_2^+, and C_2H_2. *J. Chem. Phys.* **2010**, *133*, 244108.

132. Fortenberry, R. C.; Lee, T. J.; Layfield, J. P. Communication: The failure of correlation to describe carbon=carbon bonding in out-of-plane bends. *J. Chem. Phys.* **2017**, *147*, 221101.

133. Fortenberry, R. C.; Novak, C. M.; Layfield, J. P.; Matito, E.; Lee, T. J. Overcoming the failure of correlation for out-of-plane motions in a simple aromatic: Rovibrational quantum chemical analysis of c-C_3H_2. *J. Chem. Theory Comput.* **2018**, *14*, 2155–2164.

134. Lee, T. J.; Fortenberry, R. C. The unsolved issue with out-of-plane bending frequencies for C=C multiply bonded systems. *Spectrochim. Acta A* **2021**, *248*, 119148.

135. Agbaglo, D.; Fortenberry, R. C. The performance of CCSD(T)-F12/aug-cc-pVTZ for the computation of anharmonic fundamental vibrational frequencies. *Int. J. Quantum Chem.* **2019**, *119*, e25899.

136. Agbaglo, D.; Fortenberry, R. C. The performance of explicitly correlated wavefunctions [CCSD(T)-F12b] in the computation of anharmonic vibrational frequencies. *Chem. Phys. Lett.* **2019**, *734*, 136720.

137. Agbaglo, D.; Lee, T. J.; Thackston, R.; Fortenberry, R. C. A small molecule with PAH vibrational properties and a detectable rotational spectrum: c-(C)C_3H_2, cyclopropenylidenyl carbene. *Astrophys. J.* **2019**, *871*, 236.

138. Agbaglo, D.; Fortenberry, R. C. Quantum chemical rovibrational characterization of CH_2ClH^+, a low-energy isomer of ionized chloromethane. *ACS Earth Space Chem.* **2019**, *3*, 1296–1301.

139. Fortenberry, R. C.; Peters, D.; Ferari, B. C.; Bennett, C. J. Rovibrational spectral analysis of CO_3 and C_2O_3: Potential sources for O_2 observed in comet 67P/Churyumov-Gerasimenko. *Astrophys. J. Lett.* **2019**, *886*, L10.

140. Valencia, E. M.; Worth, C. J.; Fortenberry, R. C. Enstatite ($MgSiO_3$) and forsterite (Mg_2SiO_4) monomers and dimers: Highly-detectable infrared and radioastronomical molecular building blocks. *Mon. Not. R. Astron. Soc.* **2020**, *492*, 276–282.

141. Westbrook, B. R.; Rio, W. A. D.; Lee, T. J.; Fortenberry, R. C. Overcoming the out-of-plane bending issue in an aromatic hydrocarbon: The anharmonic vibrational frequencies of c-(CH)$C_3H_2{}^+$. *Phys. Chem. Chem. Phys.* **2020**, *22*, 12951–12958.

142. Inostroza-Pino, N.; Palmer, C. Z.; Lee, T. J.; Fortenberry, R. C. Theoretical rovibrational characterization of the *cis/trans*-HCSH and H_2SC isomers of the known interstellar molecule thioformaldehyde. *J. Mol. Spectrosc.* **2020**, *369*, 111273.

143. Gardner, M. B.; Westbrook, B. R.; Fortenberry, R. C. Anharmonic vibrational frequencies for small clusters of silicon and oxygen: SiO_2, SiO_3, Si_2O_3, & Si_2O_4. *Planet Space Sci.* **2020**, *193*, 105176.

144. Westbrook, B. R.; Valencia, E. M.; Rushing, S. C.; Tschumper, G. S.; Fortenberry, R. C. Anharmonic vibrational frequencies of ammonia borane (BH_3NH_3). *J. Chem. Phys* **2021**, *154*, 041104.

145. Watrous, A. G.; Davis, M. C.; Fortenberry, R. C. Pathways to detection of strongly-bound inorganic species: The vibrational and rotational spectral data of AlH_2OH, HMgOH, AlH_2NH_2, and $HMgNH_2$. *Front. Astron. Space Sci.* **2021**, *8*, 17.

146. Watrous, A. G.; Westrbook, B. R.; Davis, M. C.; Fortenberry, R. C. Vibrational and rotational spectral data for possible interstellar detection of AlH_3OH_2, SiH_3OH, and SiH_3NH_2. *Mon. Not. R. Astron. Soc.* **2021**, *508*, 2613–2619.

147. Westbrook, B. R.; Fortenberry, R. C. Anharmonic frequencies of $(MO)_2$ & related hydrides for M = Mg, Al, Si, P, S, Ca, & Ti and heuristics for predicting anharmonic corrections of inorganic oxides. *J. Phys. Chem. A* **2020**, *124*, 3191–3204.

148. Cernicharo, J.; Guèlin, M.; Agundez, M.; McCarthy, M. C.; Thaddeus, P. Detection of C_5N^- and vibrationally excited C_6H in IRC+10216. *Astrophys. J.* **2008**, *688*, L83–L86.

149. Turner, B. E. Detection of vibrationally excited SiS in IRC+10216. *Astron. Astrophys.* **1987**, *183*, L23–L26.

150. Lacy, J. H.; Evans II, N. J.; Achtermann, J. M.; Bruce, D. E.; Arens, J. F.; Carr, J. S. Discovery of interstellar acetylene. *Astrophys. J.* **1989**, *342*, L43–L46.

151. d'Hendecourt, L. B.; Jourdain de Muizon, M. The discovery of interstellar carbon dioxide. *Astron. Astrophys.* **1989**, *223*, L5–L8.

152. Cami, J.; Bernard-Salas, J.; Peeters, E.; Malek, S. E. Detection of C_{60} and C_{70} in young planetary nebulea. *Science* **2010**, *329*, 1180–1192.

153. McGuire, B. A. Census of interstellar, circumstellar, extragalactic, protoplanetary disk, and exoplanetary molecules. *Astrophys. J. Suppl. Ser.* **2018**, *239*, 17.

154. Mills, I. M. Vibration-Rotation Structure in Asymmetric- and Symmetric-Top Molecules. In *Molecular Spectroscopy — Modern Research*; Rao, K. N., Mathews, C. W., Eds.; Academic Press: New York, 1972; pp. 115–140.

155. Papousek, D.; Aliev, M. R. *Molecular Vibration-Rotation Spectra*. Elsevier: Amsterdam, 1982.

156. Thaddeus, P.; Cummins, S. E.; Linke, R. A. Identification of the SiCC radical toward IRC +10216: The first molecular ring in an astronomical source. *Astrophys. J.* **1984**, *283*, L45–L48.

157. Cernicharo, J.; McCarthy, M. C.; Gottlieb, C. A.; Agúndez, M.; Prieto, L. V.; Baraban, J. H.; Changala, P. B.; Guéin, M.; Kahane, C.; Martin-Drumel, M. A. *et al.* Discovery of SiCSi in IRC+10216: A missing link between gas and dust carriers of Si-C Bonds. *Astrophys. J. Lett.* **2015**, *806*, L3.

158. Fortenberry, R. C.; Lee, T. J.; Müller, H. S. P. Excited vibrational level rotational constants for SiC_2: A sensitive molecular diagnostic for astrophysical conditions. *Mol. Astrophys.* **2015**, *1*, 13–19.

159. Butenhoff, T. J.; Rohlfing, E. A. Laser-induced fluorescence spectroscopy of jet-cooled SiC_2. *J. Chem. Phys.* **1991**, *95*, 1–8.

160. Campbell, E. K.; Holz, M.; Gerlich, D.; Maier, J. P. Laboratory confirmation of C_{60}^+ as the carrier of two diffuse interstellar bands. *Nature* **2015**, *523*, 322–324.

161. Cordinder, M. A.; Linnartz, H.; Cox, N. L. J.; Cami, J.; Najarro, F.; Proffitt, C. R.; Lallement, R.; Ehrenfreund, P.; Foing, B. H.; Gull, T. R. *et al.* Confirming interstellar C_{60}^+ using the *Hubble Space Telescope*. *Astrophys. J. Lett.* **2019**, *875*, L28.

162. Heger, M. L. The spectra of certain class B stars in the regions 5630 Å–6680 Å and 3280 Å–3380 Å. *Lick Observatory Bulletin* **1922**, *10*, 146.

163. Merrill, P. W. Unidentified interstellar lines. *Publ. Astron. Soc. Pac.* **1934**, *46*, 206–207.

164. Merrill, P. W. Stationary lines in the spectrum of the binary star boss 6142. *Astrophys. J.* **1936**, *83*, 126–128.

165. McKellar, A. Evidence for the molecular origin of some hitherto unidentified interstellar lines. *Publ. Astron. Soc. Pac.* **1940**, *52*, 187–192.

166. Douglas, A. E. Origin of diffuse interstellar lines. *Nature* **1977**, *269*, 130–132.

167. Sarre, P. J. The diffuse interstellar bands: A major problem in astronomical spectroscopy. *J. Mol. Spectrosc.* **2006**, *238*, 1–10.

168. McCall, B. J.; Griffin, R. E. On the discovery of the diffuse interstellar bands. *Proc. R. Soc. A* **2013**, *469*, 20120604.

169. Swings, P.; Rosenfield, L. Considerations regarding interstellar molecules. *Astrophys. J.* **1937**, *86*, 483–486.

170. Adams, W. S. Some results with the COUDÉ spectrograph of the Mount Wilson observatory. *Astrophys. J.* **1941**, *93*, 11–23.

171. Douglas, A. E.; Herzberg, G. Note on CH^+ in interstellar space and in the laboratory. *Astrophys. J.* **1941**, *94*, 381.

172. Huggins, W. Preliminary note on the photographic spectrum of comet b 1881. *Proc. R. Soc. Lond.* **1881**, *33*, 1–3.

173. Fowler, A. Investigations relating to the spectra of comets. *Mon. Not. R. Astron. Soc.* **1910**, *70*, 484–496.

174. Swings, P.; Page, T. The spectrum of comet Bester (1947k). *Astrophys. J.* **1950**, *111*, 530–554.

175. Herzberg, G.; Lew, H. Tentative identffication of the H_2O^+ ion in comet kohoutek. *Astron. Astrophys.* **1974**, *31*, 123–124.

176. Pierce, D. M.; A'Hearn, M. F. An analysis of CO production in cometary comae: Contributions from gas-phase phenomena. *Astrophys. J.* **2010**, *718*, 340–347.

177. Fortenberry, R. C.; Pierce, D. M.; Bodewits, D. Knowledge gaps in the emission spectra of oxygen-bearing molecular cations. *Astrophys. J. Suppl. Ser.* **2021**, *256*, 6.

178. Fortenberry, R. C.; Huang, X.; Crawford, T. D.; Lee, T. J. The 1 $^3A'$ HCN and 1 $^3A'$ HCO$^+$ vibrational frequencies and spectroscopic constants from quartic force fields. *J. Phys. Chem. A* **2013**, *117*, 9324–9330.

179. Fortenberry, R. C.; Crawford, T. D.; Lee, T. J. Vibrational frequencies and spectroscopic constants for 1 $^3A'$ HNC and 1 $^3A'$ HOC$^+$ from high-accuracy quartic force fields. *J. Phys. Chem. A* **2013**, *117*, 11339–11345.

180. Inostroza, N.; Huang, X.; Lee, T. J. Accurate *ab initio* Quartic force fields of cyclic and bent HC_2N isomers. *J. Chem. Phys.* **2011**, *135*, 244310.

181. Czekner, J.; Cheung, L. F.; Johnson, E. L.; Fortenberry, R. C.; Wang, L.-S. A high resolution photoelectron imaging and theoretical study of CP$^-$ and C_2P^-. *J. Chem. Phys.* **2018**, *148*, 044301.

182. Morgan, W. J.; Fortenberry, R. C. Additional diffuse functions in basis sets for dipole-bound excited states of anions. *Theor. Chem. Acc.* **2015**, *134*, 47.

183. Stanton, J. F.; Bartlett, R. J. The equation of motion coupled-cluster method — a systematic biorthogonal approach to molecular excitation energies, transition-probabilities, and excited-state properties. *J. Chem. Phys.* **1993**, *98*, 7029–7039.

184. Krylov, A. I. Equation-of-motion coupled cluster methods for open-shell and electronically excited species: The Hitchiker's guide to fock space. *Annu. Rev. Phys. Chem.* **2007**, *59*, 433–463.

185. Fortenberry, R. C.; King, R. A.; Stanton, J. F.; Crawford, T. D. A benchmark study of the vertical electronic spectra of the linear chain radicals C_2H and C_4H. *J. Chem. Phys.* **2010**, *132*, 144303.

186. Christiansen, O.; Koch, H.; Jørgensen, P. Response functions in the CC3 iterative triple excitation model. *J. Chem. Phys.* **1995**, *103*, 7429–7441.

187. Koch, H.; Christiansen, O.; Jørgensen, P.; de Meràs, A. M. S.; Helgaker, T. The CC3 model: An iterative coupled cluster approach including connected triples. *J. Chem. Phys.* **1997**, *106*, 1808–1818.

188. Smith, C. E.; King, R. A.; Crawford, T. D. Coupled cluster excited methods including triple excitations for excited states of radicals. *J. Chem. Phys.* **2005**, *122*, 054110-1–8.

189. Morgan, W. J.; Fortenberry, R. C. Quartic force fields for excited electronic states: Rovibronic reference data for the 1 $^2A'$ and 1 $^2A''$ states of the isoformyl radical, HOC. *Spectrochim. Acta A* **2015**, *135*, 965–972.

190. Morgan, W. J.; Fortenberry, R. C. Theoretical rovibronic treatment of the $\tilde{X}\,^2\Sigma^+$ and $\tilde{A}\,^2\Pi$ states of C_2H & $\tilde{X}\,^1\Sigma^+$ state of C_2H^- from quartic force fields. *J. Phys. Chem. A* **2015**, *119*, 7013–7025.

191. Bassett, M. K.; Fortenberry, R. C. Symmetry breaking and spectral considerations of the surprisingly floppy c-C_3H radical and the related dipole-bound excited state of c-C_3H^-. *J. Chem. Phys.* **2017**, *146*, 224303.

192. Davis, M. C.; Fortenberry, R. C. (T)+EOM quartic force fields for theoretical vibrational spectroscopy of electronically excited states. *J. Chem. Theory Comput.* **2021**, *17*, 4374–4382.

193. McGuire, B. A.; Burkhardt, A. M.; Kalenskii, S.; Shingledecker, C. N.; Remijan, A. J.; Herbst, E.; McCarthy, M. C. Detection of the aromatic molecule benzonitrile (c-C_6H_5CN) in the interstellar medium. *Science* **2018**, *359*, 202–205.

194. McGuire, B. A.; Loomis, R. A.; Burkhardt, A. M.; Lee, K. L. K.; Shingledecker, C. N.; Charnley, S. B.; Cooke, I. R.; Cordiner, M. A.; Herbst, E.; Kalenskii, S. *et al.* Detection of two interstellar polycyclic aromatic hydrocarbons via spectral matched filtering. *Science* **2021**, *371*, 1265.

195. Burkhardt, A. M.; Lee, K. L. K.; Changala, P. B.; Shingledecker, C. N.; Cooke, I. R.; Loomis, R. A.; Wei, H.; Charnley, S. B.; Herbst, E.; McCarthy, M. C. *et al.* Discovery of the pure polycyclic aromatic hydrocarbon indene (c-C_9H_8) with GOTHAM observations of TMC-1. *Astrophys. J. Lett.* **2021**, *913*, L18.

196. Cernicharo, J.; Agúndez, M.; Cabezas, C.; Tercero, B.; Marcelino, N.; Pardo, J. R.; de Vicente, P. Pure hydrocarbon cycles in TMC-1: Discovery of ethynyl cyclopropenylidene, cyclopentadiene, and indene. *Astron. Astrophys.* **2021**, *649*, L15.

197. Cernicharo, J.; Heras, A. M.; Tielens, A. G. G. M.; Pardo, J. R.; Herpin, F.; Guélin, M.; Waters, L. B. F. M. Infrared space observatory's discovery of C_4H_2, C_6H_2, and benzene in CRL 618. *Astrophys. J.* **2001**, *546*, L123–L126.

198. Langhoff, S. R. Theoretical infrared spectra for polycyclic aromatic hydrocarbon neutrals, cations, and anions. *J. Phys. Chem.* **1996**, *100*, 2819–2841.

199. Bauschlicher, Jr., C. W.; Ricca, A. The infrared spectra of polycylic aromatic hydrocarbons with some or all hydrogen atoms removed. *Astrophys. J.* **2013**, *776*, 102.

200. Boersma, C.; Bauschlicher, Jr., C. W.; Ricca, A.; Mattioda, A. L.; Cami, J.; Peeters, E.; de Armas, F. S.; Saborido, G. P.; Hudgins, D. M.; Allamandola, L. J. The NASA Ames PAH IR spectroscopic database version 2.00: Updated content, web site, and on(off)line tools. *Astrophys. J. Suppl. Ser.* **2014**, *211*, 8.

201. Ricca, A.; Bauschlicher, Jr., C. W.; Boersma, C.; Tielens, A. G. G. M.; Allamandola, L. J. The infrared spectroscopy of compact polycyclic aromatic hydrocarbons containing up to 384 carbons. *Astrophys. J.* **2012**, *754*, 75.

202. Martin, J. M. L.; Taylor, P. R. Accurate *ab Initio* quartic force field for *trans*-HNNH and treatment of resonance polyads. *Spectrochim. Acta A* **1997**, *53*, 1039–1050.

203. Mackie, C. J.; Candian, A.; Maltseva, E.; Petrignani, A.; Oomens, J.; Lee, W. J. B. T. J.; Tielens, A. G. G. M. The anharmonic quartic force field infrared spectra of three polycyclic aromatic hydrocarbons: Naphthalene, anthracene, and tetracene. *J. Chem. Phys.* **2015**, *143*, 224314.

204. Maltseva, E.; Petrignani, A.; Candian, A.; Mackie, C. J.; Huang, X.; Lee, T. J.; Tielens, A. G. G. M.; Oomens, J.; Buma, W. J. High-resolution IR absorption spectroscopy of polycyclic aromatic hydrocarbons: The realm of anharmonicity. *Astrophys. J.* **2015**, *814*, 23.

205. Hamprecht, F. A.; Cohen, A. J.; Tozer, D. J.; Handy, N. C. Development and assessment of new exchange-correlation functionals. *J. Chem. Phys.* **1998**, *109*, 6264–6271.

206. Frisch, M. J.; Trucks, G. W.; Schlegel, H. B.; Scuseria, G. E.; Robb, M. A.; Cheeseman, J. R.; Scalmani, G.; Barone, V.; Petersson, G. A.; Nakatsuji, H. *et al.* Gaussian 16 Revision C.01. 2016; Gaussian Inc. Wallingford CT.

207. Mackie, C. J.; Candian, A.; Huang, X.; Maltseva, E.; Petrignani, A.; Oomens, J.; Mattioda, A. L.; Buma, W. J.; Lee, T. J.; Tielens, A. G. G. M. The anharmonic quartic force field infrared spectra of five non-linear polycyclic aromatic hydrocarbons: Benz[a]anthracene, chrysene, phenanthrene, pyrene, and triphenylene. *J. Chem. Phys.* **2016**, *145*, 084313.

208. Mackie, C. J.; Candian, A.; Huang, X.; Maltseva, E.; Petrignani, A.; Oomens, J.; Buma, W. J.; Lee, T. J.; Tielens, A. G. G. M. The anharmonic quartic force field infrared spectra of hydrogenated and methylated PAHs. *Phys. Chem. Chem. Phys.* **2018**, *20*, 1189–1197.

209. Becke, A. D. Density-functional thermochemistry. III. The role of exact exchange. *J. Chem. Phys.* **1993**, *98*, 5648–5652.

210. Yang, W. T.; Parr, R. G.; Lee, C. T. Various functionals for the kinetic energy density of an atom or molecule. *Phys. Rev. A* **1986**, *34*, 4586–4590.

211. Lee, C.; Yang, W. T.; Parr, R. G. Development of the Colle-Salvetti correlation-energy formula into a functional of the electron density. *Phys. Rev. B* **1988**, *37*, 785–789.

212. Barone, V.; Cimino, P.; Stendardo, E. Development and validation of the B3LYP/N07D computational model for structural parameter and magnetic tensors of large free radicals. *J. Chem. Theory Comput.* **2008**, *4*, 751–764.

213. Peeters, E.; Allamandola, L. J.; Hudgins, D. M.; Hony, S.; Tielens, A. G. G. M. The Unidentified InfraRed Features after ISO. In *Astrophysics of Dust, ASP Conference Series*, Vol. 309; Witt, A. N., Clayton, G. C., Draine, B. T., Eds.; Astronomical Society of the Pacific: San Francisco, CA, 2004.

214. Peeters, E.; Mackie, C.; Candian, A.; Tielens, A. G. G. M. A spectroscopic view on cosmic PAH emission. *Acc. Chem. Res.* **2021**, *54*, 1921–1933.

215. Huang, X.; Schwenke, D. W.; Lee, T. J. What it takes to compute highly accurate rovibrational line lists for use in astrochemistry. *Acc. Chem. Res.* **2021**, *54*, 1311–1321.

216. Tennyson, J.; Zobov, N. F.; Williamson, R.; Polyansky, O. L. Experimental energy levels of the water molecule. *J. Phys. Chem. Ref. Data* **2001**, *30*, 735.

217. Rey, M.; Nikitin, A. V.; Babikov, Y. L.; Tyuterev, V. G. TheoReTS — An information system for theoretical spectra based on variational predictions from molecular potential energy and dipole moment surfaces. *J. Mol. Spectrosc.* **2016**, *327*, 138–158.

218. Partridge, H.; Schwenke, D. W. The determination of an accurate isotope dependent potential energy surface for water from extensive *ab initio* calculations and experimental data. *J. Chem. Phys.* **1997**, *106*, 4618–4639.

219. Fortney, J. J.; Robinson, T. D.; Domagal-Goldman, S.; Genio, A. D. D.; Gordon, I. E.; Gharib-Nezhad, E.; Lewis, N.; Sousa-Silva, C.; Airapetian, V.; Drouin, B. *et al.* The need for laboratory measurements and ab initio studies to aid understanding of exoplanetary atmospheres. 2019; http://arxiv.org/abs/1905.07064.

220. Huang, X.; Schwenke, D. W.; Tashkun, S. A.; Lee, T. J. An isotopic-independent highly accurate potential energy surface for CO_2 isotopologues and *ab initio* $^{12}C^{16}O_2$ infrared line list. *J. Chem. Phys.* **2012**, *136*, 124311.

221. Huang, X.; Freedman, R. S.; Tashkun, S. A.; Schwenke, D. W.; Lee, T. J. Semi-empirical $^{12}C^{16}O_2$ IR line lists for simulations up to 1500 K and 20,000 cm^{-1}. *J. Quant. Spectrosc. Radiat. Transf.* **2013**, *130*, 134–146.

222. Huang, X.; Gamache, R. R.; Freedman, R. S.; Schwenke, D. W.; Lee, T. J. Reliable infrared line lists for 13 CO_2 isotopologues up to E'= 18,000 cm^{-1} and 1500 K, with line shape parameters. *J. Quant. Spectrosc. Radiat. Transf.* **2014**, *147*, 134–144.

223. Huang, X.; Schwenke, D. W.; Freedman, R. S.; Lee, T. J. Ames-2016 line lists for 13 isotopologues of CO_2: Updates, consistency, and remaining issues. *J. Quant. Spectrosc. Radiat. Transf.* **2017**, *203*, 224–241.

224. Huang, X.; Schwenke, D. W.; Lee, T. J. Isotopologue consistency of semi-empirically computed infrared line lists and further improvement for rare isotopologues: CO_2 and SO_2 case studies. *J. Quant. Spectrosc. Radiat. Transf.* **2019**, *230*, 222–246.

225. Gamache, R. R.; Roller, C.; Lopes, E.; Gordon, I. E.; Rothman, L. S.; Polyansky, O. L.; Zobov, N. F.; Kyuberis, A. A.; Tennyson, J.; Yurchenko, S. N. *et al.* Total internal partition sums for 166 isotopologues of 51 molecules important in planetary atmospheres: Application to HITRAN2016 and beyond. *J. Quant. Spectrosc. Radiat. Transf.* **2017**, *203*, 70–87.

226. Tarczay, G.; Császár, A. G.; Klopper, W.; Szalay, V.; Allen, W. D.; Schaefer, III, H. F. The barrier to linearity of water. *J. Chem. Phys.* **1999**, *110*, 11971–11981.

227. Polyansky, O. L.; Zobov, N. F.; Mizus, I. I.; Lodi, L.; Yurchenko, S. N.; Tennyson, J.; Császár, A. G.; Boyarkin, O. V. Global spectroscopy of the water monomer. *Phil. Trans. R. Soc. A* **2012**, *370*, 2728–2748.

228. Cheung, A. C.; Rank, D. M.; Townes, C. H.; Thornton, D. D.; Welch, W. J. Detection of NH_3 molecules in the interstellar medium by their microwave emission. *Phys. Rev. Lett.* **1968**, *21*, 1701–1705.

229. V. N. Salinas, M. R. H.; Bergin, E. A.; Cleeves, L. I.; Brinch, C.; Blake, G. A.; Lis, D. C.; Melnick, G. J.; Panić, O.; Pearson, J. C.; Kristensen, L. *et al.* First detection of gas-phase ammonia in a planet-forming disk. NH_3, N_2H^+, and H_2O in the disk around TW hydrae. *Astron. Astrophys.* **2016**, *591*, 122.

230. Fink, U.; Larson, H. P.; Bjoraker, G. L.; Johnson, J. R. The NH_3 spectrum in saturn's 5 micron window. *Astrophys. J.* **1983**, *268*, 880–888.

231. Huang, X.; Schwenke, D. W.; Lee, T. J. Rovibrational spectra of ammonia. I. unprecedented accuracy of a potential energy surface

used with nonadiabatic corrections. *J. Chem. Phys.* **2011**, *134*, 044320.

232. Huang, X.; Schwenke, D. W.; Lee, T. J. An accurate global potential energy surface, dipole moment surface, and rovibrational frequencies for NH_3. *J. Chem. Phys.* **2008**, *129*, 214304.

233. Müller, H. S.; Schlöder, F.; Stutzki, J.; Winnewisser, G. The cologne database for molecular spectroscopy, CDMS: A useful tool for astronomers and spectroscopists. *J. Mol. Struct.* **2005**, *742*, 215–227.

234. Huang, X.; Schwenke, D. W.; Lee, T. J. Rovibrational spectra of ammonia. II. Detailed analysis, comparison, and prediction of spectroscopic assignments for $^{14}NH_3$, $^{15}NH_3$, and $^{14}ND_3$. *J. Chem. Phys.* **2011**, *134*, 044321.

235. Sung, K.; Brown, L. R.; Huang, X.; Schwenke, D. W.; Lee, T. J.; Coy, S. L.; Lehmann, K. K. Extended line positions, intensities, empirical lower state energies and quantum assignments of NH_3 from 6300 to $7000\,cm^{-1}$. *J. Quant. Spectrosc. Radiat. Transf.* **2012**, *113*, 1066–1083.

236. Huang, X.; Schwenke, D. W.; Lee, T. J. Ames $^{32}S^{16}O^{18}O$ Line list for high-resolution experimental IR analysis. *J. Mol. Spectrosc.* **2016**, *330*, 101–111.

237. Stewart, A. I.; Anderson, D. E.; Esposito, L. W.; Barth, C. A. Ultraviolet spectroscopy of venus: Initial results from the pioneer venus orbiter. *Science* **1979**, *203*, 777–779.

238. Hintze, P. E.; Kjaergaard, H. G.; Vaida, V.; Burkholder, J. B. Vibrational and electronic spectroscopy of sulfuric acid vapor. *J. Phys. Chem. A* **2003**, *107*, 1112–1118.

239. Snyder, L. E.; Hollis, J. M.; Ulich, B. L.; Lovas, F. J.; Johnson, D. R.; Buhl, D. Radio detection of interstellar sulfur dioxide. *Astrophys. J.* **1975**, *198*, L81–L84.

240. Huang, X.; Schwenke, D. W.; Lee, T. J. Empirical infrared line lists for five SO_2 isotopologues: $^{32/33/34/36}S^{16}O_2$ and $^{32}S^{18}O_2$. *J. Mol. Spectrosc.* **2015**, *311*, 19–24.

241. Underwood, D. S.; Tennyson, J.; Yurchenko, S. N.; Huang, X.; Schwenke, D. W.; Lee, T. J.; Clausen, S.; Fateev, A. ExoMol molecular line lists — XIV. The rotation-vibration spectrum of hot SO_2. *Mon. Not. R. Astron. Soc.* **2016**, *459*, 3890–3899.

242. Huang, X.; Schwenke, D. W.; Lee, T. J. Exploring the limits of the data-model-theory synergy: "Hot" MW transitions for rovibrational IR studies. *J. Mol. Spectrosc.* **2020**, *1217*, 128260.

243. Sousa-Silva, C.; Seager, S.; Ranjan, S.; Petkowski, J. J.; Zhan, Z.; Hu, R.; Bains, W. Phosphine as a biosignature gas in exoplanet atmospheres. *Astrobiology* **2020**, *20*, 235–268.

244. Tielens, A. G. G. M. *Molecular Astrophysics.* Cambridge University Press: Cambridge, UK, 2021.

245. Allamandola, L. J.; Tielens, A. G. G. M.; Baker, J. R. Interstellar polycyclic aromatic hydrocarbons: The infrared emission bands, the excitation/emission mechanism, and the astrophysical implications. *Astrophys. J. Suppl. Ser.* **1989**, *71*, 733–775.

246. Puget, J. L.; Leger, A. A new component of the interstellar matter: Small grains and large aromatic molecules. *Annu. Rev. Astron. Astrophys.* **1989**, *27*, 161–198.

247. Mackie, C. J.; Chen, T.; Candian, A.; Lee, T. J.; Tielens, A. G. G. M. Fully anharmonic infrared cascade spectra of polycyclic aromatic hydrocarbons. *J. Chem. Phys.* **2018**, *149*, 134302.

248. Bauschlicher, Jr., C. W.; Boersma, C.; Ricca, A.; Mattioda, A. L.; Cami, J.; Peeters, E.; Sánchez de Armas, F.; Puerta Sabroido, G.; Hudgins, D. M.; Allamandola, L. J. The NASA ames polycyclic aromatic hydrocarbon infrared spectroscopic database: The computed spectra. *Astrophys. J. Suppl. Ser.* **2010**, *189*, 341–351.

249. Bauschlicher, Jr., C. W.; Ricca, A.; Boersma, C.; Allamandola, L. J. The NASA ames PAH IR spectroscopic database: Computational version 3.00 with updated content and the introduction of multiple scaling factors. *Astrophys. J. Suppl. Ser.* **2018**, *234*, 32.

250. Mattioda, A. L.; Hudgins, D. M.; Boersma, C.; Bauschlicher, Jr., C. W.; Ricca, A.; Cami, J.; Peeters, E.; Sánchez de Armas, F.; Puerta Saborido, G.; Allamandola, L. J. The NASA ames PAH IR spectroscopic database: The laboratory spectra. *Astrophys. J. Suppl. Ser.* **2020**, *251*, 22.

251. Hoy, A. R.; Mills, I. M.; Strey, G. Anharmonic force constant calculations. *Mol. Phys.* **1972**, *24*, 1265–1290.

252. Califano, S. *Vibrational States.* Wiley: London, 1976.

253. Boersma, C.; Bauschlicher, Jr., C. W.; Ricca, A.; Mattioda, A. L.; Peeters, E.; Tielens, A. G. G. M.; Allamandola, L. J. Polycyclic aromatic hydrocarbon far-infrared spectroscopy. *Astrophys. J.* **2011**, *729*, 64.

254. Tielens, A. G. G. M. Interstellar polycyclic aromatic hydrocarbon molecules. *Annu. Rev. Astron. Astrophys.* **2008**, *46*, 289–337.

255. Maragkoudakis, A.; Peeters, E.; Ricca, A. Probing the size and charge of polycyclic aromatic hydrocarbons. *Mon. Not. R. Astron. Soc.* **2020**, *494*, 642–664.

256. Maltseva, E.; Petrignani, A.; Candian, A.; Mackie, C. J.; Huang, X.; Lee, T. J.; Tielens, A. G. G. M.; Oomens, J.; Buma, W. J. High-resolution IR absorption spectroscopy of polycyclic aromatic

hydrocarbons in the $3\,\mu$m region: Role of hydrogenation and alkylation. *Astrophys. J.* **2016**, *831*, 58.

257. Maltseva, E.; Mackie, C. J.; Candian, A.; Petrignani, A.; Huang, X.; Lee, T. J.; Tielens, A. G. G. M.; Oomens, J.; Buma, W. J. High-resolution IR absorption spectroscopy of polycyclic aromatic hydrocarbons in the $3\,\mu$m region: Role of periphery. *Astron. Astrophys.* **2018**, *610*, A65.

258. Boese, A. D.; Martin, J. M. L. Vibrational spectra of the azabenzenes revisited: Anharmonic force fields. *J. Phys. Chem. A* **2004**, *108*, 3085–3096.

259. Barone, V.; Cimino, P.; Stendardo, E. Development and validation of the B3LYP/N07D computational model for structural parameter and magnetic tensors of large free radicals. *J. Chem. Theory Comput.* **2008**, *4*, 751–764.

260. Martin, J. M. L.; Lee, T. J.; Taylor, P. R.; François, J.-P. The anharmonic force field of ethylene, C_2H_4, by means of accurate *ab initio* calculations. *J. Chem. Phys.* **1995**, *103*, 2589–2602.

261. Basire, M.; Parneix, P.; Calvo, F.; Pino, T.; Bréchignac, P. Temperature and anharmonic effects on the infrared absorption spectrum from a quantum statistical approach: Application to naphthalene. *J. Phys. Chem. A* **2009**, *113*, 6947–6954.

262. Basire, M.; Parneix, P.; Calvo, F. Quantum anharmonic densities of states using the Wang–Landau method. *J. Chem. Phys.* **2011**, *129*, 081101.

263. Calvo, F.; Basire, M.; Parneix, P. Temperature effects on the rovibrational spectra of pyrene-based PAHs. *J. Phys. Chem. A* **2011**, *115*, 8845–8854.

264. Chen, T.; Mackie, C. J.; Candian, A.; Lee, T. J.; Tielens, A. G. G. M. Anharmonicity and the IR emission spectrum of highly excited PAHs. *Astron. Astrophys.* **2018**, *618*, A49.

265. Wagner, D.; Kim, H.; Saykally, R. Peripherally hydrogenated neutral polycyclic aromatic hydrocarbons as carriers of the 3 micron interstellar infrared emission complex: Results from single-photon infrared emission spectroscopy. *Astrophys. J.* **2000**, *545*, 854–860.

266. Mackie, C. J.; Candian, A.; Lee, T. J.; Tielens, A. G. G. M. Modelling the infrared cascade spectra of anharmonic interstellar PAHs: The $11.2\,\mu$m band. *Theor. Chem. Acc.* **2021**, *140*, 124.

Chapter 8

MULTIMODE, The n-Mode Representation of the Potential and Illustrations to IR Spectra of Glycine and Two Protonated Water Clusters

Qi Yu[*,**], Chen Qu[†,††], Paul L. Houston[‡,§], Riccardo Conte[¶],
Apurba Nandi[∥], and Joel M. Bowman[∥,‡‡]

*Department of Chemistry, Yale University,
New Haven, Connecticut, 06520, USA
†Department of Chemistry & Biochemistry,
University of Maryland, College Park, Maryland 20742, USA
‡Department of Chemistry and Chemical Biology,
Cornell University, Ithaca, New York 14853, USA
§Department of Chemistry and Biochemistry,
Georgia Institute of Technology, Atlanta, Georgia 30332, USA
¶Dipartimento di Chimica, Università degli Studi di Milano,
via Golgi 19, 20133 Milano, Italy
∥Department of Chemistry and Cherry L. Emerson Center
for Scientific Computation, Emory University,
Atlanta, Georgia 30322, USA
**q.yu@yale.edu
††szquchen@gmail.com
‡‡jmbowma@emory.edu

8.1. Introduction

This chapter describes rigorous quantum approaches to vibrational dynamics, with an emphasis on infrared (IR) spectroscopy, of molecules and also molecular clusters. This is a very active area of research, as attested to by recent themed issues of *Phys. Chem.*

Chem. Phys.[1] and *J. Phys. Chem. A.*[2] and a perspective that two of us wrote in 2019.[3] In this chapter, the focus is on the challenges to theory to provide accurate quantum simulations of these vibrational dynamics as well as useful interpretations of the simulations.

Although the study of vibrational and rovibrational dynamics is pervasive in chemical physics, it is fair to assert that IR spectroscopy is the major area where theory meets experiment. High-resolution rovibrational spectroscopy is of course an important part of this field; however, here we consider medium-resolution IR spectroscopy (*i.e.*, not rotationally resolved). As such, the focuses are band positions, intensities, bandwidths and band sub-structures. Challenges to theory to reproduce such features range from "routine" to very difficult. Sharp bands usually indicate that a single vibrationally excited eigenstate is dominantly responsible for the band. Often, a standard double-harmonic analysis is sufficient to assign the band; however, to improve quantitative accuracy, vibrational perturbation theory is now typically employed. Sharp bands separated by a few wavenumbers may indicate a relatively simple resonance interaction, *e.g.*, Fermi and/or Darling–Dennison resonances. These are usually successfully treated by degenerate perturbation theory.[4] The analysis of a complex band, that is a broad band with sub-structure, is more challenging both qualitatively and quantitatively. In this case, a vibrational configuration interaction (VCI) or vibrational coupled cluster (VCC) approach is almost certainly needed.

These approaches, starting with the harmonic one and ending with the VCI approach (which is "exact", in principle), provide the critical assignment and interpretation of the spectral bands. While the harmonic approach provides the simplest interpretation, it may not be correct for bands that are inherently not captured by this approach, *e.g.*, combination and overtone bands, *etc.* A more rigorous approach may provide acceptable agreement for a complex experimental band, but then the matter of interpretation may also become complex.

The most rigorous, and thus computationally intensive, methods are the VCI and VCC ones. These methods are described in this book. These and many other methods work with given potential energy surfaces (PESs). The form of the PES is often an important aspect of the computational method used in VCI and VCC theories. So-called global PESs can describe large-amplitude motion, perhaps across multiple minima and even dissociation to ones that are more restrictive to describing molecular motion about a minimum on the PES. Our focus is on PESs that describe large-amplitude motion of molecules with 10 or more atoms. However, we use the Watson Hamiltonian, which is a general Hamiltonian in mass-scaled rectilinear normal modes. The trade-off in using this Hamiltonian is the use of rectilinear normal modes, which are not universally optimum for the broad class of molecular vibrations. They are generally better for high-frequency stretch modes than low-frequency, large-amplitude ones, which often are better described using curvilinear coordinates.

To go significantly beyond 10-atom systems and still use a VCI approach requires some major modifications of a straightforward increase in the size of the basis, owing to the well-known "curse of dimensionality", which refers to the highly non-linear increase in the size of the basis.

The chapter is organized as follows. In the next section, we give a brief derivation of normal coordinates and zero-order Hamiltonians, including a brief digression on the Eckart frame in the context of a semi-classical method termed Adiabatic Switching. Then we introduce the VCI method and relate it to perturbation theory and vibrational self-consistent field (VSCF) theory. The Watson Hamiltonian is then introduced along with the n-mode representation of the general potential and the interaction between this representation and the VCI excitation space.

Finally, two examples of VSCF/VCI calculations are given, namely VSCF/VCI calculations of IR spectra of glycine and protonated water clusters, $H_7O_3^+$ and $H_9O_4^+$. Comparisons with experiment are done for both. These examples make use of high-dimensional *ab initio* potential energy surfaces and comments about these are also given.

8.2. Normal Modes and Zero-Order Hamiltonians

The simplest and most general model of molecular vibrations is uncoupled harmonic oscillators. But where does this model come from? The rigorous answer is normal mode analysis, which is briefly reviewed below and is done from a purely classical perspective.

For N atoms, there are $3N$ Cartesian coordinates, $(x_1, y_1, z_1, \ldots, x_N, y_N, z_N)$. Next, multiply the coordinates of the ith atom by $\sqrt{m_i}$, where m_i is mass of atom i. There are a total of $3N$ of these new coordinates, and we denote them all by the column vector q of length $3N$. The classical Hamiltonian of these coordinates is given by

$$H = \frac{1}{2}p^\top p + V(q), \tag{8.1}$$

where the momentum vector p is just the derivative of q with respect to time. Next, expand the potential about a stationary point, *i.e.*, where $\nabla V = 0$, to second order. In this case, H is given by

$$H = \frac{1}{2}p^\top p + \frac{1}{2}q^\top F q, \tag{8.2}$$

where F is the matrix of second partial derivatives of V (in the mass-scaled coordinates) evaluated at the stationary point. Next, introduce an orthogonal transformation between q and a new set of coordinates Q, namely $q = CQ$. Owing to a key property of an orthogonal transformation, *i.e.*, $C^{-1} = C^\top$, we note that $p^\top p = P^\top P$ and F is transformed to $C^\top F C$. There exists a matrix C such that this transformed matrix is diagonal. With this transformation, the above Hamiltonian is given by

$$H = \frac{1}{2}P^\top P + \frac{1}{2}Q^\top \Lambda Q, \tag{8.3}$$

where Λ is a diagonal matrix of the eigenvalues of F, and C is the matrix of eigenvectors. Q are the mass-scaled normal coordinates. Note that six eigenvalues are zero, as these correspond to the three translational and three rotational modes. The former are rigorously decoupled from the rotational and vibrational normal coordinates. However, the rotational modes are decoupled from vibrational modes only for infinitesimal displacements in the rotational normal modes.

Thus, there exists a rotation–vibration coupling for these modes, and indeed for any set of internal modes. Ignoring this coupling is typically done for zero-order classical and quantum Hamiltonians. Finally, we note that for the remaining $3N - 6$ eigenvalues, the ith harmonic frequency, ω_i, is simply given by the $\sqrt{\Lambda_i}$.

The exact quantum version of the exact classical Hamiltonian given by Eq. (8.3) with the harmonic potential replaced by the exact one (in normal coordinates) is known as the Watson Hamiltonian and is given in the next section.

To continue we give the zero-order quantum version of the classical Hamiltonian given by Eq. (8.3). This Hamiltonian, denoted H_0, is just in the vibrational normal coordinates and neglect vibration–rotation coupling. This Hamiltonian is given by a separable sum

$$H_0 = \sum_{i=1}^{3N-6} h_i. \tag{8.4}$$

Classically, h_i is given by

$$h_i = \frac{P_i^2}{2} + \frac{1}{2}\omega_i^2 Q_i^2 \tag{8.5}$$

and quantum mechanically h_i is given by

$$h_i = -\frac{1}{2}\frac{\partial^2}{\partial Q_i^2} + \frac{1}{2}\omega_i^2 Q_i^2. \tag{8.6}$$

Quantum mechanically and semi-classically, the zero-order eigenvalues are identical and given by the familiar expression

$$E_0(n_1, \ldots, n_F) = \sum_{i=1}^{F} \hbar\omega_i(n_i + 1/2), \tag{8.7}$$

where now $F = 3N - 6$ and the zero-order eigenfunctions are given by $\prod_i \phi_{n_i}^{(0)}(Q_i)$, where each $\phi_{n_i}^{(0)}(Q_i)$ is an eigenfunction of h_i.

To provide some insight into vibration–rotation coupling, which is ignored in these Hamiltonians, we make a brief pedagogical digression to describe a semi-classical method, called Adiabatic Switching (AS), to obtain vibrational energies.

8.2.1. *Brief digression on adiabatic switching and the Eckart conditions*

AS is a method to evolve a zero-order Hamiltonian to the "exact one" and thereby obtain the exact energy eigenvalues of that Hamiltonian. The assumption in AS is that the semi-classical quantum numbers (actions) n_i in Eq. (8.7) do not change during this evolution; however, the energy does and evolves to the exact semi-classical energy.

Specifically, in the AS method, a time-dependent Hamiltonian is constructed as

$$H(t) = H_0 + s(t)(H - H_0), \qquad (8.8)$$

where $s(t)$ is a switching function that slowly turns on the remaining part of the Hamiltonian, $H - H_0$. In brief, at $t = 0$ the classical vibrational normal coordinates and momenta are given by standard expressions for each mode with assigned values of the quantum numbers (the normal coordinates associated with translation and rotation are set to zero) and the total angular momentum is rigorously set to zero. From these the $3N$ Cartesian coordinates and momenta are obtained and the equations of motion are solved.

The difference $H - H_0$ is approximated accurately by just the difference in the full (V) and harmonic (V_0) potentials, $V - V_0$. In order to evaluate V_0 at each step a transformation back to the vibrational normal modes is necessary. Ideally this is done such that rotational normal coordinates are zero; however, a straightforward transformation back to normal coordinates from the propagated Cartesian coordinates shows that this is not the case (owing to vibration-rotation coupling). A rotation of the Cartesian coordinates is done to the Eckart frame to produce a minimized value of the rotational normal coordinates. This is done via the equation

$$\sum_i m_i \vec{r}_i \times \vec{R}_i = \vec{0}, \qquad (8.9)$$

where $\vec{r}_i$ is the Cartesian coordinates of atom i after the rotation, and $\vec{R}_i$ is the coordinates of atom i in the reference geometry. The

Cartesian coordinates in the Eckart frame are then transformed to the normal coordinates, and V_0 can be easily evaluated.

The details of this procedure are given in Ref. 5, where citations to other important papers are given. This procedure produces much improved energies for the test case of CH_4 than previous application of AS which did not minimize the vibration–rotation coupling.

8.3. Fundamentals of Vibrational Configuration Interaction

Now, considering an exact quantum Hamiltonian H, a basic postulate of quantum mechanics is that the exact eigenfunctions of H, which we denote by Ψ_L, can be written as

$$\Psi_L = \sum_K c_K^{(L)} \Phi_K, \tag{8.10}$$

where Φ_K are a complete, orthonormal set of functions. For example, these are the eigenfunctions of the zero-order, separable Hamiltonian given above, Eq. (8.6). The goal is to obtain the expansion coefficients $c_K^{(L)}$. Once these are known, the problem is solved; however, the challenge remains to extract information from these coefficients that can be used to interpret the answer. Obviously, if a single coefficient dominates, *e.g.*, has a magnitude of say 0.9 or greater, the interpretation is essentially the one from the familiar normal mode analysis and the experimental band should be simple to interpret. The challenge comes from those interesting cases where there is not one or even several dominant coefficients. In this case, one could conclude that the harmonic normal mode analysis has broken down and that there is strong mode-mode coupling. This is usually manifested experimentally by a broad and complex band, which is perforce difficult to interpret. The goal in this case theoretically is to at least determine which modes are strongly coupled and ultimately to understand the source of the coupling.

Another major challenge with this approach is that it is rarely the case that one is interested in a single or even a few eigenstates, Ψ_L. One exception is the ground rovibrational state and associated zero-point energy, as this energy is important in thermochemistry, *e.g.*, the dissociation energy, D_0, of a molecule, the relative energies of isomers, *etc.* For spectroscopy, even spanning a region of several hundred to roughly $1,000\,\mathrm{cm}^{-1}$, there can be hundreds or more transitions just from the ground vibrational state. Obviously, this number of states depends on the size of the molecule/cluster.

The most widely used VCI methods ultimately lead to finding eigenvalues of the matrix representation of H, denoted $\boldsymbol{H}$. The advantage of this approach is that many energy eigenvalues and eigenfunctions are obtained at once. And these can be used to obtain the spectrum of interest. For the IR spectrum the dipole moment must also be known. As discussed in detail in the next section the size of the $\boldsymbol{H}$ must be controlled by a selection of zero-order states to include in the above expansion and there has been a resurgence of interest in this recently. The procedures used in MULTIMODE will be given in the next section.

Perturbation theory, together with the so-called quartic force field (QFF), is a widely used method to obtain the expansion coefficients in Eq. (8.10) approximately and we review this briefly below.

8.3.1. *Second-order perturbation theory*

The second-order perturbation theory correction for energies comes from the first-order correction to the wavefunction. This is not normally considered a "CI" method. However, it is possible to regard it as such, since the goal is to obtain expressions for the expansion coefficients $c_K^{(L)}$ in Eq. (8.10).

This is clear from the first-order correction to the wavefunction given by

$$\Psi_K \approx \Phi_K + \sum_{L \neq K}^{\infty} \frac{\langle L| V' |K\rangle}{E_K^{(0)} - E_L^{(0)}} \Phi_L. \qquad (8.11)$$

In this equation, $|L\rangle$ and $|K\rangle$ denote zero-order wavefunctions and $E_L^{(0)}$ and $E_K^{(0)}$ are the corresponding zero-order energies. These could be those for any zero-order Hamiltonian; however, here we mean the separable harmonic oscillator one given above. And V' is the difference between the full potential and the harmonic oscillator one.

The usual assumption is that the zero-order term in the expansion dominates (and is given the coefficient 1.0) and the other terms are (hopefully) small corrections. There are two issues with this expression. One is computational, *i.e.*, the evaluation of the potential matrix element in the high-dimensional $(3N - 6)$ space of the wavefunction. The second is small denominators, known as resonances (Fermi and Darling-Dennison resonances being well-known examples).[4] These resonances can result in large corrections which invalidate the assumption that these coefficients are small. This breakdown of standard perturbation theory is actually of interest, as it may signal something unusual in the dynamics. Degenerate perturbation theory, *i.e.*, basically a small VCI, is able to handle these resonances.[6] This approach, however, still has a limited range of validity.

The other issue mentioned, *i.e.*, the computational cost of matrix elements of the perturbation potential is basically trivial if one follows the conventional approach and expands V in a QFF, as follows formally from conventional second-order perturbation theory, which is typically denoted as VPT2.[7]

To be more specific, the QFF is given schematically by

$$V = \sum_{i,j,k,l} F_{i,j,k,l} Q_i Q_j Q_k Q_l, \tag{8.12}$$

where the indices refer to all $3N - 6$ normal modes. Note V' in Eq. (8.11) is just this V minus the quadratic terms. (Recall there are no mixed bi-linear terms since these have been eliminated by design in making the transformation to normal coordinates.) Using this form for the potential, the matrix elements of V are analytical using a basis of harmonic oscillator functions.

There are some important comments to be made about this representation of V. First, while it is formally justified for VPT2 theory, it is just a Taylor series expansion of V at a reference configuration truncated at fourth order. For general applications, especially to X–H stretches, this is not sufficient for high accuracy when used in "exact", *e.g.*, basis expansion or grid, methods. Nevertheless, QFFs have been used in a number of recent papers that propose methods to efficiently select terms in the above expansion of the wavefunction. A recent study, guided by perturbation theory and a QFF with an extensive review of this recent literature is recommended.[8] Since the performance of these methods (to achieve sub cm^{-1} convergence) does depend significantly on the truncation of the force field, this re-examination seems worthwhile.

A simple way to examine this is to compare a QFF for a single mode, q, obtained as a fourth-order Taylor series expansion of a Morse potential. Specifically, consider the Morse potential that is an approximation for the H_2 potential. Thus, the dissociation energy is 4.7 eV, the range parameter a equals 1.0 $bohr^{-1}$ and r_e equal to 1.4 bohr. This is shown in Figure 8.1. As indicated, the mean absolute error in this simple QFF for energies below the fundamental

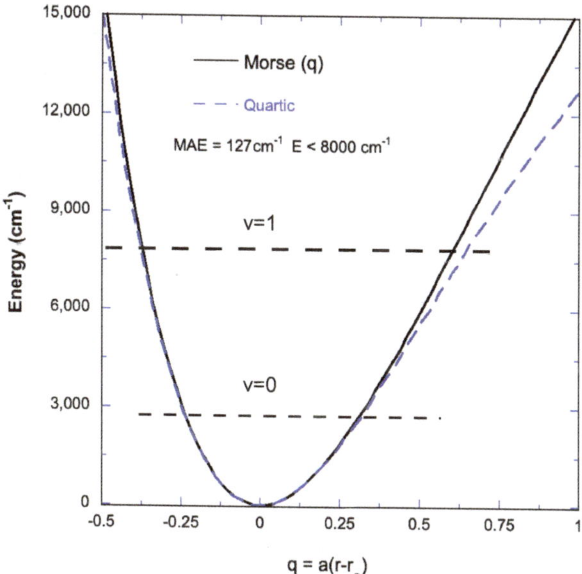

Figure 8.1. Morse potential and fourth order Taylor series of it and the mean absolute error in fourth order expansion. The vibrational levels 0 and 1 indicated are typical for an X–H stretch.

excitation is obviously very large compared to a convergence target of sub cm^{-1}. Clearly, the error is much larger for energies above the fundamental.

This inaccuracy of a QFF in normal coordinates (or in any coordinates used in a zero-order Hamiltonian based on a separable harmonic oscillator model) is well known and has been for many years. So fully rigorous applications of a variational theory to say an experimental IR spectrum aiming for high (several cm^{-1} or less) accuracy do not use such QFFs. It is also known that a happy result of VPT2 theory is that it yields exact Morse vibrational energies using the quartic expansion of that potential.

This focus on the inaccuracy of the QFF (some might say "beating a dead horse") is really a caution. The methods proposed for selecting an optimum expansion of the wavefunction may in fact be good ones. However, testing them on more accurate (non-QFF) potentials seems important to us.

A final note about QFF; it is a simple example of a *sum-of-products* (SOP) representation of V. The generalization of this for a general potential has been done very effectively as explained in chapters in this book. This makes the evaluation of matrix elements just a sum of products of 1D elements. In the former, multidimensional quadratures of the V can be written as the SOP of 1D-quadratures. This is an extremely effective, albeit not universal strategy. The example of the QFF is relevant here. In early vibrational CI calculations, this form of the SOP was used, *e.g.*, in the code POLYMODE[9] and ANHAR.[10] SOP approaches that are not Taylor series expansions of the potential, can in principle overcome this, *e.g.*, the "POTFIT" method in the Multi-configuration time-dependent Hartree (MCTDH) approach.[11–13]

8.3.2. *VSCF and VSCF+VCI*

Returning to Eq. (8.10), the rigorous approach is to determine the exact expansion coefficients, where we have used the harmonic-oscillator wavefunctions as the example basis. Another, more sophisticated basis consists of the virtual states of a VSCF

Hamiltonian, analogous to the basis used in electronic structure theory. In brief the single-mode functions in the product $\prod_i \phi_{n_i}^{(0)}(Q_i)$ are variationally determined and this leads to the usual self-consistent field equations for each single-mode function.[14, 15]

We will give more details of the VSCF/VCI approach used in MULTIMODE in the next section.

8.3.3. *The Watson Hamiltonian, the n-mode representation of potential and the excitation space*

Quantum approaches to vibrational dynamics are generally performed using a set of $3N - 6$ internal vibrational coordinates, as these can be done for specified values of the total angular momentum. (An important exception is quantum diffusion Monte Carlo (DMC) calculations, which are routinely done in all the Cartesian coordinates.[16] Also, see Chapter 4.)

The Hamiltonian operator is transformed to a matrix using a basis representation of the wavefunction, *c.f.*, Eq. (8.2) for example. The details depend on the choice of internal coordinates,[17] but in a basis of zero-order functions, *e.g.*, harmonic oscillator or virtual states, the potential matrix generally requires multidimensional quadrature. This is one, if not the major, computational bottleneck in applying VCI approaches, for two reasons. First, a strictly numerical quadrature scales exponentially with dimensionality. For illustration, if there are 15 quadrature points per degree of freedom, each potential matrix element effort scales as $\mathcal{O}(15^{3N-6})$. Second, the size of the H-matrix also scales exponentially with the number of degrees of freedom. For example, if there are 10 basis functions per degree of freedom, the H-matrix is $10^{3N-6} \times 10^{3N-6}$. So, this is basically hopeless even for a non-linear tetraatomic molecule with 6 degrees of freedom. Clearly, strategies are needed to deal with these two aspects of exponential scaling. This is an active area of research[17–24] and also chapters in this book and as noted much recent work based on truncated force fields.[8]

Here, we describe our approaches to deal with these bottlenecks, for the case where the internal coordinates are mass-scaled normal coordinates, $\boldsymbol{Q}$. The corresponding Hamiltonian (for non-linear molecules), known as the Watson Hamiltonian,[25] is given by

$$\hat{H} = \frac{1}{2} \sum_{\alpha\beta} (\hat{J}_\alpha - \hat{\pi}_\alpha) \mu_{\alpha\beta} (\hat{J}_\beta - \hat{\pi}_\beta) - \frac{1}{2} \sum_{k}^{F} \frac{\partial^2}{\partial Q_k^2} - \frac{1}{8} \sum_{\alpha} \mu_{\alpha\alpha} + V(\boldsymbol{Q}),$$

$$(8.13)$$

where $\alpha(\beta)$ represent the x, y, z coordinates, $\hat{J}_\alpha$ and $\hat{\pi}_\alpha$ are the components of the total and vibrational angular momenta respectively, $\mu_{\alpha\beta}$ is the inverse of effective moment of inertia, and $V(\boldsymbol{Q})$ is the potential. The number of normal modes is denoted by F, and for non-linear molecules F equals $3N - 6$.

In many applications, the vibrational angular momentum terms are neglected. This approximate Hamiltonian cannot lead to exact results, but the errors are generally not large, otherwise it would not be widely used. Nevertheless, for molecules with hydrogen atoms, errors can be of the order of $10\,\mathrm{cm}^{-1}$. A particularly glaring case is the ground state tunneling splitting in H_3O^+. Using the full Watson Hamiltonian, with an accurate PES, the calculated splitting is $46\,\mathrm{cm}^{-1}$,[26] in good agreement with experiment, versus $21.9\,\mathrm{cm}^{-1}$ without the vibrational angular momentum terms with the same PES.[27] We do not neglect these terms in the MULTIMODE software.

Returning to the issue of exponential scaling, we begin with the evaluation of matrix elements. One strategy to deal with this scaling is to use the SOP form for the PES, of which the QFF is one example, as noted already. Another very different one is based on the n-mode representation of the potential $V(\boldsymbol{Q})$. In normal coordinates, this representation is given by[28]

$$V(Q_1, Q_2, \ldots, Q_F) = \sum_{i} V_i^{(1)}(Q_i) + \sum_{i,j} V_{ij}^{(2)}(Q_i, Q_j)$$

$$+ \sum_{i,j,k} V^{(3)}_{ijk}(Q_i, Q_j, Q_k)$$

$$+ \sum_{i,j,k,l} V^{(4)}_{ijkl}(Q_i, Q_j, Q_k, Q_l) + \cdots, \qquad (8.14)$$

where $V^{(1)}_i(Q_i)$ is the one-mode potential, *i.e.*, the 1D cut through the full-dimensional PES in each mode, one-by-one, $V^{(2)}_{ij}(Q_i, Q_j)$ is the intrinsic 2-mode potential among all pairs of modes, *etc.* Here, intrinsic means that any n-mode term is zero if any of the arguments is zero. Also, each term in the representation is in principle of infinite order in the sense of a Taylor series expansion. So, for example $V^{(1)}(Q)$ could be a Morse potential.

There are two impacts of this representation of the potential. One is that matrix elements of the potential, which are done by numerical quadrature, are hierarchical, and obviously increase in computational effort with n. Second, the excitation space can be tightly coupled to each term in the n-mode expansion. For example, consider the 3-mode representation. The highest dimensional matrix element, independent of the total number of modes, is given by

$$\langle n'_i, n'_j, n'_k | V^{(3)}_{ijk} | n_i, n_j, n_k \rangle. \qquad (8.15)$$

Clearly in this element modes other than i, j, k result in factors of Kronecker delta functions, meaning no excitations in those other modes. So, for these elements the excitation space can be fairly large. For two-mode terms the excitation space can be very large for each mode. Indeed this tailoring the excitation space has been a fundamental part of the MULTIMODE software from the earliest version.[29] Examples of this excitation protocol are given below for the examples of glycine and two protonated water clusters.

After years of experience with many molecules, the 4MR of the full-dimensional PES is typically used as it generally gives results that are converged to within a few cm^{-1} or less.[18] Of course tests with a 5MR of the potential can be done, and this is always recommended. Also, for the zero-point energy, for which benchmark full dimensional

DMC calculations can be performed, tests for this energy can be done and we almost always do this. However, DMC energies typically have statistical uncertainties of a few cm^{-1} so this only provides a benchmark test within this level of uncertainty.

The n-mode representation and generalization to the full Hamiltonian is now widely used.[20,28–33] Each term in this representation can be obtained exactly if a full-dimensional PES is available. Alternatively, these terms can be obtained directly by calls to an electronic structure code. This is an option in MOLPRO.[34] MULTIMODE also can make use of direct calculations of *ab initio* energies to evaluate each term in the n-mode representation, as described in an application to N-methyl acetamide.[35]

Some comments on the limitations of rectilinear normal coordinates and thus the Watson Hamiltonian are in order. It is generally thought that these are suitable only for "semi-rigid" molecules. This is certainly correct for the important class of molecules with internal rotors, of which methyl rotors are a widespread example. However, we have successfully used these coordinates with the Watson Hamiltonian to obtain tunneling splittings for umbrella motion in NH_3,[36] H_3O^{+}[26] and tunneling between equivalent minima in the vinyl radical.[27] In these cases the reference geometry is the relevant saddle point separating the two minima and the rectilinear imaginary frequency normal mode is the large-amplitude coordinate. The basis for the MULTIMODE calculations are numerical ones determined by relaxed 1D cuts through the multidimensional potential. Details can be found elsewhere;[26,37] however, it is sufficient to note that these are not harmonic oscillator bases.

For molecules with low-energy torsional modes or internal rotors, a more general approach is needed. The reaction path version of MULTIMODE[38] is able to describe these problems; however, it is beyond the scope of this chapter to delve into this class of molecular vibrational dynamics. MULTIMODE has many features that space does not permit describing in detail here. A summary of features of MULTIMODE is given in Figure 8.2.

MULTIMODE FEATURES

- Eigenvalues and eigenfunctions of the Watson Hamiltonian

- VSCF and VSCF/VCI

- n-mode representation of the potential, maximum $n = 6$

- General user-supplied potential

- Force field in normal coordinates

- Option for direct calculation of the potential on n-mode grids

- User-supplied dipole moment (n-mode rep)

- Calculation of rovibrational spectra at non-zero temperature

- Approximate vibrational spectra at T = 0 K and J = 0

- Expectation values

- Torsional motion via the "Reaction Path" version of MULTIMODE

- MULTIMODE 4.9 information and download link can be found at https://scholarblogs.emory.edu/bowman/softwares/multimode/

Figure 8.2. A list of features of MULTIMODE and url for download of version 4.9.

A typical workflow for a MULTIMODE calculation of the IR spectrum is given in Figure 8.3. The calculation is done in two steps. First, the eigenvalues and eigenfunctions of the Watson Hamiltonian are obtained and written to disk. Then dipole matrix elements are obtained for select initial states using numerical quadrature and the n-mode representation of a user-supplied dipole moment surface. Then the IR spectrum is obtained. The temperature-dependent rovibrational, line-list spectrum can also be calculated with MULTIMODE; however, this is clearly more involved than a "vibration-only" IR spectrum. Details of such calculations have been reported for ethylene[39,40] and the interested reader can consult those papers. Also, see recent work by Rauhut and co-workers, who also use a very similar "MULTIMODE" approach to obtain rovibrational spectra within MOLPRO.[41]

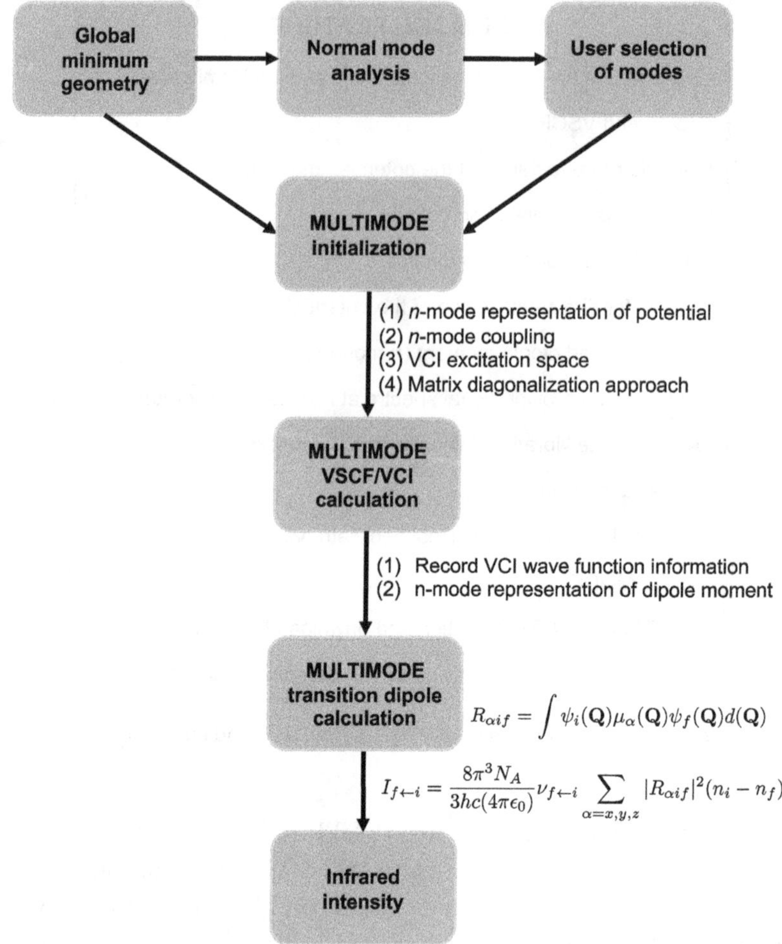

$$R_{\alpha if} = \int \psi_i(\mathbf{Q})\mu_\alpha(\mathbf{Q})\psi_f(\mathbf{Q})d(\mathbf{Q})$$

$$I_{f\leftarrow i} = \frac{8\pi^3 N_A}{3hc(4\pi\epsilon_0)}\nu_{f\leftarrow i}\sum_{\alpha=x,y,z}|R_{\alpha if}|^2(n_i - n_f)$$

Figure 8.3. Flow chart of MULTIMODE VSCF/VCI and IR spectrum calculation.

8.3.4. *IR spectra of glycine*

Glycine is the smallest amino acid. Its side chain is made of a single hydrogen atom. In spite of the simplicity of this amino acid, the electronic ground state PES of glycine has been intensively investigated to learn about the several low-energy conformers.[42]

We recently reported a permutationally invariant polynomial (PIP) PES for glycine at the DFT/B3LYP/aug-cc-pVDZ level of theory.[43] This PES describes the eight low-lying conformers. The optimized structures are in agreement both with pioneering calculations by Császár,[44] who predicted eight conformers but focused only on two, and recent CCSD(T) calculations by Czakó and co-workers.[45] In addition, we obtained the zero-point energies of all glycine conformers by means of DMC and adiabatically switched[46] semi-classical initial value representation (AS-SCIVR) molecular dynamics. AS-SCIVR results are in agreement with a previous SCIVR study which adopted *ab initio* "on-the-fly" dynamics at the same level of electronic structure theory and basis set.[47] More details about SCIVR can be found in Chapter 10. The two approaches, DMC and SCIVR, led to results in excellent agreement with one another, allowing us to conclude that the eight conformers could be described as four pairs of separated asymmetric double wells instead of eight distinct conformers in eight isolated wells.

Given the importance of glycine, there have been several calculations of the vibrational energies of the various conformers. They include harmonic calculations,[44,48] self-consistent field theory,[49] model Hamiltonians,[50] second-order vibrational perturbation theory,[51] and semi-classical initial value representation.[47]

Experimental data on the spectra are more difficult to come by, primarily because glycine is a solid at room temperature and decomposes at its melting point. Experimental studies of the rotational spectrum in the millimeter-wave region were motivated by the 2003 report that glycine spectral lines were observed from interstellar sources,[52] but these observations were not reproduced by several other investigators.[53–55] It now appears that the presence or absence of glycine in the interstellar medium is currently unresolved and that the previous positive identification was likely in error. This issue is of key importance for theories about the origin of life on Earth because it may clarify whether glycine could have arrived from the outer space upon impacts with asteroids or it is the result of more and more complex terrestrial synthetic processes.

The vibrational spectrum has been studied using matrix isolation spectroscopy,[48,56–59] helium nanodroplets,[59] FTIR in liquids and solids,[60,61] and molecular beam jet spectroscopy.[62,63]

The most comprehensive results have come from matrix isolation spectroscopy.[48] The jet cooling technique has provided an excellent Raman spectra of glycine in the 180–500 cm^{-1} region.[62,63]

As mentioned, we recently reported a PES for glycine and used it subsequently in MULTIMODE calculations of the IR spectra of two conformers. We present the highlights of those calculations here and also give some details of the excitation space for the VSCF/VCI calculations.

8.3.5. *Potential energy surface*

The glycine potential energy surface used here has been described previously.[43] Briefly, it is derived from a data set of energies at approximately 70,000 geometries calculated using density functional theory (B3LYP) with Dunning's aug-cc-pVDZ basis set. The fit is to 20,000 energies and their associated gradients as well as to 50,000 additional energies using a permutationally invariant polynomial basis[64–66] in 45 Morse variables with maximum polynomial order of four. Investigation of the surface located eight conformers of glycine as well as 15 saddle points connecting them. The harmonic frequencies calculated for the global minimum on the surface (Conformer I) were found to agree well with those of the direct-DFT optimized structure, with a mean absolute error of 4.3 cm^{-1}.

8.3.6. *Aspects of the glycine calculations*

Calculations were performed using Version 5.1.4 of MULTIMODE. The calculations used a 4-mode representation of the potential in mass-scaled normal coordinates and a 3MR representation of the effective inverse moment of inertia for the vibrational angular momentum terms in the exact Watson Hamiltonian.[25] We explored reduced mode models as well as a full, 24-mode model. With a 9-mode model with CI blocks of size 7869 and 5064 for the A' and A'' symmetries, we calculated 578 and 474 CI vibrational levels of these symmetries up to the energy of 5,000 cm^{-1}. The model gave encouraging results for intensities, but several levels were significantly shifted from the experimental results. A 10-mode model was not much of an improvement. Thus, a full, 24-mode was used to obtain 3,098 vibrational levels up to the energy of $3,700\,\mathrm{cm}^{-1}$.

We use this full-mode model to present a section of the full MULTIMODE input file to illustrate the parameters of the excitation space. This is shown below.

```
C**NATOM   CONV   ICOUPL ICOUPC ISCFCI MCHECK INORM
    10     1.D-2    4       3     8000     1      1
C**NMAX    CUT    MATSIZ
    -4     3.7D+3    0
C**MAXSUM
    5 5 5 5
C**MAXBAS
    5 5 5 5 5 5 5 5 5 5 5 5 5 5 5 5 5 5 5 4 4 4 4 4
    4 4 4 4 4 4 4 4 4 4 4 4 4 4 4 4 4 4 3 3 3 3 3
    3 3 3 3 3 3 3 3 3 3 3 3 3 3 3 3 3 3 3 2 2 2 2 2
    2 2 2 2 2 2 2 2 2 2 2 2 2 2 2 2 2 2 1 1 1 1 1
C**NSYM ISYM (ORDER A1,B2,B1,A2)
    15    3   4   6   8   9  11  12  14  16  17  18  19  20  22  24
     9    1   2   5   7  10  13  15  21  23
```

Here is a brief explanation of each key parameter.

- NATOM: Number of atoms.
- CONV: Criteria of convergence, in cm^{-1}.
- ICOUPL: n-mode representation of the potential; max $= 6$.
- ICOUPC: n-mode representation of the Coriolis coupling; max $= 4$.
- ISCFCI: Number of states to calculate in VCI.
- MCHECK: MULTIMODE rotates the molecule to principal-axis frame (which is required in the calculation) when this parameter is set to 1.
- INORM: MULTIMODE performs normal-mode analysis when it is 1; otherwise the user must supply harmonic frequencies and normal-mode vectors in the input file.
- NMAX: Number of modes allowed to be excited simultaneously in VCI; -4 means up to quadruple excitation.
- CUT: Cutoff energy for reported CI eigen-energies.

- **MATSIZ**: When set to 1, MULTIMODE just calculates the size of the CI matrix; always encouraged before actually performing the VCI calculation.

- **MAXSUM**: Sum-over-quanta in single, double, triple, and quadruple excitation basis; the number of parameters equals to |NMAX|. In the example input above, the first "5" means that the sum-over-quanta in the single-excitation basis cannot exceed 5; the second "5" means the sum-over-quanta in double-excitation basis cannot exceed 5, *etc.*

- **MAXBAS**: Maximum excitation quanta for each mode; the number of lines in this input block equals to |NMAX|, and the first line is for single-excitation basis, the second line is for double-excitation basis, *etc.* In the example shown above, the first line of **MAXBAS** means that in the single-excitation basis, the first 19 modes can at most be excited to the fourth excited state (the fifth state when the ground state is included), while the last 5 modes can be excited up to the third excited state. The second line means that in double-excitation basis, the first 19 modes can be excited to the third excited state while the last 5 modes can be excited to the second excited state at most.

- **NSYM** and **ISYM**: Number of modes that belong to this symmetry, and a list of specific modes in this symmetry; the number of lines in this input block equals the number of irreducible representations.

Three parameters, **NMAX**, **MAXSUM**, **MAXBAS**, determine the size of the VCI matrix. If the molecule has symmetry, *e.g.*, Conformer I of glycine, it is always encouraged to take advantage of this by specifying **NSYM** and **ISYM**. As a result, MULTIMODE only computes symmetry blocks rather than the full CI matrix.

For IR calculations a dipole moment surface is needed. The dipole moment surface of glycine is approximated by a separable form, which in normal coordinates is given by

$$\mu(Q_1, \ldots, Q_{24}) = \mu_e + \sum_{i=1}^{24} \Delta\mu_i(Q_i), \qquad (8.16)$$

where $\boldsymbol{\mu}_e$ is the dipole moment of glycine at the equilibrium configuration, and $\Delta\boldsymbol{\mu}_i(Q_i)$ is the change of dipole caused by the change in the ith normal mode, obtained by a fourth-order polynomial fit to 10 points on the 1D DFT (B3LYP/aug-cc-pVDZ) dipole cut along this mode. Note that this expansion was calculated separately for Conformer I and Conformer II.[67] Figures of these cuts have been published.[67]

The output from MM provided the energies of the CI states, the mixing coefficient corresponding to each basis state, and the intensity of the dipole transition between the zero-point level and each CI state. The corresponding spectrum was constructed using the CI energies and transition dipole intensities and then simulated using Gaussian peaks with full width at half maximum (FWHM) of $5\,\mathrm{cm}^{-1}$.

8.3.7. *MULTIMODE spectra*

Figure 8.4 shows a comparison between the experimental spectrum of Stepanian *et al.*,[48] the spectrum calculated by the Barone group using GVPT2, and that for transitions from the ground state calculated using MM and our potential energy surface. In order to emphasize the relative intensities, we have used two adjustable amplitude parameters to compare the two calculated spectra with the experimental spectrum. There is good agreement between the three spectra on the intensities of most of the peaks, but the MM ones appear to be shifted somewhat to higher energy than the experimental ones. The GVPT2 ones are a bit closer in energy to the experimental ones than the MM peaks. Nonetheless, the general agreement between all three spectra is good. Note that the experimental peak at $3{,}200\,\mathrm{cm}^{-1}$ has been shown to be due to the presence of small concentrations of Conformer II.[56]

Figure 8.5 shows a comparison between the experimental peaks,[60] the GVPT2 calculations,[51] and the MM results for the spectra starting from Conformer II. Again, there are two arbitrary adjustable parameters for setting the intensities. The general agreement is excellent, showing that the GVPT2 and MM calculations are performing well. Of particular interest is the fact that both the GVPT2 and MM calculations produce a peak near $3{,}200\,\mathrm{cm}^{-1}$, known to be

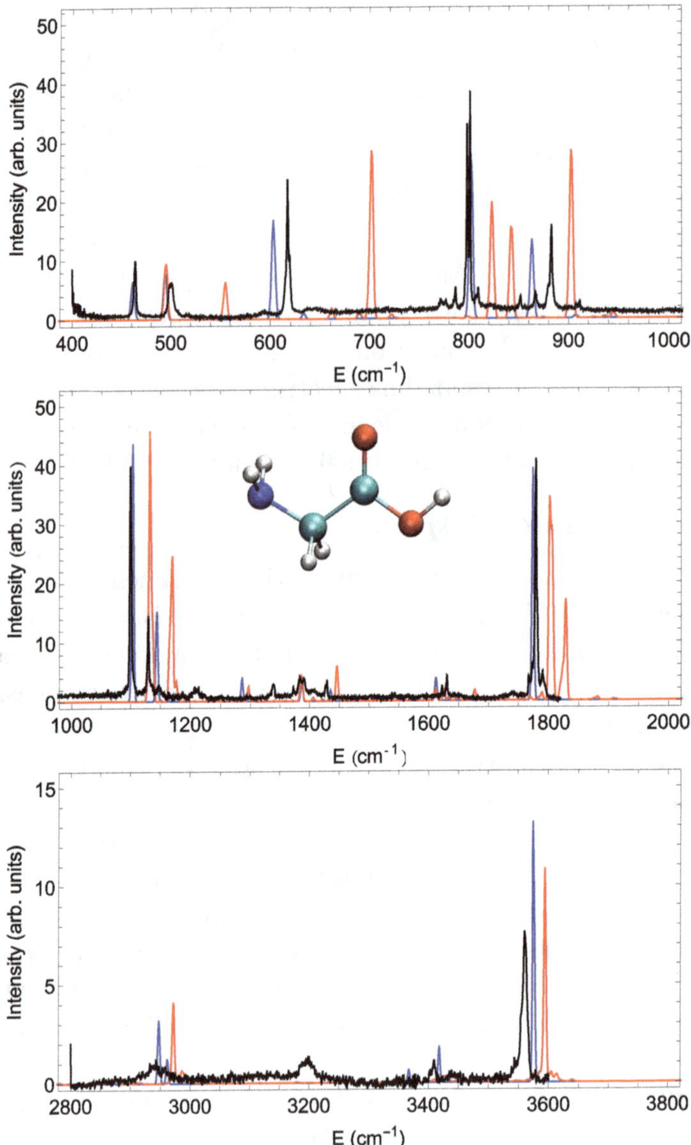

Figure 8.4. The ground state spectrum of glycine, including the experimental spectrum[48] (black), the GVPT2 results[51] (blue), and the calculated spectrum using MULTIMODE and the glycine PES (red). For the calculated spectra, a FWHM linewidth of $5\,\mathrm{cm}^{-1}$ was used. The experimental peak at $3{,}200\,\mathrm{cm}^{-1}$ is due to a small amount of Conformer II.

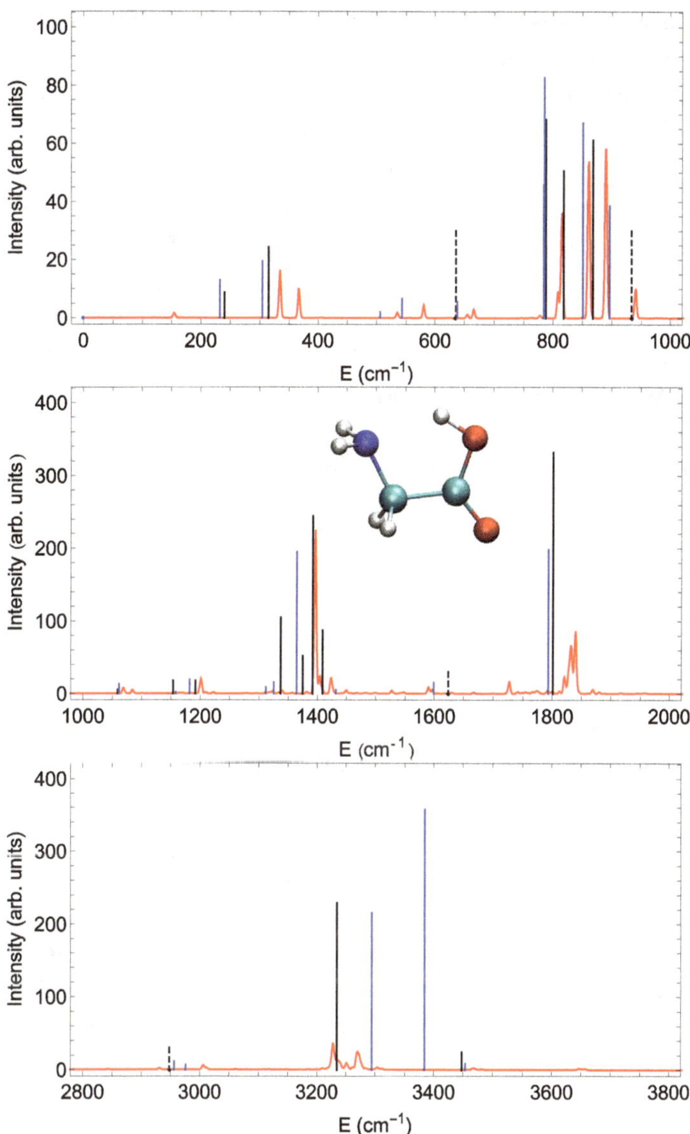

Figure 8.5. The spectrum of Conformer II, including the experimental stick spectrum[60] (black), the GVPT2 stick spectrum[51] (blue) and the spectrum calculated using MULTIMODE and the glycine PES (red). A FWHM linewidth of $5\,\text{cm}^{-1}$ was used for the MULTIMODE spectrum. The dashed black lines represent peaks in the spectrum for which the intensity is not reported.

attributable to absorption from Conformer II.[56] The MM peak is reasonably broad due to substantial mixing of levels, but its breadth is comparable to the experimental peaks shown in Figure 3(b), peak h of Ref. 60 or Figure 2(b) of Ref. 56.

10-atom glycine, described by a global potential energy surface, is certainly a major challenge for any state-of-the-art vibrational method. The approach taken above, albeit done for all 24 vibrational modes, is not totally general as the low-frequency torsional modes were not rigorously described. So, for example, inter-confomer coupling is not described. Fortunately, this coupling is evidently small for the case of Conformers I and II for which independent IR spectra align well with experiment.

Going beyond 10 atoms is of course even more challenging. We consider this next when we review our calculations on $H_9O_4{}^+$ and $H_7O_3{}^+$ using MULTIMODE and a many-body *ab initio* PES and DMS.

8.4. IR Spectra of of $H_7O_3{}^+$ and $H_9O_4{}^+$

Vibrational spectra of cold protonated water clusters, $H^+(H_2O)n$, have provided important insights about the structure(s) and vibrational dynamics of the excess proton in water.[68–73] The well-known Zundel and Eigen structural motifs have been found to have distinctive spectral signatures in various protonated water clusters. The smallest Zundel cation is $H_5O_2{}^+$. This has the proton located equidistant between two flanking water molecules. The smallest Eigen cation, $H_9O_4{}^+$, has central hydronium equally shared by three solvated water molecules. We have referred to the bare hydronium cation, H_3O^+, as "baby Eigen".

Various post-harmonic theoretical methods have been applied to investigate the IR spectra of different sizes of protonated water clusters, examples include VPT2,[7,74,75] VSCF/VCI[14,15,76–79] and MCTDH method.[11,80,81] These quantum methods attempt to take account of the strong anharmonicity and mode coupling within the molecular systems and provide accurate predictions of vibrational states.

Molecular dynamics (MD)-based methods have also been widely used for IR spectra calculation of molecular systems from gas phase to condensed phase. At present the *ab initio* MD (AIMD) approach is the workhorse for such simulations.[82] However, for smaller clusters semi-classical,[83,84] semi-quantum ring-polymer molecular dynamics (RPMD),[85,86] and centroid molecular dynamics (CMD)[87–89] have also been applied to the IR spectra calculation. These methods do capture the significant effects of anharmonicity of the system, like O–H stretch in water,[90,91] and in this respect are more accurate than pure classical MD simulations of the IR spectrum. However, these alternatives to the MD approach are significantly more computationally intensive except at high temperature where they merge with MD simulations. Also, since these methods are somewhat heuristic, it is not clear how they can be made systematically more accurate.

We recently reported a systematic assessment of the MD and RPMD approaches for the IR spectra of of the $H_7O_3^+$ and $H_9O_4^+$ clusters.[92] This was done by applying those methods using our many-body PES and DMS and comparing the results to our VSCF/VCI calculations of those spectra and experiment. We highlight those comparisons here, but first we briefly review the MULTIMODE VSCF/VCI calculations of those spectra.[78,79,93]

8.4.1. *Summary of MULTIMODE VSCF/VCI calculations*

We have reported a high-level *ab initio* PES and DMS that can be used to obtain the IR spectrum of general protonated water

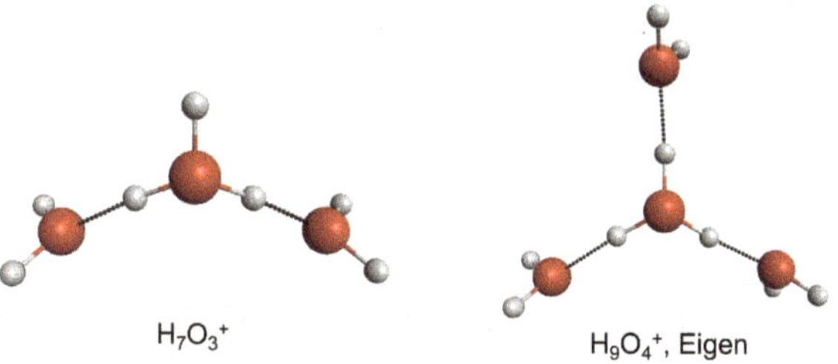

Figure 8.6. The global minimum structures of $H_7O_3^+$ and $H_9O_4^+$ (Eigen).

clusters.[94,95] The potential energy of the system is represented in a many-body expression with truncation order of 4. The dipole moment is also represented using a many-body expression but is truncated at 2-body level. Details of both the PES and DMS are given elsewhere.[95]

The VSCF/VCI spectra of $H_7O_3^+$ and $H_9O_4^+$ were calculated using MULTIMODE software.[30] Standard normal mode analyses were conducted first using global minimum geometries of $H_7O_3^+$ and $H_9O_4^+$. Due to consideration of computational cost, a subset of normal modes for each molecule was selected for reduced-dimensional VSCF/VCI calculations. Specifically, for $H_7O_3^+$, 18 normal modes (above $350\,cm^{-1}$) were selected and they were coupled using a 4-mode representation of the potential. The excitation space in VCI calculation consists of single, double, triple and quadruple excitations. The Hamiltonian matrix is in the order of 150,000 and was diagonalized using a block Davidson method. As to $H_9O_4^+$, we carried out 15-mode calculation and these modes included two OO stretches, OO umb, hydronium rot, two hydronium wag, hydronium umb, three water bends, two hydronium bends and three hydronium stretches. The 4-mode representation (4MR) of the potential was used and the final VCI matrix is 138,661. For water stretches, we did a separate calculation that included all water bends and stretches with also 4MR representation of the potential. To obtain IR spectra of the two molecules, we recorded the final VCI wavefunction for each system and calculated the transition dipole moment with 3-mode representation (3MR) of the dipole moment. A flow chart of MULTIMODE VSCF/VCI and IR spectrum calculation is shown in Figure 8.3. More details of VSCF/VCI calculations of these two molecules are given in our previous papers.[79,96]

For both molecules, we systematically selected different mode coupling options (2-mode coupling, 3-mode coupling, *etc.*) and also investigated different excitation spaces to obtain the converged anharmonic vibrational frequencies and spectra. In Figure 8.7, we show how the total VCI matrix size grows when increasing the excitation space through MULTIMODE variable MAXSUM, taking

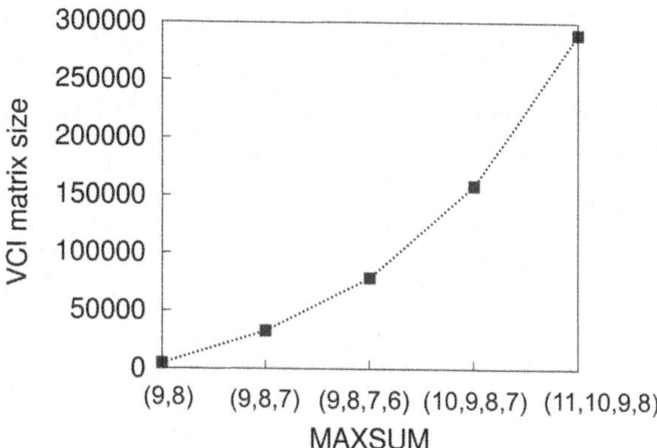

Figure 8.7. The VCI matrix sizes as a function of different choices of excitation space (variable MAXSUM) for 18-mode system calculations. The indices (i,j,k,l) represent a maximum quanta of i for 1-mode excitation, a maximum sum of j for 2-mode excitations, a maximum sum of k for 3-mode excitations, and a maximum sum of l for 4-mode excitations.

the example of 18-mode VCI calculation. As seen, the VCI matrix size is small when only 1 and 2-mode excitations are allowed. It grows exponentially once 3-mode or 4-mode excitations are activated. With large 4-mode excitation space, like (11,10,9,8), the matrix size becomes extremely large, close to 300,000. Thus, it is computationally expensive to store and diagonalize this VCI matrix. As already been mentioned above, MULTIMODE software is able to enable large system calculation and conduct efficient matrix diagonalization algorithms, like block Davidson method.

8.4.2. *Summary of quasi-classical MD, classical MD and TRPMD simulations*

The VSCF/VCI approach has its limit to calculate the IR spectra of $H_7O_3{}^+$ and $H_9O_4{}^+$ in the region 0 to $1200\,cm^{-1}$, since the large-amplitude, torsional motion of the flanking waters cannot be accurately described by the Watson Hamiltonian. As an alternative, we conducted quasiclassical molecular dynamics (QCMD) simulation of two molecules with 100 independent trajectories for each molecule. For each trajectory, the zero-point energy was assigned to all intermolecular modes and umbrella mode of H_3O^+. Each trajectory was run for a total time of 12 ps with step size 0.06 fs.

Classical molecular dynamics simulations were also performed for two clusters at 20 K and 100 K. Langevin thermostat was used in the NVT calculations and each set of calculations contains 10 independent trajectories. Each trajectory ran for 100 ps, and the dipole moment was recorded every 0.25 fs. As to the Thermostatted Ring Polymer Molecular Dynamics (TRPMD) calculations, the temperature was set as 100 K since lower temperature requires more beads (replicas) and the influence of temperature will be discussed from classical MD. We first ran a 50 ps Path Integral Molecular Dynamics (PIMD) calculation to obtain the equilibrium structures that can be used as the starting structures for the TRPMD calculations. Two different thermostat methods were used in TRPMD simulations. These included the path integral Langevin equation (PILE) thermostat and Generalized Langevin equation (GLE) thermostat to the normal mode representation. For each cluster with one chosen thermostat, eight independent trajectories were run with the initial structure from previous PIMD simulations. Each TRPMD trajectory was run for 50 ps with 64 beads. Both classical MD and TRPMD simulations were run using i-PI[97] software combining with our developed potential energy surface/dipole moment surface.

For QCMD and classical MD, the dipole moment along the trajectory was recorded and the IR intensity was calculated through Fourier transform of the dipole autocorrelation function. For TRPMD, we recorded the dipole moment of each bead along the trajectory and calculated the final estimator for the dipole moment at each step as an average over all replicas (beads). The final IR intensity in TRPMD simulation is also obtained through Fourier transform of the dipole autocorrelation function.

8.5. Results and Discussion

In this section, we present calculated IR spectra of two important protonated water clusters $H_7O_3^+$ and $H_9O_4^+$ from approaches including classical MD, TRPMD with two thermostats, Quasi-Classical MD (QCMD) and VSCF/VCI. The theoretical spectra are compared to the experiment along with detailed discussions. As shown in Figure 8.6, the global minimum energy structure of $H_7O_3^+$ has a C_s symmetry that the hydronium core is shared by two water molecules. The lowest energy isomer of H_9O_4 is called Eigen ion where the hydronium is equally shared by three water molecules.

8.5.1. $H_7O_3^+$

The harmonic frequencies of $H_7O_3^+$ have been reported at CCSD(T)-F12/aVTZ level of theory.[95] For the central hydronium, at harmonic level, its asymmetric OH stretch is located at $2520\,cm^{-1}$ while the symmetric OH stretch is at $2688\,cm^{-1}$. The free OH stretch ($3839\,cm^{-1}$) is located in the water OH stretch region. For the water monomers, the asymmetric OH stretches are around $3820\,cm^{-1}$ and the symmetric stretches are around $3910\,cm^{-1}$. Figure 8.8(a) shows that the classical MD spectrum at $20\,K$ generally reproduces all harmonic features including hydronium stretches, water stretches, bends and several intense signatures below $1200\,cm^{-1}$. This comes from the small amount of energy that is added to the molecular system at $20\,K$. Under such low temperature conditions, the molecule remains in its equilibrium structure and not enough anharmonic effects can be revealed in classical MD simulation. At higher temperature, $100\,K$, no change of central vibrational peaks is observed. The spectrum shows broader bands along the spectral range, especially for the hydronium stretches. When compared to the experimental spectra shown in Figure 8.8(g-h), there exists a 500–$600\,cm^{-1}$ red-shift of the hydronium stretches and a 100–$200\,cm^{-1}$ red-shift for water stretches. This huge downshift phenomenon cannot be explained by the classical MD simulation.

The TRPMD spectra with different thermostats show significant improvement on the spectral peak positions, see Figure 8.8(c-d). Specifically, both TRPMD spectra predict the water stretch band at $3750\,cm^{-1}$. This successfully reproduces the 100–$200\,cm^{-1}$ red shift comparing with water OH stretches' harmonic frequencies and agrees well with the experimental measurement. As to the prominent hydronium stretches, the TRPMD spectrum with a generalized

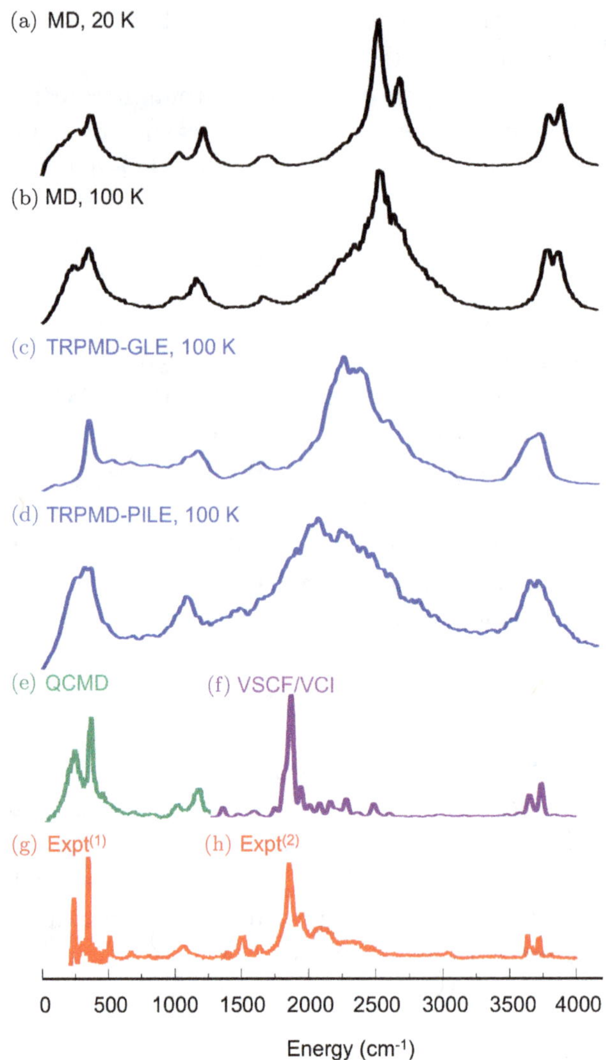

Figure 8.8. The vibrational spectrum of $H_7O_3{}^+$ for different calculation methods, as listed. For the experimental spectra, (1) is from Ref. 96 and (2) is from Ref. 93.

Langevin equation (GLE) thermostat shows a 200–300 cm^{-1} red shift relative to the peak positions in the classical MD spectrum. However, the calculated hydronium OH stretches are still 200–300 cm^{-1} higher than that measured in experiment. So even though the TRPMD approach has included some nuclear quantum effects like zero-point energy and hydrogen tunneling effects, it is evident that the vibration spectrum of $H_7O_3^+$ is beyond such effects and more sophisticated methods should be applied in vibrational spectrum calculation. Finally, as we shown in Figure 8.8(f), the high-level quantum VSCF/VCI calculations of $H_7O_3^+$ spectrum successfully reproduces all the prominent bands along the spectrum range of 1000 to 4000 cm^{-1}. Both VSCF/VCI and experimental spectra predict the intense peak at around 1900 cm^{-1} and that the hydronium stretches contribute most to this signature band. As to other less intense bands at around 1950–2500 cm^{-1}, according to VSCF/VCI analyses, the hydronium asym/sym-stretches have strong coupling with different combinations bands. Because of the highly mixing feature, those combination bands borrow the intensity from hydronium stretches and show large IR intensity. Besides of the peak position in calculated spectra, classical MD spectrum has more broadening than the VSCF/VCI and experimental results. As the temperature increases from 20 K to 100 K, the classical MD spectrum becomes broader. The two TRPMD spectra have the hydronium stretch band even broader than the classical MD ones. This stretch band covers the 300 cm^{-1} region in the TRPMD-GLE spectrum and the 500 cm^{-1} region in the TRPMD-PILE spectrum.

Figure 8.8 also shows the spectrum of $H_7O_3^+$ in the far-IR region, 200–1200 cm^{-1}. The experimental spectrum has been reported together with quasi-classical molecular dynamics simulation results.[96] As seen in Figure 8.8(e) and 8.8(g), two intense peaks appear below 500 cm^{-1}. The feature around 240 cm^{-1} is assigned as the water wagging motion while the embedded H_3O^+ hydrogen bond stretching mode contributes to the band at around 350 cm^{-1}. Another small feature at 508 cm^{-1} is captured by both experiment and QCMD calculation which comes from the combination band of low-frequency modes. The classical MD spectra at 20 K and 100 K predict both peaks below 500 cm^{-1}. Thus, these two bands are not influenced by the anharmonic effects too much and are still close to the harmonic numbers. A major difference appears in the TRPMD spectrum using GLE thermostat. The band at 240 cm^{-1}

disappears and only one $350\,\mathrm{cm}^{-1}$ band is seen below $500\,\mathrm{cm}^{-1}$. As a comparison, with PILE thermostat, the corresponding TRPMD spectrum has a broad feature along 240–$400\,\mathrm{cm}^{-1}$ and two peaks are buried in this broad band. The TRPMD-PILE spectrum in this far-IR spectrum range is reasonable considering the MD spectrum at higher temperature will introduce broadening of different vibrational bands. The reason for the lack of the important $240\,\mathrm{cm}^{-1}$ peak in the TRPMD-GLE spectrum is that this method will introduce unphysical description of the rotational and librational modes.[98] The TRPMD-GLE spectrum is not as reliable as TRPMD-PILE when simulating the spectrum in low-frequency regions.

8.5.2. $H_9O_4{}^+$ (Eigen)

The minimum energy isomer of $H_9O_4{}^+$ is the modeled "Eigen" molecule with the hydronium surrounded by three water molecules. The signature band of this molecule is the broad band centered around $2650\,\mathrm{cm}^{-1}$ in experiment.[74,99] Similar to $H_7O_3{}^+$, it is hard to explain the origin or the large red-shift of this hydronium stretch band. The harmonic analyses based on CCSD(T)-F12/aVTZ and constructed PES predict the hydronium stretches are around $3000\,\mathrm{cm}^{-1}$ and carry large intensity.[95] Thus, there exists around $400\,\mathrm{cm}^{-1}$ red shift for the anharmonic spectrum.

As shown in Figure 8.9, the classical MD simulation cannot resolve the problem at two different temperatures. The two classical MD spectra predict the hydronium stretches at around $3000\,\mathrm{cm}^{-1}$ which are almost the harmonic numbers. Other positions of intense bands also agree with the harmonic results. Two TRPMD spectra successfully reproduce right band positions of water stretches comparing with experiment where there exists 100–$200\,\mathrm{cm}^{-1}$ red-shift from the harmonic numbers. Back to the hydronium stretches, the TRPMD spectrum with GLE thermostat predicts the hydronium stretch band at $2800\,\mathrm{cm}^{-1}$ while with PILE thermostat, TRPMD calculation shows this band at around $2750\,\mathrm{cm}^{-1}$. The red-shift of the hydronium stretch band is prominent with 200–$250\,\mathrm{cm}^{-1}$ from harmonic predictions. Similar to the $H_7O_3{}^+$ case, the TRPMD spectra cannot fully explain the huge red-shift of hydronium

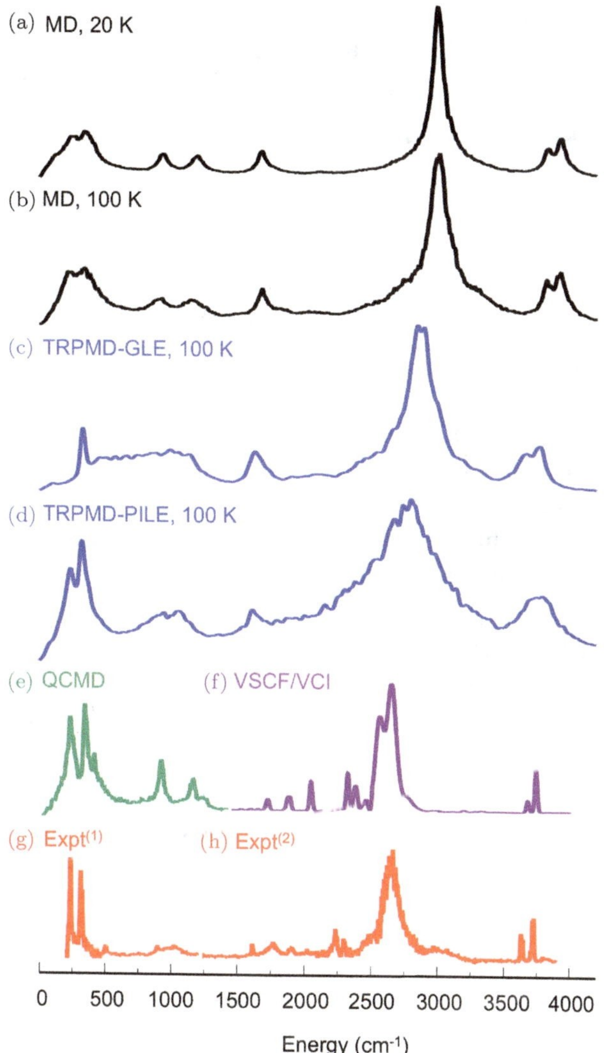

Figure 8.9. The vibrational spectrum of $H_9O_4^+$ for different calculation methods, as listed. For the experimental spectra, (1) is from Ref. 96 and (2) is from Ref. 79.

stretches in the experiment which is up to $400\,\mathrm{cm}^{-1}$. Besides, from experimental measurement (see Figure 8.9(h)), several additional peaks appear from $1750\,\mathrm{cm}^{-1}$ to $2400\,\mathrm{cm}^{-1}$. Both classical MD and TRPMD calculations capture these important features. All the above gaps are filled and explained by our VSCF/VCI calculation as shown in Figure 8.9(f). The fully quantum VSCF/VCI calculation successfully predicts the hydronium stretches at around $2650\,\mathrm{cm}^{-1}$ where they are highly coupled with other vibrational states including several complex combination bands. The VSCF/VCI spectrum also provides explanations of all small featured bands in 1750–$2400\ \mathrm{cm}^{-1}$. These bands originate mainly from the important combination bands involving hydronium core rotation, wagging and umbrella motions. Each combination band has coupling with hydronium stretches (asym- or sym-stretches) and thus they carry significant intensity in the spectrum.

The vibrational spectrum of the $H_9O_4^+$ Eigen isomer in the far-IR region $(0$–$1250\,\mathrm{cm}^{-1})$ is also investigated using different approaches. Similar to $H_7O_3^+$, the experimental spectrum has two observed bright states below $500\,\mathrm{cm}^{-1}$. The $236\,\mathrm{cm}^{-1}$ band is assigned as the fundamental transition of water wagging motion, while the hydrogen bond stretch contributes to the $316\,\mathrm{cm}^{-1}$ band. Both TRPMD with PILE thermostat and QCMD capture the above features. However, with GLE thermostat, the TRPMD simulation only shows one band around $300\,\mathrm{cm}^{-1}$. This comes from the same reason we have discussed in the $H_7O_3^+$ case. Even though TRPMD-GLE method provides a better resolved spectrum in high-frequency regions, it may cause some unphysical descriptions in low-frequency regions, especially for some motions like water librations.

8.6. Conclusions

The vibrational dynamics of molecules and molecular clusters continues to be an active research area both experimentally and theoretically. In this chapter, we have presented one theoretical approach, based on the widely used normal-coordinate framework and the associated rigorous quantum mechanical Hamiltonian. Advantages and limitation of this approach were given, along with general computational challenges that are present in all matrix-based approaches that require set-up of an H-matrix and determination of eigenvalues and eigenvectors of that matrix.

The code MULTIMODE, which is based on this approach, was described in some detail, with emphasis on the strategies to deal with these computational challenges.

A brief discussion of the centrally important potential energy surface was given and for the calculation of IR spectra the need for a general expression for the dipole moment was also given.

Two recent applications of MULTIMODE to obtain IR spectra of two protonated water clusters and two conformers of glycine were presented. These are systems of great interest and complexity and we have shown how the spectra obtained with MULTIMODE contribute to unraveling this complexity.

Acknowledgments

JMB thanks the Army Research Office, DURIP grant (W911NF-14-1-0471), for funding a computer cluster where most of the calculations were performed. JMB and AN acknowledge current support from NASA grant (80NSSC20K0360).

References

1. Hochlaf, M. Editorial of the PCCP themed issue Spectroscopy and dynamics of medium-sized molecules and clusters. *Phys. Chem. Chem. Phys.* **2013**, *15*, 9967–9969.
2. Senent, M. L.; Hochlaf, M.; Carvajal, M. Spectroscopy and dynamics of medium-sized molecules and clusters: Theory, experiment, and applications. *J. Phys. Chem. A* **2016**, *120*, 475–476.
3. Qu, C.; Bowman, J. M. Quantum approaches to vibrational dynamics and spectroscopy: is ease of interpretation sacrificed as rigor increases? *Phys. Chem. Chem. Phys.* **2019**, *21*, 3397–3413.
4. Wilson, E. B.; Decius, J. C.; Cross, P. C. *Molecular Vibrations: The Theory of Infrared and Raman Vibrational Spectra*. Dover Publ.: New York, NY, 1980.
5. Qu, C.; Bowman, J. M. Revisiting adiabatic switching for initial conditions in quasi-classical trajectory calculations: Application to CH_4. *J. Phys. Chem. A* **2016**, *120*, 4988–4993.
6. Yagi, K.; Otaki, H. Vibrational quasi-degenerate perturbation theory with optimized coordinates: Applications to ethylene and *trans*-1,3-butadiene. *J. Chem. Phys.* **2014**, *140*, 084113.

7. Nielsen, H. H. The vibration-rotation energies of molecules. *Rev. Mod. Phys.* **1951**, *23*, 90–136.

8. Fetherolf, J. H.; Berkelbach, T. C. Vibrational heat-bath configuration interaction. *J. Chem. Phys.* **2021**, *154*, 074104.

9. Romanowski, H.; Bowman, J. M.; Harding, L. B. Vibrational energy levels of formaldehyde. *J. Chem. Phys.* **1985**, *82*, 4155–4165.

10. Dunn, K. M.; Boggs, J. E.; Pulay, P. Vibrational energy levels of hydrogen cyanide. *J. Chem. Phys.* **1986**, *85*, 5838–5846.

11. Meyer, H.-D.; Manthe, U.; Cederbaum, L. S. The multi-configurational time-dependent Hartree approach. *Chem. Phys. Lett.* **1990**, *165*, 73–78.

12. Manthe, U.; Meyer, H.-D.; Cederbaum, L. S. Wave-packet dynamics within the multiconfiguration hartree framework: General aspects and application to NOCl. *J. Chem. Phys.* **1992**, *97*, 3199–3213.

13. Worth, G. A.; Beck, M. H.; Jäckle, A.; Vendrell, O.; Meyer, H.-D. The MCTDH Package, Version 8.2, (2000). H.-D. Meyer, Version 8.3 (2002), Version 8.4 (2007). O. Vendrell and H.-D. Meyer Version 8.5 (2013). Version 8.5 contains the ML-MCTDH algorithm. Current versions: 8.4.12 and 8.5.5 (2016). See http://mctdh.uni-hd.de/.

14. Bowman, J. M. Self-consistent field energies and wavefunctions for coupled oscillators. *J. Chem. Phys.* **1978**, *68*, 608.

15. Bowman, J. M. The self-consistent-field approach to polyatomic vibrations. *Acc. Chem. Res.* **1986**, *19*, 202–208.

16. McCoy, A. B. Diffusion Monte Carlo approaches for investigating the structure and vibrational spectra of fluxional systems. *Int. Rev. Phys. Chem.* **2006**, *25*, 77–107.

17. Carrington, T. Perspective: Computing (ro-)vibrational spectra of molecules with more than four atoms. *J. Chem. Phys.* **2017**, *146*, 120902:1–10.

18. Bowman, J. M.; Carrington, T.; Meyer, H.-D. Variational quantum approaches for computing vibrational energies of polyatomic molecules. *Mol. Phys.* **2008**, *106*, 2145–2182.

19. Roy, T. K.; Gerber, R. B. Vibrational self-consistent field calculations for spectroscopy of biological molecules: new algorithmic developments and applications. *Phys. Chem. Chem. Phys.* **2013**, *15*, 9468–9492.

20. Christiansen, O. Selected new developments in vibrational structure theory: potential construction and vibrational wave function calculations. *Phys. Chem. Chem. Phys.* **2012**, *14*, 6672–6687.

21. Oschetzki, D.; Rauhut, G. Pushing the limits in accurate vibrational structure calculations: anharmonic frequencies of lithium fluoride clusters $(LiF)_n$, n = 2–10. *Phys. Chem. Chem. Phys.* **2014**, *16*, 16426–16435.

22. Császár, A. G.; Fabri, C.; Szidarovszky, T.; Matyus, E.; Furtenbacher, T.; Czakó, G. The fourth age of quantum chemistry: molecules in motion. *Phys. Chem. Chem. Phys.* **2012**, *14*, 1085–1106.

23. Wang, X.; Carter, S.; Bowman, J. M. Pruning the Hamiltonian matrix in MULTIMODE: test for C_2H_4 and application to CH_3NO_2 using a new ab initio potential energy surface. *J. Phys. Chem. A* **2015**, *119*, 11632–11640.

24. Tennyson, J. Perspective: accurate ro-vibrational calculations on small molecules. *J. Chem. Phys.* **2016**, *145*, 120901:1–8.

25. Watson, J. K. G. Simplification of the molecular vibration-rotation Hamiltonian. *Mol. Phys.* **1968**, *15*, 479–490.

26. Huang, X.; Carter, S.; Bowman, J. Ab initio potential energy surface and rovibrational energies of H_3O^+ and its isotopomers. *J. Chem. Phys.* **2003**, *118*, 5431–5441.

27. Kamarchik, E.; Wang, Y.; Bowman, J. Reduced-dimensional quantum approach to tunneling splittings using saddle-point normal coordinates. *J. Phys. Chem. A* **2009**, *113*, 7556–7562.

28. Carter, S.; Culik, S. J.; Bowman, J. M. Vibrational self-consistent field method for many-mode systems: a new approach and application to the vibrations of CO Adsorbed on Cu(100). *J. Chem. Phys.* **1997**, *107*, 10458–10469.

29. Carter, S.; Bowman, J. M.; Handy, N. C. Extensions and tests of "MULTIMODE": a code to obtain accurate vibration/rotation energies of many-mode molecules. *Theor. Chem. Acc.* **1998**, *100*, 191–198.

30. Bowman, J. M.; Carter, S.; Huang, X. MULTIMODE: a code to calculate rovibrational energies of polyatomic molecules. *Int. Rev. Phys. Chem.* **2003**, *22*, 533.

31. Ostrowski, L.; Ziegler, B.; Rauhut, G. Tensor decomposition in potential energy surface representations. *J. Chem. Phys.* **2016**, *145*, 104103:1–9.

32. Ziegler, B.; Rauhut, G. Efficient generation of sum-of-products representations of high-dimensional potential energy surfaces based on multimode expansions. *J. Chem. Phys.* **2016**, *144*, 114114:1–11.

33. König, C.; Christiansen, O. Automatic determination of important mode-mode correlations in many-mode vibrational wave functions. *J. Chem. Phys.* **2015**, *142*, 144115:1–19.

34. Werner, H.-J.; Knowles, P. J.; Manby, F. R.; Black, J. A.; Doll, K.; Heelmann, A.; Kats, D.; Khn, A.; Korona, T.; Kreplin, D. A.; *et al.* The molpro quantum chemistry package. *J. Chem. Phys.* **2020**, *152*, 144107:1–24.

35. Kaledin, A. L.; Bowman, J. M. Full dimensional quantum calculations of vibrational energies of N-methyl acetamide. *J. Phys. Chem. A* **2007**, *111*, 5593–5598.

36. The vibrational levels of ammonia. *Spectrochim. Acta A* **2002**, *58*, 825–838.

37. Yu, Q.; Bowman, J. M. Ab initio potential for $H_3O^+ \rightarrow H^+ + H_2O$: a step to a many-body representation of the hydrated proton? *J. Chem. Theory Comput.* **2016**, *12*, 5284.

38. Bowman, J. M.; Huang, X.; Handy, N. C.; Carter, S. Vibrational levels of methanol calculated by the reaction path version of MULTIMODE, using an ab initio, full-dimensional potential. *J. Phys. Chem. A* **2007**, *111*, 7317–7321.

39. Carter, S.; Sharma, A. R.; Bowman, J. M.; Rosmus, P.; Tarroni, R. Calculations of rovibrational energies and dipole transition intensities for polyatomic molecules using MULTIMODE. *J. Chem. Phys.* **2009**, *131*, 224106.

40. Carter, S.; Sharma, A. R.; Bowman, J. M. First-principles calculations of rovibrational energies, dipole transition intensities and partition function for ethylene using MULTIMODE. *J. Chem. Phys.* **2012**, *137*, 154301:1–19.

41. Erfort, S.; Tschöpe, M.; Rauhut, G. Toward a fully automated calculation of rovibrational infrared intensities for semi-rigid polyatomic molecules. *J. Chem. Phys.* **2020**, *152*, 244104:1–14.

42. Barno, V.; Biczysko, M.; Bloino, J.; Puzzarini, C. Characterization of the elusive conformers of glycine from state-of the-art structural, thermodynamic, and spectroscopic computations: Theory complements experiment. *J. Chem. Theory Comput.* **2013**, *9* 1533–1547.

43. Conte, R.; Houston, P. L.; Qu, C.; Li, J.; Bowman, J. M. Full-dimensional, ab initio potential energy surface for glycine with characterization of stationary points and zero-point energy calculations by means of diffusion Monte Carlo and semiclassical dynamics. *J. Chem. Phys.* **2020**, *153*, 244301:1–11.

44. Császár, A. G. Conformers of gaseous glycine. *J. Am. Chem. Soc.* **1992**, *114*, 9568–9575.

45. Orjan, E. M.; Nacsa, A. B.; Czakó, G. Conformers of dehydrogenated glycine insomers. *J. Comput. Chem.* **2020**, *41*, 2001–2014.

46. Conte, R.; Parma, L.; Aieta, C.; Rognoni, A.; Ceotte, M. Improved semiclassical dynamics through adiabatic switching trajectory sampling. *J. Chem. Phys.* **2019**, *151*, 214107.

47. Gabas. F.; Conte, R.; Ceotte, M. On-the-fly ab initio semiclassical calculation of glycine vibrational spectrum. *J. Chem. Theory Comput.* **2017**, *13*, 2378–2388.

48. Stepanian. S. G.; Reva, I. D.; Radchenko, E. D.; Rosado, M. T. S.; Durate, M. L. T. S.; Fausto, R.; Adamowicz, L. Matrix-isolation

infrared and theoretical studies of the glycine conformers. *J. Phys. Chem. A* **1998**, *102*, 1041–1054.

49. Brauer, B.; Chaban, G. M.; Gerber, R. B. Spectroscopically-tested, improved, semi-empirical potentials for biological molecules: calculations for glycine, alanine and proline. *Phys. Chem. Chem. Phys.* **2004**, *6*, 2543–2556.

50. Senent, M.; Villa, M.; Dominguez-Gomez, R.; Fernandez-Clavero, A. Ab initio study of the far infrared spectrum of glycine. *Int. J. Quantum Chem.* **2020**, *104*, 551–561.

51. Barno, V.; Biczysko, M.; Blonio, J.; Puzzarini, C. Accurate structure, thermodynamic and spectroscopic parameters from CC and CC/DFT schemes: the challenge of the conformational equilibrium in glycine. *Phys. Chem. Chem. Phys.* **2013**, *15*, 10094–10111.

52. Kuan, Y-J.; Charnley, S. B.; Huang, H.-C.; Tseng, W.-L.; Kisiel, Z. Interstellar glycine. *Astrophys. J.* **2003**, *593*, 848–867.

53. Snyder, L. E.; Lovas, F. J.; Hollis, J. M.; Friedel, D. N.; Jewell, P. R.; Raemijan, A.; Ilyushin, V. V.; Alekseev, E. A.; Dyubko, S. F. A rigorous attempt to verify inter-stellar glycine. *Astrophys. J.* **2005**, *619*, 914–930.

54. Jones, P. A.; Cunningham, M. R.; Godfrey, P. D.; Cragg, D. M. A Search for biomolecules in Sagittarius B2 (LMH) with the Australia Telescope Compact array. *Mon. Not. R. Astron. Soc.* **2007**, *374*, 579–589.

55. Cunningham, M. R.; Jones, P. A.; Godfrey, P. D.; Cragg, D. M.; Bains, I.; Burton, M. G.; Calisse, P.; Crighton, N. J. M; Curran, S. J.; Davis, T. M.; *et al.* A search for propylene oxide and glycine in Sagittarius B2 (LMH) and Orion. *Mon. Not. R. Astron. Soc.* **2007**, *376*, 1201–1210.

56. Bazsó, G.; Magyarfalvi, G.; Tarczay, G. Near-infrared laser induced conformational change and UV laser photolysis of glycine in low-temperature matrices: observation of a short-lived conformer. *J. Mol. Struct.* **2012**, *1025*. 33–42.

57. Bazsó, G.; Magyarfalvi, G.; Tarczay, G. Tunneling lifetime of the ttc/VIp conformer of glycine in low temperature matrices. *J. Phys. Chem. A* **2012**, *116*, 10539–10547.

58. Coussan, S.; Tarczay, G. Infrared laser induced conformational and structural changes of glycine and glycine water complex in low-temperature matrices. *Chem. Phys. Lett.* **2016**, *644*, 189–194.

59. Huisken, F.; Werhahn, O.; Ivanov, A. Y.; Krasnokutski, S. A. The O-H stretching vibrations of glycine trapped in rare gas matrices and helium clusters. *J. Chem. Phys.* **1999**, *111*, 2978–2984.

60. Ivanov, A. Y.; Sheina, G.; Blagoi, Y. P. FTIR spectroscopic study of the UV-induced rotamerization of glycine in the low temperature matrices (Kr, Ar, Ne). *Spectrochim. Acta* **1999**, *55*, 219–228.

61. Kumar, S.; Rai, A. R.; Singh, V. B.; Rai S. B. Vibrational spectrum of glycine molecule. *Spectrochim. Acta* **2005**, *61*, 2741–2746.

62. Balabin, R. M. Conformational equilibrium in glycine: experimental jet-colled Raman spectrum. *J. Phys. Chem. Lett.* **2010**, *1*, 20–23.

63. Balabin, R. M. Experimental thermodynamics of free glycine conformations: the first Raman experiment after twenty years of calculations. *Phys. Chem. Chem. Phys.* **2012**, *14*, 99–103.

64. Braams, B. J.; Bowman, J. M. Permutationally invariant potential energy surfaces in high dimensionally. *Int. Rev. Phys. Chem.* **2009**, *28*, 577–606.

65. Nandi, A.; Qu, C.; Bowman, J. M. Using gradients in permutationally invariant polynomial potential fitting: a demonstration for CH4 using as few as 100 configuration. *J. Chem. Theory Comput.* **2019**, *15*, 2826–2835.

66. Conte, R.; Qu, C.; Houston, P. L.; Bowman, J. M. Efficient generation of permutationally invariant potential energy surfaces for large molecules. *J. Chem. Theory Comput.* **2020**, *16*, 3264–3272.

67. Qu, C.; Houston, P. L.; Conte, R.; Nandi, A.; Bowman, J. M. MULTIMODE calculations of vibratonal spectroscopy and 1d interconformer tunneling dynamics in glycine using a full-dimensional potential energy surface. *J. Phys. Chem. A* **2021**, *125*, 5346–5354.

68. Heberle, J.; Riesle, J.; Thiedemann, G.; Oesterhelt, D.; Dencher, N. A. Proton migration along the membrane surface and retarded surface to bulk transfer. *Nature* **1994**, *370*, 379.

69. Stowell, M. H. B.; McPhillips, T. M.; Rees, D. C.; Soltis, S. M.; Abresch, E.; Feher, G. Light-induced structural changes in photosynthetic reaction center: implications for mechanism of electron-proton transfer. *Science* **1997**, *276*, 812–816.

70. Luecke, H.; Richter, H.-T.; Lanyi, J. K. Proton transfer pathways in bacteriorhodopsin at 2.3 angstrom resolution. *Science* **1998**, *280*, 1934–1937.

71. Rini, M.; Magnes, B.-Z.; Pines, E.; Nibbering, E. T. J. Real-time observation of bimodal proton transfer in acid-base pairs in water. *Science* **2003**, *301*, 349–352.

72. de Grotthuss.; T. C. J. Sur la décomposition de l'eau et des corps qu'elle tient en dissolution à l'aide de l'électricité galvanique. *Ann. Chim.* **1806**, *58*, 54.

73. Agmon, N. The Grotthuss mechanism. *Chem. Phys. Lett.* **1995**, *244*, 456.

74. Fournier, J. A.; Wolke, C. T.; Johnson, M. A.; Odbadrakh, T. T.; Jordan, K. D.; Kathmann, S. M.; Xantheas, S. S. Snapshots of proton accommodation at a microscopic water surface: Understanding the vibrational spectral signatures of the charge defect in cryogenically cooled $H^+(H_2O)_n$=2–28 clusters. *J. Phys. Chem. A* **2015**, *119*, 9425–9440.

75. Wang, H.; Agmon, N. Reinvestigation of the infrared spectrum of the gas-phase protonated water tetramer. *J. Phys. Chem. A* **2017**, *121*, 3056–3070.

76. Christoffel, K.; Bowman, J. Investigations of self-consistent field, SCF CI and virtual state configuration interaction vibrational energies for a model three-mode system. *Chem. Phys. Lett.* **1982**, *85*, 220–224.

77. McCoy, A. B.; Huang, X.; Carter, S.; Landeweer, M. Y.; Bowman, J. M. Full-dimensional vibrational valculations for $H_5O_2^+$ using an ab initio potential energy surface. *J. Chem. Phys.* **2005**, *122*, 061101.

78. Yu, Q.; Bowman, J. M. Communication: VSCF/VCI vibrational spectroscopy of H_7O_{3+} and H_9O_{4+} using high-level, many-body potential energy surface and dipole moment surfaces. *J. Chem. Phys.* **2017**, *146*, 121102.

79. Yu, Q.; Bowman, J. M. High-level quantum calculations of the IR spectra of the Eigen, Zundel, and ring isomers of $H^+(H_2O)_4$ find a single match to experiment. *J. Am. Chem. Soc.* **2017**, *139*, 10984–10987.

80. Vendrell, O.; Gatti, F.; Meyer, H. D. Dynamics and infrared spectroscopy of the protonated water dimer. *Angew. Chem. Int. Ed.* **2007**, *46*, 6918–6921.

81. Vendrell, O.; Gatti, F.; Meyer, H. D. Strong isotope effects in the infrared spectrum of the Zundel cation. *Angew. Chem. Int. Ed.* **2009**, *48*, 352–355.

82. Thomas, M.; Brehm, M.; Fligg, R.; Vohringer, P.; Kirchner, B. Computing vibrational spectra from ab initio molecular dynamics. *Phys. Chem. Chem. Phys.* **2013**, *15*, 6608–6622.

83. Miller, W. H. The semiclassical initial value representation: a potentially practical way for adding quantum effects to classical molecular dynamics simulations. *J. Phys. Chem. A* **2001**, *105*, 2942–2955.

84. Liu, J.; Miller, W. H.; Paesani, F.; Zhang, W.; Case, D. A. Quantum dynamical effects in liquid water: a semiclassical study on the diffusion and the infrared absorption spectrum. *J. Chem. Phys.* **2009**, *131*, 164509.

85. Habershon, S.; Fanourgakis, G. S.; Manolopoulos, D. E. Comparison of path integral molecular dynamics methods for the infrared absorption spectrum of liquid water. *J. Chem. Phys.* **2008**, *129*, 074501.

86. Rossi, M.; Ceriotti, M.; Manolopoulos, D. E. How to remove the spurious resonances from ring polymer molecular dynamics. *J. Chem. Phys.* **2014**, *140*, 234116.

87. Paesani, F.; Voth, G. A. A quantitative assessment of the accuracy of centroid molecular dynamics for the calculation of the infrared spectrum of liquid water. *J. Chem. Phys.* **2010**, *132*, 014105.

88. Medders, G. R.; Paesani, F. Infrared and Raman spectroscopy of liquid water through "first-principles" many-body molecular dynamics. *J. Chem. Theory Comput.* **2015**, *11*, 1145–1154.

89. Medders, G. R.; Paesani, F. On the interplay of the potential energy and dipole moment surfaces in controlling the infrared activity of liquid water. *J. Chem. Phys.* **2015**, *142*, 212411.

90. Liu, H.; Wang, Y.; Bowman, J. M. Quantum calculations of intramolecular IR spectra of ice models using ab initio potential and dipole moment surfaces. *J. Phys. Chem. Lett.* **2012**, *3*, 3671–3676.

91. Rossi, M.; Liu, H.; Paesani, F.; Bowman, J.; Ceriotti, M. Communication: On the consistency of approximate quantum dynamics simulation methods for vibrational spectra in the condensed phase. *J. Chem. Phys.* **2014**, *141*, 181101.

92. Yu, Q.; Bowman, J. M. Classical, Thermostated ring polymer, and quantum VSCF/VCI calculations of IR spectra of $H_7O_3^+$ and $H_9O_4^+$ (Eigen) and comparison with experiment. *J. Phys. Chem. A* **2019**, *123*, 1399–1409.

93. Duong, C. H.; Gorlova, O.; Yang, N.; Kelleher, P. J.; Johnson, M. A.; McCoy, A. B.; Yu, Q.; Bowman, J. M. Disentangling the complex vibrational spectrum of the protonated water trimer, $H^+(H_2O)_3$, with two-color IR-IR photodissociation of the bare ion and anharmonic VSCF/VCI theory. *J. Phys. Chem. Lett.* **2017**, *8*, 3782–3789.

94. Qu, C.; Yu, Q.; Bowman, J. M. Permutationally invariant potential energy surfaces. *Annu. Rev. Phys. Chem.* **2018**, *69*, 6.1–6.25.

95. Heindel, J. P.; Yu, Q.; Bowman, J. M.; Xantheas, S. S. Benchmark electronic structure calculations for $H_3O^+(H_2O)_n$, n=0-5 clusters and tests of an existing 1,2,3-body potential energy surface with a new 4-body correction. *J. Chem. Theory Comput.* **2018**, *14*, 4553–3566.

96. Esser, T. K.; Knorke, H.; Asmis, K. R.; Schöllkopf, W.; Yu, Q.; Qu, C.; Bowman, J. M.; Kaledin, M. Deconstructing prominent bands in the terahertz spectra of $H_7O_3^+$ and $H_9O_4^+$: Intermolecular modes in Eigen clusters. *J. Phys. Chem. Lett.* **2018**, *9*, 798–803

97. Ceriotti, M.; More, J.; Manolopoulos, D. E. i-PI: A Python interface for ab initio path integral molecular dynamics simulations. *Comput. Phys. Commun.* **2014**, *185*, 1019–1026.

98. Rossi, M.; Kapil, V.; Ceriotti, M. Fine tuning classical and quantum molecular dynamics using a generalized Langevin equation. *J. Chem. Phys.* **2018**, *148*, 102301.

99. Wolke, C. T.; Fournier, J. A.; Dzugan, L. C.; Fagiani, M. R.; Odbadrakh, T. T.; Knorke, H.; Jordan, K. D.; McCoy, A. B.; Asmis, K. R.; Johnson, M. A. Spectroscopic snapshots of the proton-transfer mechanism in water. *Science* **2016**, *354*, 1131.

https://doi.org/10.1142/9789811237911_0009

Chapter 9

Vibrational Spectra of Flexible Systems using the MCTDH Approach

H.-D. Meyer, M. Schröder, and O. Vendrell*

*Theoretische Chemie, Physikalisch-Chemisches Institut,
Universität Heidelberg Im Neuenheimer Feld 229,
69120 Heidelberg, Germany
oriol.vendrell@pci.uni-heidelberg.de

9.1. Introduction

In this chapter, we discuss how to solve problems of (molecular) quantum dynamics with a time-dependent method. A time-dependent approach has the obvious disadvantage compared to a time-independent one that there is one more variable, the time. However, a time-dependent approach offers advantages: it can treat continuum states in a more natural way, important for dissociating systems, and it is able to work with time-dependent Hamiltonians, *e.g.*, it can easily include a laser field. More importantly, the time-dependent Schrödinger equation poses an initial value problem rather than an eigenvalue problem. The former is simpler in structure and, as only the quantum states present in the initial wavefunction are propagated, allows us to concentrate on relevant features. This is important when treating large systems with a high density of states.

A time-dependent method consists of three steps:

(1) Generation of an initial wave function;
(2) Propagation of the wave function;

(3) Analysis of the time-dependent wavefunction to obtain the desired observables like spectra or cross-sections.

The last step is often performed by a Fourier-transform of a correlation function. The most elaborate task is, of course, the propagation. Here we employ the Multi-configuration time-dependent Hartree (MCTDH) approach and its multi-layer extension, ML-MCTDH. The (ML-)MCTDH theory is briefly reviewed in Section 9.9.2. In Section 9.3, we discuss the sum-of-products form of potential energy surfaces, in Section 9.4, proton transfer processes in malonaldehyde, and in Section 9.5, transient infrared absorption spectroscopy of the Zundel cation.

9.2. Multi-Configuration Time-Dependent Hartree

9.2.1. *Standard method*

As explained in Section 9.1, the aim is to solve the *time-dependent* Schrödinger equation. The most direct way to solve it by a basis set method is to expand the wavefunction into a direct product basis and to solve the resulting equations of motion. The f-dimensional wavefunction is hence expanded as

$$\Psi(q_1, \ldots q_f, t) = \sum_{l_1=1}^{N_1} \cdots \sum_{l_f=1}^{N_f} C_{l_1 \ldots l_f}(t)\, \chi_{l_1}^{(1)}(q_1) \cdots \chi_{l_f}^{(f)}(q_f), \quad (9.1)$$

where N_κ denotes the number of basis functions employed for the κth degree of freedom (DOF), $\{\chi^{(\kappa)}\}$ denotes a set of 1D orthonormal basis functions for the κ-th DOF, and $\{q_\kappa\}_{\kappa=1}^{f}$ is a set of internal coordinates. Often a grid representation, *e.g.*, a *Discrete Variable Representation* (DVR),[1–5] is used. In this case $C_{l_1 \ldots l_f}$ is the amplitude of the wavefunction at the grid point $(l_1 \cdots l_f)$, multiplied with DVR-weights. The C-tensor then directly represents the wavefunction on the grid.

Applying the Dirac-Frenkel variational principle[6,7] yields the well-known equations of motion

$$i\dot{C}_{j_1, \ldots j_f} = \sum_{l_1, \ldots l_f} \langle \chi_{j_1}^{(1)} \cdots \chi_{j_f}^{(f)} | H | \chi_{l_1}^{(1)} \cdots \chi_{l_f}^{(f)} \rangle\, C_{l_1, \ldots l_f}. \quad (9.2)$$

The approach defined by Eqs. (9.1) and (9.2) is often called *Standard Method*; it is easy to implement and works well. However, the standard method is plagued by the *curse of dimensionality*; it shows a strong exponential increase of both memory and computation time with the number of DOFs. If all N_κ are alike, the C-tensor has N^f entries. This makes the standard method unsuitable for problems with more than 6 DOFs and in general unusable for more than 9 DOFs.

9.2.2. *MCTDH*

To overcome the limitations of the standard method we introduce an intermediate basis set of time-dependent functions $\varphi^{1;\kappa}$ and write the wave function as[8-13]

$$\Psi(q_1,\ldots,q_f,t) = \sum_{j_1=1}^{n_1} \cdots \sum_{j_f=1}^{n_f} A^1_{j_1\ldots j_f}(t) \prod_{\kappa=1}^{f} \varphi^{1;\kappa}_{j_\kappa}(q_\kappa,t)$$

$$= \sum_{J} A^1_J \, \Phi^1_J. \tag{9.3}$$

The second line implicitly defines the composite index J and the *Hartree product* or *configuration* Φ^1_J. (The meaning of the seemingly needless upper index 1 will become clear when Eq. (9.8) is discussed.) The *single particle functions* (SPFs) φ in turn are expanded in a primitive basis $\{\chi\}$

$$\varphi^{1;\kappa}_{j_\kappa}(q_\kappa,t) = \sum_{l=1}^{N_\kappa} A^{2;\kappa}_{j_\kappa,l}(t)\, \chi^{(\kappa)}_l(q_\kappa), \tag{9.4}$$

i.e., one uses a standard method expansion for the SPFs. The SPFs are propagated according to a variational scheme and follow the wave packet. The number of SPFs, n_κ, needed to converge the expansion of the wave function is in general considerably smaller than the number of primitive functions, N_κ, needed to converge the expansion of the SPFs. MCTDH does not break the exponential scaling of the effort, but the scaling is to a lower base, n^f rather than N^f.

An important step forward towards reducing the numerical effort was the introduction of mode combination.[14,15] A SPF may not be a 1D but a low-dimensional function, which depends on a combined or logical coordinate

$$Q_\kappa = (q_{\kappa,1}, \ldots, q_{\kappa,d_\kappa}) \tag{9.5}$$

where d_κ denotes how many physical coordinates are combined to yield the logical coordinate Q_κ. The index κ no longer runs over the DOFs from 1 to f but rather over the combined particles from 1 to p. The expansion of the wavefunction is essentially still given by Eq. (9.3), one merely has to replace f with p and q with Q. The size of the A^1 tensor is considerably reduced, because the tensor has fewer indices as the number of particles p is smaller than the number of DOFs f ($p = f/d$ if all d_κ are alike). The propagation of the A^1 tensor is thus faster, but the propagation of the combined SPFs

$$\varphi_{j_\kappa}^{1;\kappa}(Q_\kappa, t) = \sum_{l_1=1}^{N_{\kappa,1}} \cdots \sum_{l_{d_\kappa}=1}^{N_{\kappa,d_\kappa}} A_{j_\kappa,l_1\cdots l_{d_\kappa}}^{2;\kappa}(t)\, \chi_{l_1}^{(\kappa,1)}(q_{\kappa,1}) \cdots \chi_{l_{d_\kappa}}^{(\kappa,d_\kappa)}(q_{\kappa,d_\kappa})$$

$$\tag{9.6}$$

becomes more elaborate as multi-dimensional functions are to be propagated. The combination orders, d_κ, are typically rather small, 1, 2, or 3, only in special cases they may take values up to 4 or 5.

MCTDH with mode combination has proven to be very powerful. Already in 1999 we could investigate a realistic vibronic model of pyrazine by propagating a 24D wavefunction on two coupled electronic states.[16]

9.2.3. *ML-MCTDH*

The mode combination trick is limited, because it becomes more and more difficult to propagate larger and larger combined SPFs when going to bigger systems. But we know a method to efficiently propagate multi-dimensional wavefunctions: MCTDH. One thus may use the MCTDH algorithm to propagate the SPFs of an underlying MCTDH expansion. This is the idea of the ML-MCTDH approach. A vivid illustration of the approach is provided by pictures of ML-trees, see Figure 9.1. A rectangular box represents a primitive basis function and a circle represents a node, *i.e.*, a coefficient tensor.

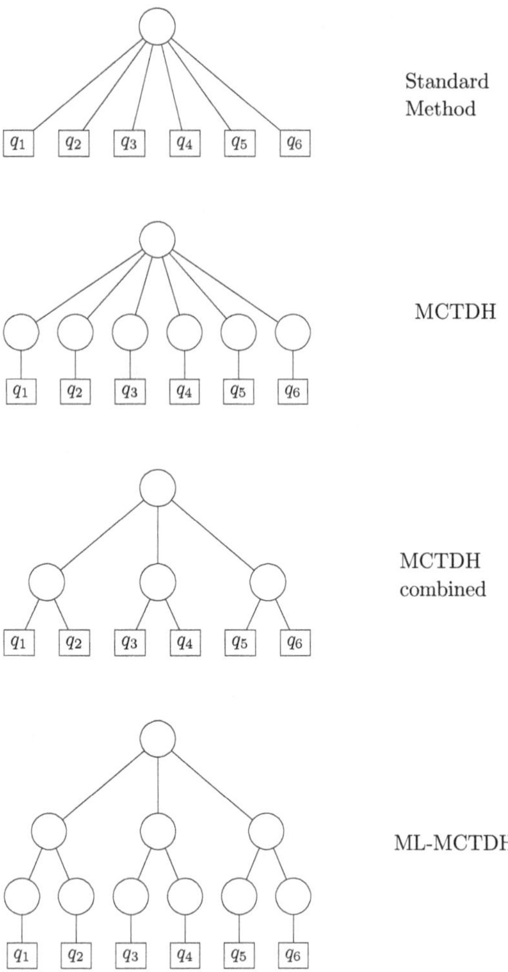

Figure 9.1. Trees of standard method, MCTDH, and ML-MCTDH.

Counting the nodes from the top down along a branch gives the layer number. The last figure thus shows a three-layers tree. A two-layers ML-MCTDH calculation is equivalent to a MCTDH one, and the standard method may be viewed as a one-layer approach. Note that ML-MCTDH can in addition make use of mode combination. For additional examples of ML-trees see Figures 9.3 and 9.4.

To formally discuss the ML-MCTDH expansion[17–20] one has to specify which node of the ML-tree is under consideration. For this it is convenient to introduce the symbol

$$z = (l; \kappa_1, \ldots, \kappa_{l-1}) \quad \text{and} \quad z-1 = (l-1; \kappa_1, \ldots, \kappa_{l-2}). \qquad (9.7)$$

Here l denotes the layer and $\kappa_1, \ldots, \kappa_{l-1}$ denote the modes of the layers above which leads to the node under consideration. Hence, z precisely defines one particular node in the ML-tree.[18,19]

The ML expansion can now be written in full generality

$$\varphi_m^{z-1,\kappa_{l-1}}(q_{\kappa_{l-1}}^{z-1}) = \sum_{j_1=1}^{n_1^z} \cdots \sum_{j_{p^z}=1}^{n_{p_z}^z} A_{m;j_1,\ldots,j_{p^z}}^z \prod_{\kappa_l=1}^{p^z} \varphi_{j_{\kappa_l}}^{z,\kappa_l}(q_{\kappa_l}^z)$$

$$= \sum_J A_{m;J}^z \cdot \Phi_J^z(q_{\kappa_{l-1}}^{z-1}), \qquad (9.8)$$

where the coordinates are combined as

$$q_{\kappa_{l-1}}^{z-1} = (q_1^z, \ldots, q_{p^z}^z). \qquad (9.9)$$

Eq. (9.8) holds also for the top layer if one defines q^0 as the combination of all DOFs and sets $\Psi(q^0, t) = \varphi_1^0(q^0, t)$, and it also holds for the bottom layer if one identifies φ^{z,κ_l} with the primitive function χ^{κ_l}, when z points to a bottom node. Hence Eq. (9.8) holds for all layers. It also includes the MCTDII expansion equations, compare with Eqs. (9.3), (9.4) and (9.6).

The equations of motion (EOM) are derived using the Dirac–Frenkel variational principle. The EOM for the top layer has a particular simple appearance.

$$i\frac{\partial A_I^1}{\partial t} = \sum_J \left\langle \Phi_I^1 \middle| \hat{H} \middle| \Phi_J^1 \right\rangle A_J^1. \qquad (9.10)$$

This equation looks similar to the EOM of the standard method, Eq. (9.2). Remember, however, that here the configurations Φ_I^1 are time-dependent.

The EOM for the propagation of the SPFs are formally the same for all layers

$$i\frac{\partial \varphi_n^{z,\kappa_l}}{\partial t} = (1 - P^{z,\kappa_l}) \sum_{j,m} (\rho^{z,\kappa_l})_{nj}^{-1} \cdot \langle \hat{H} \rangle_{jm}^{z,\kappa_l} \varphi_m^{z,\kappa_l}, \qquad (9.11)$$

where

$$P^{z,\kappa_l} = \sum_j |\varphi_j^{z,\kappa_l}\rangle\langle\varphi_j^{z,\kappa_l}| \qquad (9.12)$$

is the projector onto the space spanned by the φ_j^{z,κ_l} SPFs, ρ^{z,κ_l} is a one-particle reduced density matrix and $\langle\hat{H}\rangle^{z,\kappa_l}$ is a matrix of mean-field operators acting on the φ_j^{z,κ_l} functions. For the definitions of the density and the mean-field operators see the MCTDH[9–13, 21] and the ML-MCTDH[17–20] literature.

The EOM for MCTDH and ML-MCTDH have the same appearance, but the computation of the density matrices and mean-fields entering the EOM is more involved in ML-MCTDH than in MCTDH. The discovery by Manthe[18] that these objects can be computed recursively, layer by layer, made the computation simpler and made it possible to write ML-MCTDH codes for an arbitrary number of layers.

The positive semi-definite density matrix, ρ, becomes singular when there are unoccupied SPFs and Eq. (9.11) ceases to be well defined. This problem is usually solved by simply regularizing ρ, *i.e.*, by replacing zero eigenvalues with small constants. This procedure works fine for MCTDH, but for ML-MCTDH, where often large combined modes appear, one may run into convergence problems in special cases.[22, 23] This has led to the development of an improved regularization scheme,[22–24] and to the derivation of transformed, singularity-free EOM.[25–28] A third way includes only occupied SPFs; new SPFs are added on the fly when needed.[29, 30]

With the ML-MCTDH method one can solve problems of high dimensionality, and in suitable cases of very high dimensionality. We refer to four examples from the literature, ranging from 186D[31] over 256D[32] and 1458D[19] to 100,000D.[23]

9.2.4. *The constant mean-field integrator and the improved relaxation algorithm*

The EOM (Eqs. (9.10), (9.11)) form a set of coupled non-linear first-order differential equations. They can be solved with a standard numerical integration method. This approach is called *variable mean-field* (VMF) integration scheme. The numerical integrator has to take rather small step sizes, because the wavefunction has oscillatory components. The mean-fields, which are rather costly to evaluate, change less rapidly. One hence may keep them constant and update them after some update-time τ_{up}, which is considerably larger than the integrator step size. Doing so, the EOM (Eqs. (9.10), (9.11)) decouple from each other and the integrator may take individual optimal step sizes for A^1 propagation and for each set of SPFs at a given node and DOF: z, κ_l. As Eq. (9.10) is now linear, a *short iterative Lanczos* (SIL)[33] integrator is used in general. The approach outlined is called *constant mean-field* (CMF) integration scheme. The actual implementation[12,34,35] is more complicated and consists of a predictor and a corrector step. This improves the accuracy and allows us to automatically determine an optimal update time τ_{up}.

(ML-)MCTDH is a time-propagation method and was not developed for computing eigenvalues and eigenvectors. However, the ground state of an Hamiltonian can be computed with (ML-)MCTDH by propagation in negative imaginary time, the so-called relaxation method.[36] Based on the CMF integration scheme an improved relaxation algorithm has been developed for MCTDH. Here the SPFs are relaxed by imaginary time propagation, but the A^1-tensor is taken as eigenvector of the matrix representation of the Hamiltonian in the top-layer configurations (see Eq. (9.10)) at the end of each update step.[12,37] By choosing the appropriate eigenvector the improved relaxation algorithm can converge to excited states as well. A block form of improved relaxation is available with which one can relax a set of states simultaneously.[38] Improved relaxation is implemented in the Heidelberg package[39] for MCTDH but not for ML-MCTDH. However, other researchers have implemented improved relaxation in their ML-codes.[40,41]

9.2.5. *(ML-)MCTDH viewed as tensor contraction method*

The (ML-)MCTDH method was derived with the idea of introducing time-dependent intermediate basis sets. But it is interesting to view it from a different angle. The standard method is impractical because the C-tensor becomes too large. However, tensor contraction methods make it possible to work with formally very large tensors. A often used tensor contraction scheme is the Tucker format. Using this scheme one expands the C-tensor as

$$C_{l_1 \ldots l_f}(t) = \sum_{j_1=1}^{n_1} \cdots \sum_{j_f=1}^{n_f} A^1_{j_1 \ldots j_f}(t)\, A^{2,1}_{j_1,l_1}(t) \cdots A^{2,\kappa}_{j_\kappa,l_\kappa}(t) \cdots A^{2,f}_{j_f,l_f}(t).$$

$$(9.13)$$

This is equivalent to Eqs. (9.3) and (9.4), hence MCTDH is in essence a Tucker-decomposition of the wavefunction tensor C, and A^1 is called *core tensor*. Similarly, ML-MCTDH can be viewed as a hierarchical Tucker-decomposition. Note however, that ML-MCTDH was derived before the mathematicians introduced the notion *hierarchical Tucker-decomposition*. Note also that the tensor-decomposition describes only how the wavefunction representation is structured. To derive the EOM is a different story.

Viewing (ML-)MCTDH as a tensor-contraction method is helpful as this builds a bridge to other methods like the well-known *density matrix renormalization group* (DMRG) approach. DMRG uses matrix product states (MPS) rather than a (hierarchical) Tucker format to achieve a tensor decomposition. A MPS tree, called tensor-train (TT) tree by the mathematicians, can be viewed as a special form of a ML-tree.

9.3. Sum-of-Products Form of High-Dimensional Potential Energy Surfaces

9.3.1. *Evaluation of high-dimensional integrals*

When solving Eq. (9.10) of the (ML-)MCTDH EOM, one has to evaluate at each time step the matrix representation of the Hamiltonian, which formally is an f-dimensional integral. Performing this integral directly by quadrature over the product grid would be very time consuming or even impossible. To make (ML-)MCTDH work, a fast algorithm is needed for computing the matrix representation of the Hamiltonian as well as the mean-fields, which are given by similar integrals. To this end Manthe has developed a correlation-DVR (CDVR) which, in essence, does the quadrature over a coarse grid build from the SPFs and then adds a Smolyak-like correction by including 1D fine grids. The advantage of this approach is that one can link a routine, which evaluates the potential energy surface (PES), directly with the (ML-)MCTDH code. The disadvantage is that this routine is called very often during each time step, which may slow down the propagation.

The Heidelberg package, on the other hand, requires that the Hamiltonian is given as a sum of products, *i.e.*,

$$H = \sum_{r=1}^{R} \prod_{\kappa=1}^{p} h_{\kappa,r} \tag{9.14}$$

where $h_{\kappa,r}$ operates on a (combined) bottom-layer coordinate Q_κ only. With this form, the Hamiltonian matrix elements turn into a sum of products of 1D or low-dimensional integrals. The latter can be done fast and the sum-of-products (SOP) form may speed up the evaluation of the integral by several orders of magnitude. (The actual speed-up obviously depends on R, the number of terms of the SOP expansion.)

Fortunately, the kinetic energy operator is in general already in SOP form; exceptions may occur when very special coordinate systems are used. The potential energy, on the other hand, is typically a high-dimensional hypersurface in coordinate representation. For simple models (vibronic coupling Hamiltonian, Henon-Heiles, spin-boson) one is able to construct a model surface in SOP form, but realistic and highly accurate surfaces lack such a simple form. These surfaces are usually available as numerical library functions that take a coordinate vector as input and return an energy value. Typically, these routines interpolate between values from high precision quantum chemistry calculations on a large number of supporting points. It is therefore important to develop methods which can transform a given multi-dimensional PES to SOP form, while introducing only negligible re-fitting errors and keeping the expansion order, R, as low as possible.

9.3.2. *PES re-fitting*

The starting point to achieve a decomposition of the PES is its representation on primitive product grid points:

$$V(Q_{i_1}^{(1)}, Q_{i_2}^{(2)}, \ldots, Q_{i_p}^{(p)}) = V_{i_1, i_2, \cdots, i_p} = V_I, \qquad (9.15)$$

where $Q_{i_\kappa}^{(\kappa)}$ denotes the coordinate of the combined bottom-layer mode Q_κ at the grid point i_κ. Like the wavefunction, the potential is stored as tensor on the primitive points.

A natural choice for a tensor decomposition of the potential energy surface is a Tucker format similar to the one used for the wavefunction. This has been implemented in the Heidelberg MCTDH package as the POTFIT algorithm[10,42,43] and its multi-layer extension.[44] POTFIT performs the decomposition through the generation of reduced one-mode densities of the potential and subsequent eigen-decompositions of the latter. The respective eigenvectors, also called single-particle potentials (SPP), serve as basis vectors for the Tucker tensor. The core-tensor is subsequently calculated by overlap of outer products of the SPP with the original potential.

One may readily see that this procedure is limited by the curse of dimensionality as the full tensor needs to be held in memory and multiple contractions over it must be performed. Recent developments tried to mitigate this by replacing full-grid quadrature by sparse integrals, most notably the multi-grid POTFIT[45] algorithm that uses regular sparse grids and the Monte Carlo (MC) POTFIT algorithm[46] and its multi-layer variant[47] that use randomly distributed points to perform the integrals.

Other methods employ machine learning techniques to achieve a SOP decomposition: specifically designed neural networks[48-54] can be trained to reproduce the potential energy given a coordinate vector as input. The networks are designed in such a way that the extraction of a SOP representation of the potential they have been trained with is directly possible. Further methods use expansions of the PES into n-particle interaction terms, also known as n-mode representation[55-57] or high-dimensional model representation (HDMR).[49,58-60] The expansion is truncated at low orders such that the n-particle terms can be further decomposed into a SOP form using traditional tensor-decomposition tools like POTFIT.

In the following a further tensor decomposition method is discussed: the canonical polyadic decomposition[61] (CPD), in the literature also referred to as canonical decomposition (CANDECOMP), or parallel factors (PARAFAC). The potential, Eq. (9.15), in CPD form can be written as

$$V_{i_1,i_2,\cdots,i_p} \approx V^{\mathrm{CPD}}_{i_1,i_2,\cdots,i_p} = \sum_{r=1}^{R} c_r\, \nu^{(1)}_{r,i_1} \nu^{(2)}_{r,i_2} \cdots \nu^{(p)}_{r,i_p} \tag{9.16}$$

where R is the expansion order, also called the *rank* of the decomposition, $\nu_r^{(\kappa)}$ are basis functions that only depend on a single physical index i_κ, and c_r are expansion coefficients. The main difference to a Tucker decomposition is that the basis functions for one physical index are not restricted to be orthogonal. This in fact is essential to achieve a compact decomposition. Often one demands the $\nu_r^{(\kappa)}$ to be normalized which gives rise to expansion coefficients which could otherwise be absorbed into one of the expansion functions.

Obtaining the decomposition Eq. (9.16) is a highly non-linear task. Furthermore, unlike for Tucker, determining the expansion order necessary to achieve a pre-defined quality of the fit is difficult.[62] In practice, one can use an alternating least squares (ALS) algorithm[63,64] to obtain an optimized CPD expansion. The ALS algorithm optimizes the basis functions for one mode while keeping those of the other modes fixed. The process iteratively sweeps over all modes using results from previous sweeps to improve the results. We found it useful to repeatedly grow the expansion order after a number of ALS optimizations until sufficient accuracy is reached.

Such an algorithm, however, requires the repeated evaluation of high-dimensional contractions over the tensor. Within the Heidelberg MCTDH package we therefore implemented[65] a MC variant of the ALS algorithm that performs the contraction using MC integrals. The algorithm uses a set of random sampling points which may be distributed according to some coordinate-dependent weight function. The sampling points comprise a small subset of the total primitive product grid. For each of the sampling points the algorithm performs a one-mode scan of the potential along each grid-index while keeping all other indices fixed such that all 1-mode scans cross at the respective sampling point. Such one-mode scans are also called *1D-cuts* or *fibers* of the tensor as they can be stored as vectors with length of one dimension of the tensor.

Within the ALS algorithm the basis functions $\nu_r^{(\kappa)}$ are then optimized along all fibers for the mode κ while evaluating integrals over the other modes using MC integrals. The sparsity of the sampling points not only mitigates the curse of dimensionality, their distribution also allows for importance sampling, *i.e.*, emphasizing regions of the potential which should be fitted with elevated accuracy by using a higher density of points in these regions. Such regions are typically low energy regions of the potential where the wavefunction resides.

In the simplest case the sampling points are uniformly distributed without importance sampling. Other choices could be Boltzmann-distributed sampling points obtained by propagating random walkers with a Metropolis algorithm or sampling obtained with a diffusion MC algorithms, Path-Integral methods, *etc.*

One shortcoming of the MC sampling, however, is that it spoils symmetries. In general, symmetries will not be retained within the fitted potential if it is obtained with a procedure outlined above. This can be cured either by using symmetrized sampling (which is numerically expensive), or by imposing symmetry constraints to the fit during the optimization. The latter is easily possible if the used symmetry operations do not mix combined modes, that is, if symmetry operations only mix coordinates within one combined mode and/or swap complete combined modes. This allows for the implementation of a symmetry-adapted ALS algorithm as outlined in Ref. 65.

9.4. Malonaldehyde

Malonaldehyde is one of the benchmark systems for modeling proton transfer processes and large amplitude motions within high-dimensional molecular systems. It consists of nine atoms

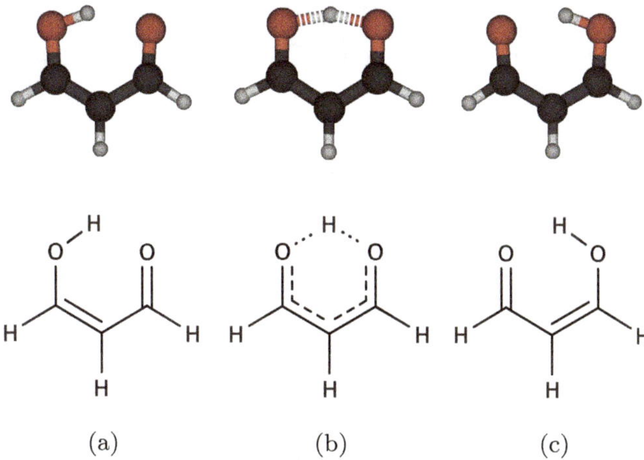

Figure 9.2.   Chemical structure of malonaldehyde. Proton transfer between the two oxygen atoms leads to re-arrangement of double bonds. Inter-atomic distances change substantially.

arranged in a horseshoe-shaped chain of three carbon atoms with conjugated double bonds as depicted in Figure 9.2. The carbon chain is terminated on each end by one oxygen atom and saturated with hydrogen. One additional proton is evenly shared between the two oxygen atoms: it can be located on either one of those atoms which gives rise to a tunneling splitting. Upon tunneling of the proton, large scale reorganizations of the double bonds take place which change the inter-atomic distances of the backbone. Due to this reorganization, the system is highly correlated, which makes an accurate quantum-mechanical description within a full, 21D, approach very difficult.

9.4.1. *Coordinates and system Hamiltonian*

9.4.1.1. *Normal mode coordinates*

In the following we repeat calculations we have performed approximately one decade ago. At this time, techniques to transform PESs directly and globally into a SOP form were not available such that we resorted to an n-mode representation of the potential where elaborate studies had to be performed to identify relevant terms of the expansion. The details of these studies can be found in Refs. 66 and 67. We use the same PES routine[68] as before, this time attempting a global SOP representation using the MC ALS method to obtain a canonical decomposition as outlined above.

We use the same set of coordinates which are normal mode coordinates $\tilde{q}_i$ with normal mode frequencies ω_i obtained at the transition state and primitive grids as in Ref. 67. To reduce the correlation and account for the changes of the inter-atomic distances the normal mode coordinates are transformed according to

$$
\begin{aligned}
q_i &= \tilde{q}_i - F_i(\tilde{q}_{21}), \quad i = 1, \dots, 20, \\
q_{21} &= \tilde{q}_{21},
\end{aligned}
\tag{9.17}
$$

with functions $F_i(\tilde{q}_{21})$ which resemble the minimum of the potential as the proton travels along the reaction coordinate (proton transfer coordinate) $\tilde{q}_{21}$. With this transformation the kinetic energy operator reads (neglecting rotation–vibration interaction)

$$
\begin{aligned}
\hat{T} = -\frac{1}{2} &\left[\sum_{i=1}^{21} \omega_i \frac{\partial^2}{\partial q_i^2} + \omega_{21} \sum_{i,j=1}^{20} F_i'(q_{21}) F_j'(q_{21}) \frac{\partial^2}{\partial q_i \partial q_j} \right. \\
&\left. - \omega_{21} \sum_{i=1}^{20} \left(F_i'(q_{21}) \frac{\partial}{\partial q_{21}} + \frac{\partial}{\partial q_{21}} F_i'(q_{21}) \right) \frac{\partial}{\partial q_i} \right].
\end{aligned}
\tag{9.18}
$$

9.4.1.2. *Effective Hamiltonian*

Note that this kinetic energy operator is approximate as vibration–rotation interaction terms ($\pi - \pi$ terms) have been neclected. Further reduction of the complexity of the system can be achieved by exploiting symmetries. At the transition state, malonaldehyde has C_{2v} symmetry with a reflection symmetry along the proton transfer coordinate. This can be used to construct two effective Hamiltonians H_+ and H_- for symmetric and antisymmetric states in the proton transfer coordinate, respectively. Effectively only one of the two equivalent potential wells needs to be modeled explicitly. For details the reader is referred to Hammer *et al.*[69] The effective time-independent Schrödinger equations for the symmetry-adapted

Hamiltonians read ($\hbar = 1$)

$$H_+ \left| \Psi_+ \right\rangle = \left(h \hat{H} h + h \hat{T} h' \mathcal{R} \right) \left| \Psi_+ \right\rangle = E_+ \left| \Psi_+ \right\rangle ,$$
$$H_- \left| \Psi_- \right\rangle = \left(h \hat{H} h - h \hat{T} h' \mathcal{R} \right) \left| \Psi_- \right\rangle = E_- \left| \Psi_- \right\rangle \tag{9.19}$$

where $\mathcal{R}$ is a reflection operator which changes the sign of proton transfer coordinate, h is a projector on one half of the transfer coordinate and $h' = 1 - h$ projects on the other half. The states $\left| \Psi_+ \right\rangle$ and $\left| \Psi_- \right\rangle$ are symmetric and anti-symmetric eigenstates, respectively. Eq. (9.19) can be solved independently for the symmetric and anti-symmetric states.

9.4.1.3. *Mode combinations*

Before transforming the PES into a SOP form the mode combinations need to be defined. We use slightly different mode combinations as in Ref. 67 as listed in Table 9.1 to better account for correlations already in the level of basis functions. The current mode combination scheme mostly resembles the mode combination scheme labeled "MC10" from Ref. 67. The most important alteration in the mode combination scheme here is that the proton transfer coordinate q_{21} is now combined with q_4 which mainly describes the O–O distance. These coordinates are naturally strongly correlated.

9.4.1.4. *PES fit*

The PES was fitted using a symmetry-adapted ALS algorithm in the above given logical coordinates with in total $R = 600$ terms using

Table 9.1. Mode combination scheme used in MCTDH and the potential fit of malonaldehyde.

Logical coord.	Physical coord.	No. of grid points
$\bar{Q}_1$	q_1, q_2, q_{18}	1573
$\bar{Q}_2$	q_3, q_{12} , q_{15}	1573
$\bar{Q}_3$	q_{13}, q_{14}, q_{16}, q_{17}	20449
$\bar{Q}_4$	q_5, q_6, q_9, q_{20}	20449
$\bar{Q}_5$	q_7, q_8, q_{10}, q_{19}	14641
$\bar{Q}_6$	q_4, q_{11}, q_{21}	6600

$N_{\text{gen}} = 2 \times 10^5$ sampling points. While the symmetry with respect to reflection of the proton transfer coordinate is already implicitely accounted for through the effective Schrödinger equations, Eq. (9.19), only the mirror-symmetry with respect to the plane of the molecule had to be explicitly implemented.

The sampling points have been obtained from a small Diffusion MC simulation which was implemented according to Ref. 70. The sampling points, therefore, resembled the density $|\Psi_0(\vec{Q})|$ of the system ground state wave function, $\Psi_0(\vec{Q})$, and hence emphasize the minimum of PES while also providing considerable sampling along the reaction path.

The obtained fit has been validated on an independent set of $N_{\text{test}} = 2 \times 10^6$ sampling points obtained with the same method. The root-mean-square fit-error

$$E_{\text{RMS}} = \sqrt{\frac{1}{N_{\text{test}}} \sum_s \left(V_s^{\text{CPD}} - V_s\right)^2} \tag{9.20}$$

with $s = (s_1, \ldots, s_p)$ a random sampling point out of the index $I = (i_1, \ldots, i_p)$ has been obtained as $62\,\text{cm}^{-1}$ and a mean fit error

$$E_{\text{mean}} = \frac{1}{N_{\text{test}}} \sum_s \left(V_s^{\text{CPD}} - V_s\right) \tag{9.21}$$

of $-0.4\,\text{cm}^{-1}$.

With the same validation set we obtain errors $E^{\text{RMS}} = 893\,\text{cm}^{-1}$ and $E^{\text{mean}} = 24.8\,\text{cm}^{-1}$ for the cluster expansion "MC13" used in Ref. 67. The error reported in Ref. 67. $E^{\text{RMS}} = 167\,\text{cm}^{-1}$ and $E^{\text{mean}} = -0.8\,\text{cm}^{-1}$ have been obtained with a much narrower sampling distribution of $|\Psi_0(\vec{Q})|^2$. We hence consider the current fit as more precise while at the same time being composed of much fewer terms than for the largest calculation "MC13" of Ref. 67 (600 terms in the present contribution versus $88,000$ in the latter case).

9.4.2. *Ground state energy and tunneling splitting*

Using Eq. (9.19) one can calculate the ground state wavefunctions and energies and hence the tunneling splitting using the *improved relaxation* algorithm[37,38] of MCTDH. Due to correlations between the different combined modes the obtained energies are not exact but are, due to the variational character of MCTDH, upper bounds of the energies. The variational character ensures that the obtained energies decrease as more basis functions are used for representing the wavefunction. The property can be exploited to extrapolate[69] the true ground state energy and tunneling splitting with a series of calculations with different numbers of basis functions.

To obtain an upper bound for the ground state energy one can use an extrapolation scheme as in Refs. 66, 67 and 71; one starts with a reasonably well converged MCTDH wavefunction that serves as a reference calculation. In this case an energy drop resulting from adding basis functions is approximately mode-local, *i.e.*, the energy drop is almost independent from the number of basis function in the other modes. Iterating through all modes and calculating the ground state energy with number of SPF doubled in the respective mode as compared to the reference calculation then results in a vector of mode-local energy drops. Adding the sum of the energy drops to the reference energy then results in a new upper bound of the true ground state energy.

For the present set of mode combinations we have performed such an extrapolation scheme as listed in Table 9.2. The first line of Table 9.2 outlines the reference calculation. In the first six columns the number of SPF for the six modes are outlined, respectively. In columns seven and nine follow the reference energies obtained by solving Eq. (9.19) for the symmetric (+) and asymmetric (−) ground state. The rightmost column contains the energy difference between the (+) and (−) states and hence the tunneling splitting.

In the following lines of Table 9.2 the same calculations are repeated but with one set of SPF doubled in size. The respective energy drops with respect to the reference calculation in the first line are listed in columns eight for (+) state and ten for the (−) state, respectively.

At the bottom of Table 9.2 the sum of the energy drops, the extrapolated ground state energies and the extrapolated splittings

Table 9.2.   Extrapolation of the ground state energy using the mode combination as outlined in table.

No. of SPF for										
Q_1	Q_2	Q_3	Q_4	Q_5	Q_6	Energy (+)	ΔE	Energy (−)	ΔE	Splitting
15	11	18	14	14	16	14670.771	—	14694.426	—	23.655
30	11	18	14	14	16	14670.520	−0.251	14694.155	−0.271	23.635
15	22	18	14	14	16	14670.694	−0.077	14694.343	−0.083	23.649
15	11	36	14	14	16	14670.483	−0.288	14694.115	−0.311	23.632
15	11	18	28	14	16	14670.519	−0.252	14694.151	−0.275	23.632
15	11	18	14	28	16	14670.374	−0.397	14694.049	−0.377	23.675
15	11	18	14	14	32	14670.498	−0.273	14694.153	−0.273	23.655
						Sum	−1.538		−1.590	
						Extrapolated energy	14669.233		14692.836	23.603

Note: The first line contains the reference calculation. In the lines below, ΔE denotes the difference to the reference state. All energies are in cm^{-1}.

can be found. Extrapolated zero point energy is obtained as $14669.2\,\mathrm{cm}^{-1}$ and the tunneling splitting as $23.6\,\mathrm{cm}^{-1}$.

In Ref. 67, a zero point energy of $14{,}664.6\,\mathrm{cm}^{-1}$ and a tunneling splitting of $22.7\ \mathrm{cm}^{-1}$ has been obtained for mode combination "MC13" while Hammer *et al.*[71] report a zero point energy of $14{,}676\,\mathrm{cm}^{-1}$ and a splitting of $23.6\,\mathrm{cm}^{-1}$ in a similar extrapolation scheme. Wang *et al.* obtained in their original publication a zero point energy of $14{,}678.3\,\mathrm{cm}^{-1}$ and a splitting of $22.6\,\mathrm{cm}^{-1}$ for this surface.[68]

The ground state energy of the previous[67] calculations is $4.6\ \mathrm{cm}^{-1}$ below the present result. This is most likely due to the fact that the n-mode representation tends to yield expansions which are slightly too low in energy. The ground state energy reported by Hammer *et al.*[71] is $6.8\ \mathrm{cm}^{-1}$ above our result. This is probably because the present MCTDH calculations are better converged.

In addition to the single-layer MCTDH calculations also ML-MCTDH calculations have been performed to obtain the ground state energy and tunneling splitting directly without extrapolation scheme. To this end, a propagaton in imaginary time has been performed. Figure 9.3 outlines the ML-tree used for both, the $(+)$ and the $(-)$ states. With this setup the ground state energy has been obtained as $14669.5\,\mathrm{cm}^{-1}$ and the tunneling splitting as $23.6\,\mathrm{cm}^{-1}$. The ground state energy is hence obtained about $0.3\,\mathrm{cm}^{-1}$ above the result of the extrapolation while the tunneling splittings are the same within $<0.1\,\mathrm{cm}^{-1}$ accuracy.

Note that all energies reported here, except those of Wang *et al.*, are obtained with the approximate kinetic energy operator

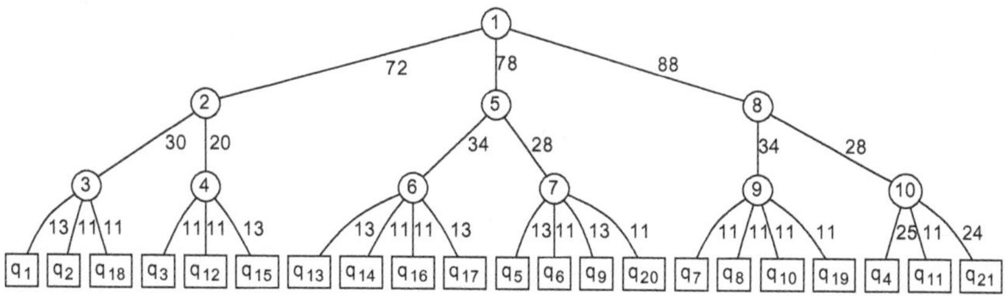

Figure 9.3.   ML-MCTDH tree for malonaldehyde.

in Eq. (9.18) without $\pi - \pi$ terms. Hammer *et al.* estimated that an inclusion of the missing terms increase the ground state energy by 2 cm^{-1} and the tunneling splitting by 0.2 cm^{-1}, respectively.

9.4.3. *Excited states*

In addition to the ground state, the state energies and tunneling splittings of the first few low-lying eigenstates have been calculated as outlined in Table 9.3 alongside with theoretical results from Refs. 67 and 72 and experimental data from Ref. 73. The state energies are given relative to the ground state energy and refer to the lower one of the respective $(+)$ or $(-)$ states which is except for the q_3 state in all cases the $(+)$ state. As above, the $\pi - \pi$ terms are neglected.

The state energies have been calculated with the block-improved relaxation scheme of MCTDH[13,38] where two blocks of four states have been used where the two blocks overlapped one state. The number of SPF in the two blocks have been set to $[24, 14, 17, 14, 14, 25]$ for the block with the first four states and to $[28, 15, 17, 15, 14, 28]$ for the second block for Q_1 to Q_6, respectively.

For the $(+)$ states a third block of three states (again with one state overlap with the second block) was included because the state energy of the $q_1 \times q_1$ state was obtained lower than the energy of the q_3 state such that these states changed positions compared to the calculations in Ref. 67. Hence, in the second $(+)$ block the q_3 state was no longer included.

Comparing the state energies listed in Table 9.3 one observes that the energies are consistently lower than the state energies obtained in Ref. 72. We believe that this is due to better converged wavefunctions in the present case. The first three excited state energies agree very well with previous state energies obtained in Ref. 67, especially for "MC10", where the energy of the q_4 state is obtained slightly lower than for both mode combinations used in Ref. 67. This most likely results from combining the highly correlated coordinates q_4 (O–O distance) and q_{21} (proton transfer) in one combined mode such that their correlation is accounted for already in the level of basis functions. This also reflects in the much lower energy of the $q_4 \times q_4$ and $q_1 \times q_4$ states.

Table 9.3. Excited states and tunneling splittings: this work and references.

	State energies				Tunneling splittings					
Ref[72]	MC10[67]	MC13[67]	This work	Exp.[73]	Ref[71]	MC10[67]	MC13[67]	This work	Exp.[73]	Assignment
0	0	0	0	0	23.5	23.5	22.4	23.7	21.6	ground state
259	254	252	249	241	64.0	64.8	64.2	63.4	57	q_4
325	271	270	271	273	6.7	17.3	13.9	17.3	$6\cdots9$	q_1
425	382	385	381	390	16.3	22.4	27.4	22.4	15	q_2
—	507	508	497	—	—	—	—	—	—	$q_4 \times q_4$
—	526	523	519	—	—	53.3	49.5	49.6	—	$q_1 \times q_4$
566	527	539	523	512	18.8	16.8	16.8	18.7	15	q_3
—	—	—	535	—	—	—	—	13.6	—	$q_1 \times q_1$

Note: Assignments are labeled according to the physical coordinate in which a node is present. All energies are given in cm^{-1} relative to the ground state and without extrapolation and $\pi - \pi$ terms.

Interestingly also the q_3 (+) state is obtained at higher energies than in the previous calculations such that for the (+) states it changes positions with the $q_1 \times q_1$ state. The reason for this is most likely due the different representation of the PES where it is currently unclear which representation delivers the correct order of states. Note, however, that the density of states increases significantly above 500 cm^{-1} such that slight change in the potential representation can lead to a change in the state order.

Compared to the experiment only the energy of the q_1 state is obtained in good agreement and about 2 cm^{-1} too low. The q_3 and q_4 state energies are obtained 11 and 8 cm^{-1} too high, respectively, while the state energy of the q_2 state is obtained 9 cm^{-1} too low.

Turning to the tunneling splittings it can be observed that for the ground state the results from the present contribution and the theoretical results from Ref. 72 and from "MC10" of Ref. 67 are in good agreement as they differ within 0.2 cm^{-1}. The result of "MC13" of Ref. 67 is approximately 1 cm^{-1} below the other theoretical results and hence closer to the experimental result which is approximately 2 cm^{-1} below the other theoretical results.

The tunneling splitting of the q_4 state is consistently obtained as approximately 64 cm^{-1} within all theoretical approaches and therefore approximately 7 cm^{-1} too large compared to the experiment.

While Hammer *et al.*[72] obtain a tunneling splitting of 6.7 cm^{-1} for the q_1 state which is close to the experimental value, it is obtained with 17.3 cm^{-1} in the present contribution. This is consistent with the result of the "MC10" mode combination of Ref. 67 while the result of the "MC13" mode combination of Ref. 67 lies between the former two.

Interestingly the splitting of $q_1 \times q_4$ in the present work is consistent with the splitting of the "MC13" mode combination of Ref. 67 while it was the other way round for the previous splittings. Furthermore, the splitting of the q_3 state is obtained with 18.7 cm^{-1}, about 2 cm^{-1} higher than for the previous calculations in Ref. 67. It is close to the results of Ref. 72 but approximately 4 cm^{-1} above the experimental value.

In conclusion, we believe that the potential fit in the present contribution is globally much more accurate than in the previous calculations in Ref. 67. The n-mode representation used in Ref. 67, however, has the quality that the 1-mode terms are represented exactly and hence also the 1-mode energies would be represented correctly if no correlation would exist. On the other hand, the current potential representation in terms of a CPD expansion contains correlations on a global level but does probably not represent the uncorrelated 1-mode terms exactly. Nevertheless we believe that the state energies and splittings reported in Table 9.3 are more accurate as compared to previous calculations. This is also due to the fact that the much smaller number of terms allows for much more SPFs and therefore much better converged MCTDH wavefunctions as much fewer terms enter the mean fields and Hamiltonian matrix elements.

9.5. Zundel Cation: Transient Infrared Absorption Spectroscopy

The linear IR absorption spectrum of the Zundel cation and several of its isotopologues in the gas phase is extremely well characterized from a theoretical perspective following a series of works by some of us using the MCTDH approach.[57, 74–78] The key to the success of these endeavors was the use of polyspherical coordinates, as well as the availability of a high-quality PES based on permutation invariant polynomials,[79] which was then transformed to product form for use within the MCTDH framework.[57] This allowed the consistent calculation of the IR absorption spectrum from 0 to 4,000 cm^{-1} from a single propagation of the dipole-operated ground state with an accuracy of 5–10 cm^{-1}, and the subsequent assignment of all lines seen in the spectrum plus a large number of IR-inactive states.[75, 78] The IR spectrum features a doublet absorption peak related to a strong Fermi resonance of the shared-proton fundamental with a combination of various modes of its environment, mainly but not exclusively the pyramidalization of the two water molecules and the hydrogen-bond stretch.[74] Although the structural aspects of this spectral feature are well understood, the real-time dynamics of the shared proton in direct connection with the couplings of the Fermi resonance are not so well characterized.

In this section we report on IR transient absorption spectrum (TAS) of the Zundel cation after excitation by a broad-band pulse that targets the shared-proton doublet structure. These calculations are inspired by recent experimental[80] and theoretical[81] work on the solvated proton in acidic solutions, where the transition energy to the first overtone of the shared proton is used as a proxy of the shape of the local potential around the proton. As we will see, the TAS is not only sensitive to the transition to the first shared-proton overtone, but it is also an excellent probe of the shared-proton dynamics and its energy relaxation to the surrounding water molecules following the pump-pulse excitation.

9.5.1. *Calculation of infrared transient absorption spectra within the MCTDH framework*

The transient absorption spectrum (TAS) (or dynamic absorption spectrum) is related to the third order polarizability of the system through the term $\langle \Phi^{(1)}(t)|\hat{\mu}|\Phi^{(2)}(t)\rangle$, where $\Phi^{(n)}(t)$ is the nth order correction term of the wavefunction Ψ in the framework of time-dependent perturbation theory.[5] After an initial pump pulse places the system in a transient state, the response of the system to a delayed probe pulse provides information about the dynamics during the pump-probe delay, while providing access to final quantum mechanical states that are not accessible via one-photon transitions because of the corresponding selection rules.

The IR-TAS spectrum is calculated for arbitrary pump and probe pulses in the time-domain representation. For this, we find it most convenient to start from the more general non-perturbative spectrum, which describes the total absorption (or emission) of the system for any incoming fields $\epsilon(t)$ (a single polarization direction is considered for simplicity) under the Hamiltonian

$$\hat{H}(t) = \hat{H}_0 - \hat{\mu}\epsilon(t). \tag{9.22}$$

The total energy exchanged between the system and the fields is

$$\Delta E = - \int_{-\infty}^{+\infty} \mu(t)\dot{\epsilon}(t)dt \qquad (9.23)$$

from the perspective of the system, *i.e.*, $\Delta E > 0$ corresponds to over-all absorption and $\Delta E < 0$ to overall induced emission. Introducing the Fourier transform of the polarization $\mu(t) = \langle \Psi(t)|\hat{\mu}|\Psi(t)\rangle$ and the fields, and the Fourier relation for the derivative of a function, one arrives at

$$\Delta E = - \int_{-\infty}^{+\infty} \omega \, \mathrm{Im}\{\mu(\omega)\epsilon^*(\omega)\}d\omega, \qquad (9.24)$$

where the integrand represents the differential of energy exchanged between the system and the fields at frequency ω. Note that this matter-centered treatment is only valid if changes to the incoming fields caused by the light-matter interaction can be neglected. Dividing the integrand in Eq. (9.24) by the differential of incident electromagnetic energy per unit area at frequency ω, $|\epsilon(\omega)|^2 c/(2\pi)$, one arrives at the expression for the non-perturbative absorption cross-section[5]

$$\sigma(\omega) = \frac{-2\pi\omega}{c} \frac{\mathrm{Im}\{\mu(\omega)\epsilon^*(\omega)\}}{|\epsilon(\omega)|^2}. \qquad (9.25)$$

Although the starting point of our treatment is the non-perturbative cross-section valid at any order of interaction, it is desirable to separate the contributions of the pump and probe pulses to the total polarizability. This can be done in practice by introducing an auxiliary degree of freedom with discrete states, analogous to how electronic states are described in non-adiabatic problems. The corresponding wavefunction reads

$$|\Psi(t)\rangle = |\Phi_0(t)\rangle \otimes |0\rangle + |\Phi_1(t)\rangle \otimes |1\rangle + |\Phi_2(t)\rangle \otimes |2\rangle. \qquad (9.26)$$

By writing the Hamiltonian as

$$\hat{H} = \hat{H}_0 \otimes \mathbf{1} - \hat{\mu}(\epsilon_{\mathrm{pu}}(t) \otimes |1\rangle\langle 0| + \epsilon_{\mathrm{pr}}(t) \otimes |2\rangle\langle 1|), \qquad (9.27)$$

$|\Phi_1(t)\rangle \otimes |1\rangle$ collects the wavepacket excited by the pump pulse, whereas $|\Phi_2(t)\rangle \otimes |2\rangle$ collects the response of the wavepacket to the probe pulse. The non-Hermitian auxiliary operators $|i + 1\rangle\langle i|$ in Eq. (9.27) ensure that only one-photon transitions are considered both for the pump and probe steps. In addition, one should note that both excited state absorption and induced emission by the probe pulse are included in $|\Phi_2(t)\rangle \otimes |2\rangle$, but that the excitation of the ground state due to the probe is excluded, *i.e.*, only the background-free part of the transient spectrum is being considered.

Finally, evaluating the expectation value of the dipole operator $\mu(t) = \langle \Psi(t)|\hat{\mu}|\Psi(t)\rangle$ with the total wave function expansion (9.26) allows one to separate the polarization components of the pump and probe pulses

$$\mu_{\mathrm{pu}}(t) = \langle \Phi_0(t)|\hat{\mu}|\Phi_1(t)\rangle + \text{c.c.}, \tag{9.28}$$

$$\mu_{\mathrm{pr}}(t) = \langle \Phi_1(t)|\hat{\mu}|\Phi_2(t)\rangle + \text{c.c.}, \tag{9.29}$$

where $\mu(t) = \mu_{\mathrm{pu}}(t) + \mu_{\mathrm{pr}}(t)$. The FT of these quantities together with the FT of the corresponding time-dependent fields enter Eq. (9.25) to obtain the TAS of the pump and probe pulses separately.

9.5.2. *Infrared transient absorption of the Zudel cation*

The anharmonicity of the proton-stretch potential in the solvated proton contains important information on the immediate environment and solvation-shell structure of the proton. Its super-harmonic character, $\omega_{12} > \omega_{01}$, which has been measured in highly acidic solutions,[80] is indicative of a Zundel-like environment under such conditions. In contrast, a sub-harmonic progression would be attributable to the Eigen form, where the excess proton is subject to a Morse-like potential.

Coherent IR spectroscopy has been applied to the study of the excess-proton overtones in the liquid phase,[80] and VCI calculations based on eigenstates and their transition dipoles were used to approximate the spectra and interpret the measurements.[81] Coherent spectroscopy relies on two pump and one probe pulse, all with different k-vectors, which coincide at the sample to generate a background free signal in the direction of their phase-matching condition. These signals can be computed from first principles in a time-dependent representation, as demonstrated for model systems,[82] but the amount of computation needed to generate them for the Zundel cation makes their direct simulation currently impractical. Instead, we simulate here the simpler TAS of the Zundel cation from first principles. The TAS is also a non-linear observable that has access to similar dynamics and to the same set of final states (*e.g.*, proton-stretch overtones) as its coherent counterparts.

Each TAS is evaluated from an MCTDH propagation starting from the ground vibrational state and using Hamiltonian (9.27), *i.e.*, with explicit inclusion of the pump and probe pulses. Both pulses are defined from the corresponding vector potentials

$$A_k(t) = \frac{E_{0,k}}{\omega_k} \exp\left[-\frac{2\ln 2}{F_k^2}(t - t_{0,k})^2\right] \cos(\omega_k(t - t_{0,k})), \qquad (9.30)$$

as $\epsilon_k(t) = -\partial A_k(t)/\partial t$, $k \in \{\mathrm{pu}, \mathrm{pr}\}$. $E_{0,k}$ is the maximal field strength, ω_k is the central frequency of the pulse, F_k is the pulse duration (fwhm of the intensity profile: $\exp\left[-\frac{4\ln 2}{F_k^2}(t - t_{0,k})^2\right]$) and $t_{0,k}$ represents the center of the k-th pulse. We consider pump and probe pulses of duration $F_k = 30$ fs and central frequencies $\omega_{\mathrm{pu}} = 1,050$ cm^{-1} and $\omega_{\mathrm{pr}} = 1,250$ cm^{-1}, respectively. Their spectral representation is shown as shaded areas in Figure 9.5(a) superimposed with the one-photon absorption spectrum of the pump pulse. The pump-probe delay $\Delta\tau = t_{0,pr} - t_{0,pu}$ is varied between 200 and 900 fs in steps of 50 fs and the pulse amplitudes $E_{0,k}$ are 0.01 a.u., chosen such that $\langle\Phi_j(t)|\Phi_j(t)\rangle \approx 0.5$ after the pump and probe pulses. Of course, the norm of the wavefunction is not conserved under Hamiltonian (9.27) but this is not a problem for the evaluation of the cross-section as long as one consistently uses the FT of the polarizabilities and fields in Eq. (9.25).

The calculations are performed using the ML-MCTDH method based on the ML tree in Figure 9.4. There, the degree of freedom corresponding to the auxiliary states is denoted as "el". Besides, a

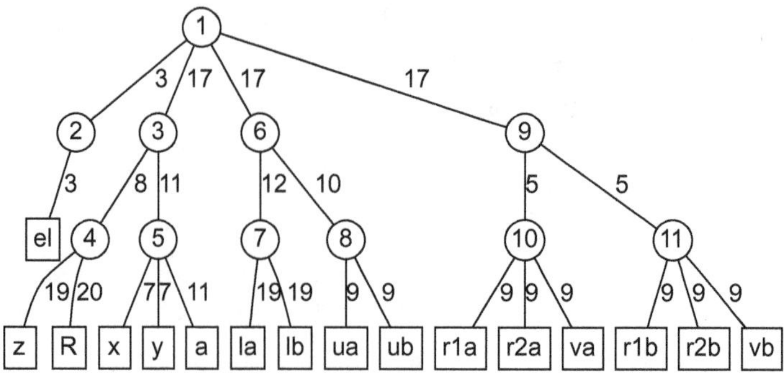

Figure 9.4.   Multilayer tree-structure for the Zundel cation transient absorption spectrum calculations.

three-layer structure is used where the second layer separates three main subgroups of DOFs: node 3: proton and O–O stretch; node 6: rocking and wagging angles; node 9: internal water coordinates. The third layer consists of the same combined modes as the "classical" standard-MCTDH calculations on the Zundel cation, as well as the same primitive grids.[77] The pump-probe calculations are based on a new potential re-fit of the original Zundel PES by Bowman[79] using the newly developed MC-CPD approach,[65] instead of the older cluster expansion strategy (see also Section 9.3).[57] Each propagation of 1500 fs (each pump-probe delay) took roughly 350 hours of wall-clock time on 8 shared-memory processors. Finally, only the μ_z component of the dipole, which is parallel to the O–O axis of the cation, is considered in the following discussions.

The pump pulse creates a coherent superposition of the two eigenstates of the proton-stretch doublet, as seen in Figure 9.5(a): the eigenstate assigned to the proton stretch fundamental $|z_1\rangle$, and the eigenstate assigned to the hydrogen-bond-stretch-pyramidalization combination $|R_1, w_3\rangle$. These two eigenstates can be understood as linear combinations of two zeroth order states of the corresponding characters, which we refer to as $|z_1\rangle_0$ and $|R_1, w_3\rangle_0$. The complete picture involves also other excitations, for example the asymmetric water bending, and a detailed account of the mixing coefficients can be found in Ref. 75. The $\hat{\mu}_z$ dipole is to a large extent

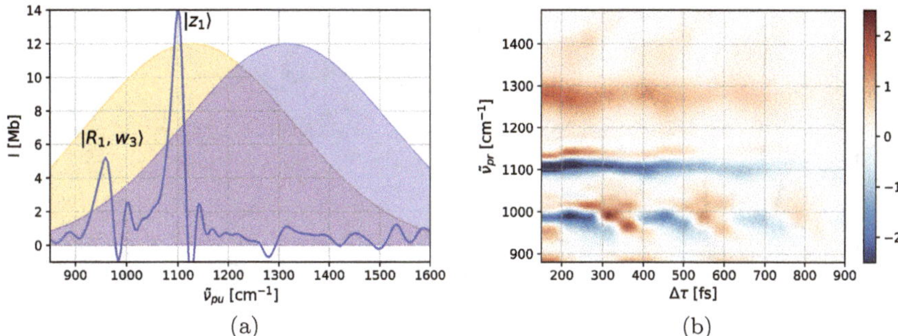

Figure 9.5. (a) One-photon absorption spectrum corresponding to the pump pulse. *Kets* indicate the final state in the absorption lines from the vibrational ground state, $|0\rangle \rightarrow |f\rangle$. $|z_1\rangle$: proton-transfer fundamental transition. $|R_1 w_3\rangle$: combination band of the oxygen-oxygen stretch fundamental with the asymmetric pyramidalization of the two water molecules. The intensity profile of the pump and probe pulses is shown in the background in orange and blue, respectively. (b) Transient absorption spectrum of the Zundel cation after exciting the system with the broadband pump pulse shown in (a). The x-axis indicates the delay between pump and probe pulses. The y-axis indicates the probe-pulse frequency (wavenumber). Blue and red regions correspond to excited state emission and absorption, respectively.

proportional to the proton displacement $\hat{z}$ along the hydrogen bond, which is the main displacement directly excited by the short pump pulse. The shared proton is strongly coupled to its environment. Thus, the pump pulse transfers energy to the shared proton, which then is transferred towards the coupled modes in a combination of periodic and dissipative dynamics. These are the dynamics that the probe pulse interrogates. Moreover, the probe pulse can excite the system to the first overtone of the shared-proton stretch, $|z_2\rangle$, which is only accessible via two photon transitions.

$\sigma_{\mathrm{pr}}(\omega_{\mathrm{pr}}, \Delta\tau)$, shown in Figure 9.5(b), features both regions of transient absorption (red) and excited state emission (blue). The main excited state emission is centered at $\omega_{\mathrm{pr}} = 1100\,\mathrm{cm}^{-1}$ and coincides with the proton-stretch fundamental of the shared-proton doublet in the linear absorption spectrum. At a higher frequency, a strong excited state absorption band centered at $\omega_{\mathrm{pr}} = 1250\ \mathrm{cm}^{-1}$ corresponds to the transition from the shared-proton fundamental to the first proton-stretch overtone, $|z_1\rangle \rightarrow |z_2\rangle$, and

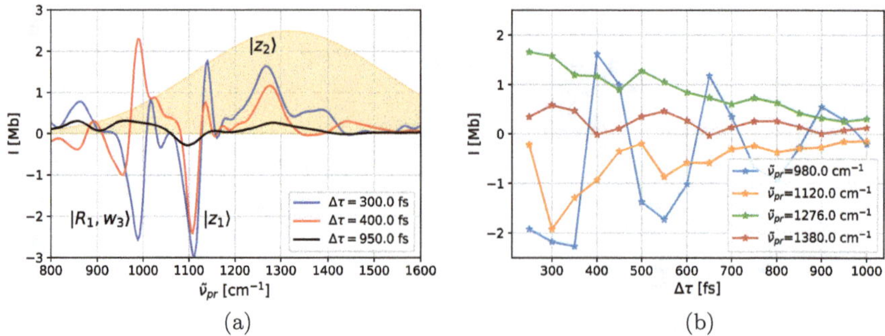

Figure 9.6. Cuts of the 2D transient spectrum in Figure 9.5(b) along the (a) probe frequency and (b) time-delay axes. *Kets* in (a) indicate the final state in excited state absorption lines (positive intensity) and the initial state in excited state emission lines (negative intensity). The intensity profile of the probe pulse is shown in the background.

it is clearly super-harmonic in agreement with the measurements in Ref. 80.

The TAS spectrum is sensitive to the wavepacket dynamics of the shared proton triggered by the pump pulse. This is seen, for example, by inspecting the TAS for the pump-probe delays 300 and 400 fs in Figure 9.6(a), where the band at 980 cm^{-1} changes from excited state emission to transient absorption, respectively. The period of this oscillation (*cf.* Figure 9.6(b)) is ≈ 250 fs and matches precisely the splitting of the shared-proton doublet.

Finally, all features seen in the TAS decay and approach the baseline of the spectrum within about one picosecond, as seen by inspecting time traces of $\sigma_{\mathrm{pr}}(\omega_{\mathrm{pr}}, \Delta\tau)$ at different probe frequencies in Figure 9.6(b). This is a consequence of the strong coupling of the shared proton to various modes of its environment, namely pyramidalization, bending and hydrogen-bond stretch.[75,76] Although the wavepacket mostly involves the shared-proton doublet, the large density of states in the spectral region covered by the bandwidth of the pump pulse leads effectively to an irreversible energy transfer from the shared proton to its surroundings. This irreversible dynamics is remarkable for such a small cluster with only seven atoms, and it deserves further consideration and comparison with similar dynamics reported in the liquid phase.[83]

Concluding, the probe transition to the shared-proton overtone $|z_2\rangle$ has been unequivocally assigned and is of super-harmonic character. It will be interesting to achieve a better characterization of the final states accessed by the probe pulse. A short pulse targeting the shared-proton spectral region of the Zundel cation creates a wavepacket consisting mainly of the states of the well-known shared-proton doublet. This excitation decays to its environment with about 1 ps. Its proving shines light on the energies of higher excited states accessible only by two-photon transitions, and about the relaxation pathways and time-scales of the shared-proton excitation towards its environment. These studies can be extended to other solvated proton clusters and complemented by experimental observations.

References

1. Light, J. C.; Hamilton, I. P.; Lill, J. V. Generalized discrete variable approximation in quantum mechanics. *J. Chem. Phys.* **1985**, *82*, 1400.
2. Lill, J. V.; Parker, G. A.; Light, J. C. The discrete variable-finite basis approach to quantum scattering. *J. Chem. Phys.* **1986**, *85*, 900–910.
3. Light, J. C. Discrete Variable Representations in Quantum Dynamics. In *Time-Dependent Quantum Molecular Dynamics*, Broeckhove, J., Lathouwers, L., Eds.; Plenum: New York, 1992; pp. 185–199.
4. Light, J. C.; Carrington Jr., T. Discrete variable representations and their utilization. *Adv. Chem. Phys.* **2000**, *114*, 263–310.
5. Tannor, D. J. *Introduction to Quantum Dynamics: A Time-Dependent Perspective.* University Science Books: Sausalito, CA, 2007.
6. Dirac, P. A. M. Note on exchange phenomena in the thomas atom. *Proc. Cambridge Philos. Soc.* **1930**, *26*, 376–385.
7. Frenkel, J. *Wave Mechanics.* Clarendon Press: Oxford, 1934.
8. Meyer, H.-D.; Manthe, U.; Cederbaum, L. S. The multi-configurational time-dependent Hartree approach. *Chem. Phys. Lett.* **1990**, *165*, 73–78.
9. Manthe, U.; Meyer, H.-D.; Cederbaum, L. S. Wave-packet dynamics within the multiconfiguration Hartree framework: General aspects and application to NOCl. *J. Chem. Phys.* **1992**, *97*, 3199–3213.
10. Beck, M. H.; Jäckle, A.; Worth, G. A.; Meyer, H.-D. The multi-configuration time-dependent Hartree (MCTDH) method: A highly efficient algorithm for propagating wave packets. *Phys. Rep.* **2000**, *324*, 1–105.

11. Meyer, H.-D.; Worth, G. A. Quantum molecular dynamics: Propagating wavepackets and density operators using the multiconfiguration time-dependent Hartree (MCTDH) method. *Theor. Chim. Acta* **2003**, *109*, 251–267.

12. Meyer, H.-D.; Gatti, F.; Worth, G. A., Eds. *Multidimensional Quantum Dynamics: MCTDH Theory and Applications*. Wiley-VCH: Weinheim, 2009.

13. Meyer, H.-D. Studying molecular quantum dynamics with the multi-configuration time-dependent Hartree method. *WIREs: Comput. Mol. Sci.* **2012**, *2*, 351–374.

14. Ehara, M.; Meyer, H.-D.; Cederbaum, L. S. Multi-configuration time-dependent Hartree (MCTDH) study on rotational and diffractive inelastic molecule-surface scattering. *J. Chem. Phys.* **1996**, *105*, 8865–8877.

15. Worth, G. A.; Meyer, H.-D.; Cederbaum, L. S. Relaxation of a system with a conical intersection coupled to a bath: A benchmark 24-dimensional wavepacket study treating the environment explicitly. *J. Chem. Phys.* **1998**, *109*, 3518–3529.

16. Raab, A.; Worth, G.; Meyer, H.-D.; Cederbaum, L. S. Molecular dynamics of pyrazine after excitation to the S_2 electronic state using a realistic 24-mode model Hamiltonian. *J. Chem. Phys.* **1999**, *110*, 936–946.

17. Wang, H.; Thoss, M. Multilayer formulation of the multiconfiguration time-dependent Hartree theory. *J. Chem. Phys.* **2003**, *119*, 1289–1299.

18. Manthe, U. A multilayer multiconfigurational time-dependent Hartree approach for quantum dynamics on general potential energy surfaces. *J. Chem. Phys.* **2008**, *128*, 164116.

19. Vendrell, O.; Meyer, H.-D. multilayer multiconfiguration time-dependent Hartree method: Implementation and applications to a Henon-Heiles Hamiltonian and to pyrazine. *J. Chem. Phys.* **2011**, *134*, 044135.

20. Wang, H. Multilayer multiconfiguration time-dependent Hartree theory. *J. Phys. Chem. A* **2015**, *119*, 7951.

21. Gatti, F.; Lasorne, B.; Meyer, H.-D.; Nauts, A. *Applications of Quantum Dynamics in Chemistry*. Vol. 98, Lectures Notes in Chemistry; Springer: Heidelberg, 2017.

22. Wang, H.; Meyer, H.-D. On regularizing the ML-MCTDH equations of motion. *J. Chem. Phys.* **2018**, *149*, 044119.

23. Wang, H.; Meyer, H.-D. Importance of appropriately regularizing the ML-MCTDH equations of motion. *J. Phys. Chem. A* **2021**, *125*, 3077.

24. Meyer, H.-D.; Wang, H. On regularizing the MCTDH equations of motion. *J. Chem. Phys.* **2018**, *148*, 124105.

25. Lubich, C. Time integration in the multiconfiguration time-dependent Hartree method of molecular quantum dynamics. *Appl. Math. Res. Express* **2015**, *2015*, 311–328.

26. Kloss, B.; Burghardt, I.; Lubich, C. Implementation of a novel projector-splitting integrator for the multi-configurational time-dependent Hartree approach. *J. Chem. Phys.* **2017**, *146*, 174107.

27. Bonfanti, M.; Burghardt, I. Tangent space formulation of the multi-configuration time-dependent Hartree equations of motion: the projector-splitting algorithm revisited. *Chem. Phys.* **2018**, *515*, 252–261.

28. Weike, T.; Manthe, U. Symmetries in the multi-configurational time-dependent Hartree wavefunction representation and propagation. *J. Chem. Phys.* **2021**, *154*, 194108.

29. Lee, K.-S.; Fischer, U. R. Truncated many-body dynamics of interacting bosons: A variational principle with error monitoring. *Int. J. Mod. Phys. B* **2014**, *28*, 1550021.

30. Mendive-Tapia, D.; Meyer, H.-D. Regularizing the MCTDH equations of motion through an optimal choice on-the-fly (i.e. spawning) of unoccupied single-particle functions. *J. Chem. Phys.* **2020**, *153*, 234114.

31. Westermann, T.; Brodbeck, R.; Rozhenko, A. B.; Schoeller, W.; Manthe, U. Photodissociation of methyl iodide embedded in a host-guest complex: a full dimensional (189D) quantum dynamics study of CH_3I@resorc[4]arene. *J. Chem. Phys.* **2011**, *135*, 184102.

32. Mendive-Tapia, D.; Mangaud, E.; Firmino, T.; de la Lande, A.; Desouter-Lecomte, M.; Meyer, H.-D.; Gatti, F. Multidimensional quantum mechanical modeling of electron transfer and electronic coherence in plant cryptochromes: the role of initial bath conditions. *J. Phys. Chem. B* **2018**, *122*, 126–136.

33. Park, T. J.; Light, J. C. Unitary quantum time evolution by iterative Lanczos reduction. *J. Chem. Phys.* **1986**, *85*, 5870.

34. Beck, M. H.; Meyer, H.-D. An efficient and robust integration scheme for the equations of motion of the multiconfiguration time-dependent Hartree (MCTDH) method. *Z. Phys. D* **1997**, *42*, 113–129.

35. Manthe, U. On the integration of the multi-configurational time-dependent Hartree (MCTDH) equations of motion. *Chem. Phys.* **2006**, *329*, 168–178.

36. Kosloff, R.; Tal-Ezer, H. A direct relaxation method for calculating eigenfunctions and eigenvalues of the Schrödinger equation on a grid. *Chem. Phys. Lett.* **1986**, *127*, 223.

37. Meyer, H.-D.; Le Quéré, F.; Léonard, C.; Gatti, F. Calculation and selective population of vibrational levels with the multiconfiguration time-dependent Hartree (MCTDH) algorithm. *Chem. Phys.* **2006**, *329*, 179–192.

38. Doriol, L. J.; Gatti, F.; Iung, C.; Meyer, H.-D. Computation of vibrational energy levels and eigenstates of fluoroform using the multiconfiguration time-dependent Hartree method. *J. Chem. Phys.* **2008**, *129*, 224109.

39. Worth, G. A.; Beck, M. H.; Jäckle, A.; Vendrell, O.; Meyer, H.-D. The MCTDH Package, Version 8.2, 2000; H.-D. Meyer, Version 8.3, 2002; Version 8.4, 2007; O. Vendrell and H.-D. Meyer Version 8.5, 2013. Version 8.5 contains the ML-MCTDH algorithm. Current versions: 8.4.20 and 8.5.13, 2020; See http://mctdh.uni-hd.de/ for a description of the Heidelberg MCTDH package.

40. Wang, H. Iterative calculation of energy eigenstates employing the multilayer multiconfiguration time-dependent Hartree theory. *J. Phys. Chem. A* **2014**, *118*, 9253.

41. Wang, H.; Shao, J. Quantum phase transition in the spin-boson model: A multilayer multiconfiguration time-dependent Hartree study. *J. Phys. Chem. A* **2019**, *123*, 1882.

42. Jäckle, A.; Meyer, H.-D. Product representation of potential energy surfaces. *J. Chem. Phys.* **1996**, *104*, 7974.

43. Jäckle, A.; Meyer, H.-D. Product representation of potential energy surfaces II. *J. Chem. Phys.* **1998**, *109*, 3772.

44. Otto, F. Multi-layer potfit: An accurate potential representation for efficient high-dimensional quantum dynamics. *J. Chem. Phys.* **2014**, *140*, 014106.

45. Peláez, D.; Meyer, H.-D. The multigrid POTFIT (MGPF) method: Grid representations of potentials for quantum dynamics of large systems. *J. Chem. Phys.* **2013**, *138*, 014108.

46. Schröder, M.; Meyer, H.-D. Transforming high-dimensional potential energy surfaces into sum-of-products form using Monte Carlo methods. *J. Chem. Phys.* **2017**, *147*, 064105.

47. Otto, F.; Chiang, Y.-C.; Peláez, D. Accuracy of Potfit-based potential representations and its impact on the performance of (ML-)MCTDH. *Chem. Phys.* **2018**, *509*, 116.

48. Manzhos, S.; Wang, X.-G.; Dawes, R.; Carrington, Jr., T. A nested molecule-independent neural network approach for high-quality potential fits. *J. Phys. Chem. A* **2006**, *110*, 5295.

49. Manzhos, S.; Carrington, Jr., T. A random-sampling high dimensional model representation neural network for building potential energy surfaces. *J. Chem. Phys.* **2006**, *125*, 084109.

50. Manzhos, S.; Carrington, Jr., T. Using neural networks to represent potential surfaces as sum of products. *J. Chem. Phys.* **2006**, *125*, 194105.

51. Koch, W.; Zhang, D. H. Communication: Separable potential energy surfaces from multiplicative artificial neural networks. *J. Chem. Phys.* **2014**, *141*, 021101.

52. Shen, X.; Chen, J.; Zhang, Z.; Shao, K.; Zhang, D. H. Methane dissociation on Ni(111): A fifteen-dimensional potential energy surface using neural network method. *J. Chem. Phys.* **2015**, *143*, 144701.

53. Pradhan, E.; Brown, A. Vibrational energies for HFCO using a neural network sum of exponentials potential energy surface. *J. Chem. Phys.* **2016**, *144*, 174305.

54. Pradhan, E.; Brown, A. Neural network exponential fitting of a potential energy surface with multiple minima: Application to HFCO. *J. Mol. Spectrosc.* **2016**, *330*, 158–164.

55. Carter, S.; Culik, S. J.; Bowman, J. M. Vibrational self-consistent field method for many-mode systems: A new approach and application to the vibrations of CO adsorbed on Cu(100). *J. Chem. Phys.* **1997**, *107*, 10458.

56. Bowman, J. M.; Carter, S.; Huang, X. MULTIMODE: a code to calculate rovibrational energies of polyatomic molecules. *Int. Rev. Phys. Chem.* **2003**, *22*, 533–549.

57. Vendrell, O.; Gatti, F.; Lauvergnat, D.; Meyer, H.-D. Full dimensional (15D) quantum-dynamical simulation of the protonated water dimer I: Hamiltonian setup and analysis of the ground vibrational state. *J. Chem. Phys.* **2007**, *127*, 184302.

58. Rabitz, H.; Alis, O. F. General foundations of high-dimensional model representations. *J. Math. Chem.* **1999**, *25*, 197.

59. Alis, O. F.; Rabitz, H. Efficient implementation of high dimensional model representations. *J. Math. Chem.* **2001**, *29*, 127–142.

60. Li, G. Y.; Hu, J. S.; Wang, S. W.; Georgopoulos, P. G.; Schoendorf, J.; Rabitz, H. Random sampling-high dimensional model representation (RS-HDMR) and orthogonality of its different order component functions. *J. Phys. Chem. A* **2006**, *110*, 2474–2485.

61. Hitchcock, F. L. The expression of a tensor or a polyadic as a sum of products. *J. Math. Phys.* **1927**, *6*, 164–189.

62. Håstad, J. Tensor rank is np-complete. *J. Algorithms* **1990**, *11*, 644.

63. Harshman, R. A. Foundations of the PARAFAC procedure: Models and conditions for an "explanatory" multi-modal factor analysis. *UCLA Work. Pap. Phon.* **1970**, *16*, 1.

64. Carroll, J. D.; Chang, J. J. Analysis of individual differences in multidimensional scaling via an n-way generalization of "Eckard-Young" decomposition. *Psychometrika* **1970**, *35*, 283.

65. Schröder, M. Transforming high-dimensional potential energy surfaces into a canonical polyadic decomposition using Monte Carlo methods. *J. Chem. Phys.* **2020**, *152*, 024108.

66. Schröder, M.; Gatti, F.; Meyer, H.-D. Theoretical studies of the tunneling splitting of malonaldehyde using the multiconfiguration time-dependent Hartree approach. *J. Chem. Phys.* **2011**, *134*, 234307.

67. Schröder, M.; Meyer, H.-D. Calculation of the vibrational excited states of malonaldehyde and their tunneling splittings with the multiconfiguration time-dependent Hartree method. *J. Chem. Phys.* **2014**, *141*, 034116.

68. Wang, Y.; Braams, B. J.; Bowman, J. M.; Carter, S.; Tew, D. P. Full-dimensional quantum calculations of ground-state tunneling splitting of malonaldehyde using an accurate ab initio potential energy surface. *J. Chem. Phys.* **2008**, *128*(22), 224314.

69. Hammer, T.; Coutinho-Neto, M. D.; Viel, A.; Manthe, U. Multiconfigurational time-dependent Hartree calculations for tunneling splittings of vibrational states: Theoretical considerations and application to malonaldehyde. *J. Chem. Phys.* **2009**, *131*, 224109.

70. Kosztin, I.; Faber, B.; Schulten, K. Introduction to the diffusion Monte Carlo method. *Am. J. Phys.* **1996**, *64*, 633–644.

71. Hammer, T.; Manthe, U. Intramolecular proton transfer in malonaldehyde: Accurate multilayer multi-configurational time-dependent Hartree calculations. *J. Chem. Phys.* **2011**, *134*, 224305.

72. Hammer, T.; Manthe, U. Iterative diagonalization in the state-averaged multi-configurational time-dependent Hartree approach: Excited state tunneling splitting in malonaldehyde. *J. Chem. Phys.* **2012**, *136*, 054105.

73. Lüttschwager, N. O. B.; Wassermann, T. N.; Coussan, S.; Suhm, M. A. Vibrational tuning of the hydrogen transfer in malonaldehyde: a combined FTIR and Raman jet study. *Phys. Chem. Chem. Phys.* **2013**, *14*, 2211.

74. Vendrell, O.; Gatti, F.; Meyer, H.-D. Dynamics and infrared spectroscopy of the protonated water dimer. *Angew. Chem. Int. Ed.* **2007**, *46*, 6918–6921.

75. Vendrell, O.; Gatti, F.; Meyer, H.-D. Full dimensional (15D) quantum-dynamical simulation of the protonated water dimer II: Infrared spectrum and vibrational dynamics. *J. Chem. Phys.* **2007**, *127*, 184303.

76. Vendrell, O.; Gatti, F.; Meyer, H.-D. Strong isotope effects in the infrared spectrum of the Zundel cation. *Angew. Chem. Int. Ed.* **2009**, *48*, 352–355.

77. Vendrell, O.; Brill, M.; Gatti, F.; Lauvergnat, D.; Meyer, H.-D. Full dimensional (15D) quantum-dynamical simulation of the protonated water dimer III: mixed Jacobi-valence parametrization and benchmark results for the zero-point energy, vibrationally excited states and infrared spectrum. *J. Chem. Phys.* **2009**, *130*, 234305.

78. Vendrell, O.; Gatti, F.; Meyer, H.-D. Full dimensional (15D) quantum-dynamical simulation of the protonated water dimer IV: Isotope effects in the infrared spectra of $D(D_2O)_2^+$, $H(D_2O)_2^+$ and $D(H_2O)_2^+$ isotopologues. *J. Chem. Phys.* **2009**, *131*, 034308.

79. Huang, X.; Braams, B. J.; Bowman, J. M. *Ab initio* potential energy and dipole moment surfaces for $H_5O_2^+$. *J. Chem. Phys.* **2005**, *122*, 044308.

80. Fournier, J. A.; Carpenter, W. B.; Lewis, N. H. C.; Tokmakoff, A. Broadband 2D IR spectroscopy reveals dominant asymmetric $H_5O_2^+$ proton hydration structures in acid solutions. *Nat. Chem.* **2018**, *10*(9), 932–937.

81. Carpenter, W. B.; Yu, Q.; Hack, J. H.; Dereka, B.; Bowman, J. M.; Tokmakoff, A. Decoding the 2d IR spectrum of the aqueous proton with high-level VSCF/VCI calculations. *J. Chem. Phys.* **2020**, *153*(12), 124506.

82. Gelin, M. F.; Egorova, D.; Domcke, W. Efficient calculation of time- and frequency-resolved four-wave-mixing signals. *Acc. Chem. Res.* **2009**, *42*, 1290–1298.

83. Kundu, A.; Dahms, F.; Fingerhut, B. P.; Nibbering, E. T. J.; Pines, E.; Elsaesser, T. Ultrafast vibrational relaxation and energy dissipation of hydrated excess protons in polar solvents. *Chem. Phys. Lett.* **2018**, *713*, 111–116.

Chapter 10

Semiclassical Vibrational Dynamics for Molecular and Supra-Molecular Systems

Riccardo Conte[*] and Michele Ceotto[†]

*Dipartimento di Chimica, Universitá degli Studi di Milano,
via Golgi 19, 20133 Milano, Italy*
[*]*riccardo.conte1@unimi.it*
[†]*michele.ceotto@unimi.it*

10.1. Introduction

Resolution of the time-independent and time-dependent Schrödinger equations has always been the central topic of quantum mechanics. The typical issue for such a task is represented by the exponential scaling of the complexity of the problem with respect to the number of degrees of freedom. This has historically limited purely quantum calculations on scattering or spectroscopy to small model systems or molecules made of a tiny number of atoms (typically three or four atoms). To overcome this issue, in the last 20 years slow progress has been achieved in the field of quantum computing.[1] However, application of this innovative computational approach to actual large dimensional chemistry problems appears still far from being routine owing to the huge effort in hardware development still needed. In a more traditional framework there are attempts to introduce quantum mechanical effects into complex and large dimensional systems starting from computationally affordable classical molecular dynamics runs.[2] Semiclassical molecular dynamics is one of them, and this chapter is devoted to present recent theoretical and methodological advances in the field with some remarkable

378

applications. The latter include quantum vibrational spectroscopy of the fluxional Zundel cation, resolution of an open experimental controversy, quantum vibrational spectroscopy of biological species, and quantum solvation.

10.1.1. *The semiclassical way to quantum spectroscopy*

As reported in some of the other chapters of this book, vibrational energies can be obtained by solving the following eigenvalue equation:

$$\hat{H}\Psi_i\left(\boldsymbol{q}\right) = E_i\Psi_i\left(\boldsymbol{q}\right), \tag{10.1}$$

where $\boldsymbol{q}$ is the collection of the nuclear positions, $\hat{H} = \hat{\boldsymbol{p}}^2/2m + V\left(\boldsymbol{q}\right)$ is the nuclear Hamiltonian for the Born–Oppenheimer potential energy surface (PES) $V\left(\boldsymbol{q}\right)$, and $\Psi_i\left(\boldsymbol{q}\right)$ is the vibrational eigenfunction of eigenvalue E_i. The transition frequency $\nu_{i\leftarrow j}$ between two vibrational states to be compared with a typical IR and/or Raman experimental signal is readily obtained by Planck's relation $(E_i - E_j) = h\nu_{i\leftarrow j}$, where $E_i > E_j$ and h is Planck's constant. Solving Eq. (10.1) to calculate the molecular spectroscopic frequency $\nu_{i\leftarrow j}$ usually implies that all the eigenvalues of the system up to E_i have to be calculated. As the dimensionality of the system increases, the number of eigenvalues within an energy range rapidly increases. However, only first (called fundamentals) and low order transitions (called combination bands and overtones) are usually populated and detected in experimental spectra.

In a time-dependent approach the spectrum is calculated by Fourier transform of a correlation function.[3] For example, the dipole–dipole correlation function is calculated to reproduce IR spectra. If one wants to reproduce all the eigenvalues as in Eq. (10.1), they will calculate the Fourier transform of the wavepacket $|\chi(t)\rangle$ autocorrelation function. Specifically, the power spectrum is

$$\begin{aligned} I(E) &= \frac{1}{2\pi\hbar} \int_{-\infty}^{+\infty} dt\, e^{\mathrm{i}Et/\hbar}\langle\chi|\chi(t)\rangle \\ &= \frac{1}{2\pi\hbar} \int_{-\infty}^{+\infty} dt\, e^{\mathrm{i}Et/\hbar}\langle\chi|e^{-\mathrm{i}\hat{H}t/\hbar}|\chi\rangle \end{aligned}$$

$$= \frac{1}{2\pi\hbar} \int_{-\infty}^{+\infty} dt\, e^{iEt/\hbar} \sum_i |\langle \chi|E_i\rangle|^2 e^{-iE_i t/\hbar}$$

$$= \sum_i |\langle \chi|E_i\rangle|^2 \delta(E - E_i). \tag{10.2}$$

Eq. (10.2) shows how the power spectrum provides the collection of eigenvalues E_i on an absolute scale with respect to the zero of $V(q)$. The computational cost is independent of the portion of the spectrum $I(E)$ chosen to simulate, but it depends on the dimensionality of the system. Specifically, the computational bottleneck is represented by the real-time quantum evolution of the reference wavepacket $|\chi(t)\rangle = e^{-i\hat{H}t/\hbar}|\chi(0)\rangle$. Feynman path integrals[4] calculate the quantum evolution by performing a sum over all possible paths connecting the initial and the final positions of the wavepacket

$$\langle \chi|e^{-i\hat{H}t/\hbar}|\chi\rangle = \int \mathcal{D}\left[q\left(t\right)\right] e^{iS_t(q)/\hbar}, \tag{10.3}$$

where each path is weighted by the complex exponential of the path action. Making the sum over all possible paths rapidly becomes an inaccessible task, even when the number of degrees of freedom is just more than a few. For this reason, several approximations have been developed over the years.[5-7] The semiclassical approximation to the Feynman path integrals consists on reducing the sum over all possible paths to the sum over all possible *classical* paths.[8-10] This is equivalent to perform the path integration using the stationary phase approximation. In fact, when the phase is stationary, *i.e.*, $\delta S_t(q)/\delta q(t) = 0$, Hamilton's equations are enforced for the path. The semiclassical approximation accounts for quantum mechanical effects not only by considering multiple paths but also by retaining the fluctuations up to the second order of the action calculated along the classical path, *i.e.*, $S_t(q) \approx S^{cl}(q(t)) + \frac{1}{2}\left(\delta^2 S_t(q)/\delta q^2(t)\right)(q - q(t))^2$.[11,12] However, finding all possible classical trajectories starting and ending at given multidimensional positions is a challenging double-boundary problem. Miller[13] suggested to avoid this task by shooting the trajectories, *i.e.*, starting several trajectories at given initial conditions (p_0, q_0) and picking up those $(\boldsymbol{p}_t, \boldsymbol{q}_t)$ pairs which satisfy the double-boundary conditions. This calculation can be conveniently done by using Monte Carlo approaches. Later, Heller[14] thought to represent the wavepacket and the semiclassical propagator in terms of coherent states ($|\boldsymbol{p}, \boldsymbol{q}\rangle$), which have the following Gaussian representation in coordinate space:

$$\langle \boldsymbol{x} | \boldsymbol{p}, \boldsymbol{q} \rangle = \left(\frac{\det(\Gamma)}{\pi^{N_{vib}}} \right)^{1/4} e^{-(\boldsymbol{x}-\boldsymbol{q})^{\mathrm{T}} \frac{\Gamma}{2}(\boldsymbol{x}-\boldsymbol{q}) + i\boldsymbol{p}^{\mathrm{T}}(\boldsymbol{x}-\boldsymbol{q})/\hbar}, \qquad (10.4)$$

where the Gaussian width is determined by the (generally diagonal) Γ width parameter matrix and N_{vib} is the number of vibrational degrees of freedom. Herman and Kluk,[15] and eventually Kay[16–18] implemented these ideas[9, 19] into the stationary phase approximation to the path integral in Eq. (10.3) obtaining the following semiclassical expression for the survival amplitude of Eq. (10.3)

$$\langle \chi | e^{-\frac{i}{\hbar} \hat{H} t} | \chi \rangle \approx \frac{1}{(2\pi\hbar)^{N_{vib}}} \int \int d\boldsymbol{q}_0 d\boldsymbol{p}_0 C_t(\boldsymbol{p}_0, \boldsymbol{q}_0) e^{\frac{i}{\hbar} S_t(\boldsymbol{p}_0, \boldsymbol{q}_0)}$$
$$\times \langle \chi | \boldsymbol{p}_t, \boldsymbol{q}_t \rangle \langle \boldsymbol{p}_0, \boldsymbol{q}_0 | \chi \rangle, \qquad (10.5)$$

where, in its most general form,[16]

$$C_t(\boldsymbol{p}_0, \boldsymbol{q}_0) = \sqrt{\left| \frac{1}{2} \left(\frac{\partial \boldsymbol{q}_t}{\partial \boldsymbol{q}_0} + \Gamma^{-1} \frac{\partial \boldsymbol{p}_t}{\partial \boldsymbol{p}_0} \Gamma - i\hbar \frac{\partial \boldsymbol{q}_t}{\partial \boldsymbol{p}_0} \Gamma + \frac{i\Gamma^{-1}}{\hbar} \frac{\partial \boldsymbol{p}_t}{\partial \boldsymbol{q}_0} \right) \right|}. \qquad (10.6)$$

The pre-exponential factor in Eq. (10.6) is obtained from the stationary phase approximation after Gaussian integration of the quadratic term, which corresponds to the second order fluctuations of the action around the classical path. Thus, $C_t(\boldsymbol{p}_0, \boldsymbol{q}_0)$ is important for reproducing quantum effects, in addition to the term given by the complex exponential of the action in Eq. (10.5).[20–30] As anticipated, the phase space integral in Eq. (10.5) is performed by Monte Carlo integration. A much detailed derivation of the semiclassical propagator and the Herman–Kluk expression of Eq. (10.5) can be found in Ref. 31. When the semiclassical initial value representation (SCIVR) propagator of Eq. (10.5) is substituted into the Fourier transform of Eq. (10.2), the following basic SCIVR power spectrum formula is obtained:

$$I\left(E\right) = \frac{1}{2\pi\hbar}\int_{-\infty}^{+\infty} dt e^{\mathrm{i}Et/\hbar}\frac{1}{\left(2\pi\hbar\right)^{N_{vib}}}\int\int d\boldsymbol{q}_0 d\boldsymbol{p}_0 C_t\left(\boldsymbol{p}_0, \boldsymbol{q}_0\right)$$
$$\times\, e^{\mathrm{i}S_t(\boldsymbol{p}_0,\boldsymbol{q}_0)/\hbar}\langle\chi|\boldsymbol{p}_t, \boldsymbol{q}_t\rangle\langle\boldsymbol{p}_0, \boldsymbol{q}_0|\chi\rangle. \tag{10.7}$$

10.1.2. *Theoretical and practical challenges for the quantum spectroscopy of high dimensional molecular and supra-molecular systems*

Unfortunately the semiclassical expression of the power spectrum $I\left(E\right)$ in Eq. (10.7) becomes quite computationally intensive as the dimensionality of the system is increased beyond that of a diatomic system. A significant step forward in the calculation of the power spectra of molecular systems has been performed in 2003 by Kaledin and Miller.[32, 33] They introduced inside Eq. (10.7) a time-averaging filter of the type $\frac{1}{T}\int_0^T dt$ that allows the reduction of the number of trajectories required for the Monte Carlo phase space integration. As a rule of thumb, one runs a thousand trajectories per degree of freedom. Their "separable" time-averaging SCIVR (TA SCIVR) expression is

$$I\left(E\right) = \left(\frac{1}{2\pi\hbar}\right)^{N_{vib}}\int\int d\boldsymbol{p}_0 d\boldsymbol{q}_0\frac{1}{2\pi\hbar T}$$
$$\times\left|\int_0^T dt e^{\frac{\mathrm{i}}{\hbar}[S_t(\boldsymbol{p}_0,\boldsymbol{q}_0)+Et+\phi_t]}\langle\chi|\boldsymbol{p}_t\boldsymbol{q}_t\rangle\right|^2. \tag{10.8}$$

To obtain this expression the pre-exponential factor has been approximated to a unitary complex number, *i.e.*, $C_t\left(\boldsymbol{p}_0, \boldsymbol{q}_0\right) \approx e^{\mathrm{i}\phi(t)/\hbar}$, where $\phi\left(t\right) = \mathrm{phase}\left[C_t\left(\boldsymbol{p}_0, \boldsymbol{q}_0\right)\right]$. Eq. (10.8) has been successfully applied to the calculations of small molecular systems, such as H_2O, CH_4 and NH_3. These calculations have been performed by employing pre-computed PES, otherwise it would not have been possible to simulate thousands of *ab initio* molecular dynamics trajectories.[34] To significantly reduce the number of classical trajectories in the phase space integration of Eq. (10.8) and allow for *ab initio on-the-fly* TA-SCIVR calculations when PESs are not available, the multiple coherent MC SCIVR method has been introduced.[35,36] The idea originated from a work by De Leon and Heller[37] who demonstrated on low dimensional model systems that accurate semiclassical eigenenergies and eigenfunctions[38–42] can be obtained employing a single classical trajectory when this is run at the correct (quantum) energy. Similarly, in the MC-SCIVR method it is assumed that reliable eigenvalues can be obtained by employing a single *ab initio* molecular dynamics trajectory with an energy in the neighborhood of the true (but unknown) quantum one. The initial phase conditions are such that the *ab initio* molecular dynamics trajectory is started from the equilibrium molecular geometry $\left(\boldsymbol{q}_{eq}\right)$ with each momentum component equal to the harmonic energy of the vibrational frequency that one wants to simulate, *i.e.*, $p_{i,\mathrm{eq}} = \sqrt{(2n+1)\hbar\omega_i}$, where ω_i is the harmonic frequency of the i-th normal mode and n is fixed according to the vibrational state of interest. In addition, one needs to choose a reference state $|\chi\rangle$ in a way to enhance the Fourier transform signal of the vibrational peak of interest. This goal is obtained by a combination of the coherent states originated by the starting trajectory conditions. For each vibrational state K between the N_{st} states to be investigated, the reference state is of the type

$$|\chi^{(K)}\rangle = \prod_{J=1}^{N_{vib}} \sum_{\alpha=1}^{N_\alpha} \varepsilon_{\alpha,J}^{(K)} \, |p_{\mathrm{eq},\alpha,J}^{(K)}, q_{\mathrm{eq},\alpha,J}^{(K)}\rangle, \quad K=1,\ldots,N_{st}, \quad (10.9)$$

where N_α is the number of coherent states and of the associated coefficients $\varepsilon_{\alpha,J}^{(K)}$, which are employed to enforce parity or molecular symmetry that allows for detecting specific mode excitation or symmetry species. By taking as many trajectories as the number of vibrational peaks of interest $(N_{tr} = N_{st})$,[38,43,44] Eq. (10.8) becomes

$$I(E) = \left(\frac{1}{2\pi\hbar}\right)^{N_{vib}} \sum_{K=1}^{N_{tr}} \frac{1}{2\pi\hbar T}$$

$$\left|\int_0^T dt\, e^{\frac{i}{\hbar}\left[S_t^{(K)}\left(p_0^{(K)},q_0^{(K)}\right)+Et+\phi_t^{(K)}\right]} \langle\chi^{(K)}|p_t^{(K)}q_t^{(K)}\rangle\right|^2. \quad (10.10)$$

Eq. (10.10) is the MC-SCIVR working formula. When a single *ab initio* classical trajectory is employed ($N_{st} = 1$), the ground (usually harmonic zero point energy) initial conditions are taken and employed for getting the full power spectrum. This turns out to be accurate for estimating fundamentals, but less accurate for overtones.[45] We briefly mention here that in an effort to improve over the harmonic sampling of trajectory initial conditions, an adiabatic switching procedure has been recently developed and interfaced to semiclassical calculations leading to more accurate and precise frequency estimates.[46]

10.2. Recent Methodological Advances

10.2.1. *The divide-and-conquer semiclassical initial value representation*

There are two main issues to face when dealing with the quantum (semiclassical) vibrational spectroscopy of large dimensional systems. One is at the theoretical level and it consists in getting a well-resolved signal, which is harder and harder to achieve as the dimensionality of the systems increases. The second issue is related to the necessity to employ an accurate PES in the calculations. While recent research efforts are trying to make analytical PESs accessible also at the level of the gold coupled-cluster standard, one generally has to rely on different approaches when treating dozens of atoms. These include *ab initio* "on-the-fly" DFT approaches in which electronic structure calculations are constantly performed along the dynamics and that are accessible thanks to the MC-SCIVR technique described above, or, mainly with limitation to qualitative estimates, polarizable force fields. For the sake of this section, we assume that an accurate PES for the system under study is available, and focus on the former issue.

The starting point of our discussion is given by the Kaledin–Miller time-averaged semiclassical initial value representation formula (see Eq. (10.8)). The vibrational spectral density, $I(E)$, is peaked in correspondence of the quantum frequencies of vibration. However, the overlap term $\langle \chi | \mathbf{p}_t, \mathbf{q}_t \rangle$ is less and less likely to provide a sensible signal as the number of degrees of freedom to treat increases.

This aspect is better clarified if we turn to the single-trajectory multiple coherent version of the Kaledin–Miller formula in separable approximation, which is employed when studying large dimensional systems. As already anticipated, in the multiple coherent approach the phase space integration of Eq. (10.8) is substituted (without significant loss in the accuracy of results) by the contribution of a single-trajectory dynamics whose initial conditions $(\mathbf{p}_{eq}, \mathbf{q}_{eq})$ and quantum reference state $|\chi\rangle = |\mathbf{p}_{eq}, \mathbf{q}_{eq}\rangle$ are chosen in a tailored way.[36,37]

$$I(E) = \left(\frac{1}{2\pi\hbar}\right)^{N_{vib}} \frac{1}{2\pi\hbar T}$$
$$\times \left| \int_0^T dt \langle \mathbf{p}_{eq}, \mathbf{q}_{eq} | \mathbf{p}_t, \mathbf{q}_t \rangle e^{i[S_t(\mathbf{p}_{eq}, \mathbf{q}_{eq}) + Et + \phi_t]/\hbar} \right|^2 \quad (10.11)$$

and

$$\langle \mathbf{p}_{eq}, \mathbf{q}_{eq} | \mathbf{p}_t, \mathbf{q}_t \rangle = \prod_{i=1}^{N_{vib}} \langle p_{eq}^{(i)}, q_{eq}^{(i)} | p_t^{(i)}, q_t^{(i)} \rangle. \quad (10.12)$$

It is clear that the larger N_{vib} the more difficult it is to have an overlap which is substantially different from zero along the whole dynamics. This observation is at the heart of the divide-and-conquer strategy.[47] In fact, the goal of the divide-and-conquer semiclassical initial value representation (DC-SCIVR) technique is to reproduce the vibrational features of high dimensional systems based on full-dimensional trajectories but reduced dimensionality semiclassical calculations. Projection of Eq. (10.11) can be readily written down as

$$\tilde{I}(E) = \left(\frac{1}{2\pi\hbar}\right)^{N_{vib}} \frac{1}{2\pi\hbar T}$$
$$\times \left| \int_0^T dt \langle \tilde{\mathbf{p}}_{eq}, \tilde{\mathbf{q}}_{eq} | \tilde{\mathbf{p}}_t, \tilde{\mathbf{q}}_t \rangle e^{i[\tilde{S}_t(\tilde{\mathbf{p}}_{eq}, \tilde{\mathbf{q}}_{eq}) + Et + \tilde{\phi}_t]/\hbar} \right|^2 \quad (10.13)$$

where the tilde symbol $(\sim)$ indicates a subspace-projected quantity. All projected terms are straightforward to obtain from the dynamics, with the exception of $\tilde{S}$ which will be discussed later. $\tilde{\phi}_t$ is the phase of the projected Herman–Kluk pre-exponential factor. In practice, the modes of vibration are partitioned into subspaces in which the semiclassical calculation is performed by means of Eq. (10.13), see Fig. 10.1. The coupling between modes belonging to different subspaces is still accounted for (in an approximate way) by means of the full-dimensional dynamics.

10.2.2. *Choosing the subspaces for DC-SCIVR calculations*

The way the vibrational modes should be best grouped into different subspaces is a problem whose resolution is far from being trivial. Notwithstanding, this aspect is a crucial one. Figure 10.2 demonstrates the dramatically different outcome of DC-SCIVR simulations when different subspace partitions are employed. Specifically, two different model systems have been considered. One system is made of eight Morse oscillators, the other one has 18 of them. Strong coupling is present between modes that belong to the same sub-unit, while sub-units are weakly coupled between them. From this example it is

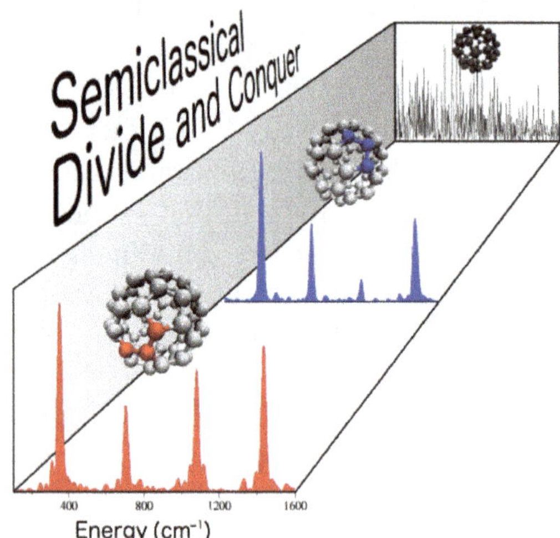

Figure 10.1. Schematic representation of the DC-SCIVR technique. A full-dimensional calculation (back) would provide noisy spectra. The DC-SCIVR method allows one to focus on a subspace of vibrational modes and collect accurate spectroscopic features (front and middle).

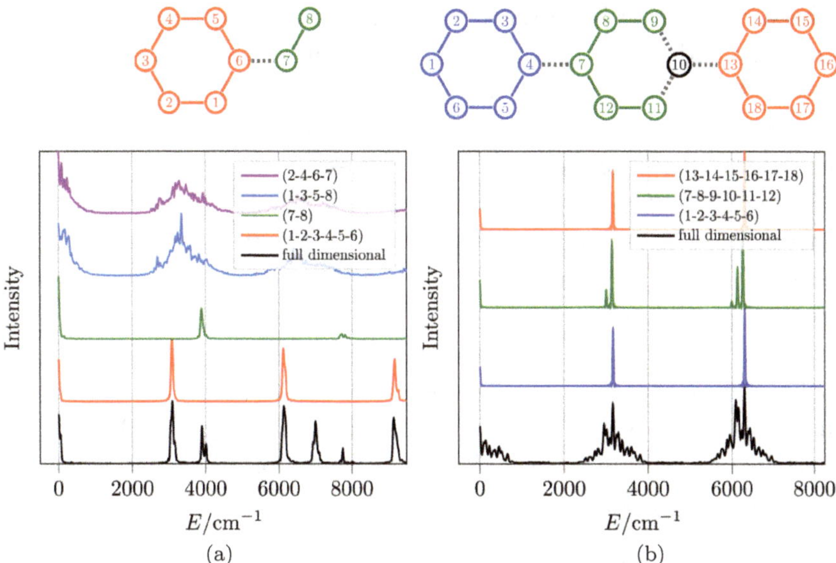

Figure 10.2. Model systems made of (a) 8 and (b) 18 Morse oscillators (upper panel) and their semiclassical full-dimensional and divide-and-conquer power spectra (bottom panel). *Source*: This figure is the synthesis of two figures reproduced from *J. Chem. Phys.* **2020**, *153*, 204104 with permission of AIP Publishing.

clear that on the one hand a non-optimized distribution of modes into subspaces leads to poor spectroscopic results but, on the other hand, a full-dimensional calculation is not always doable. In fact (panel (a)) a random or non-sensible choice of the subspace (blue and violet lines) leads to a poor spectral signal, while an increase in the system dimensionality (panel (b)) leads to an unresolved spectral signal.[48]

A few techniques have been developed and are available to determine the subspaces accurately.[49] One method is based on evaluating the off-diagonal elements of the normal-mode Hessian matrix of the potential. Off-diagonal elements of the Hessian matrix provide a way to estimate the coupling between normal modes. However, DC-SCIVR is a dynamical approach and thousands of different molecular geometries (with associated different Hessian matrices) are calculated during a simulation. For this reason the subspace partition is better achieved based on the averaged Hessian along the trajectory. To apply this procedure, a threshold parameter ε must be selected. After that, all modes whose coupling is higher than ε are collected in the same subspace. This is an effective procedure, which has been applied successfully numerous times.[49–52] However, it is dependent on an arbitrary threshold and there is no rigorous way to select it apart from the necessity to keep the size of the largest subspace small enough to be able to collect a well-resolved spectroscopic signal.

A rigorous way to determine the subspace partition is provided by Liouville's theorem. The theorem states that, along the dynamics, the phase space volume is conserved ($d\mathbf{p}_t d\mathbf{q}_t = d\mathbf{p}_0 d\mathbf{q}_0$). The direct consequence of this is that the determinant of the monodromy or stability matrix (*i.e.*, the Jacobian J) is always equal to 1, *i.e.*,

$$
M = \begin{pmatrix} M_{\mathbf{qq}} & M_{\mathbf{qp}} \\ M_{\mathbf{pq}} & M_{\mathbf{pp}} \end{pmatrix}
$$
$$
= \begin{pmatrix} \partial\mathbf{q}_t/\partial\mathbf{q}_0 & \partial\mathbf{q}_t/\partial\mathbf{p}_0 \\ \partial\mathbf{p}_t/\partial\mathbf{q}_0 & \partial\mathbf{p}_t/\partial\mathbf{p}_0 \end{pmatrix} \equiv J \qquad \det(\mathcal{M}) = 1 \quad \text{at all times.}
$$

$$(10.14)$$

In the ideal case of a separable potential, it also stands that $\prod_{i=1}^{N_{\text{sub}}} \det \tilde{J}_i(t) = 1$, where N_{sub} is the total number of subspaces, but in a real case the potential energy is not separable and the previous condition is no longer verified. However, a sensible choice of subspaces is one for which the product is as close as possible to 1. This search is not doable by testing all possible subspace combinations in a brute force approach. For this purpose an evolutionary algorithm has been recently developed.[48]

10.2.3. *Non-separability of the potential energy and ease of CPU times in DC-SCIVR simulations*

As anticipated, the projected action $\tilde{S}$ is not straightforward to be evaluated due to the non-separability of the potential. To overcome this issue we have come up with a formula for the projected potential, which is still exact in the case of a separable potential and provides an approximate but reasonable guess for the non-separable case.[47, 49] The relevant formula for the projected potential downgrades the instantaneous degrees of freedom not belonging to the subspace of interest to parameters

$$V_s(t) = V(\tilde{\mathbf{q}}_M(t); \mathbf{q}_{N_{vib}-M}(t)) - V(\tilde{\mathbf{q}}_M^{eq}; \mathbf{q}_{N_{vib}-M}(t)), \qquad (10.15)$$

where the subscript s indicates a generic M-dimensional subspace. Consequently, considering that the momentum is separable and the kinetic energy $(\tilde{T})$ easily projected onto the subspace, the formula for the projected action relative to the subspace s is

$$\tilde{S}_t = \int_0^t dt' \, (\tilde{T} - V_s). \qquad (10.16)$$

Eq. (10.16) has been adopted for several DC-SCIVR simulations with very good accuracy. The additional cost is represented by the necessity to calculate the second term in Eq. (10.15). However, this is much less computationally intensive than calculation of the Hessian matrix of the potential necessary to evolve the monodromy matrix elements and collect second-order quantum effects.

The computational bottleneck of the DC-SCIVR calculations is represented by the need to evaluate the Hessian matrices instantaneously. In principle, this should be done at every step of the dynamics. To ease the computational overhead, we have recently elaborated a strategy based on the creation of a dynamical database of representative Hessian matrices.[53] The idea at the heart of the approach is that similar geometries feature similar Hessian matrices. In practice one estimates the "distance" of each instantaneous geometry from those already present in the database of geometries by means of

$$\Theta(\mathbf{q}, \mathbf{q}_{db}) = |q_l - q_{l,db}| \qquad l = 1, \ldots, N_{vib}, \qquad (10.17)$$

where the calculation is performed in the normal mode space. The value of $\Theta(\mathbf{q}, \mathbf{q}_{db})$ is compared to an arbitrary cutoff to restrict the possible candidates among Hessian matrices in the database through an iterative procedure. If, at the end of this process, more than one database entry is still suitable, then the one associated to the smaller value of $\Theta(\mathbf{q}, \mathbf{q}_{db})$ is chosen. The Hessian matrix thus extracted from the database will be employed in the DC-SCIVR simulations avoiding calculation of a new Hessian matrix. Conversely, if no Hessian database entry is available for association to the instantaneous geometry, then a new Hessian matrix is calculated and inserted in the database.

The Hessian database approach is based on a cutoff choice. The cutoff is generally chosen according to the trade-off between accuracy of results and computational costs. Several applications have demonstrated that it is possible to slash CPU times up to a factor of 10–15 without significant loss in the accuracy of DC-SCIVR outcomes. The procedure for constructing the Hessian database is reported in Figure 10.3.

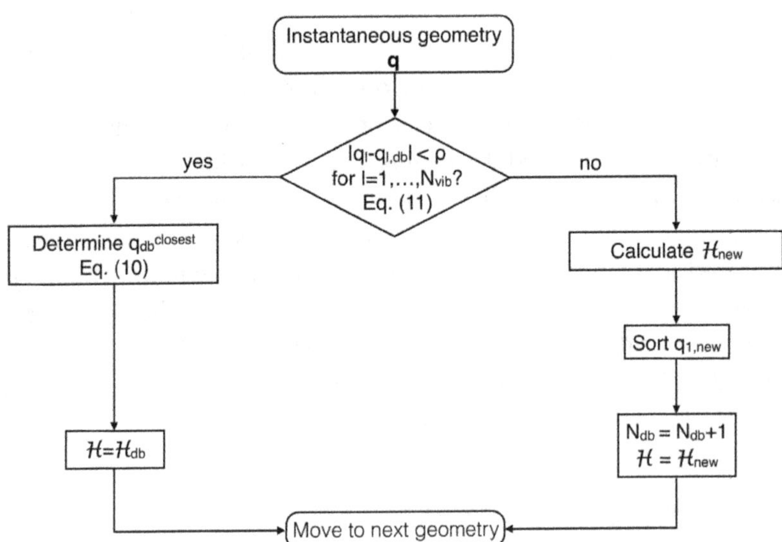

Figure 10.3.   Flow chart for the Hessian database method. *Source*: Reproduced from *J. Chem. Phys.* **2019**, *150*, 244118 with permission of AIP Publishing.

10.2.4. *A practical workflow for DC-SCIVR implementation*

DC-SCIVR can be implemented exploiting *ab initio* molecular dynamics or an analytical PES, if available. In Figure 10.4, we report a workflow chart to describe the complete implementation of a DC-SCIVR simulation. When an analytical expression for the potential is available, the dynamics, Hessian database, and Hessian calculations are all performed (or updated in the case of the Hessian database) step after step. In the case of an *ab initio* "on-the-fly" dynamics, first the entire trajectory is run, then the Hessian database exploits the trajectory to determine the Hessian matrices to be calculated. In all cases the method proceeds with the determination of the subspaces and, upon calculation of the projected potential, the DC-SCIVR vibrational spectrum is eventually obtained.

10.2.5. *Semiclassical wavefunctions and IR spectra*

Differently from other molecular dynamics techniques, semiclassical (SC) calculations for power spectra are not linked to a specific temperature. This is a direct consequence of the Schrödinger equation. There is no temperature dependence in the Schrödinger equation, so quantum vibrational energy levels are temperature independent too. An SC vibrational power spectrum basically returns the eigenenergies of the vibrational Hamiltonian (or the frequencies of vibration as difference between eigenenergies) and, consequently, it is temperature independent. However, there are observables in quantum mechanics that do depend on the population of energy levels and indeed on the temperature. This is the case of infrared (IR) spectra, which are directly comparable to the experiment not only in terms of frequencies of vibration, but also in terms of transition intensities (an additional feature with respect to power spectra). In SC dynamics thermal effects can be added *a posteriori* without running new simulations.

To get to the IR spectrum, similarly to other quantum mechanical approaches, we start from the calculation of the eigenfunctions. Once the SC eigenfunctions are known, the IR spectrum $(S(\omega, T))$ can

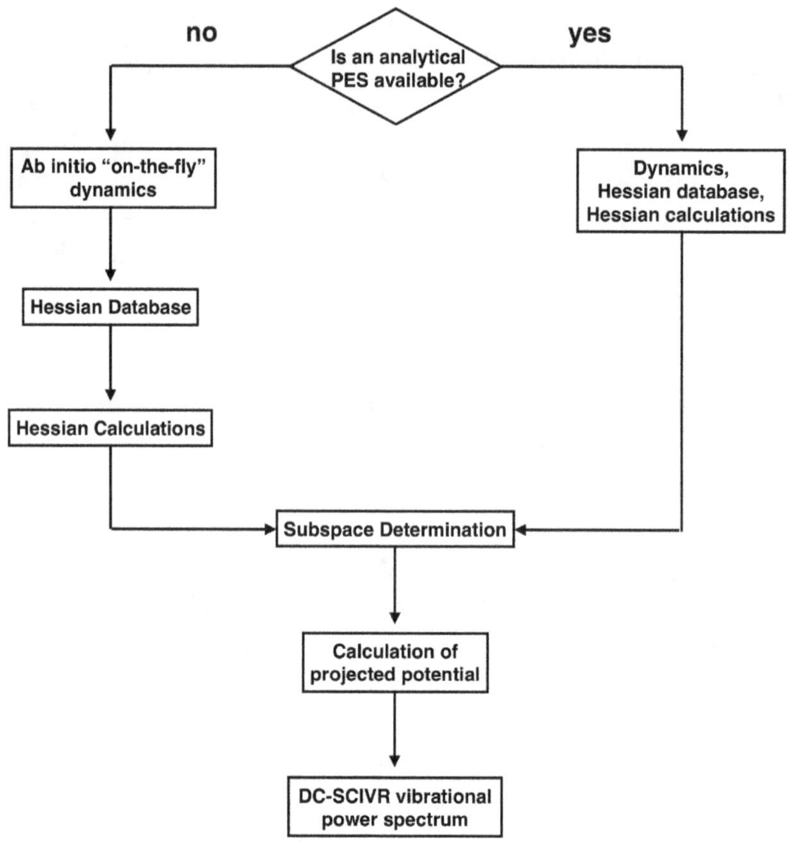

Figure 10.4.   Workflow for DC-SCIVR simulations.

be calculated within the framework of quantum linear response in its sum-over-states version for isotropic and homogeneous molecular systems,[54] *i.e.*,

$$S(\omega, T) = \sum_{n \neq m} [P_n(T) - P_m(T)] F_{nm} \delta(\omega - \Omega_{nm}), \qquad (10.18)$$

where $\Omega_{nm} = E_m - E_n$ is the energy difference between vibrational energies; $P_n(T)$ is the population of the n-th state at temperature T, *i.e.*, $P_n(T) = e^{-E_n/k_B T}/Z$ where k_B is the Boltmann constant and $Z = \sum_n e^{-E_n/k_B T}$ is the canonical partition function; $F_{nm} \propto \Omega_{nm} \langle \Psi_n | \hat{\mu} | \Psi_m \rangle$ are the oscillator strengths, which depend on the system eigenfunctions Ψ_n and dipole moment operator $\hat{\mu}$. Therefore, it is clear that knowledge of the eigenfunctions allows one to calculate the IR spectrum.

Before moving to the actual strategy for the SC determination of the eigenfunctions, we spend a few words about the possible approximations commonly employed when calculating IR spectra. The clue lies in the transition dipole element $\langle \Psi_n | \hat{\mu} | \Psi_m \rangle$, which is the difficult part of the calculation. It is difficult for two main reasons. The first one is that even a few wavefunctions are commonly difficult to calculate with the exclusion of model or low dimensional systems. Secondly, the dipole moment operator is trivial to calculate in its nuclear components but this is not the case for the electronic component. For the latter either an analytical dipole surface is available, or one should rely on an "on-the-fly" Monte Carlo sampling for evaluating the electronic dipole contribution. This has been done for instance in the case of water,[39] but it gets more and more complicated to do as the dimensionality of the system increases. To overcome these major issues, some approximations are commonly employed. First of all, the dipole moment is linearized and the wavefunctions substituted by their harmonic counterparts. In this case a harmonic "electrical" approximation is invoked. This is often coupled with a harmonic "mechanical" approximation counterpart, *i.e.*, frequencies of vibration are estimated harmonically upon diagonalization of the Hessian matrix of the potential at the equilibrium geometry. The two together are known under the name of double harmonic approximation. However, anharmonicities are fundamental and they can be included both at the level of vibrational frequencies and at the level of wavefunctions.

To have the anharmonic eigenfunctions available, an SC approach has been introduced.[39] The idea is to construct each eigenfunction $|\Psi_n\rangle$ as a linear combination of harmonic eigenstates $|\phi_{\mathbf{K}}\rangle$

$$|\Psi_n\rangle = \sum_{\mathbf{K}} C_{n,\mathbf{K}} |\phi_{\mathbf{K}}\rangle, \tag{10.19}$$

and the problem is reduced to the calculation of coefficients. The square modulus of the coefficients can be easily related to the SC power spectrum. In fact

$$|C_{n,\mathbf{K}}|^2 = |\langle \phi_{\mathbf{K}} | \Psi_n \rangle|^2, \qquad (10.20)$$

and

$$\langle \phi_{\mathbf{K}} | e^{-\frac{i}{\hbar}\hat{H}t} | \phi_{\mathbf{K}} \rangle = \sum_m e^{-\frac{i}{\hbar}E_m t} |\langle \phi_{\mathbf{K}} | \Psi_m \rangle|^2. \qquad (10.21)$$

From Eq. (10.21) one readily gets, by means of the properties of Fourier transforms, the following working formula:

$$|C_{n,\mathbf{K}}|^2 = \frac{1}{\pi T} Re \left[\int_0^T dt \, \langle \phi_{\mathbf{K}} | e^{-\frac{i}{\hbar}\hat{H}t} | \phi_{\mathbf{K}} \rangle e^{-\frac{i}{\hbar}E_n t} \right] = \bar{I}_{\phi_{\mathbf{K}}}(E_n), \qquad (10.22)$$

which basically states that the square modulus of the coefficient for the n-th state on the $\mathbf{K}$-th element of the harmonic basis set $\{|\phi_{\mathbf{K}}\rangle\}$ is equal to the intensity of the power spectrum calculated using $|\phi_{\mathbf{K}}\rangle$ as reference state at the eigenenergy E_n. In this way, one can decompose the eigenfunctions into the harmonic basis set. The only aspect which remains to be clarified is the sign of the coefficients. This requires a bit of workaround. It can be worked out that[39]

$$C_{n,\mathbf{K}} = \mathrm{sign}(C_{0,\mathbf{K}}) \frac{\bar{I}_{0\mathbf{K}}(E_n) - \bar{I}_{\phi_0}(E_n) - \bar{I}_{\phi_{\mathbf{K}}}(E_n)}{2\sqrt{\bar{I}_{\phi_0}(E_n)}}, \qquad (10.23)$$

where $\bar{I}_{0\mathbf{K}}(E_n)$ is the value of the SC power spectrum obtained using a reference state which is the sum of the harmonic ground state and the $\mathbf{K}$th harmonic state, i.e., $|\phi_0\rangle + |\phi_{\mathbf{K}}\rangle$; $\bar{I}_{\phi_0}(E_n)$ employs just $|\phi_0\rangle$ as a reference state, and $\bar{I}_{\phi_{\mathbf{K}}}(E_n)$ adopts just $|\phi_{\mathbf{K}}\rangle$ as a reference state; $\mathrm{sign}(C_{0,\mathbf{K}})$ sets a global sign, and therefore it is irrelevant.

Eventually it is possible to calculate the eigenfunctions by simulating a set of power spectra characterized by different reference states and evaluating them at the eigenenergy corresponding to the eigenfunction. The method has been successfully applied not only to model systems and a small molecule like water,[39] but it has been also adopted for the protonated glycine amino acid.[41] However, calculation of the relevant eigenfunctions for larger molecules may turn out to be a computationally unfeasible task.

Another semiclassical approach, which is less accurate but able to avoid calculation of eigenfunctions, has been also developed quite recently.[40]

10.3. Some Remarkable Applications

In this section, we present a few representative applications of semiclassical dynamics. They all involve complex systems of different dimensionality. We start with the study of the vibrational spectroscopy of the fluxional Zundel cation, a strict test for any theoretical method in quantum spectroscopy that aims at being considered reliable. Then we move to the resolution of an experimental controversy involving a biological species, the protoned glycine dimer, and get to larger biological systems like nucleosides. Finally, we have a look at a new and promising field of research: quantum spectroscopy in solution. In our simulations we make use of short-time dynamics (less than 1 ps long) as required by semiclassical approaches, and different kinds of PES representations.

10.3.1. *A small but challenging molecule: The Zundel cation*

The first study we present is about the small protonated water cluster $H_5O_2^+$, *i.c.*, the Zundel cation. This molecule is a prototypical example of protonated water clusters, which has been tackled by various approaches given the great biological relevance of the charge transport mechanism in aqueous solutions.[55–58] From the vibrational point of view, at high frequency the molecule features two bands associated to the O–H stretching modes. Furthermore, it is characterized by strongly anharmonic dynamics for the proton shared by the two water monomers, a motion which spectroscopically leads to the signature proton transfer doublet. This feature is quite complicated. It involves both proton transfer modes, wagging of the two water moieties and stretching modes along the O–O axis. These motions are strongly coupled together. The investigation is made particularly difficult by the topology of the PES. In fact, the presence of several low-frequency barriers between equivalent global minima make classical trajectories particularly unstable, which is challenging for theoretical methods in general and particularly for semiclassical ones, because they include quantum corrections starting from the evaluation of the stability (monodromy) matrix. The first successful simulations and explanations of the proton transfer motion in the Zundel cation were performed by means of the multi-configuration time-dependent Hartree (MCTDH) approach. Before that, the doublet origin was unclear and thought to be related to an asymmetry in the spectra due to the tagging atom.[59–61] We show in the following that similar highly accurate results can be obtained by means of SC dynamics too.

The Zundel cation is characterized by 15 vibrational degrees of freedom. This is a number which is still treatable semiclassically in a full-dimensional fashion by means of an ensemble of trajectories to get Monte Carlo convergence of the phase-space integral in Eq. (10.13). However, the complexity of the simulations requires adoption of some devices concerning the way some of the degrees of freedom not directly involved in the calculation are initialized. In other words, the study of the Zundel cation serves as a link between the standard time-averaged full-dimensional semiclassical calculations and DC-SCIVR ones.

We focus on the study of the proton transfer motion, which well exemplifies all the aspects mentioned above. From the experimental point of view, the proton transfer motion shows a neat doublet at 928 cm^{-1} and 1,047 cm^{-1}.[56] Several studies have been undertaken to explain this doublet.[59,61] One hypothesis is that the doublet is the result of a tunneling splitting process similar to the one that can be sometimes found in hydrogen bonding. However, this possibility has been ruled out[62] and from semiclassical trajectories it is clear that in most of the relevant configuration space close to equilibrium, the shared proton only visits a shallow single minimum.[63] This points out another key feature of semiclassical dynamics: the fact of being based on classical trajectories allows one to get a physical interpretation of the phenomena which is more intuitive than other quantum mechanical methods.

Eventually the accepted explanation of the proton transfer doublet is that it arises from a Fermi resonance between the bare proton transfer mode and a combination of O–O stretching and wagging modes which is close in frequency to the bare proton transfer mode.[60,61] To deal with such a complex situation it requires a careful choice of reference state $|\chi\rangle$ and starting dynamics conditions. In particular, for this simulation, the starting geometry was chosen from a Husimi distribution around the equilibrium one ($\mathbf{q}_{eq}$), while an initial velocity centered at one quantum of harmonic excitation was given just to the proton transfer mode ($p_7^{exc}, p_{0...6,8...15}^{eq}$, the proton transfer mode being the seventh upon sorting of the harmonic frequencies in increasing order). The initial velocity of all other modes was set to zero, but obviously the modes get excited along the trajectories because of the quick energy transfer. In Figure 10.5, besides experimental values, three different power spectra are reported. These are obtained employing three different reference states. As anticipated, $|\chi\rangle$ can be chosen to enhance specific features of the spectrum exploiting parity and symmetry. In Figure 10.5, the green line is obtained with a reference state centered on the proton transfer mode and symmetrized in a way that only features related to this mode get enhanced, *i.e.*,

$$|\chi\rangle = |\mathbf{p}_7^{eq}, \mathbf{q}_7^{eq}\rangle$$

$$= \prod_{i=1}^{15} \left(|\mathbf{p}_i^{eq}, \mathbf{q}_i^{eq}\rangle + \varepsilon_i| - \mathbf{p}_i^{eq}, \mathbf{q}_i^{eq}\rangle \right), \qquad \varepsilon_7 = -1, \varepsilon_i = 0 \text{ if } i \neq 7.$$

$$(10.24)$$

The result is that the two peaks due to the doublet are well identified at frequencies of 891 cm^{-1} and 1,062 cm^{-1}, which are not far from the experimental results (928 cm^{-1} and 1047 cm^{-1}, respectively). To better identify which modes contribute to the proton transfer motion it is possible to build a reference state starting from the harmonic ground state and apply the $\hat{z}$ operator to it, *i.e.*, the projection operator along the O–O axis $|\chi\rangle = \hat{z}|0\rangle$. This is the axis along which the proton transfer motion takes place. The result of such simulation is given by the orange dashed-dotted line in

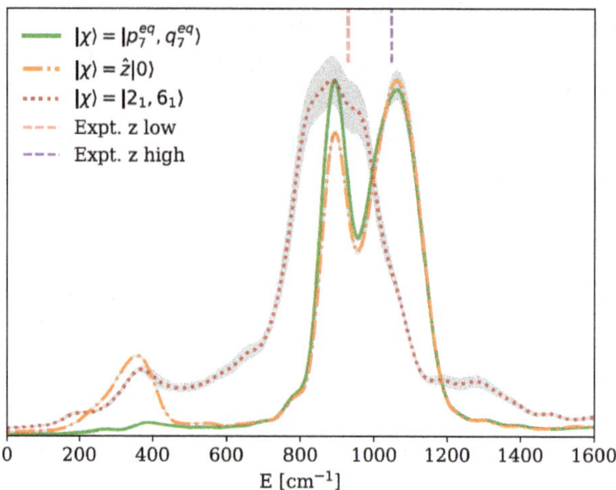

Figure 10.5. Several semiclassical simulations describing the proton transfer motion by means of different reference states. Reference state centered at the hydrogen transfer motion mode (green solid line); reference state obtained from the projection of the ground harmonic state to the O–O axis (orange dashed-dots); reference state centered at the first combined excitation of the wagging and O–O stretch modes (red dots). *Source*: Reproduced from *J. Chem. Phys.* **2019**, *151*, 114307 with permission of AIP Publishing.

Figure 10.5. In addition to the doublet a peak at 358 cm^{-1} is found, compatible with the wagging mode and corroborating the view that the wagging mode is intimately linked to proton transfer. Finally, it remains to determine within the Fermi resonance originating the doublet which peak has a major component due to the wagging O–O stretch combination and which one has a major component due to the proton transfer mode. To this end a reference state centered on the combination is employed ($|\chi\rangle = |2_1, 6_1\rangle$), providing the red dotted spectrum. It is clear that this spectrum confirms the wagging band and suggests that the signal due to the wagging O–O stretch combination is the one at lower frequency in the doublet. MCTDH calculations came to the same results,[61] and an MCTDH application to deuterated isotopologues of the Zundel cation showed that the identity of the doublet peaks reverses upon deuteration.[64]

This example demonstrates the ability of SC dynamics to tackle with high accuracy difficult spectroscopic problems that only the most refined quantum theoretical approaches are able to treat.[61] When moving to larger systems, though, other quantum methods start facing the issue of exponential scaling of computational costs. Conversely, semiclassical techniques like DC-SCIVR combined with the devices described above are characterized by a mild increase of CPU costs, which permits application to much larger systems. This will be the focus of the next examples.

10.3.2. *Beyond experimental controversy: The case of the protonated glycine dimer*

The protonated glycine dimer represents a building block of more complex biological species characterized by protonation. Since these systems are ubiquitous in biochemistry and the proton has the lightest mass among ions, an important question to answer is whether accurate quantum mechanical investigations, even if difficult to perform, can be crucial for such kinds of systems or not. In this regard, a well-known experimental controversy has involved the protonated glycine dimer (Gly_2H^+). Studies conducted in 2005 by McLafferty *et al.* and in 2007 by McMahon and co-workers were clearly at odds.[65, 66] They tried to determine which conformer of the protonated glycine dimer was the main contributor to the measured IR spectrum. Gly_2H^+ has two low-energy conformers in gas phase, which are denoted as CS01 and CS02. For CS01 there is a $O \cdots H^+N$ interaction, while an $N \cdots H^+N$ one is typical of CS02. McLafferty concluded that CS02 was dominating the experimental spectrum, while McMahon deemed that CS01 was mainly responsible for the detected spectrum. Both of them substantiated their conclusion by presenting calculations based on a scaled harmonic approach, which indeed was not able to provide a final answer.

Development and implementation of DC-SCIVR has given us the opportunity to investigate the problem by means of a much refined and accurate technique.[67] We based our study on a single-trajectory *ab initio* "on-the-fly" dynamics. The energy of the trajectory was equal to the harmonic zero point energy of Gly_2H^+, and its length equal to about 0.6 ps or 25,000 atomic time units. The idea at the

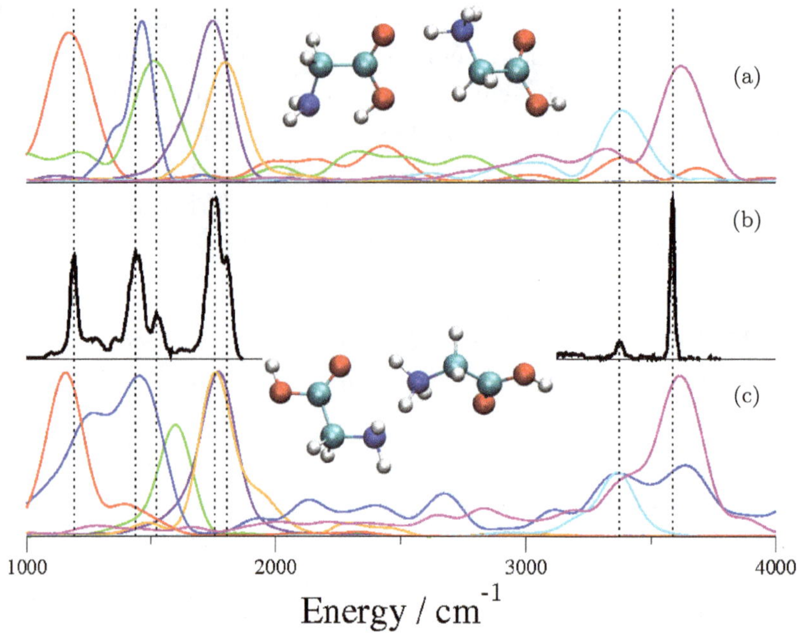

Figure 10.6. Power spectra for (a) Conformer CS01 and (c) Conformer CS02 of Gly_2H^+. The measured IR spectrum is reported in panel (b). *Source*: Reproduced from Ref. 67 with permission from the Royal Society of Chemistry.

heart of the study was to calculate a semiclassical power spectrum for each of the two conformers and then compare the frequencies of vibration thus obtained to the experimental spectrum. For this reason, we focused on the vibrational features above $1,000$ cm^{-1}. Subspaces were determined on the basis of the averaged Hessian approach.

A comparison between computational and experimental spectra in Figure 10.6 clearly shows that in panel (a) all spectroscopic features of the experiment are accurately reproduced, while in panel (c) there are some differences. In particular, the two carbonyl stretch frequencies at around $1,800$ cm^{-1} are correctly described in panel (a) but they are basically degenerate in panel (c). Furthermore, there seems to be better overall agreement between panels (a) and (b) rather than (b) and (c). This is confirmed by looking at the mean absolute error (MAE) of DC-SCIVR frequencies vs experimental ones. For Conformer CS01 the MAE is 14 cm^{-1}, while for Conformer CS02 the MAE is 32 cm^{-1}. This discrepancy in the mean absolute error values is pretty significant and conclusive even upon weighing in the typical accuracy of SC calculations.

This study demonstrates how an accurate quantum spectroscopic investigation can assist and clarify experimental outcomes. Also, the study involves a biological species and this is a promising area of development for the DC-SCIVR technique. More details on this topic will be given in the application presented next.

10.3.3. *Going big: Quantum spectroscopy of biological species*

When interested in the computational study of biological systems, dimensionality represents probably the biggest issue to face. On the one hand, dealing with a large number of degrees of freedom is a difficult task for quantum theories. DC-SCIVR allows one to overcome this issue. On the other hand, it is difficult to find a PES which is both accurate and fast to evaluate for such big systems.

One peculiar feature of SC methods in general is that they treat quantum mechanically (in semiclassical approximation, of course) all degrees of freedom and not just a subset or portion of them. Therefore, it is necessary to face the issue of the PES to be employed when applying DC-SCIVR to biological molecules. We take advantage of this aspect for introducing a very brief description of possible ways to deal with the PES issue.

One possible approach is represented by the *ab initio* "on-the-fly" molecular dynamics. In this approach an *ab initio* electronic structure calculation is performed at each step of the dynamics to get the system energy and gradient commonly by means of one of the several scientific softwares for electronic structure calculations available. NWCHEM is one of such software.[68] It is free to use and we have employed it successfully in many simulations. The advantage of this technique is twofold. First of all, at least in principle, it is applicable to any kind of system. Secondly, the calculated energy and gradients are exact for the level of electronic structure theory and basis set employed. The drawback is represented by the large amount of CPU time that these calculations require, limiting in practice the size of treatable systems and, most of all, the accuracy of the level of electronic structure theory to be used. In practice, *ab initio* "on-the-fly" simulations are usually performed by means of density functional theory.

Another possible approach is represented by the fitting of accurate *ab initio* PESs. This is a hot field in theoretical and computational chemistry involving several techniques like permutationally invariant fitting[69–74] and machine learning methods.[75–82] One starts by generating a dataset of *ab initio* energies at a high level of electronic structure theory. The dataset must be representative of the several features of the PES, which include all conformers and dissociation channels, at both low and high energy. Once the PES has been built, the great advantage over "on-the-fly" methods is that the PES can be reused as many times as needed. Furthermore, potential energy calls are faster and applications to more complex systems are in principle possible. The drawbacks are that a fitting error is inevitable[83] and that construction of a precise analytical surface for large systems is difficult even if great progress has been achieved recently.[84] Furthermore, when the molecular size gets large, the potential calls to this PESs become costly too.

Finally, a third way to deal with the PES issue is to make use of classical force fields. These are built as the sum of several covalent and non-covalent terms based on a set of so-called natural values and parameters. The big advantage of force fields is that potential calls are very fast and this allows for application to very large systems. However, speed in evaluation comes at the cost of accuracy, and many times force fields can be employed just for qualitative rather than quantitative investigations. Many types of force fields of different accuracy have been developed over the years, specializing in tackling particular kinds of systems.

On this point it is possible to employ DC-SCIVR to better evaluate the opportunity to adopt force fields in vibrational

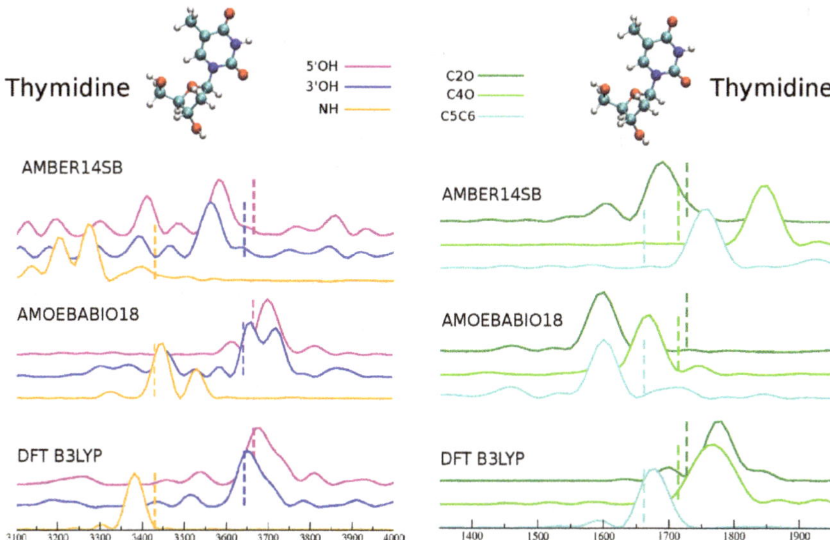

Figure 10.7. Vibrational power spectra for thymidine using *ab initio* "on-the-fly" DFT and force fields. The dashed vertical lines show the experimental findings. *Source*: Reproduced from *J. Chem. Theory Comput.* **2020**, *16*, 3476–3485.

investigations. To this end we have studied the vibrational spectroscopy of nucleosides.[85] The results for thymidine are reported in Figure 10.7.

The good agreement with experimental data obtained using DFT demonstrates that the DC-SCIVR method is an adequate approach for systems of this size and that the B3LYP functional may be appropriate for studying even biological systems. The best estimates of the popular but not polarizable AMBER14SB force field are harmonic ones,[86] while application of the semiclassical recipe worsens the prediction for almost all the simulated signals. Conversely, we obtained a reasonable set of vibrational frequencies when AMOEBABIO18,[87] a polarizable force field, was used for the semiclassical analysis. As anticipated, this opens up the possibility to employ polarizable force fields for qualitative and semi-quantitative quantum descriptions of biological systems in connection with DC-SCIVR.

10.3.4. *How many water molecules are (spectroscopically) needed to solvate one?*

Another topic of great interest in physical chemistry is represented by solvation. DC-SCIVR offers the opportunity to undertake a quantum mechanical investigation of solvation and microsolvation. While this is a general topic for which some perspectives will be presented further below, it is interesting to study the solvation of water itself. Since IR and Raman experimental spectra of liquid water are known, we asked ourselves how many water molecules are needed to solvate the central one in a spectroscopical sense.[88] In other words, one can start by building a larger and larger network of water molecules around a central one. The DC-SCIVR spectrum of the target central molecule varies as its surroundings change and can be compared to the experimental spectra of liquid water. The question is how many water molecules should be added before the DC-SCIVR spectrum is able to reproduce all the spectroscopic features of liquid water?

A portion of interest of the IR spectrum of liquid water is reported in Figure 10.8. We focus on the region between 1,500 and 4,000 cm^{-1}, which is characterized by three bands. Around 1,600 cm^{-1} is the band of bendings, followed by a large band centered at about 2,100 cm^{-1} due to the combination of bendings and low frequency librational motions. We will refer to this in short as the combination band. Finally, mostly above 3,000 cm^{-1} is the band of stretches made of symmetric and asymmetric stretching signals.

Following the idea to build a more and more complex network of water molecules around the target one, it is possible to employ the highly accurate MB-Pol[89] or WHBB[90] water potentials to calculate the corresponding DC-SCIVR spectra. In our work we employed MB-Pol. The smaller clusters (dimer, trimer, hexamer) are known to not reproduce the spectral features of the bulk, because even the signals corresponding to fundamental vibrational transitions are outside the experimental range of liquid water. However, in liquid water tetrahedral coordination is playing the major role and there are two small clusters — the pentamer and the heptamer — which are characterized by tetrahedral coordination.

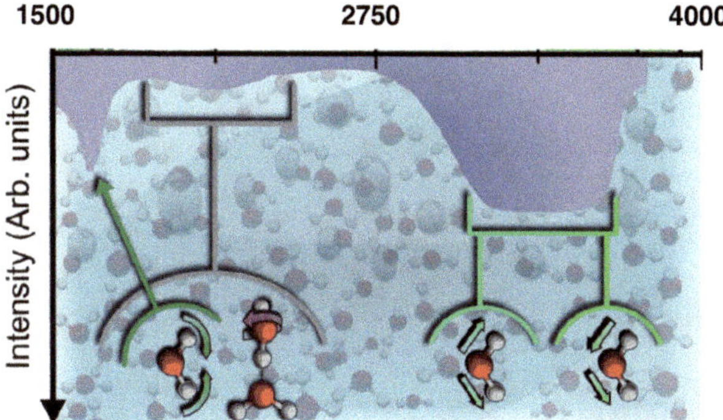

Figure 10.8. IR spectrum of liquid water. The principal vibrational features are highlighted and explained in the text. *Source*: Reproduced from Ref. 88 with permission from the Royal Society of Chemistry.

Figure 10.9(a) shows these two conformers and their DC-SCIVR power spectra superimposed onto the IR spectrum of liquid water. We notice that fundamentals are within the experimental bands. This would lead to the conclusion that a structure made of a central water molecule tetrahedrally coordinated to other four water molecules is representative of solvation. Actually, this would be the answer in a classical world, one where only classical frequencies of vibration, their higher harmonics and classical beatings are accounted for. In a quantum molecular world, which is what the experiment reproduces, there is another crucial aspect: combination bands. Combination bands are representative of mixed excitations and are purely quantum features. On this point it is evident the importance of employing a technique, like DC-SCIVR, able to reproduce quantum effects. In fact, from Figure 10.9, it is clear that the combination band signal for these small clusters is not matching the one for liquid water. The signal we present for the combination band is the most representative one, since it is due to the strongest coupling between a libration mode and the bending mode of the central water monomer, as determined by means of the Hessian approach described above. Therefore, differently from the classical case, the quantum answer to water solvation is not the pentamer or heptamer. One needs to look at bigger structures.

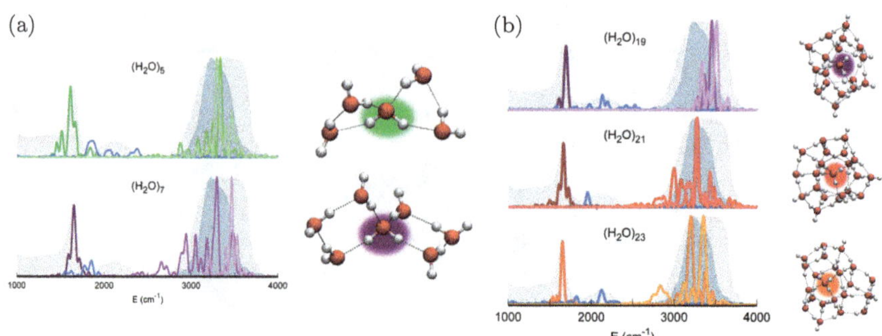

Figure 10.9. DC-SCIVR vibrational power spectra for (a) $(H_2O)_5$, and $(H_2O)_7$, and (b) $(H_2O)_{19}$, $(H_2O)_{21}$, $(H_2O)_{23}$. The bending-libration combined signal is drawn in blue. *Source*: Reproduced from Ref. 88 with permission from the Royal Society of Chemistry.

Figure 10.9(b) shows the DC-SCIVR vibrational power spectra for clusters made of 19, 21, and 23 water monomers respectively. We notice that such clusters have a truly "3D" structure, in the sense that they have not an "all-monomer surface" conformation as it is in the case of smaller clusters. In particular, from Figure 10.9, it is clear that the 19-mer has some issues in the reproduction of the fundamental transition features, while it is the 21-mer that displays the desired vibrational pattern in its entirety. This result is confirmed by investigation of the larger 23-mer. The final quantum spectroscopic answer to the problem of water solvation is indeed very different from its classical counterpart. Quantum mechanically to solvate a target water molecule a double shell made of 20 water monomers is necessary. The target molecule is tetrahedrally coordinated to four molecules in the first solvation shell. Each of these molecules is in turn tetrahedrally coordinated to other 4 molecules, which build up the second solvation shell.

10.4. A Quick Look at Some Perspectives

DC-SCIVR has been shown to be a powerful technique for the quantum spectroscopy of medium and large size systems. This feature opens the way to the study of more and more complex molecules and supra-molecular aggregates. There are three fields of application and research we can envisage for the near future: larger biological systems; supra-molecular aggregates involving molecules adsorbed on a surface; solvated systems with explicit solvent treatment.[91–94]

As for biological species DC-SCIVR will allow for an *in silico* description of the dynamics of several building blocks of life. These range from simple amino acids to more complex peptides, from nucleobases to elaborated structures like quadruplexes and nucleic acid strands. Currently the vast majority of simulations involving these systems are classical ones or limited to electronic structure calculations. The possibility to provide a dynamical description of the molecular events involving biological species at the quantum level is expected to improve the description of the many hydrogen bonds which characterize these systems, leading to a more accurate description of their structural characterization and functionalization.

Molecules adsorbed on the solid surface of TiO_2 or other semiconductors represent typical cases of interest in material science. An extension of the DC-SCIVR method to plane wave electronic structure codes has already been proposed,[95] to allow for vibrational characterization of adsorbates. Accurate studies of these systems are expected to provide an alternative way to rationalize catalytic processes for better functional material engineering.

In the field of solvation, DC-SCIVR opens up the possibility to perform calculations on solvated and microsolvated species by means of an explicit solvent treatment. This is certainly the case for water, for which a number of highly precise potentials are available. The procedure will allow one not only to overcome the limitations of implicit solvation models, but also to bring the high accuracy of gas-phase simulations to the realm of condensed phase systems. This will permit a better and more frequent comparison to experiments.

References

1. Cao, Y.; Romero, J.; Olson, J. P.; Degroote, M.; Johnson, P. D.; Kieferová, M.; Kivlichan, I. D.; Menke, T.; Peropadre, B.; Sawaya, N. P. *et al.* Quantum chemistry in the age of quantum computing. *Chem. Rev.* **2019**, *119*, 10856–10915.

2. Parrinello, M.; Rahman, A. Crystal structure and pair potentials: A molecular-dynamics study. *Phys. Rev. Lett.* **1980**, *45*, 1196.

3. Heller, E. J. The semiclassical way to molecular spectroscopy. *Acc. Chem. Res.* **1981**, *14*, 368–375.

4. Feynman, R. P.; Hibbs, A. R. *Quantum Mechanics and Path Integrals.* McGraw-Hill: New York, 1965.

5. Makri, N.; Miller, W. H. Monte Carlo path integration for the real time propagator. *J. Chem. Phys.* **1988**, *89*, 2170–2177.

6. Habershon, S. Automated prediction of catalytic mechanism and rate law using graph-based reaction path sampling. *J. Chem. Theory Comput.* **2016**, *12*, 1786–1798.

7. Habershon, S.; Manolopoulos, D. E.; Markland, T. E.; Miller III, T. F. Ring-polymer molecular dynamics: quantum effects in chemical dynamics from classical trajectories in an extended phase space. *Annu. Rev. Phys. Chem.* **2013**, *64*, 387–413.

8. Berry, M. V.; Mount, K. Semiclassical approximations in wave mechanics. *Rep. Prog. Phys.* **1972**, *35*, 315.

9. Baranger, M.; de Aguiar, M. A.; Keck, F.; Korsch, H.-J.; Schellhaass, B. Semiclassical approximations in phase space with coherent states. *J. Phys. A* **2001**, *34*, 7227.

10. Grossmann, F.; Xavier, A. L. From the coherent state path integral to a semiclassical initial value representation of the quantum mechanical propagator. *Phys. Lett. A* **1998**, *243*, 243–248.

11. Miller, W. H. The semiclassical initial value representation: a potentially practical way for adding quantum effects to classical molecular dynamics simulations. *J. Phys. Chem. A* **2001**, *105*, 2942–2955.

12. Heller, E. J. Cellular dynamics: a new semiclassical approach to time dependent quantum mechanics. *J. Chem. Phys.* **1991**, *94*, 2723–2729.

13. Miller, W. H. Classical S matrix: numerical application to inelastic collisions. *J. Chem. Phys.* **1970**, *53*, 3578–3587.

14. Heller, E. J. Frozen Gaussians: A very simple semiclassical approximation. *J. Chem. Phys.* **1981**, *75*, 2923–2931.

15. Herman, M. F.; Kluk, E. A semiclassical justification for the use of non-spreading wavepackets in dynamics calculations. *Chem. Phys.* **1984**, *91*, 27–34.

16. Kay, K. G. Integral expressions for the semiclassical time-dependent propagator. *J. Chem. Phys.* **1994**, *100*, 4377–4392.

17. Kay, K. G. Numerical study of semiclassical initial value methods for dynamics. *J. Chem. Phys.* **1994**, *100*, 4432–4445.

18. Kay, K. G. Semiclassical propagation for multidimensional systems by an initial value method. *J. Chem. Phys.* **1994**, *101*, 2250–2260.

19. Weissman, Y. Semiclassical approximation in the coherent states representation. *J. Chem. Phys.* **1982**, *76*, 4067–4079.

20. Bonella, S.; Montemayor, D.; Coker, D. F. Linearized path integral approach for calculating nonadiabatic time correlation functions. *Proc. Natl. Acad. Sci.* **2005**, *102*, 6715–6719.

21. Kay, K. G. Semiclassical initial value treatments of atoms and molecules. *Annu. Rev. Phys. Chem.* **2005**, *56*, 255–280.

22. Zhuang, Y.; Siebert, M. R.; Hase, W. L.; Kay, K. G.; Ceotto, M. Evaluating the accuracy of Hessian approximations for direct dynamics simulations. *J. Chem. Theory Comput.* **2012**, *9*, 54–64.

23. Martin-Fierro, E.; Pollak, E. Forward-backward semiclassical initial value series representation of quantum correlation functions. *J. Chem. Phys.* **2006**, *125*, 164104.

24. Pollak, E.; Martin-Fierro, E. New coherent state representation for the imaginary time propagator with applications to forward-backward semiclassical initial value representations of correlation functions. *J. Chem. Phys.* **2007**, *126*, 164107.

25. Zhang, S.; Pollak, E. Hybrid prefactor semiclassical initial value series representation of the quantum propagator. *J. Chem. Theory Comput.* **2005**, *1*, 345–352.

26. Conte, R.; Pollak, E. Comparison between different Gaussian series representations of the imaginary time propagator. *Phys. Rev. E* **2010**, *81*, 036704.

27. Antipov, S. V.; Ye, Z.; Ananth, N. Dynamically consistent method for mixed quantum-classical simulations: A semiclassical approach. *J. Chem. Phys.* **2015**, *142*, 184102.

28. Church, M. S.; Ananth, N. Semiclassical dynamics in the mixed quantum-classical limit. *J. Chem. Phys.* **2019**, *151*, 134109.

29. Grossmann, F. *Theoretical Femtosecond Physics: Atoms and Molecules in Strong Laser Fields.* Springer: Berlin/Heidelberg, 2013.

30. Ceotto, M.; Zhuang, Y.; Hase, W. L. Accelerated direct semiclassical molecular dynamics using a compact finite difference Hessian scheme. *J. Chem. Phys.* **2013**, *138*, 054116.

31. Conte, R.; Ceotto, M. Semiclassical Molecular Dynamics for Spectroscopic Calculations. In *Quantum Chemistry and Dynamics of Excited States: Methods and Applications*; González, L., Lindh, R., Eds.; John Wiley and Sons, Inc., 2020; pp. 595–628.

32. Kaledin, A. L.; Miller, W. H. Time averaging the semiclassical initial value representation for the calculation of vibrational energy levels. *J. Chem. Phys.* **2003**, *118*, 7174–7182.

33. Kaledin, A. L.; Miller, W. H. Time averaging the semiclassical initial value representation for the calculation of vibrational energy levels. II.

Application to H_2CO, NH_3, CH_4, CH_2D_2. *J. Chem. Phys.* **2003**, *119*, 3078–3084.

34. Ma, X.; Di Liberto, G.; Conte, R.; Hase, W. L.; Ceotto, M. A quantum mechanical insight into SN2 reactions: semiclassical initial value representation calculations of vibrational features of the Cl-CH_3Cl pre-reaction complex with the VENUS suite of codes. *J. Chem. Phys.* **2018**, *149*, 164113.

35. Ceotto, M.; Atahan, S.; Shim, S.; Tantardini, G. F.; Aspuru-Guzik, A. First-principles semiclassical initial value representation molecular dynamics. *Phys. Chem. Chem. Phys.* **2009**, *11*, 3861–3867.

36. Ceotto, M.; Atahan, S.; Tantardini, G. F.; Aspuru-Guzik, A. Multiple coherent states for first-principles semiclassical initial value representation molecular dynamics. *J. Chem. Phys.* **2009**, *130*, 234113.

37. De Leon, N.; Heller, E. J. Semiclassical quantization and extraction of eigenfunctions using arbitrary trajectories. *J. Chem. Phys.* **1983**, *78*, 4005–4017.

38. Ceotto, M.; Valleau, S.; Tantardini, G. F.; Aspuru-Guzik, A. First principles semiclassical calculations of vibrational eigenfunctions. *J. Chem. Phys.* **2011**, *134*, 234103.

39. Micciarelli, M.; Conte, R.; Suarez, J.; Ceotto, M. Anharmonic vibrational eigenfunctions and infrared spectra from semiclassical molecular dynamics. *J. Chem. Phys.* **2018**, *149*, 064115.

40. Micciarelli, M.; Gabas, F.; Conte, R.; Ceotto, M. An Effective semiclassical approach to IR spectroscopy. *J. Chem. Phys.* **2019**, *150*, 184113.

41. Aieta, C.; Micciarelli, M.; Bertaina, G.; Ceotto, M. Anharmonic quantum nuclear densities from full dimensional vibrational eigenfunctions with application to protonated glycine. *Nat. Commun.* **2020**, *11*, 1–9.

42. Aieta, C.; Bertaina, G.; Micciarelli, M.; Ceotto, M. Representing molecular ground and excited vibrational eigenstates with nuclear densities obtained from semiclassical initial value representation molecular dynamics. *J. Chem. Phys.* **2020**, *153*, 214117.

43. Ceotto, M.; Tantardini, G. F.; Aspuru-Guzik, A. Fighting the curse of dimensionality in first-principles semiclassical calculations: non-local reference states for large number of dimensions. *J. Chem. Phys.* **2011**, *135*, 214108.

44. Ceotto, M.; Dell' Angelo, D.; Tantardini, G. F. Multiple coherent states semiclassical initial value representation spectra calculations of lateral interactions for CO on Cu (100). *J. Chem. Phys.* **2010**, *133*, 054701.

45. Gabas, F.; Conte, R.; Ceotto, M. On-the-fly *ab* initio semiclassical calculation of glycine vibrational spectrum. *J. Chem. Theory Comput.* **2017**, *13*, 2378.

46. Conte, R.; Parma, L.; Aieta, C.; Rognoni, A.; Ceotto, M. Improved semiclassical dynamics through adiabatic switching trajectory sampling. *J. Chem. Phys.* **2019**, *151*, 214107.

47. Ceotto, M.; Di Liberto, G.; Conte, R. Semiclassical "divide-and-conquer" method for spectroscopic calculations of high dimensional molecular systems. *Phys. Rev. Lett.* **2017**, *119*, 010401.

48. Gandolfi, M.; Rognoni, A.; Aieta, C.; Conte, R.; Ceotto, M. Machine learning for vibrational spectroscopy via divide-and-conquer semiclassical initial value representation molecular dynamics with application to N-methylacetamide. *J. Chem. Phys.* **2020**, *153*, 204104.

49. Di Liberto, G.; Conte, R.; Ceotto, M. "Divide and conquer" semiclassical molecular dynamics: A practical method for spectroscopic calculations of high dimensional molecular systems. *J. Chem. Phys.* **2018**, *148*, 014307.

50. Gabas, F.; Di Liberto, G.; Ceotto, M. Vibrational investigation of nucleobases by means of divide and conquer semiclassical dynamics. *J. Chem. Phys.* **2019**, *150*, 224107.

51. Di Liberto, G.; Conte, R.; Ceotto, M. "Divide-and-conquer" semiclassical molecular dynamics: an application to water clusters. *J. Chem. Phys.* **2018**, *148*, 104302.

52. Conte, R.; Aspuru-Guzik, A.; Ceotto, M. Reproducing deep tunneling splittings, resonances, and quantum frequencies in vibrational spectra from a handful of direct ab initio semiclassical trajectories. *J. Phys. Chem. Lett.* **2013**, *4*, 3407–3412.

53. Conte, R.; Gabas, F.; Botti, G.; Zhuang, Y.; Ceotto, M. Semiclassical vibrational spectroscopy with Hessian databases. *J. Chem. Phys.* **2019**, *150*, 244118.

54. McQuarrie, D. A. *Statistical Mechanics.* Harper-Collins: New York, 1976.

55. Huang, X.; Habershon, S.; Bowman, J. M. Comparison of quantum, classical, and ring-polymer molecular dynamics infra-red spectra of Cl-(H2O) and H^+(H2O)2. *Chem. Phys. Lett.* **2008**, *450*, 253–257.

56. Hammer, N. I.; Diken, E. G.; Roscioli, J. R.; Johnson, M. A.; Myshakin, E. M.; Jordan, K. D.; McCoy, A. B.; Huang, X.; Bowman, J. M.; Carter, S. The vibrational predissociation spectra of the $H_5O_2^+ \cdot$ RGn(RG=Ar,Ne) clusters: correlation of the solvent perturbations in the free OH and shared proton transitions of the Zundel ion. *J. Chem. Phys.* **2005**, *122*, 244301.

57. McCoy, A. B.; Huang, X.; Carter, S.; Landeweer, M. Y.; Bowman, J. M. Full-dimensional vibrational calculations for H5O2$^+$ using an ab initio potential energy surface. *J. Chem. Phys.* **2005**, *122*, 061101.

58. Rossi, M.; Ceriotti, M.; Manolopoulos, D. E. How to remove the spurious resonances from ring polymer molecular dynamics. *J. Chem. Phys.* **2014**, *140*, 234116.

59. Vendrell, O.; Gatti, F.; Meyer, H.-D. Dynamics and infrared spectroscopy of the protonated water dimer. *Angew. Chem. Int. Ed.* **2007**, *46*, 6918–6921.

60. Vendrell, O.; Gatti, F.; Lauvergnat, D.; Meyer, H.-D. Full-dimensional (15-dimensional) quantum-dynamical simulation of the protonated water dimer. I. Hamiltonian setup and analysis of the ground vibrational state. *J. Chem. Phys.* **2007**, *127*, 184302.

61. Vendrell, O.; Gatti, F.; Meyer, H.-D. Full dimensional (15-dimensional) quantum-dynamical simulation of the protonated water dimer. II. Infrared spectrum and vibrational dynamics. *J. Chem. Phys.* **2007**, *127*, 184303.

62. Schran, C.; Brieuc, F.; Marx, D. Converged colored noise path integral molecular dynamics study of the Zundel cation down to ultralow temperatures at coupled cluster accuracy. *J. Chem. Theory Comput.* **2018**, *14*, 5068–5078.

63. Bertaina, G.; Di Liberto, G.; Ceotto, M. Reduced rovibrational coupling Cartesian dynamics for semiclassical calculations: application to the spectrum of the Zundel cation. *J. Chem. Phys.* **2019**, *151*, 114307.

64. Vendrell, O.; Gatti, F.; Meyer, H.-D. Full dimensional (15 dimensional) quantum-dynamical simulation of the protonated water-dimer IV: Isotope effects in the infrared spectra of D (D 2 O) 2+, H (D 2 O) 2+, and D (H 2 O) 2+ isotopologues. *J. Chem. Phys.* **2009**, *131*, 034308.

65. Oh, H.-B.; Lin, C.; Hwang, H. Y.; Zhai, H.; Breuker, K.; Zabrouskov, V.; Carpenter, B. K.; McLafferty, F. W. Infrared photodissociation spectroscopy of electrosprayed ions in a Fourier transform mass spectrometer. *J. Am. Chem. Soc.* **2005**, *127*, 4076–4083.

66. Wu, R.; McMahon, T. B. Infrared multiple photon dissociation spectra of proline and glycine proton-bound homodimers. Evidence for zwitterionic structure. *J. Am. Chem. Soc.* **2007**, *129*, 4864–4865.

67. Gabas, F.; Di Liberto, G.; Conte, R.; Ceotto, M. Protonated glycine supramolecular systems: the need for quantum dynamics. *Chem. Sci.* **2018**, *9*, 7894–7901.

68. Valiev, M.; Bylaska, E.; Govind, N.; Kowalski, K.; Straatsma, T.; Dam, H. V.; Wang, D.; Nieplocha, J.; Apra, E.; Windus, T. *et al.* NWChem: A comprehensive and scalable open-source solution for large scale molecular simulations. *Comput. Phys. Commun.* **2010**, *181*, 1477–1489.

69. Braams, B. J.; Bowman, J. M. Permutationally invariant potential energy surfaces in high dimensionality. *Int. Rev. Phys. Chem.* **2009**, *28*, 577–606.

70. Conte, R.; Houston, P. L.; Bowman, J. M. Communication: A benchmark-quality, full-dimensional ab initio potential energy surface for Ar-HOCO. *J. Chem. Phys.* **2014**, *140*, 151101.

71. Conte, R.; Qu, C.; Bowman, J. M. Permutationally invariant fitting of many-body, non-covalent interactions with application to three-body methane-water-water. *J. Chem. Theory Comput.* **2015**, *11*, 1631–1638.

72. Conte, R.; Houston, P. L.; Qu, C.; Li, J.; Bowman, J. M. Full-dimensional, ab initio potential energy surface for glycine with characterization of stationary points and zero-point energy calculations by means of diffusion Monte Carlo and semiclassical dynamics. *J. Chem. Phys.* **2020**, *153*, 244301.

73. Houston, P. L.; Conte, R.; Qu, C.; Bowman, J. M. Permutationally invariant polynomial potential energy surfaces for tropolone and H and D atom tunneling dynamics. *J. Chem. Phys.* **2020**, *153*, 024107.

74. Qu, C.; Conte, R.; Houston, P. L.; Bowman, J. M. Full-dimensional potential energy surface for acetylacetone and tunneling splittings. *Phys. Chem. Chem. Phys.* **2020**..

75. Bartok, A. P.; De, S.; Poelking, C.; Bernstein, N.; Kermode, J. R.; Csanyi, G.; Ceriotti, M. Machine learning unifies the modeling of materials and molecules. *Sci. Adv.* **2017**, *3*, e1701816.

76. Dral, P. O.; Owens, A.; Yurchenko, S. N.; Thiel, W. Structure-based sampling and self-correcting machine learning for accurate calculations of potential energy surfaces and vibrational levels. *J. Chem. Phys.* **2017**, *146*, 244108.

77. Zaspel, P.; Huang, B.; Harbrecht, H.; von Lilienfeld, O. A. Boosting quantum machine learning models with a multilevel combination technique: pople diagrams revisited. *J. Chem. Theory Comput.* **2019**, *15*, 1546–1559.

78. Sauceda, H. E.; Chmiela, S.; Poltavsky, I.; Müller, K.-R.; Tkatchenko, A. Molecular force fields with gradient-domain machine learning: Construction and application to dynamics of small molecules with coupled cluster forces. *J. Chem. Phys.* **2019**, *150*, 114102.

79. Bogojeski, M.; Vogt-Maranto, L.; Tuckerman, M. E.; Müller, K.-R.; Burke, K. Quantum chemical accuracy from density functional approximations via machine learning. *Nat. Commun.* **2020**, *11*, 5223.

80. Unke, O. T.; Koner, D.; Patra, S.; Käser, S.; Meuwly, M. High-dimensional potential energy surfaces for molecular simulations: from empiricism to machine learning. *Mach. Learn. Sci. Technol.* **2020**, *1*, 013001.

81. Dral, P. O.; Owens, A.; Dral, A.; Csnyi, G. Hierarchical machine learning of potential energy surfaces. *J. Chem. Phys.* **2020**, *152*, 204110.

82. Qu, C.; Houston, P. L.; Conte, R.; Nandi, A.; Bowman, J. M. Breaking the coupled cluster barrier for machine-learned potentials of large molecules: the case of 15-atom acetylacetone. *J. Phys. Chem. Lett.* **2021**, *12*, 4902–4909.

83. Conte, R.; Botti, G.; Ceotto, M. Sensitivity of semiclassical vibrational spectroscopy to potential energy surface accuracy: a test on formaldehyde. *Vib. Spectrosc.* **2020**, *106*, 103015.

84. Conte, R.; Qu, C.; Houston, P. L.; Bowman, J. M. Efficient generation of permutationally invariant potential energy surfaces for large molecules. *J. Chem. Theory Comput.* **2020**, *16*, 3264–3272.

85. Gabas, F.; Conte, R.; Ceotto, M. Semiclassical vibrational spectroscopy of biological molecules using force fields. *J. Chem. Theory Comput.* **2020**, *16*, 3476–3485.

86. Abraham, M. J.; Murtola, T.; Schulz, R.; Páll, S.; Smith, J. C.; Hess, B.; Lindahl, E. GROMACS: high performance molecular simulations through multi-level parallelism from laptops to supercomputers. *SoftwareX* **2015**, *1*, 19–25.

87. Rackers, J. A.; Wang, Z.; Lu, C.; Laury, M. L.; Lagardere, L.; Schnieders, M. J.; Piquemal, J.-P.; Ren, P.; Ponder, J. W. Tinker 8: software tools for molecular design. *J. Chem. Theory Comput.* **2018**, *14*, 5273–5289.

88. Rognoni, A.; Conte, R.; Ceotto, M. How many water molecules are needed to solvate one? *Chem. Sci.* **2021**, *12*, 2060.

89. Babin, V.; Medders, G. R.; Paesani, F. Development of a first principles water potential with flexible monomers. II: trimer potential energy surface, third virial coefficient, and small clusters. *J. Chem. Theory Comput.* **2014**, *10*, 1599–1607.

90. Wang, Y.; Huang, X.; Shepler, B. C.; Braams, B. J.; Bowman, J. M. Flexible, ab initio potential, and dipole moment surfaces for water. I. Tests and applications for clusters up to the 22-mer. *J. Chem. Phys.* **2011**, *134*, 094509.

91. Buchholz, M.; Grossmann, F.; Ceotto, M. Mixed semiclassical initial value representation time-averaging propagator for spectroscopic calculations. *J. Chem. Phys.* **2016**, *144*, 094102.

92. Buchholz, M.; Grossmann, F.; Ceotto, M. Application of the mixed time-averaging semiclassical initial value representation method to complex molecular spectra. *J. Chem. Phys.* **2017**, *147*, 164110.

93. Buchholz, M.; Grossmann, F.; Ceotto, M. Simplified approach to the mixed time-averaging semiclassical initial value representation for the

calculation of dense vibrational spectra. *J. Chem. Phys.* **2018**, *148*, 114107.

94. Rognoni, A.; Conte, R.; Ceotto, M. Caldeira - Leggett model vs ab initio potential: a vibrational spectroscopy test of water solvation. *J. Chem. Phys.* **2021**, *154*, 094106.

95. Cazzaniga, M.; Micciarelli, M.; Moriggi, F.; Mahmoud, A.; Gabas, F.; Ceotto, M. Anharmonic calculations of vibrational spectra for molecular adsorbates: a divide-and-conquer semiclassical molecular dynamics approach. *J. Chem. Phys.* **2020**, *152*, 104104.

Chapter 11

Direct Dynamics for Vibrational Spectroscopy: From Large Molecules in the Gas Phase to the Condensed Phase

Sana Bougueroua[*], Vladimir Chantitch[*], Wanlin Chen[*], Simone Pezzotti[*,†], and Marie-Pierre Gaigeot[*,‡]

[*]*Université Paris-Saclay, Univ Evry, CNRS, LAMBE UMR8587, 91025, Evry-Courcouronnes, France*
[†]*Department of Physical Chemistry II, Ruhr University Bochum, D-44801, Bochum, Germany*
[‡]*mgaigeot@univ-evry.fr*

11.1. Introduction

In this chapter, we turn our attention to vibrational spectroscopy calculations performed by molecular dynamics (MD) simulations, that we can also call "finite temperature dynamical vibrational spectroscopy". For reasons that are discussed in the chapter, dynamical spectroscopy is mainly achieved with an electronic representation of the interactions, mostly with the density functional theory (DFT) *ab initio* representation. This approach has indeed proven to be a rather good compromise between accuracy of the DFT-based spectroscopy and computational cost of the MD simulation. Examples taken from our own research illustrate the rather excellent agreement of DFT-MD-based spectroscopy.

We start this chapter by presenting the basic background of molecular dynamics. We do not review the DFT or force field representations for the calculation of the interactions and forces.

416

We encourage the readers who are beginners in MD and/or DFT and force field formalisms to go to some of the reference textbooks and reviews of the domain for more details.[1–6]

We then review the theoretical formalism for the calculation of vibrational spectra from MD simulations based on the Fourier transformation of a time-correlation function. As will be seen, the formalism of dynamical spectroscopy is versatile and hence can be applied to the gas phase molecules and clusters of primary interest in this book, as well as to liquids and interfaces (solid–liquid and liquid–air interfaces) in the condensed phase. This is one of the strength of dynamical spectroscopy. We hence review the MD-based spectroscopy calculations for Infrared (IR) spectroscopy (probing gas phase and condensed phase molecular systems), Raman (gas phase and condensed phase) and Sum Frequency Generation (SFG) (probing non-centrosymmetric media in the condensed phase). Other spectroscopies are not presented. We will then present a hybrid formalism that we have recently developed that takes advantage of both the dynamical formalism presented in this chapter and of the more usual static formalism of spectroscopy calculations presented in other chapters of this book.

We then present a discussion on some issues inherent to MD-based spectroscopy that beginners should have in mind before starting similar investigations and that all readers should keep in mind when assessing MD-based spectroscopy data in the literature. We especially emphasize the time-length of the MD trajectories for the convergence of spectroscopic signals, the equipartition of energy and its consequence on the intensity of the bands, the choice of temperature in the MD and its influence for the positions of the bands, and the quantum effects of nuclei in the position of the bands. The relevance of the level of representation of the interactions and forces (electronic versus force field) in the MD simulations is then discussed.

An important aspect of dynamical spectroscopy is the assignment of the vibrational modes. This is reviewed in this chapter, in particular with a presentation of a recent method that we have developed based on algorithmic graph theory. This method has the advantage to extract anharmonic modes and their couplings, which is an achievement in dynamical spectroscopy.

We will then illustrate some applications of DFT-MD-based dynamical spectroscopy taken from our own investigations. Section 11.8 reviews the IR spectroscopy of floppy peptides for which the MD-based approach to spectroscopy was essential in order to achieve a comprehension of the peptidic conformations that were responsible for the recorded IR signatures at finite temperature. Section 11.9 shows applications of DFT-MD-based dynamical spectroscopy in the THz domain, for revealing the 3D-structures of peptides, as well as for providing a global comprehension of the motions activated in this domain. Such mapping of vibrational modes can be used further in order to automatically dissect experimental spectra. In Section 11.10, the dynamical IR spectroscopy in the THz domain is used to quantify vibrational anharmonicities, especially the couplings between modes for a hydrogen bonded β-sheet peptidic model structure. With Section 11.11, we highlight the versatility of dynamical spectroscopy by showing the spectroscopy–structure relationships for complex inhomogeneous interfaces between a silica oxide and liquid water. We not only show how the DFT-MD dynamical spectrum of such interface matches the experiment but also how one can obtain the detailed molecular comprehension of the spectral fingerprints of such complex inhomogeneous molecular systems.

We finish this chapter by presenting our opinions on where the dynamical spectroscopy field is evolving, by discussing some exciting and promising perspectives.

11.2. Basic Background on Molecular Dynamics

We provide here the basic background on MD. For a more detailed description, any beginner with MD should refer to some textbooks.[1–6]

MD predicts the motion of all the atoms of a given molecular system, over an interval of time that can be of a few tens of pico-seconds when electronic representations are used for the calculation of the interactions and forces, up to hundreds of nano-seconds and even beyond when a classical or coarse-grain representation of the interactions and forces are used. See later in this chapter more description on the representations for the calculation of the interactions and forces that are relevant in the context of the modeling of vibrational spectroscopy based on MD/direct dynamics. When an electronic representation is used, the most common implementation of MD is based on the Born–Oppenheimer (BO) approximation and hence on the separation in time-scale of the motions of nuclei and electrons. In other words, electrons adapt instantaneously to the displacement of the nuclei. In the BO approximation, one therefore solves the motions in time of the nuclei, while the time-independent Schrödinger equation is solved in order to find the electronic wavefunction that corresponds to the positions of the nuclei at a given time. Moreover, nuclei are heavy in comparison to the electrons, so that the motion of any nucleus can be described classically without loss of accuracy for structural and dynamical properties of the molecular systems. Of course, the classical representation for light atoms like the hydrogens is far less accurate, their quantum nature might be of importance to certain properties. In particular, vibrations can be affected by the quantum nature of hydrogen atoms. This will be discussed later in this chapter.

All MD simulations presented in this chapter rely on the BO approximation with the motions of the nuclei being treated with classical mechanics. The motion of each nucleus i of a given molecular system composed of N nuclei is thus obtained by solving the Newton's equation of motion:

$$m_i \frac{\partial^2 \vec{r}_i}{\partial t^2} = \vec{F}_i \quad \forall i, \tag{11.1}$$

where m_i is the mass of nucleus i, $\vec{r}_i$ is the position vector (expressed in Cartesian coordinates) of the nucleus, $\vec{F}_i$ is the total force acting on nucleus i and arising from the interactions with all other nuclei ($\forall j \neq i$) in the molecular system, $\vec{F}_i$ being the negative of the gradient of the potential energy $V(\vec{r}_1, \vec{r}_2, \ldots, \vec{r}_N)$ of the system at the given positions in space of all the nuclei:

$$\vec{F}_i = -\frac{\partial V}{\partial \vec{r}_i} \quad \forall i.$$

As stated above, V is either calculated quantum mechanically through the resolution of the time-independent Schrödinger equation or classically through the use of a force field (it can also be with a coarse-grain representation). See later in this chapter more discussion for V in the context of the modeling of vibrational spectroscopy based on MD/direct dynamics.

For a given molecular system composed of N nuclei, the N coupled equations of motions are solved numerically. This requires the discretization of the time in small time-intervals δt, called the time-step. Numerous algorithms have been developed to that end, generally based on a Taylor expansion of the position and velocity of each nucleus i. All most commonly used algorithms derive from the seminal work of L. Verlet. The initial Verlet algorithm is obtained by writing two Taylor expansions for the position of nucleus i, one Taylor expansion being for a time-forward $(t + \delta t)$, the other being for a time-backward $(t - \delta t)$, with the following expressions:

$$\vec{r}_i(t + \delta t) = \vec{r}_i(t) + \vec{v}_i(t)\delta t + \frac{\vec{F}_i(t)}{2m_i}\delta t^2 + O(\delta t^3) + \cdots, \quad (11.2)$$

$$\vec{r}_i(t - \delta t) = \vec{r}_i(t) - \vec{v}_i(t)\delta t + \frac{\vec{F}_i(t)}{2m_i}\delta t^2 - O(\delta t^3) + \cdots. \quad (11.3)$$

By summing these two equations, one obtains:

$$\vec{r}_i(t + \delta t) = 2\vec{r}_i(t) - \vec{r}_i(t - \delta t) + \frac{\vec{F}_i(t)}{m_i}\delta t^2 + O(\delta t^4). \quad (11.4)$$

Knowing the position of nucleus i at times t and $t - \delta t$ as well as the total force acting on the nucleus at time t, one is therefore capable to calculate the position in time of nucleus i at the subsequent time $t + \delta t$. The estimate of this new position hence contains an error of the order of δt^4. Note that this algorithm does not use any knowledge on the velocity of the nucleus in order to compute the position.

The velocity $\vec{v}_i(t)$ of the nucleus at time t is then obtained by

$$\vec{v}_i(t) = \frac{\vec{r}_i(t + \delta t) - \vec{r}_i(t - \delta t)}{2\delta t}. \quad (11.5)$$

Note that with Eqs. (11.4)–(11.5), the position of nucleus i is known at time $t + \delta t$ while the velocity of the nucleus is known at time t (*i.e.*, the previous moment in time).

The Verlet algorithm can be cast into a form that computes both the position $\vec{r}_i(t)$ and the velocity $\vec{v}_i(t)$ of nucleus i at the same $t + \delta t$ time. This is the so-called velocity Verlet algorithm, which equations are

$$\vec{r}_i(t + \delta t) = \vec{r}_i(t) + \vec{v}_i(t)\delta t + \frac{\vec{F}_i(t)}{2m_i}\delta t^2, \tag{11.6}$$

$$\vec{v}_i(t + \delta t) = \vec{v}_i(t) + \frac{\delta t^2}{2}\left(\frac{\vec{F}_i(t)}{2m_i} + \frac{\vec{F}_i(t + \delta t)}{2m_i}\right). \tag{11.7}$$

There is another algorithm, the so-called Leap-Frog Verlet algorithm, that evaluates the velocity at $t + \delta t/2$ and uses this velocity to compute the position at time $t + \delta t$. See for instance Refs. 1 and 4 for more details on all these algorithms.

All common algorithms derived from the Verlet algorithms are symplectic, thus capable to conserve the total energy E of the system. Note that this statement is depending on the choice of δt being sufficiently small to ensure that the total energy can be a constant in time. An MD simulation performed with these symplectic algorithms will thus always provide the time evolution of the simulated system in the microcanonical NVE ensemble where the number of particles N, the volume V and the total energy E of the system are constant/conserved. Other ensembles can be simulated, such as the canonical NVT ensemble where the number of particles N, the volume V and the temperature T of the system are constant, or such as the isobaric-isothermal NPT ensemble where the number of particles N, the pressure P and the temperature T of the system are constant. For example, it can be desirable to reproduce the structural and dynamical properties of the investigated system at constant room temperature, and one way to ensure that this constraint is satisfied is to perform MD directly in the canonical NVT ensemble. Specific algorithms commonly known as thermostat/barostat have been developed to fix temperature/pressure in the MD. Please refer to the reference textbooks of the MD literature to gain more knowledge on the modified algorithms for the resolution of the equations of Newton in order to simulate these thermodynamic conditions.[1–6]

Following the equipartition theorem, the average internal kinetic energy of a system is related to the macroscopic temperature T of the system through the following equation:

$$E_{kin}(t) = \sum_{i=1}^{N} \frac{1}{2} m_i \vec{v}_i^{2}(t) = (3N - 6)\frac{k_B T(t)}{2}, \qquad (11.8)$$

where k_B is the Boltzmann constant, $(3N-6)$ the number of internal degrees of freedom of the system, and $\vec{v}_i$ is the velocity of nucleus i which has the mass m_i. The equipartition theorem states that each degree of freedom in a molecular system has an *average* kinetic energy equal to $k_B T(t)/2$, or alternatively that each atom of a molecular system has an *average* kinetic energy of $3k_B T(t)/2$. As the velocities of the nuclei are known at each time-step of the dynamics, this equation is used to calculate the instantaneous temperature $T(t)$ of the system at each time t. In an NVT molecular dynamics, there is a heat bath which reference temperature is set to T_0 that serves as a thermostat. Because the instantaneous temperature is directly related to the atomic internal velocities, maintaining the temperature constant for a dynamics in the NVT ensemble requires imposing some control on the rate of change of these velocities, and thus thermostat algorithms require a modification of the Newton's equations of motions.

One comment is here needed on the perturbations on the velocities induced by algorithms for imposing a thermostat. If the average distribution of atomic velocities obtained from dynamics in the NVT ensemble is correct with respect to the target temperature, the time evolution of the velocities of all atoms is not correct, as these velocities are affected by the coupling with the thermostat. Therefore, when dynamics are used to calculate properties which are directly dependent on the evolution with time of the atomic velocities, as is the case for the calculation of vibrational spectroscopic signals, the microcanonical NVE ensemble has to be chosen as the ensemble for the dynamics.

At the initial time $t = 0$ of a MD simulation, both the positions and the velocities of all nuclei have to be assigned. The atomic positions are dictated by the structure of the system investigated, they are therefore chosen accordingly at $t = 0$. The atomic velocities are generated from a Maxwell–Boltzmann distribution centered at the desired temperature of the dynamics. This distribution is generated from random numbers, according to the following equation:

$$p(\vec{v}_i) = \sqrt{\frac{m_i}{2\pi k_B T}} \exp\left(-\frac{m_i v_i^2}{2k_B T}\right),$$

where $p(\vec{v}_i)$ is the probability of the ith nucleus to have $\vec{v}_i$ velocity, m_i the mass of this particle and T the targeted temperature of the system. This equation is actually written for each direction of the Cartesian space.

An important control parameter for a MD is the time-step δt chosen for the Verlet algorithms. The maximum time-step that can be taken is determined by the rate of the fastest process/motion in the molecular system. δt is thus typically chosen an order of magnitude smaller than this fastest process. Molecular motions, such as rotations and vibrations, occur with periods of time in the range 10^{-11}–10^{-14} s. A time-step of the order of a femto-second (10^{-15} s) or less is thus required to model such motions with sufficient accuracy. In particular, fast vibrational motions as the OH-stretching requires a time-step of the order of 0.1–0.5 fs to be correctly sampled. Any dynamics obtained within the BO approximation requires δt of this order for a proper conservation of the total energy in the NVE ensemble. All illustrations taken from our works presented later in this chapter of direct dynamics for vibrational spectroscopy in the BO approximation, and all using the DFT electronic representation for the calculation of interactions and forces, were done with a time-step of 0.4 fs.

11.3. Review of the Formalism for MD-based Dynamical Anharmonic Vibrational Spectroscopy

There are a few reviews on (DFT-)MD simulations for vibrational spectroscopy.[7-11] We also refer to a brief selection of papers applying finite temperature MD to gas phase IR, Raman and vibrational circular dichroism (VCD) spectroscopies,[7-10] to IR/Raman spectroscopies of liquids,[7,12,13] to SFG spectroscopy of (aqueous) interfaces,[14-18] that show the versatility of MD-based theoretical spectroscopy. We review in this chapter the formalism for dynamical spectroscopy, discuss a few key points, and address some possible issues that one should have in mind when applying the dynamical formalism for vibrational spectroscopy.

Within the well-known time-correlation function formalism from linear response theory,[19,20] an IR absorption spectrum is calculated by

$$I(\omega) = \frac{2\pi\beta\omega^2}{3cV} \int_{-\infty}^{+\infty} dt\, e^{i\omega t} \langle \delta\boldsymbol{\mu}(t) \cdot \delta\boldsymbol{\mu}(0) \rangle$$

$$= \frac{2\pi\beta}{3cV} \int_{-\infty}^{+\infty} dt\, e^{i\omega t} \left\langle \delta\left(\frac{d\boldsymbol{\mu}(t)}{dt}\right) \cdot \delta\left(\frac{d\boldsymbol{\mu}(0)}{dt}\right) \right\rangle, \quad (11.9)$$

where $\beta = 1/kT$, ω is the frequency of the absorbed light, c is the speed of light in vacuum, V the volume of the system, $\boldsymbol{\mu}(t)$ is the instantaneous dipole moment vector of the system at the time t, $\frac{d\boldsymbol{\mu}(t)}{dt}$ its time derivative, $\delta\boldsymbol{\mu}(t) = \boldsymbol{\mu}(t) - \langle\boldsymbol{\mu}\rangle$ is the fluctuation with respect to the mean value and $\langle...\rangle$ is the equilibrium time-correlation function. This formula is for a classical representation of the nuclei, therefore the quantum correction factor $\beta\hbar/(1 - \exp(-\beta\hbar\omega))$ that is included in order to correct the classical line shape[12] in Eq. (11.9). See, *e.g.*, Refs. 12 and 21 for other empirical corrections and for discussions on the exact introduction of quantum nuclei effects. Eq. (11.9) hence provides the whole IR spectrum of the molecular system, *i.e.*, the active bands, the band-shapes and the band-intensities. There are no further model(s) to apply in order to obtain the band-shapes and intensities.

The advantage of the linear response theory for theoretical spectroscopy is that there are no assumptions on the calculation of the dipole moment that enters into Eq. (11.9). Coupled with MD simulations at finite temperature, the fluctuations of the dipole moment over time reflect the subtle changes in the electronic distribution over the molecule(s) as the 3D-conformations evolve with time and with the changes in the interactions. Therefore, both the exploration of the potential energy surface (PES) for the sampling of the conformations over time and for the fluctuations of the dipole moment take into account anharmonicities in the final vibrational calculated spectrum. The shape and broadening of the IR bands is the direct result of the underlying conformational dynamics at a given temperature, and of the anharmonicities in the mode-couplings. Whenever isomerization and/or proton transfers occur over time, these events, together with all intermediate/transient conformations explored over time, are naturally taken into account into the final calculation of the IR spectrum in Eq. (11.9). This participates to the final shapes of the IR bands.

Equivalent formula based on the linear response theory can be written for the calculation of a Raman signal, VCD, SFG (see, *e.g.*, Refs. 7–9, 15, 22 and 23), where polarizability tensors (Raman, SFG) and magnetic moments (VCD) enter into the correlation function. We report below the dynamical Raman and SFG signals, as they are respectively useful for characterizing gas phase molecules, clusters and liquids (Raman), interfaces between *e.g.*, solids and liquids (SFG). Applications of DFT-MD simulations for such spectroscopies are illustrated later in this chapter.

An MD-based Raman spectrum is calculated through the following equations within the time-correlation formalism[19]:

$$I(\omega) = I_{\text{iso}}(\omega) + I_{\text{ani}}(\omega), \tag{11.10}$$

$$I_{\text{iso}} = \int_{-\infty}^{+\infty} dt \, e^{i\omega t} \langle \delta \bar{\alpha}(t) \delta \bar{\alpha}(0) \rangle, \tag{11.11}$$

$$I_{\text{ani}} = \frac{2}{15} \int_{-\infty}^{+\infty} dt \, e^{i\omega t} \langle \text{Tr} \boldsymbol{\delta\beta}(t) \boldsymbol{\delta\beta}(0) \rangle, \tag{11.12}$$

where I_{iso} and I_{ani} are, respectively, the isotropic and anisotropic parts of the Raman signal, in which $\boldsymbol{\alpha}$ is the polarizability tensor of the molecular system, Tr is the trace operator, and $\bar{\alpha} = 1/3(\alpha_{xx} + \alpha_{yy} + \alpha_{zz})$, $\boldsymbol{\beta} = \boldsymbol{\alpha} - \bar{\alpha}\boldsymbol{I}$. $\boldsymbol{I}$ is the identity matrix. As in Eq. (11.9), δ stands for the fluctuations of the observable, *e.g.*, $\delta\bar{\alpha}(t) = \bar{\alpha}(t) - \langle\bar{\alpha}\rangle$ is the fluctuation of the polarizability tensor with respect to its mean value, and $\langle...\rangle$ is the equilibrium time-correlation function. Introducing the time-derivatives of $\bar{\alpha}$ and $\boldsymbol{\beta}$ in Eqs. (11.11)–(11.12), as done for the IR dynamical signal in Eq. (11.9), one obtains:

$$I_{\text{iso}} = \frac{1}{\omega^2} \int_{-\infty}^{+\infty} dt \, e^{i\omega t} \left\langle \delta \left(\frac{d\bar{\alpha}}{dt}(t) \right) \delta \left(\frac{d\bar{\alpha}}{dt}(0) \right) \right\rangle, \tag{11.13}$$

$$I_{\text{ani}} = \frac{2}{15} \frac{1}{\omega^2} \int_{-\infty}^{+\infty} dt \, e^{i\omega t} \left\langle \text{Tr}\delta \left(\frac{d\boldsymbol{\beta}}{dt}(t) \right) \delta \left(\frac{d\boldsymbol{\beta}}{dt}(0) \right) \right\rangle. \tag{11.14}$$

At interfaces between two media, for instance between a solid and a liquid, there is breaking of the centrosymmetry of the liquid and therefore a spectroscopic signal can be recorded with the SFG vibrational scheme. See, *e.g.*, Refs. 24–27 for the basics on SFG spectroscopy. The SFG signal is due to the total resonant electric dipole non-linear susceptibility $\chi^{(2)}(\omega)$ (with real and imaginary components). Following the time-dependent approach developed by Morita *et al.*,[28,29] $\chi^{(2)}(\omega)$ is calculated by the following equation:

$$\chi^{(2)}_{pqr}(\omega) = \frac{i\omega}{k_B T} \int_0^\infty dt \exp(i\omega t)\langle\delta\left(A_{pq}(t)\right)\delta\left(M_r(0)\right)\rangle, \tag{11.15}$$

$$= \frac{i}{k_B T \omega} \int_0^\infty dt \exp(i\omega t) \left\langle \delta\left(\frac{dA_{pq}}{dt}(t) \right) \delta\left(\frac{dM_r}{dt}(0) \right) \right\rangle, \tag{11.16}$$

where A_{pq} and M_r are, respectively, the components of the polarizability tensor and dipole moment of the whole system, (p,q,r) any direction amongst (x,y,z), k_B and T are the Boltzmann constant and temperature, $\langle...\rangle$ denotes an equilibrium average over the MD trajectory, and δ is the fluctuation of the observable with respect to its mean value (see Eqs. (11.9)–(11.12)). Corrections for the quantum nature of the nuclei are included in the pre-factor.

In the condensed phase (liquids and interfaces), the dipole moment and polarizability tensor that enter into Eqs. (11.9), (11.11, 11.12) and (11.15), respectively for IR, Raman and SFG signals, are the ones of the *whole* molecular system. However, this might not be useful for a more detailed comprehension of the signal. For instance, in liquids one might be interested in extracting the IR/Raman signal of the molecules that solvate an ion, and hence distinguish this signal from the whole response of the liquid. At solid–liquid interfaces, one needs to separate the SFG contribution arising from the top-surface of the solid from the one arising from the water (if water is the liquid) in contact with the top-surface, one also needs to separate the contributions to the SFG signal arising from the various layers of the water. We have in particular given the molecular definitions[15, 22] of the binding interfacial layer (BIL) and of the diffuse layer (DL) that are the only two layers (each one of them being several Å to nm thickness) contributing to the SFG signal. Such decomposition of IR/Raman/SFG signals into specific molecular and layering contributions is the strength of MD-based spectroscopies, out of range of experiments (even with, *e.g.*, deuteration schemes).

One has therefore to decompose the total polarizability tensor $A(t)$ and total dipole moment $M(t)$ of the whole system into individual molecular components at each time-step of the dynamics:

$$M(t) = \sum_i^N \mu_i(t) \quad A(t) = \sum_i^N \overline{\alpha}_i(t), \tag{11.17}$$

$\mu_i(t)$ and $\overline{\alpha}_i(t)$ being, respectively, the dipole moment vector and polarizability tensor of molecule i within the whole molecular system composed of N molecules, at time t.

Taking the example of the IR spectroscopy signal of a liquid composed of N molecules, Eq. (11.9) now becomes (we only take the first equation without the time derivatives of the dipole moment for illustration):

$$I(\omega) = \frac{2\pi\beta\omega^2}{3cV} \int_{-\infty}^{+\infty} dt\, e^{i\omega t} \left\langle \sum_{i=1}^N \delta\boldsymbol{\mu}_i(t) \cdot \sum_{j=1}^N \delta\boldsymbol{\mu}_i(0) \right\rangle$$

$$= \frac{2\pi\beta\omega^2}{3cV} \sum_{i=1}^{N} \int_{-\infty}^{+\infty} dt \, e^{i\omega t} \langle \delta\boldsymbol{\mu}_i(t) \cdot \delta\boldsymbol{\mu}_i(0) \rangle$$

$$+ \frac{2\pi\beta\omega^2}{3cV} \sum_{i=1}^{N} \int_{-\infty}^{+\infty} dt \, e^{i\omega t} \langle \delta\boldsymbol{\mu}_i(t) \cdot \sum_{j=1, j\neq i}^{N} \delta\boldsymbol{\mu}_j(0) \rangle.$$

$$(11.18)$$

In this latter equation, the first term is the sum of "self-correlation" terms, *i.e.*, correlations between the dipole of the same molecule i at the two times t and $t = 0$, while the second summation is made of "cross-correlation" terms, *i.e.*, correlations between the dipole moment of two different molecules i and j at the two times t and $t = 0$. As discussed in Ref. 12, cross-correlation terms require long enough MD trajectories in order for them to converge, tens of pico-seconds trajectories were shown in this work insufficient. Cross-terms are indeed small terms, they fluctuate a lot in time, therefore the difficulty for their convergence. The more molecules within the system, the harder for convergence, as more terms are added up in the cross-term summation. Thus, most of the dynamical spectra calculations obtained in the literature from DFT-MD trajectories over 10–100 ps do not (and should not!) include such terms. Note that Nagata and Mukamel[30] introduced a truncation scheme in which the cross-correlation terms are calculated over a radius around molecule i, this radius likely being somehow system dependent. With this, the cross-correlation term for the calculation of a dynamical signal seems to converge over shorter time-scales. See more results on that topic in a work by Nagata and Kuhne.[31] It is important to have in mind that while the cross-correlation terms are very seldom included in the dynamical spectra calculations, the correlations between the dipole/polarizability of the molecules are however included in these calculations. In *ab initio* MD simulations and in classical simulations including polarization effects, each molecular dipole moment and polarizability tensor is indeed calculated as the response to all inter-molecular interactions, thus including the correlations between all molecules within the system. Therefore, even if the cross-terms are neglected in signals calculations, these terms are implicitly taken into account in the actual values of the molecular dipoles and polarizabilities in a mean-field way. We will come back into the discussion over cross-terms in correlation functions in the next section that will report a new hybrid formalism for MD dynamical spectroscopy calculation.

Taking the other example of the SFG signal and dismissing the cross-correlation terms from the final dynamical spectroscopic signal because of the issue of their convergence over the rather limited time-length of DFT-MD simulations (the convergence for the SFG signal is even harder because it involves cross-correlation between two different observables calculated for two different molecules), the SFG signal of the whole system from Eq. (11.15) becomes[28, 29, 32]:

$$\chi_{pqr}^{(2),self}(\omega) = \frac{i\omega}{k_B T} \sum_{i=1}^{N} \int_0^\infty dt \exp(i\omega t) \langle \bar{\alpha}_{pq}^i(t) \mu_r^i(0) \rangle, \qquad (11.19)$$

$$= \frac{i}{k_B T \omega} \sum_{i=1}^{N} \int_0^\infty dt \exp(i\omega t) \left\langle \frac{d\bar{\alpha}_{pq}^i}{dt}(t) \frac{d\mu_r^i}{dt}(0) \right\rangle. \quad (11.20)$$

References 7 and 12 describe in details the Wannier localization and Voronoi tessellation schemes that can be applied *a posteriori* to the dynamics, in order to extract the molecular dipole moments $\mu_i(t)$ and molecular polarizability tensors $\bar{\alpha}_i(t)$ at each time-step of DFT-MD simulations. These schemes are however computationally costly, as discussed in these papers. In Ref. 32, these observables were extracted from the DFT-MD trajectories, by applying the scheme of the Wannier localization of the wavefunction[12, 33] in order to get the individual molecular dipole moments, and by applying the scheme of an external field to the whole system to get the individual molecular response in terms of molecular polarizability tensors (calculations involving finite electric fields of 0.0001 a.u. intensity were performed independently along the x, y and z directions at each time-step). This procedure is extremely costly, roughly multiplying by four the computational cost of a DFT-MD simulation. Proof-of-concept and accuracy of the procedure for the theoretical SFG signal was given in Ref. 32 for the prototype air–liquid water interface. A slightly less expensive method has been developed by Kuhne *et al.*[34] where the Wannier localization and Wannier volumes around the atoms are used to build the molecular polarizabilities. Though it still relies on the Wannier localization of the wavefunction, it is avoiding the computationnally expensive stage that we applied in Ref. 32 where the molecular polarizabilities were obtained by the application of an external field. In classical MD simulations, the individual molecular dipoles and polarizabilities are obtained through classical models that require parameterization. See the works by Morita *et al.* for accurate models.[35–37] The transferability of these models from the air–liquid–water interface to electrolytic air-liquid water and to silica–liquid–water interfaces is however not straightforward. Changing molecular systems requires re-parameterization of the models (which is both computationally and time costly).

11.4. A Hybrid Formalism for Dynamical Spectroscopy

11.4.1. *Demonstration for the IR spectroscopic signal*

As reviewed above, any MD-based vibrational spectrum is calculated from the Fourier transform of a time-correlation function. Taking the example of IR spectroscopy (Eqs. (11.9) and (11.18), respectively for a gas phase molecule or a molecular liquid) the dynamical IR signal is the Fourier transform of the time-correlation of the molecular dipole moment(s). While the strength of an electronic-based MD method is to calculate these dipole moments without pre-established/parameterized models, calculating dipole moments in "on-the-fly" Born–Oppenheimer MD (BOMD) simulations remains however computationally expensive. This is true for the two most popular theoretical strategies in the literature for extracting molecular dipoles, *i.e.*, the Wannier localization of the delocalized wavefunction,[12,38–40] or the Voronoi tesselation.[41,42] We have developed[43,44] an alternative method, which formalism for IR dynamical spectroscopy is now presented. Its generalization to Raman and SFG spectroscopic signals is straightforward. Only the extension to the SFG signal will be presented in this text.

The basis for the new formalism is that the derivative with time of the dipole moment vector in Eq. (11.9) can be expressed as the summation over the $3N - 6$ internal coordinates (ICs) of the system of the derivatives of the dipole moment vector with respect to ICs times the derivative of IC with respect to time, as expressed in Eq. (11.21). N is the total number of atoms in the molecular system. ICs are labeled R_m and are the components of vector $\boldsymbol{R} = [R_1, R_2, R_3, \ldots, R_m, \ldots, R_{3N-6}]^T$. The choice for the ensemble of $3N - 6$ non-redundant ICs (*i.e.*, bond distances, angles, dihedral angles) is natural in vibrational spectroscopy, as the vibrational motions (stretchings, bendings, rocking motions, torsional & wagging motions) are naturally expressed into ICs. However, the same theoretical strategy can be employed by using the $3N$ Cartesian coordinates instead of the ICs.[43]

The derivative with time of a dipole moment can be expressed as in Eq. (11.21) (for the total dipole moment of the whole system for a gas phase molecule and for the molecular dipole moment of one molecule within the whole system as in liquids, alike):

$$\frac{d\boldsymbol{\mu}(t)}{dt} = \sum_{m=1}^{3N-6} \frac{\partial \boldsymbol{\mu}(t)}{\partial R_m} \cdot \frac{\partial R_m(t)}{\partial t} = \sum_{m=1}^{3N-6} \frac{\partial \boldsymbol{\mu}(t)}{\partial R_m} \cdot \dot{R}_m(t), \qquad (11.21)$$

where $\dot{R}_m(t)$ is the velocity of the mth $R_m(t)$ IC at time t. With this in hand, one can rewrite Eq. (11.9) into:

$$I(\omega) = \frac{2\pi\beta}{3cV} \sum_m \sum_l \int_{-\infty}^{+\infty} dt \, e^{i\omega t} \left\langle \frac{\partial \boldsymbol{\mu}(t)}{\partial R_m} \cdot \dot{R}_m(t) \frac{\partial \boldsymbol{\mu}(0)}{\partial R_l} \cdot \dot{R}_l(0) \right\rangle.$$

$$(11.22)$$

The dynamical IR spectrum is thus calculated through the Fourier transform of the time-correlation function of the velocities of the internal coordinates modulated by derivatives of the dipole moment with respect to the internal coordinates. These latter are the components of atomic polar tensors (APTs), here expressed in internal coordinates within a transformation from the APT components that are naturally defined in Cartesian coordinates (*i.e.*, components: $\frac{\partial \boldsymbol{\mu}}{\partial r_m}$ with $r_m = x_m, y_m, z_m$ being Cartesian coordinates).

As vibrational motions have small displacements, the $\frac{\partial \boldsymbol{\mu}(t)}{\partial R_m}$ elements can be considered constant with time, hence leading to Eq. (11.23):

$$I(\omega) = \frac{2\pi\beta}{3cV} \sum_m \sum_l \frac{\partial \boldsymbol{\mu}}{\partial R_m} \frac{\partial \boldsymbol{\mu}}{\partial R_l} \int_{-\infty}^{+\infty} dt \, e^{i\omega t} \langle \dot{R}_m(t) \cdot \dot{R}_l(0) \rangle. \quad (11.23)$$

The dynamical IR spectrum is thus now calculated from the Fourier transform of the time-correlation functions of the velocities of the ICs. At variance with the time-correlation function of dipoles, velocity time-correlation functions are known to reach their zero-limit within far less than $\sim$5 ps thanks to their isotropic character. With such a fast time-scale, several short DFT-MD trajectories can be employed to calculate a converged (average) DFT-MD-IR spectrum of a given molecular system. One important key-point in Eq. (11.23), also discussed in Ref. 43, is that the final IR intensity $I(\omega)$ includes all the self- (when $m = l$) and cross- (when $m \neq l$) correlations between internal coordinates. All motions in the dynamics are correlated, these correlations are naturally included in the calculation of the IR spectrum through Eqs. (11.9) and thus Eq. (11.23). One remark here. We have said earlier in this chapter that the cross-correlation terms that should enter dynamical spectra, see Eq. (11.18) for the IR signal of a condensed phase, are usually not included into the calculation because they are too slow to converge (especially with DFT-MD simulations). This does not hold true for the cross-correlations to be calculated in Eq. (11.23) as these ones involve velocities instead of dipole moments/polarizability tensors. The intrinsic isotropic character of velocities allows velocity time-correlations to converge in far shorter time-scales than dipoles (or polarizabilities) time-correlation functions. This holds for self-correlation and cross-correlation time-functions. This is (as will be rediscussed later in this chapter) one of the strengths of the formalism presented in this section.

One can hence rewrite Eq. (11.23) into two sums, one related to the self-correlation contributions ($\langle \dot{R}_m(t) \dots \dot{R}_m(0) \rangle$, $\forall \; m = 1, \dots, 3N-6$), and one to the cross-correlation contributions ($\langle \dot{R}_m(t) \cdot$

$\dot{R}_l(0)\rangle$, $\forall\, m \neq l$):

$$I(\omega) = \frac{2\pi\beta}{3cV} \sum_m \frac{\partial\boldsymbol{\mu}}{\partial R_m} \frac{\partial\boldsymbol{\mu}}{\partial R_m} \int_{-\infty}^{+\infty} dt\, e^{i\omega t} \langle \dot{R}_m(t) \cdot \dot{R}_m(0)\rangle$$

$$+ \frac{2\pi\beta}{3cV} \sum_m \sum_{l\neq m} \frac{\partial\boldsymbol{\mu}}{\partial R_m} \frac{\partial\boldsymbol{\mu}}{\partial R_l} \int_{-\infty}^{+\infty} dt\, e^{i\omega t} \langle \dot{R}_m(t) \cdot \dot{R}_l(0)\rangle.$$

$$(11.24)$$

This is the formula for MD-APT-IR dynamical spectroscopy calculation. The $\frac{\partial\boldsymbol{\mu}}{\partial R_m}$ pre-factors in Eq. (11.24) are obtained by transforming the Cartesian APTs components into ICs components through the following route: if vector $\boldsymbol{\xi} = [x_1, y_1, z_1, x_2, \ldots, z_N]^T$ collects the $3N$ Cartesian displacements of the N atoms of the system, it is always possible to find a linear transformation between Cartesian and internal coordinates such as $\boldsymbol{R} = \boldsymbol{B} \cdot \boldsymbol{\xi}$ and therefore

$$\frac{\partial\mu_u}{\partial R_i} = \sum_j A_{ji} \frac{\partial\mu_u}{\partial \xi_j}, \qquad (11.25)$$

where $B_{ij} = \frac{\partial R_i}{\partial \xi_j}$ and $A_{ji} = \frac{\partial \xi_j}{\partial R_i}$, and μ_u is the uth Cartesian coordinate of the dipole moment vector $\boldsymbol{\mu}$. One can adopt the Wilson definitions of internal coordinates[45] to describe the vibrational subspace and the transformation $\boldsymbol{A}$ and $\boldsymbol{B}$ matrices, hence defining the relationships between internal ($\boldsymbol{R}$) and Cartesian displacements coordinates ($\boldsymbol{\xi}$). Note that APTs can be calculated from standard quantum chemical packages such as Gaussian,[46] largely used in the gas phase community.

This is why our newly developed method is a "hybrid formalism": it indeed uses the dynamical perspective of spectroscopic signal calculations through the Fourier transform of time-correlation functions together with pre-factors that modulate this Fourier transform, that are obtained from static calculations.

In Ref. 43, we have tested several schemes for adapting the (Cartesian) APTs to the change in conformations of molecules over time (along the MD trajectory) into the DFT-MD-APT-IR modeling, hence taking into account, *e.g.*, isomerization, rotation of certain parts of molecules, proton transfers, large amplitude motions, *etc.*

The weighted linear combination of a few selected APTs adapted for molecular rotation has been found the most robust to accurately reproduce the "exact" DFT-MD-IR spectrum from Eq. (11.9). If we note $\boldsymbol{P}(t)$ the Cartesian APT tensor of the molecule at time t (which components are transformed into internal coordinates through Eq. (11.25)), a reasonable hypothesis is that $\boldsymbol{P}(t)$ can be described by a linear combination of the atomic polar tensors $\boldsymbol{P_j}^{ref}$ of a set of appropriately chosen reference structures $\{j\}$:

$$\boldsymbol{P}(t) = \sum_j w_j(t)\boldsymbol{R_j}(\text{ref} \rightarrow t)\boldsymbol{P_j^{ref}}(\text{ref})\boldsymbol{R_j}^{\boldsymbol{T}}(\text{ref} \rightarrow t), \qquad (11.26)$$

where $w_j(t)$ is the weight of the j-th reference structure in the whole combination at time t. The rotation matrix $\boldsymbol{R}(\text{ref} \rightarrow t)$ has been introduced in the equation to ensure the consistency of the internal reference system of the $\{j\}$ structures with the one of the molecule at time t along the trajectory. In other words, the structure at time t and the reference structure(s) have their three principal moments of inertia known in each molecule's own referential of Cartesian coordinates. To assign a weight in the above equation, one has to recognize "how much" the structure at time t and the reference structure(s) are alike. To do this, the principal moments of inertia of the two structures have to be expressed in the same referential of Cartesian coordinates. This is achieved by rotating one molecular framework (known by its three moments of inertia) with respect to the other one. Mathematically, this is achieved by finding the rotational matrix between the two systems of coordinates (*i.e.*, $R_j(\text{ref} \rightarrow t)$ in the equation above) and by applying this rotational matrix to the APT matrix of the reference structure (*i.e.*, $R_j P_j^{ref} R_j^T$ in the equation). See Ref. 43 for definition(s) that can be chosen for the weight in the equation above.

The reference structures can be chosen following different strategies. In the gas phase, if the potential energy surface of the molecular system is known, it is for instance reasonable to use the equilibrium conformations of the molecule as references, to which some transition points along known transition pathways can be added. A much simpler possibility is to randomly choose some structures explored along the trajectory and to use them as reference. This latter strategy

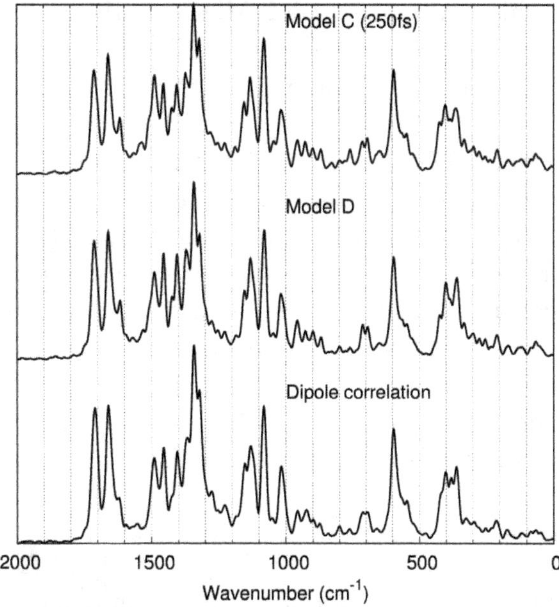

Figure 11.1. The hybrid DFT-MD-APT-IR spectroscopy calculation illustrated for the floppy Ala_2H^+ gas phase peptide (all details for dynamics and spectroscopy of this peptide are in Section 11.8 in this chapter). Model D corresponds to Eq. (11.26) using only two pre-computed APTs, each one calculated on the two lower energy conformers of the peptide. The label "Dipole correlation" corresponds to the reference DFT-MD-IR calculation through Eq. (11.9). *Source*: Reproduced with permission from Ref. 43.

was typically found the most successful in gas phase DFT-MD-APT-IR spectroscopy.[43] Figure 11.1 shows an illustration of DFT-MD-APT-IR for the floppy Ala_2H^+ gas phase peptide (which dynamics and spectroscopy will be detailed in Section 11.8 later in this chapter) taken from Ref. 43. This peptide continuously changes conformation over time, with in particular a rotational motion of the C-Terminal tail of the chain. The "exact" DFT-MD-IR spectrum is reported at the bottom of the figure, while two models for the DFT-MD-APT-IR signal calculation are applied. Model D labeled in this figure is illustrating the weighted linear combination of APTs reported in Eq. (11.26) using only two pre-computed APTs, each one calculated on the two lower energy conformers of the peptide. One hence can see that this strategy for DFT-MD-APT-IR spectroscopy is in excellent agreement with the reference DFT-MD-IR signal calculated from the time-correlation function of the dipole moment of the molecule.

See our Refs. 17, 47 and 48 for an equivalent strategy for interfacial systems. These papers describe in details how to obtain the relevant APTs and Raman tensors for a liquid (at the interface with another medium) and for solid top-surface species, to use in SFG signal calculation through the same hybrid formalism. We also refer the reader to Ref. 43 for more details for the expressions to be employed for the weights w_j in Eq. (11.26). Moreover, velocities in all MD codes are obtained in Cartesian coordinates, velocities of the internal coordinates are thus calculated by numerical derivation with typically a five points central difference algorithm that we have found of excellent accuracy for DFT-MD trajectories accumulated with a 0.4 fs time-step: $\dot{\boldsymbol{R}}(t) = \frac{-\boldsymbol{R}(t+2\delta t)+8\boldsymbol{R}(t+\delta t)-8\boldsymbol{R}(t-\delta t)+\boldsymbol{R}(t-2\delta t)}{12\delta t}$.

In Ref. 43, we have shown that a very small number of reference structures and thus of reference APTs is necessary for a high level accuracy of Eqs. (11.23)–(11.24) for DFT-MD-APT–IR spectroscopy. For instance, the IR spectroscopy of a very dynamical dipeptide that continuously changes conformation over time is accurately captured with only two reference APTs. Another example is the IR spectrum of the $Cl^- \cdots$ methanol cluster in the low frequency domain where large amplitude intermolecular motions are probed: it is accurately obtained when using only six reference APTs. Also, a large peptide like Gramicidin requires only six reference APTs to capture the correct IR spectrum in the THz domain.

11.4.2. *Advantages of this hybrid formalism*

Therefore, one can see that this methodology has many advantages among which being computationally cheap, efficient and accurate. A restricted number of APTs is indeed required, without any specific choice of the reference structures (*i.e.*, minimum, transition state, any structure explored along the trajectory, reasonable clusters for intermolecular interactions). The APT calculations are done *once and for all*, outside of the MD simulations. Short time-scale MD trajectories can then be performed over which the VDOS (Cartesian coordinates)/ICDOS (internal coordinates) Fourier transform of the velocities time-correlation functions is converged. This allows a statistical calculation of IR spectra over several trajectories, therefore a good statistical representation of the spectroscopy, that could not be achieved with a calculated signal based on the dipole time-correlation function (usually through one single "long enough" trajectory). The band-intensities are now entirely in the hands of the APTs that modulate the VDOS/ICDOS intensities, by including the charge fluxes that are mandatory for accurate IR intensities. As a consequence of this later point: if the molecular system is thermalized by MD simulations in the NVT ensemble, one can expect a rather reasonable equipartition of the energy within all the modes to be achieved, even for smallish systems. The subsequent tens of pico-seconds of NVE MD simulations for the DFT-MD-APT-IR spectroscopy would then keep this equipartition, which together with the APT modulation of the VDOS/ICDOS would provide band-intensities with an accuracy that could not be achieved through Eq. (11.9). The final, and not the least, advantage of Eq. (11.23) for MD-IR spectroscopy is that one can choose, at will, the level of representation for the calculation of the APTs. The higher *ab initio* representation the more accurate the APTs, thus the more accurate the charge fluxes for the IR calculations, thus the more accurate the dynamical IR band-intensities. Such accuracy is achieved at low computational cost as the number of atomic polar tensors is reduced to a low number, as discussed above.

This hybrid formulation of MD-based IR spectroscopy calculation is thus opening the path for hybrid theoretical methods for dynamical IR spectroscopy. The MD trajectory can be obtained at any level of theory, from low computational cost that uses force fields or semi-empirical representations, to medium computational cost with a DFT representation, while the APTs can be obtained at a very high level of electronic representation, for instance CCSD(T) for the highest affordable.

11.4.3. *Extension to SFG spectroscopy*

The same general formalism is applied in our DFT-MD-SFG spectro-scopic investigations of aqueous interfaces,[14, 15, 17, 22] where a param-eterization of the water APTs and Raman tensors has been made. An equivalent parameterization of these tensors has been conducted for the contribution of the O–H silanols at silica top-surfaces in order to include their contribution to the SFG spectrum of aqueous silica interfaces.[47, 48] This hybrid formulation allows for fast spectroscopic calculations of very complex molecular systems in the condensed phase, at low enough computational cost, *i.e.*, with rather limited time-lengths of the DFT-MD trajectories of such complex molecular systems.

Applying the hybrid formalism that we have detailed above for an IR spectrum,[49] one can re-write the SFG signal in the following way. For the water contribution to $\chi^{(2)}(\omega)$ to which we restrict our equations for the time being (transferring to, *e.g.*, surface silanol O–H groups is straightforward, see Refs. 47 and 48), the individual molecular dipole moments and polarisability tensors are calculated with the model proposed by Khatib *et al.*,[50] that supposes that in the high frequency region ($> 3,000$ cm^{-1}) only the O–H stretching motions are contributing to the spectrum. Neglecting intermolecular cross-correlation terms, as discussed earlier in this chapter, $\chi^{(2)}(\omega)$ becomes:

$$\chi^{(2)}_{PQR}(\omega) = \sum_{m=1}^{M} \sum_{n1=1}^{2} \sum_{n2=1}^{2} \frac{i}{k_b T \omega} \int_{0}^{\infty} dt\, e^{(-i\omega t)} \langle \dot{\bar{\alpha}}_{PQ}^{m,n1}(t) \dot{\mu}_{R}^{m,n2}(0) \rangle,$$

$$(11.27)$$

where (P,Q,R) are any x, y, z direction in the laboratory frame. $\bar{\alpha}_{PQ}(t)$ and $\mu_R(t)$ are respectively the individual O–H bond con-tribution to the total polarization and dipole moment of the system and $\dot{\bar{\alpha}}_{PQ}(t)$ and $\dot{\mu}_R(t)$ their time derivatives. M is the number of water molecules and n_1 and n_2 two indices that identify each of the two O–H oscillators per molecule. Using the direction cosine matrix (D) projecting the molecular frame (x, y, z) onto the fixed laboratory frame (P, Q, R), and assuming that the O-H stretching is much faster than the modes involving a bond reorientation, one can write:

$$\dot{\alpha}_{PQ}(t) \simeq \sum_{i}^{x,y,z} \sum_{j}^{x,y,z} D_{Pi}(t) D_{Qj}(t) \frac{d\bar{\alpha}_{ij}}{dr_z} v_z(t), \qquad (11.28)$$

$$\dot{\mu}_R(0) \simeq \sum_{i}^{x,y,z} D_{Ri}(0) \frac{d\mu_i}{dr_z} v_z(0). \qquad (11.29)$$

The D matrix and the projection of the velocities on the O–H bond axis (v_z) can be readily obtained from the DFT-MD trajectory, while $\frac{d\bar{\alpha}_{ij}}{dr_z}$ and $\frac{d\mu_i}{dr_z}$ are parameterized.[50,51] In most of our works, *ssp* SFG signals are calculated (this is also the most recorded in SFG experiments), *i.e.*, $P = x$, $Q = x$ and $R = z$ in Eqs. (11.27)–(11.29).

One thus sees that the time-correlation functions in Eq. (11.27) are based on velocities (here the Cartesian velocities of the atoms of a given water molecule obtained in the MD are projected on each O–H bond axis) modulated by the components of both the APT (as in the IR signal seen above) and of the Raman tensor. For SFG, the Cartesian coordinates of space are employed as the signal is calculated along the z direction of space (*i.e.*, perpendicular to the top-surface of the solid) which is the main contributor to the spectroscopic signal at interfaces. This SFG formulation based on APTs and Raman tensors modulating the velocity time-correlation functions has been shown to provide excellent agreement with respect to experimentally measured HD-SFG spectra on a vast range of water-solid and water-vapor interfaces at different pH, electrolytes and surface hydrophilicity conditions.[14,48,52–56] As an illustration, we report in Figure 11.2 the theory-experiment comparison for the air-water interface[14] (top) and an amorphous silica-water interface[15] (bottom).

The figure not only shows that the decomposition scheme described above to calculate SFG spectra from weighted velocity-velocity correlations works extremely well, but also that explicitly treating water self-correlations, with inter-molecular cross correlations implicitly accounted for in a mean field approach within the parameterization of $\frac{d\alpha_{lm}}{dr_z}\,\frac{d\mu_n}{dr_z}$, is a sufficiently good approximation for the prediction of SFG spectra in the OH-stretching frequency range.

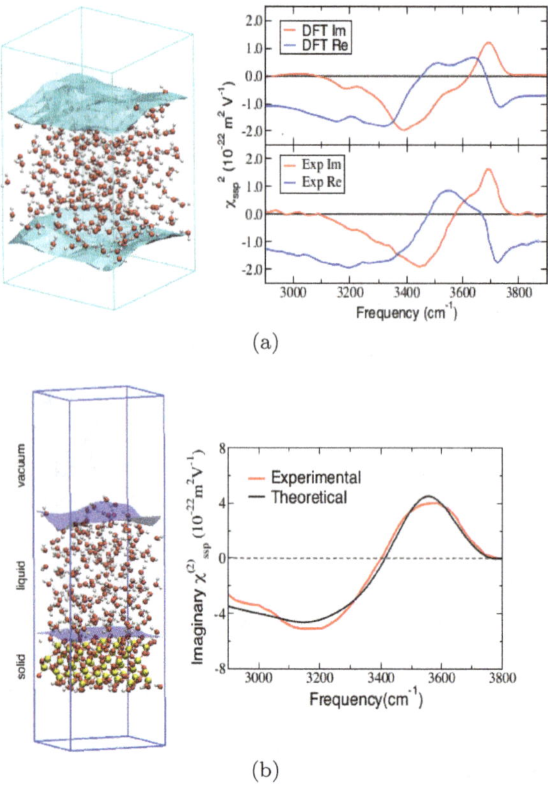

Figure 11.2. Comparison of the theoretical HD-SFG spectra obtained using our methodology (Eqs. (11.27)–(11.29), with DFT-MD trajectories) with the experimental spectra from Refs. 57 and 58, respectively for the (a) air–water and (b) amorphous silica–water interfaces.

11.5. Discussions on Some Possible Issues with MD-based Vibrational Spectroscopy

11.5.1. *Length of MD trajectories for spectroscopic signals*

As seen above, within the linear response formalism any dynamical vibrational spectrum is calculated through a Fourier transform of a time-correlation function of a given property (dipole, polarizability, *etc.*). MD is, by construction, the modeling tool to accumulate the time evolution of the properties entering into these time-correlation functions. One central element is thus the length of the trajectory over which the time-correlation function is calculated in order to ensure convergence of the calculated spectrum. By construction, a time-correlation function has to tend to zero at "long time scales" which can be hard to achieve over rather short time-scale DFT-MD trajectories. This is especially crucial for isolated gas phase molecules and their DFT-MD-based spectroscopy. The zero-limit of any time-correlation function of a property $\mathcal{A}(t)$ measures the time needed for the loss of the information in $\mathcal{A}(0)$ known at the initial time $t=0$ of the dynamics. This decorrelation time is dependent on the $\mathcal{A}(t)$ intrinsic property, and is molecular system dependent. It is much easier to get this zero limit for condensed phase systems, simply because the time-correlation is averaged over all the molecules of, *e.g.*, the liquid. To recover such statistical sampling for an isolated molecule one has to either accumulate a long trajectory (typically over at least ~100 ps time-scale), which is computationally costly in DFT-MD, or simulate a sufficient number of short time-length trajectories, each trajectory differing for its initial conditions (positions and velocities of the atoms); one then averages the time-correlation functions or their Fourier transforms over these trajectories. This is typically what we have done for DFT-MD-IR of the THz spectroscopy of small peptides.[44, 59–61] See also a related discussion[15] for aqueous solid interfaces and convergence of the DFT-MD-SFG spectra over the length of the trajectories.

In this convergence issue of the MD-spectrum, the most critical point is the convergence of the band-intensities and band-shapes. Band-positions are usually converged within 2–5 ps trajectories, even for isolated molecules, but the difficulty in reaching equipartition of energy for gas phase molecules necessitates typically at least 30–50 ps trajectories together with statistics over several trajectories for the intensities and shapes of the bands to be converged (*i.e.*, not changing anymore with the increase in trajectory time-length and/or statistical sampling). The lower number of degrees of freedom in the molecular system, the longer these trajectories have to be accumulated. The difficulties in converging band-intensities and band-shapes can hinder the theory-experiment match. Such difficulties are released in condensed phase systems thanks to the statistical sampling of properties that are now averaged over a large number of molecules, which also ensures a much better equipartition of energy within all degrees of freedom. See the subsequent discussion.

11.5.2. *Equipartition of energy*

We want to illustrate further the issue of equipartition of energy for gas phase molecules and the consequences for dynamical spectroscopy. The definition of equipartition has been given earlier in this chapter when discussing Eq. (11.8) for the calculation of the temperature T of a molecular system. We take three examples of gas phase molecules of increasing size for which we assess whether equipartition of energy can be achieved in "short" time-scale DFT-MD trajectories (over which DFT-MD-IR spectra are calculated). The molecules are phenol (13 atoms), the cytosine-guanine H-bonded DNA base-pair (26 atoms), and the gramicidin peptide (273 atoms). The dynamics and IR spectroscopy of isolated phenol is presented at several temperatures from $\sim$50 to $\sim$250 K. See Ref. 62 for the theoretical and experimental details on these investigations for several phenol derivative molecules and the synergy with IR-UV ion dip THz-IR experiments. The DFT-MD simulations of gramicidin are done at $\sim$50 K, both 50 K and room temperature DFT-MDs are presented for the C–G base pair. For all three molecules, 50 ps DFT–MD trajectories were accumulated in the NVE ensemble (over which the averages and fluctuations discussed afterwards are calculated), subsequent to 50 ps dynamics in the NVT ensemble where thermostats were employed to thermalize the molecular systems (*i.e.*, to reach the target of temperature and equipartition in all degrees of freedom). All data discussed hereafter are reported in Figures 11.3 and 11.4. To assess equipartition, the temperature and its fluctuation per atom is reported in the tables in Figure 11.3 for phenol and the base pair, and per residue for the gramicidin molecule.

The data clearly show that none of these gas phase molecules is at equipartition. In the case of phenol, the carbon atoms are systematically at lower temperatures than the average value while the hydrogen atoms are systematically at higher temperature than the average (with two exceptions). Carbons are thus "too cold", hydrogens are "too hot". This will have a direct consequence on the

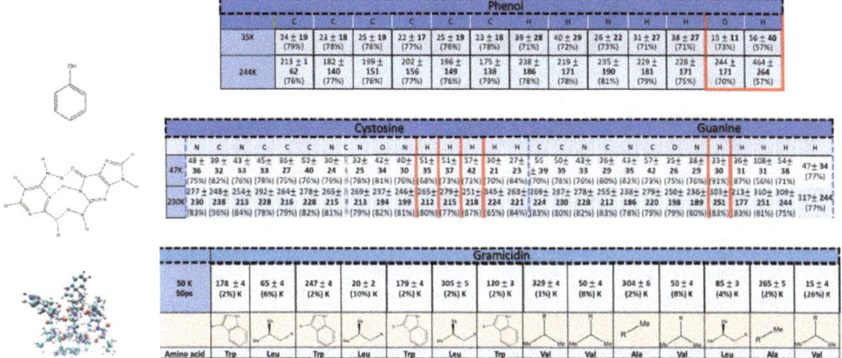

Phenol

	C	C	C	C	C	C	H	H	H	H	H	O	H
35K	24 ± 19 (79%)	23 ± 18 (78%)	25 ± 19 (76%)	22 ± 17 (77%)	25 ± 19 (76%)	23 ± 18 (78%)	39 ± 28 (71%)	40 ± 29 (72%)	26 ± 22 (73%)	31 ± 27 (71%)	38 ± 27 (71%)	15 ± 11 (73%)	56 ± 40 (57%)
244K	213 ± 62 (76%)	182 ± 140 (77%)	199 ± 151 (76%)	202 ± 156 (77%)	196 ± 149 (76%)	175 ± 138 (79%)	238 ± 186 (78%)	219 ± 171 (78%)	235 ± 190 (81%)	229 ± 181 (79%)	228 ± 171 (75%)	244 ± 171 (70%)	464 ± 264 (57%)

Cystosine

	N	C	N	C	C	C	N	C	N	O	N	H	H	H	H
47K	48 ± 36 (75%)	39 ± 32 (82%)	43 ± 33 (76%)	45 ± 33 (78%)	36 ± 27 (75%)	52 ± 40 (76%)	30 ± 24 (79%)	32 ± 25 (78%)	42 ± 34 (81%)	40 ± 30 (76%)	51 ± 35 (68%)	51 ± 37 (73%)	57 ± 42 (73%)	30 ± 21 (70%)	27 ± 23 (84%)
290K	277 ± 230 (83%)	248 ± 238 (96%)	254 ± 213 (84%)	292 ± 228 (78%)	264 ± 216 (79%)	278 ± 228 (82%)	265 ± 215 (81%)	269 ± 213 (79%)	237 ± 194 (82%)	246 ± 199 (81%)	265 ± 212 (80%)	279 ± 215 (77%)	251 ± 218 (87%)	345 ± 224 (65%)	263 ± 221 (84%)

Guanine

	C	C	N	C	N	C	O	N	H	H	H	H	H
47K	55 ± 39 (70%)	50 ± 39 (78%)	43 ± 33 (76%)	36 ± 29 (80%)	43 ± 35 (82%)	57 ± 42 (73%)	35 ± 26 (75%)	38 ± 29 (76%)	33 ± 30 (91%)	36 ± 31 (87%)	108 ± 31 (56%)	54 ± 38 (71%)	47 ± 34 (77%)
290K	269 ± 224 (83%)	287 ± 230 (80%)	278 ± 228 (82%)	255 ± 212 (83%)	238 ± 186 (78%)	279 ± 220 (79%)	250 ± 198 (79%)	236 ± 189 (80%)	303 ± 251 (83%)	213 ± 177 (83%)	310 ± 251 (81%)	309 ± 244 (75%)	317 ± 244 (77%)

Gramicidin

50 K 50ps	178 ± 4 (2%) K	65 ± 4 (6%) K	247 ± 4 (2%) K	20 ± 2 (10%) K	179 ± 4 (2%) K	305 ± 5 (2%) K	120 ± 3 (2%) K	329 ± 4 (1%) K	50 ± 4 (8%) K	304 ± 6 (2%) K	50 ± 4 (8%) K	85 ± 3 (4%) K	265 ± 5 (2%) K	15 ± 4 (26%) K
Amino acid	Trp	Leu	Trp	Leu	Trp	Leu	Trp	Val	Val	Ala	Val	Leu	Ala	Val

Figure 11.3. Tables of temperatures per atom (Phenol, C-G base pair) and per residue (gramicidin) obtained in 50 ps DFT-MD simulations in the NVE ensemble (subsequent to 50 ps NVT DFT-MD). Various temperatures are presented, one single trajectory is used for these average values. Temperatures and their fluctuations by standard deviations are reported. The values highlighted in the red boxes are for specific atoms discussed in the text. Illustrations of the molecules are reported on the left side of the table.

intensity of the IR bands in which these atoms are involved: "too hot atoms" will contribute to "too high intensity" of the bands related to these atoms' motions, "too cold atoms" will contribute to "too low intensity" of the associated bands. Thus, over-/under-estimations of band intensities will likely be induced by the shifts in temperature in the heavy/light atoms. Another interesting conclusion is that each average atomic temperature has large fluctuations, of the order of more than 70%, which shows the large variations in the dynamics of these atoms over time. Interestingly, the O–H group of the phenol molecule that has torsional and rotational large amplitude motions[62] at all temperatures, also has the "hottest" hydrogen atom within the molecule (see the red box). This hydrogen is clearly not thermalized within the whole system. Any IR spectral band related to the O–H motion (*e.g.*, torsional motion in the THz domain[62]) will thus likely suffer in reaching a good agreement in intensity with respect to experiment. Same overall conclusions can be drawn for the C–G base pair at the two investigated temperatures.

Equipartition arises from exchanges of energy between degrees of freedom, *i.e.*, through intra- and inter-atomic interactions.

Bond-stretches and angular/torsional motions within the molecule together with intermolecular interactions are the driving forces for energy exchanges. There are only intra-molecular interactions in the phenol molecule that can allow energy exchanges, it is thus not too surprising that these motions are not enough for reaching equipartition in such a small molecular system over rather limited time-lengths of trajectories (despite NVT thermalization prior to the NVE dynamics). On the other hand, the C–G base pair has three H–bonds over which intermolecular exchange of energy can be more easily achieved. This is however not enough for reaching equipartition over the 50 ps time-scale of the DFT–MD simulation. This is also not enough for the Hs involved in these H-bonds to be at equipartition (although it seems to help), see the temperatures highlighted in the red boxes.

Interestingly, the fluctuations of temperature per residue of the gramicidin peptide are now very low, presumably as the result of compensations between "too hot"/"too cold" atoms and their compensating fluctuations. Note that most of the residues have a number of atoms that is comparable to the phenol/C–G pair. Equipartition per residue is however still not reached over 50 ps DFT-MD. While the average temperature of the peptide is 50 K, some residues can have an average temperature as low as 15–20 K, others as high as 305 K. Only two residues are thermalized/at equipartition (two valine residues) with the average 50 K temperature of the peptide. Thus, the average 50 K of the peptide reflects very disparate temperatures of the individual residues. Even a large molecule such as gramicidin is hard to thermalize correctly with rather short time-scales DFT-MD trajectories. As discussed above, one consequence is that some of the DFT-MD-IR bands will likely have too low/too high intensity in comparison to the experiment. This is however to be tempered for a large molecule. Indeed, the vibrational modes of a large molecule are due to combinations of motions and of residues. Therefore, there will likely be compensations between some "hot motions" and some "cold motions" within each of the modes, that might in the end lead to reasonable band-intensities in the IR spectrum (by compensation), despite equipartition not being reached.

For phenol, we have furthermore accumulated a total of six trajectories (20 ps each) at ∼200 K (averaged over all trajectories) that differ by the initial conditions in positions and velocities of the atoms at the initial time of the dynamics. They can be used as a statistical ensemble for averages made over a total of 120 ps, especially for the DFT-MD-APT-IR spectrum. Though equipartition is still not achieved for this small molecule, the average DFT-MD-IR spectrum obtained over the six trajectories (black line in Figure 11.4)

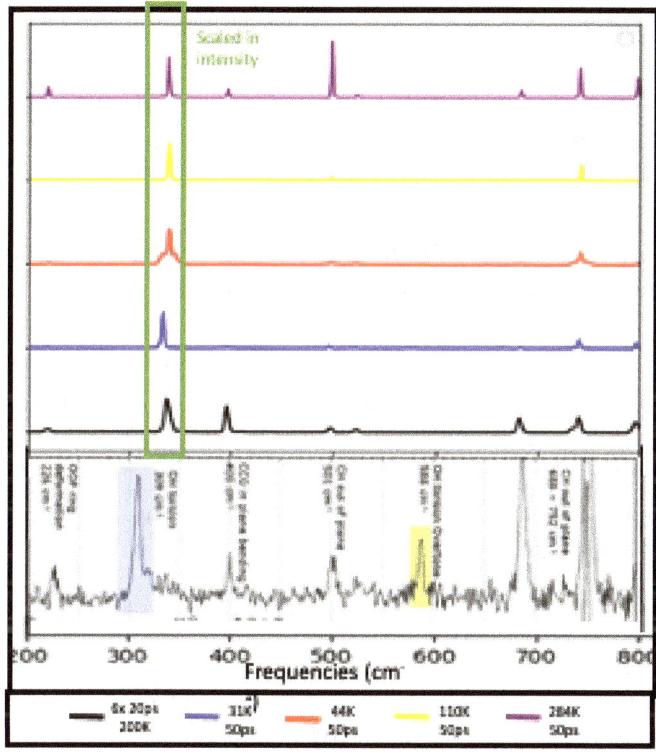

Figure 11.4. DFT-MD-APT-IR spectroscopy of gas phase phenol in the THz spectral domain. IR-UV ion-dip action spectroscopy experiment at the bottom.[62] Colored lines report the DFT-MD-APT-IR calculated at various increasing temperatures. All but the black line are calculated over one single 50 ps DFT-MD trajectory. The black-line spectrum is averaged over six DFT-MD trajectories of 20 ps each. Ref. 62 has all details for the DFT-MD set-up used for these trajectories.

is now in much better agreement with the experiment (bottom line in Figure 11.4) in terms of ratio of band-intensities than the spectrum calculated at a similar temperature of $\sim$284 K over one trajectory only (violet line in Figure 11.4). In particular, the black-line spectrum is the only one showing a relevant intensity for the $\sim$400 cm^{-1} band, also for the equal intensity of the $\sim$700–800 cm^{-1}doublet. This can be seen in Figure 11.4, where the intensity of the $\sim$300 cm^{-1} torsional O–H band has been scaled to the same intensity in all spectra, and is thus used as the reference peak. Note that the experimental THz-IR spectrum[62] has been recorded at low temperature ($\sim$50 K). The influence of "hot"/"cold" atoms is nicely seen in all DFT-MD-IR spectra calculated over one single trajectory only. Each peak in the phenol spectrum is indeed due to very simple molecular modes that involve very few atoms, each one being either "too hot" or "too cold". The 31 K, 44 K and 110 K DFT-MD-IR spectra have only two peaks of relevant intensity, *i.e.*, the $\sim$300 and $\sim$700 cm^{-1} bands. There is no doublet in the 700–800 cm^{-1} range. All bands have substantial intensities in the 284 K DFT-MD-IR spectrum, however with either too low or too high intensity compared to the 200 K spectrum averaged over six trajectoires, and compared to the experiment.

We make one last illustration of equipartition of energy for a condensed phase molecular system, with the example of the water molecules that are located in the BIL directly in contact with a silica oxide surface (the BIL is $\sim$3 Å thickness above the solid top-surface). This example is taken from the DFT-MD simulations of the silica-water interface in Ref. 15; an image of the simulation box is in Figure 11.2. The DFT-MD-SFG spectrum of the BIL has been presented in Figure 11.2, showing an excellent agreement with the experiment. There are 20 water molecules in the BIL in this simulated system, their individual average temperatures and associated fluctuations are presented in Figure 11.5 where the BIL-water molecules H-bonded to the quartz surface are in the red bars and the ones not H-bonded to the quartz surface are in the blue bars. Interestingly, while the whole solid–water system is thermalized at 300 ± 9 K, the water molecules in the BIL are thermalized around 400 K with rather large fluctuations around each individual mean

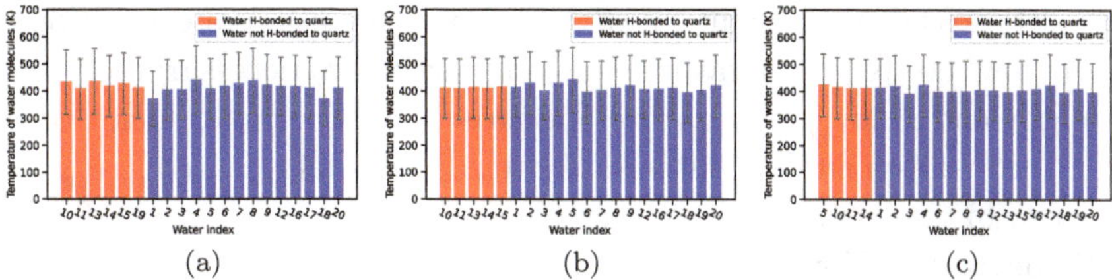

Figure 11.5. Average temperature and fluctuation per water molecule that belongs to the BIL of a silica–liquid water interface. DFT-MD trajectories are accumulated for (a) 10 ps, (b) 40 ps, and (c) 70 ps. The numbers in the x-axis report the label of each water molecule within the 20 molecules located in the BIL. Red bars are for water molecules H-bonded to the silica top-surface, blue bars for water molecules not H-bonded to the silica surface.

temperature, whatever their H-bonded status to the silica surface. There is inhomogeneity in the individual molecular temperatures for a 10 ps DFT-MD trajectory (left panel in Figure 11.5). Very nicely, one can see that the increase in the time-length of the dynamics from 10 to 70 ps (from (a) to (c)) leads to a more homogeneous distribution of the individual temperatures. This shows that time is essential for the exchange of energy between the water molecules and their surrounding in order to achieve equipartition, even in condensed phase systems where exchange of energy is easily processed through inter-molecular interactions. Within 70 ps of dynamics, one can see that all the water molecules in the BIL are at equipartition within roughly ±10 K.

11.5.3. *Choice of temperature in the MD*

The choice of the temperature in the MD trajectory has an influence on the positions of the vibrational bands. To make a 1-to-1 comparison to an experiment, the temperature in the MD simulation has to match the experimental one. In our past investigations for the IR spectroscopy of gas phase molecules and clusters, we have either made room-temperature DFT-MD trajectories to calculate IR spectra for the interpretation of room-temperature IR-MPD and IR-PD experiments,[9,10,63–67] or 50 K temperature DFT-MD trajectories for the interpretation of low-temperature IR-UV ion dip experiments.[44,59–61,68] These dynamical spectra have shown rather good to excellent agreements with the experimental band-positions. The low 50 K temperature was in particular necessary for the assignment of low temperature spectroscopic experiments, not only for the band-positions but also for the band-shapes that are very narrow in these experiments. Increasing the temperature in the DFT-MD beyond 50 K would broaden the theoretical bands: mechanically, there are "larger amplitude motions" made by the atoms with the increase in temperature, directly reflected in the band-shapes. Broadening would impede the theory-experiment assignment. The increase in temperature in the DFT-MD could also induce conformational isomerizations that are not probed at the lower temperature in the experiment. One has therefore to be careful in the choice of the temperature in the MD simulations for vibrational spectroscopy of gas phase molecules. The same is true for the condensed phase, *i.e.*, liquids and interfaces between liquids and other media (solids, liquids, air). There again, the temperature has to be concomitant with the experimental one, very often room temperature.

For a given 3D-structure of a molecule (our argument stands either for a single gas phase molecule or for any molecule within a condensed phase medium), each vibrational oscillator in the molecule will explore/sample more anharmonic motions as temperature is increased. The immediate consequence is the red-shift of the frequency of that oscillator (shift towards a lower frequency). The choice of the temperature thus controls the extent of vibrational anharmonicity probed in each oscillator. It controls the absolute value of the oscillator's vibrational frequency, therefore the careful choice to be made for the temperature in the MD simulation. On the other hand, vibrational anharmonicities also arise from modes couplings. Couplings are less affected by the choice of temperature in the (classical nuclei) MD simulation. Mode couplings being the essence of the far-IR/THz vibrational modes, our 50 K trajectories of gas phase peptides have typically shown that these anharmonicities are rather well accounted for.[44,59–61,68] See also the part later in this chapter about the vibrational graph modes (Vib-Graphs) that can measure couplings between the modes.

11.5.4. *Zero point energy and quantum nuclei*

An issue that is however remaining with classical nuclei (DFT-) MD simulations is the "classical temperature" versus the "quantum temperature" of the nuclei, *i.e.*, the ZPE-zero point energy contained in the oscillators. The classical treatment of the nuclei at, *e.g.*, room temperature will not lead to the same sampling of the anharmonicities in the motions of the X–H oscillators as in the quantum treatment of the nuclei, to the extent that the amount of "classical vibrational red-shift" will be insufficient in comparison to the "quantum red-shift". The difference between classical and quantum vibrational red-shift will depend on each X–H vibrator, and is therefore hard to quantify *a priori*. At low classical (nuclei) MD temperatures, such as the 50 K trajectories accumulated by us for THz-IR spectroscopy of isolated peptides,[44, 59–61] classical and quantum temperatures of the oscillators are rather comparable. The advantage of the (classical) MD-based vibrational spectroscopy at low temperature is that the couplings between the modes are still naturally accounted for, therefore not only the large amplitude anharmonic motions are sampled but also the anharmonic couplings between the modes, which allows DFT-MD-IR in the THz/far-IR domain to be sufficiently quantitatively accurate for the band-positions. We have hence found the dynamical vibrational bands to be within less than 10 cm^{-1} from their experimental counterparts.[44, 59–61] See, *e.g.*, Refs. 69, 70 for similar discussions on gas phase molecules IR spectroscopy. See also Bowman *et al.*[71] for some comparisons of anharmonic VSCF, VCI and dynamical MD spectra of small molecules using the same potential energy and dipole surfaces in all theoretical methods, and their discussion on band-positions from the MD-based trajectories. See also Ref. 44 for discussions on temperature effects in classical MD trajectories and their influence on band-positions.

In our works and in this text, the nuclei are treated as classical particles in the MD simulations. The trajectories thus miss the quantum nature of light atoms as well as the ZPE in the vibrational modes. There are several methods in the literature that include the quantum nature of the nuclei in the MD simulations. I refer the readers to some recent papers.[69,70,72–75] These methods however usually have a computational cost that hampers their widespread application.

The lack of ZPE can be a critical issue for the high-frequency motions/modes of O–H, N–H, C–H groups, as the classical energy put into these motions at room temperature is (too) small in comparison to the ZPE of these motions. As a consequence, the classical motions of these groups might be too small-amplitude, which hence shifts the position of the calculated stretching bands with respect to their experimental counterpart. Such an issue disappears for the large amplitude motions in the THz, where, *e.g.*, we obtained agreements within less than 10 cm^{-1} for band-positions compared to experimental values.[44,59–61]

Semi-classically *prepared* MD trajectories that include ZPE into the modes at the initial time of the classical nuclei dynamics can be done, hence taking into account the quantum representation of the nuclei in the preparation of the trajectory while propagating classically the equations of motions of the nuclei. See, *e.g.*, one of our work on that topic[76] and Refs. 71, 77 for discussions. However, one issue with semi-classicaly prepared MD is the ZPE-leakage, well documented in the literature by, *e.g.*, Hase, Bowman, Manolopoulos, and collaborators. Quantum mechanically each internal molecular mode must contain an amount of energy at least equal to its ZPE, but classical mechanics may allow vibrational energy to flow freely between all or a subset of the modes and, hence, does not preserve any ZPE constraint.[78] This is an error inherent to classical mechanics. Consequently, no matter how accurately one can initially (*i.e.*, at time zero of the MD) assign the ZPE to each normal mode of a molecule, after a number of steps, the energies in these modes may fluctuate. The energy fluctuation between modes for a multimode Hamiltonian is caused by the couplings between the modes. In the case of separable modes, the energies for these modes would be conserved. Without any control of the coupling term, it is possible for one mode to transfer its energy to other modes and to lose energy, and thus to be of an energy less than the ZPE.[79] Bowman and co-workers[80] and Miller and co-workers[81] proposed a method to constrain the ZPE by changing the sign of the momentum when the energy of any mode reaches the ZPE. This method did prevent energy from going below the ZPE, however since the momentum change occurs instantaneously, it is equivalent to an infinite impulse that is perhaps too abrupt, and can cause noise in a classical correlation function. This can be problematic in the context of time-dependent correlation functions for vibrational spectroscopy. Another method has been implemented by Bowman and co-workers[82] to constrain the ZPE by smoothly eliminating the coupling terms in the Hamiltonian as the energy of any mode falls below a specified value. This still introduces errors in the correlation functions that are needed to calculate spectroscopic signals.

Semi-classical MD, described in another chapter of this book by Ceotto *et al.*, is intermediate between quantum nuclei MD and semi-classically prepared MD, intermediate not only in terms of theoretical representation but also in terms of computational cost. Semi-classical MD retains the quantum nature of the nuclei into the propagator of the dynamics and hence includes the ZPE in each vibrational mode of the system over the MD. For computational reasons, the trajectories are accumulated over short time scales (ps time-scale), the final vibrational spectrum of a molecular system is averaged over multiple (short) trajectories. We let the readers go to this chapter and learn more about this methodology for its strengths and limitations.

11.6. Forces in MD Simulations: What Level to Choose for Vibrational Spectroscopy?

Beyond the temperature and sampling/convergence issues discussed above, the accuracy of a MD dynamical vibrational spectrum is primarily due to the quality of the representation used for the calculation of the interactions. The accuracy of the MD vibrational band-positions is arising from the internal motions of the (anharmonic) oscillators and from the interactions and couplings between these oscillators. These can be roughly speaking qualified as the intra-molecular interactions. In contrast, intensities are directly related to the (intermolecular) charge fluxes between the atoms. The choice of *ab initio*/semi-empirical/force field representations for the calculations of these intra- and inter-molecular interactions is therefore crucial. In BO MD, forces can be calculated "on-the-fly" (mainly with semi-empirical or DFT electronic representations) or from precomputed potential energy surfaces (which can be obtained with a higher level of electronic representation). Similarly, dipoles and polarizabilities needed for the calculation of the spectroscopic signal can be calculated either "on-the-fly" or on precomputed surfaces. Precomputed surfaces are generally obtained at a very high quantum level. See for instance works by Bowman *et al.*,[71] also presented in another chapter of this book.

While any quantum representation can be used for "on-the-fly" *ab initio* MD, the DFT electronic representation is certainly an excellent compromise between computational efficiency and electronic accuracy.[3] It is therefore the most used in the literature for dynamical spectroscopy of large molecular systems (nowadays a few 1000' atoms is reasonably possible) over 100' ps time-scale trajectories. DFT-MD is hence applied to the dynamics and spectroscopy of gas phase molecules and clusters, liquids and solid-liquid interfaces. Any higher level of electronic representation will be applicable only to the MD of gas phase molecules containing roughly a maximum 10–50 atoms, for trajectories accumulated over few picosecond time-scales. One consequence of the DFT-MD "smallish time-scale" is that there might be some difficulties in achieving convergence of the bands' intensities in the dynamical spectrum, as discussed earlier in this chapter. However, the hybrid method for spectroscopy based on APTs and Raman tensors discussed earlier in this chapter is one way to circumvent such issue. GGA functionals like BLYP or PBE are certainly the most used in DFT-MD simulations. Hybrid functionals as well as meta-GGA functionals would probably be of better accuracy for vibrational spectroscopy, however at a higher computational cost.

While accumulating the trajectories of large molecular systems (*e.g.*, 100–1000 atoms) over $\sim$50–100 ps in DFT-MD is already an achievement, the calculation of the properties that enter into the time-correlation functions for IR, Raman, and SFG spectroscopies, *i.e.*, the molecular dipole moments and molecular polarizability tensors, is even more challenging. This has already been discussed in this chapter. In the case of gas phase molecules, no (computationally costly) localization procedure is required. We also refer the reader to an earlier section of this chapter where we showed an alternative hybrid formalism for the calculation of IR, Raman and SFG spectra that does not require the actual evaluation of dipole moments or polarizability tensors.

The alternative to a full QM electronic representation in MD simulations for spectroscopy is the semi-empirical (SE) representation, which can come in various flavors. To our best knowledge, most of the SE-MD simulations of vibrational spectroscopy using PM methods as the SE representation are for the liquid phase, especially the IR spectroscopy of liquid water or biomolecules immersed in water.[83,84] Several IR spectroscopic investigations have been led by Rapacioli *et al.* using DFT Tight Binding (DFTB) SE-MD dynamics for gas phase systems.[85–87] Other papers can be found for static harmonic spectra calculations based on SE representations.[88,89]

On the other hand, FF-MD classical simulations are based on analytical expressions for the intra- and inter-molecular interactions (FF) with pre-established parameters. Examples of FFs used in the literature for various molecular systems (gas phase or condensed matter) are AMBER, CHARMM, AMOEBA, CLAYFF, and INTERFACE. The intramolecular part of these FFs is generally modeled by harmonic forms of the bond and angular motions that represent two- and three-body terms, an anharmonic torsional term for four-body dihedral motions, while the intermolecular part is generally modeled by two-body Coulomb and Lennard-Jones terms. Roughly speaking, the intra-molecular interactions and the parameters entering into these are mostly responsible for the vibrational band-positions while the inter-molecular interactions/parameters are mostly responsible for the band-intensities. Intermolecular parameters also modulate the band-positions through the interactions with the surrounding environment, typically for vibrational bands related to hydrogen bonds. No matter how carefully any FF is parameterized, its accuracy for spectroscopy will be limited by the choice of the functional forms entering into the analytical expressions, and then by the actual parameterization. As said above, most of the common FFs in use for gas phase molecules and condensed phases are based on harmonic oscillators for the internal motions, which is of course a huge approximation for vibrational spectroscopy. At least Morse anharmonic potentials should be used to model these motions for improved spectroscopic accuracy, cross-terms between the internal motions should also be included. All these terms would increase the complexity of the parameterization, which is therefore very seldom done. Also polarization should be included in the electrostatic part of the FFs for spectroscopic accuracy, while higher order multipolar terms to model electrostatic interactions should also be in principle taken into account for improved accuracy. This is also very seldom done, as it is extremely complex and cumbersome to parameterize such FFs. Even more importantly, electrostatic terms that explicitly take care of charge and dipole fluxes[90] should be added in FFs for spectroscopic accuracy: such fluxes have been shown to be the most important ingredients in vibrational spectroscopy analyses.[91,92] Such force fields are extremely rare, as they are complex to develop and parameterize, they furthermore become computationally rather costly.[93,94]

Very few FFs are therefore of spectroscopic accuracy, *i.e.*, very few are able to predict reliable band-positions, band-intensities and band-shapes. In Ref. 95, we have for instance shown the extent of agreement/disagreement between DFT-MD and FF-MD SFG spectra for a series of aqueous interfaces using simple and slightly more sophisticated FFs. Developments around the AMOEBA FF have been for instance performed by Clavaguera *et al.* in the gas phase community, leading to a FF of better spectroscopic accuracy.[93,96] Developing a FF is a time-consuming task, while the transferability of the parameters from one molecular system to another is always questionable.

State-of-the-art machine learning techniques are now increasingly applied for the development of FFs. The functional forms of the intra- and inter-molecular interactions are thus not needed anymore, no need for an actual parameterization. Machine learning is also applied to the learning of dipole and polarizability properties that enter into the spectroscopic theoretical signals. These ML approaches will be discussed at the end of this chapter within the perspectives to the field.

11.7. From the Peaks to Their Molecular Assignments: Revealing Vibrational Modes and Their Couplings

It is one achievement to calculate dynamical vibrational spectra, it is another task to assign the vibrational bands to molecular motions and hence reveal and rationalize the movements that give rise to the active spectroscopic bands. There are a certain number of theoretical methods that have been developed to that end.[41,44,97–104] The easiest way is to calculate the vibrational density of states (VDOS) or internal coordinates density of states (ICDOS). Both are unfortunately only qualitative decompositions of the vibrational spectra. While the atoms involved in the motions are known by these decompositions, their percentage participations are not quantitatively known. The intensity of the VDOS/ICDOS bands do not correspond to the active IR (or Raman, SFG) intensities, which limits the global assignment. "Effective normal modes" have been developed,[100,101,104–107] which are the equivalent of the normal mode analysis in static harmonic calculations, however including the mode couplings and temperature of the underlying MD trajectory. Though appealing, these methods are in practice not easy to converge, preventing their widespread use. We have developed an alternate route to modes assignments that we present below in the context of the IR spectroscopy of gas phase molecules. The method can be generalized to Raman and SFG spectroscopies also discussed in this chapter, with adjustments to be made for condensed phases.

11.7.1. *Assignments of modes by VDOS or ICDOS*

The VDOS is obtained through the Fourier transform of the *Cartesian* atomic velocity auto-correlation functions:

$$\text{VDOS}(\omega) = \sum_{i=1,N} \int_{-\infty}^{\infty} \langle \mathbf{v_i}(t)\cdot\mathbf{v_i}(0) \rangle \, \exp(i\omega t) \, dt, \quad (11.30)$$

where i runs over all atoms of the investigated system. $\mathbf{v_i}(t)$ is the velocity vector (in the Cartesian space) of atom i at time t. As in Eq. (11.9), the angular brackets in Eq. (11.30) represent the statistical average of the correlation function. The VDOS spectrum provides all vibrational modes of the molecular system. However, only some of these modes will be active in the chosen spectroscopy (following the selection rules), so VDOS spectra can by no means directly substitute IR, Raman or SFG spectra discussed in this chapter. Traditionally, the cross-correlations between the velocities of different atoms are not included. They can be included, as was done in Eq. (11.23) for the MD-APT-IR hybrid formalism for dynamical spectroscopy shown earlier in this chapter (two summations over i and j then enter Eq. (11.30)).

The strength of correlation functions based on summations over atoms is that they can be decomposed (or restricted) according to, *e.g.*, atom types, or groups of chosen atoms, or chemical groups of interest, *etc.*, in order to get a detailed assignment of the vibrational bands in terms of these specific atomic motions. This is hence done by restraining the sum over i in Eq. (11.30) to the atoms of interest only. With these individual signatures in hands, one can rather easily interpret a vibrational peak in terms of the molecular movements that are responsible for its activity. This is easy to interpret for localized vibrational modes, as they generally involve only a few atoms, *i.e.*, typically couples of atoms involved in bond-stretching motions in the 3000–4000 cm^{-1} domain. A similar interpretation of the VDOS peaks is more complex for delocalized and collective modes, and for highly coupled motions, because of numerous atoms (and motions) involved in such collective modes. Without a quantitative measure of these motions within the active mode, it is almost impossible to provide the exact atomic and molecular vibrational motions. See examples illustrated in Refs. 60, 105.

Delocalized and collective modes are activated in the far-IR/THz domain below 800 cm^{-1} (see Section 11.10). For these modes, the Fourier transform of ICs time-correlation functions is better suited to describe the molecular motions. The ICDOS spectrum is calculated as: $\text{ICDOS}(\omega) = \sum_{i=}^{3N-6} \int_{-\infty}^{\infty} \langle IC_i(t) \cdot IC_i(0) \rangle \exp(i\omega t) \, dt$, where $IC_i(t)$ is the value of the internal coordinate i at time t of the MD. This formalism requires to describe the molecular system by an ensemble of non-redundant internal coordinates.[43, 60, 61, 108] Here again, the total $\text{ICDOS}(\omega)$ spectrum can be decomposed into IC_i individual components by restricting the summation over i. Here again, the lack of quantitative % of participation of each motion into the peaks/modes limits the use of ICDOS.

11.7.2. *Graph theory for modes assignments (Vib-Graph)*

We have developed in Ref. 44 a new method for vibrational band assignments that combines the DFT-MD-APT formulation described earlier in this chapter (see, *e.g.*, Eq. (11.24) for IR spectroscopy) with algorithmic Graph Theory. In the case of IR, Eq. (11.24) has shown how to reconstruct any IR spectrum as a combination of the IR participation of the (non-redundant) ICs that describe the molecular system of interest. We thus directly obtain the measure of the IR participation of each intra- and inter-molecular motion in each IR band. This is the direct measure of the vibrational mode responsible for each IR peak. This knowledge is then coupled with Graph Theory from computer sciences. One graph is associated to each spectroscopic peak, and represents a vibrational mode. A graph is composed by an ensemble of vertices which contain the knowledge of the ICs involved in the mode (given by the self-terms in Eq. (11.24)), and by an ensemble of edges (here, non-directed) that provide the knowledge of the couplings between the ICs (given by the cross-terms in Eq. (11.24)). The graph thus provides the connectivity between the ICs that are responsible for the IR activity of each spectroscopic band, including their percentage of participation and the couplings between ICs. Each anharmonic mode probed in the finite temperature MD will thus be quantitatively described by such a graph. The different elements of the theory were detailed in Ref. 44, there is no space in this chapter to repeat them.

Interestingly, as soon as a Vib-Graph is plotted, one can immediately conclude whether a vibrational mode is made of coupled or uncoupled internal motions. By simply counting the number of connected components in the graph one sees whether a mode is made of localized (typically one main single component in the graph) or delocalized/collective motions (several components in the graph). By using the weights on the vertices of the graph (a weight is the percentage of participation of the associated IC into the molecular mode described by the graph), one can furthermore get the information on the "electronic" *vs* "mechanical" couplings involved in the motions, in a very efficient way. ICs can indeed be mechanically correlated without participating to the final IR intensity, the latter arises from the "electronic" coupling/fluxes contained in the APTs in Eqs. (11.24) and (11.25). In that case, an edge on the graph would have a high value of the weight but low/even zero weight on one of the connected vertices.

We now describe some applications. We especially want to emphasize the strength of the Vib-Graphs for measuring similarities and couplings between vibrational modes. One straightforward illustration of the Vib-Graph theory concerns the three vibrational modes of a gas phase water molecule, here obtained from a 15 ps DFT-MD simulation at 50 K (hence avoiding the rotational motion of the molecule).[44] The three Vib-Graphs for each of the three IR active vibrational modes of the isolated water molecule are presented in Figure 11.6. As can be immediately seen, the two higher frequency modes are the in-phase and out-of-phase O–H stretching motions of H_2O, while the 1609 cm^{-1} band is the bending motion. This latter is completely decoupled from the stretching motions, as expected, *i.e.*, the bending IC is never showing up in the graphs of the 3,650

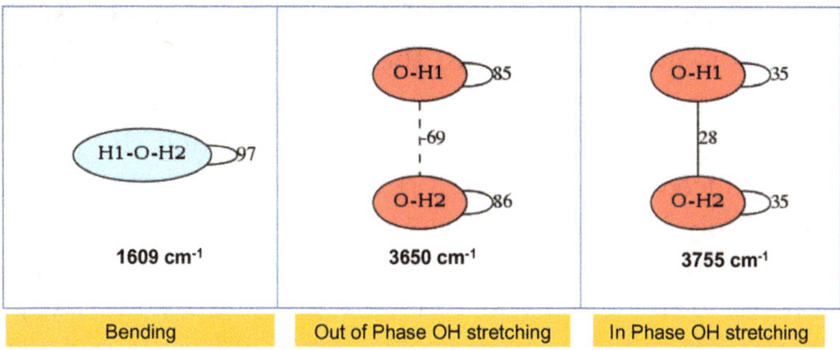

Figure 11.6. Graphs of the three active IR modes for a water gas phase molecule at 50 K. *Source*: Reproduced with permission from Ref. 44.

and 3,755 cm^{-1} stretching bands, and vice versa. One can see that hence the 1609 cm^{-1} bending mode has only one single vertex in the graph representation. For the two stretching modes, one can observe that the two O–H groups systematically equally participate to each vibrational mode, the percentage of contribution is readily seen on the vertices of the graphs. Very nicely, the in-phase/out-of-phase nature of the stretching modes is given by the sign of the weight on the edges. The 3,755 cm^{-1} mode has a positive (+28%) value of the weight for the cross-contribution between the two O–H stretching motions, thus an in-phase motion of the two stretchings, while the 3,650 cm^{-1} mode has a negative value (−69%), therefore the out-of-phase stretching motion. One can also see that the values of the weights on the vertices (self-contribution) and the ones on the edges (cross-contribution) are of the same order of magnitude, showing that both contributions are equally relevant for the final IR activity of the stretching bands. With these graphs, one can also see that a band assignment based on the self-contributions only (first term in Eq. (11.24)) would be incorrect. More illustrations of Vib-Graphs for delocalized modes in the THz domain are given in Section 11.10 later in the chapter.

We want to end this section by showing the strength of graphs in measuring the couplings between vibrational modes. One advantage of graphs is indeed the natural capability for comparing graphs and hence extract similarities between graphs. This is exactly what we need for extracting the degree of similarity between modes, which in other words is nothing else than the measure of couplings between vibrational modes. Isomorphism techniques (finding a bijection between two graphs[109]) or graph matching (finding the maximum common induced subgraph (MCIS) between two graphs)[110] can be used. We apply here graph matching by finding the sub-graph that contains the maximum number of ICs that are commonly involved in two graphs. The sub-graph gives us the ICs that are shared by two IR modes, therefore the coupled motions between the two modes. The MCIS problem is classified as a non-polynomial (NP) problem, *i.e.*, hard to solve. One traditional solution is through the reduction to a maximum clique problem, using auxiliary product graph.[111,112] For the given illustrations below, undirected unweighted graphs are used, *i.e.*, not taking into account the weights on the vertices and the edges that are known in the Vib-Graphs.

Given two graphs $G_i = (V_i, E_i)$ and $G_j = (V_j, E_j)$, where V_i (V_j) and E_i (E_j) are respectively the vertices and edges in graphs G_i and G_j, the product of the two graphs gives rise to a graph $G_i \triangledown G_j = \{(v, v') \in V_i x V_j : (v, v) \in E_i \Leftrightarrow (v', v') \in E_j\}$. The maximum clique in the graph $G_i \triangledown G_j$ can be searched, giving the maximum common induced sub-graph $G_{ij} = (V_{ij}, E_{ij})$. The degree of similarity between G_i and G_j can then be calculated by $\mathrm{sim}(G_i, G_j) = \frac{(|V_{ij}|+|E_{ij}|)^2}{(|V_i|+|E_i|)*(|V_j|+|E_j|)}$.

With such a measure in hands, one can for instance provide the similarity between two vibrational modes within the same vibrational spectrum of a given molecular system, or provide the similarity of a given mode in the IR spectra of a molecular system recorded at different temperatures, or provide the similarity between two modes that are active in two different molecular systems. This latter identifies how much a vibrational mode can be preserved/modified in going from one molecular system to another, for instance in going from a di-peptide to a tri-peptide, and to longer peptidic chains or to H-bonded chains.

We now illustrate this last point, by taking the example of the Ac-Phe-OMe monomer molecule and its H-bonded dimeric form that serves as model for a β-sheet structure. The THz-IR spectroscopy of Ac-Phe-OMe and (Ac-Phe-OMe)$_2$ has been presented in Refs. 44, 68 by coupling IR-UV ion dip experiments and DFT-MD-based spectroscopy. See also Section 11.10 later in this chapter. Figure 11.7 reports the Ac-Phe-OMe monomer and atomic nomenclature, together with the 3D-structures of the optimized structures of the monomer and dimer, and their energetics. As described in these works, the Ac-Phe-OMe monomer can adopt a $\beta_L(g^+)$ (g^+ in short notation), or $\beta_L(a)$ (a in short), or $\beta_L(g^-)$ (g^- in short) 3D-conformation, which differ by the orientation in space of the phenylalanine moiety with respect to the backbone chain (χ_1 dihedral angle).

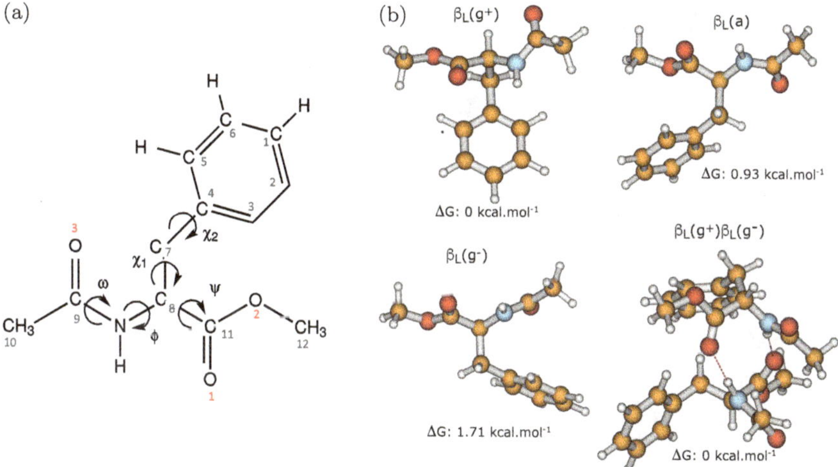

Figure 11.7. (a) Chemical scheme of the Ac-Phe-OMe monomer with the labels used in the text for the atoms and the definition of dihedral angles (ϕ, ψ, ω and χ) from the Ramachandran notation.[113] The caps "Ac" and "OMe" correspond, respectively, to an acetyl group CH_3–CO and to a O–CH_3 group. (b) Ac-Phe-OMe and (Ac-Phe-OMe)$_2$ peptides optimized geometries in the β_L organization. The "$\beta_L(g^+)$" conformation corresponds to the assigned one for the monomer (in our experimental conditions). The "$\beta_L(g^+)$-$\beta_L(g^-)$" dimer structure corresponds to the assigned one (in our experimental conditions). Free energies have been calculated at 50K at the BLYP-D3/6-311+G(d,p) level, see Tables 1–2. *Source:* Reproduced with permission from Ref. 44.

The g^+ and g^- conformers of the Ac-Phe-OMe monomer have for instance an active IR band at 276 cm^{-1} (g^+) and 265 cm^{-1} (g^-) in the THz domain, which degree of similarity we wish to quantify. Figure 11.8 shows the Vib-Graphs of these two modes. The blue color for the vertices stands for the bending internal motions that constitute these two modes. The labels of the atoms are the same in the two conformers. One hence recognizes vibrational modes made of C–O–C and C–C–O bending motions, N–C–C and C–N–C bending motions, along each monomeric chain. By applying the MCIS method to these two graphs, which is directly seen in Figure 11.8 by the common edges and nodes/vertices that are circled in red, one can immediately see that the Vib-Graph of the g^+ conformer is entirely included in the Vib-Graph of g^-. The degree of similarity measured between these two graphs is 0.53, which is a rather high degree of similarity. When g^+ is H-bonded to g^- to form a dimeric $g^+\cdots g^-$

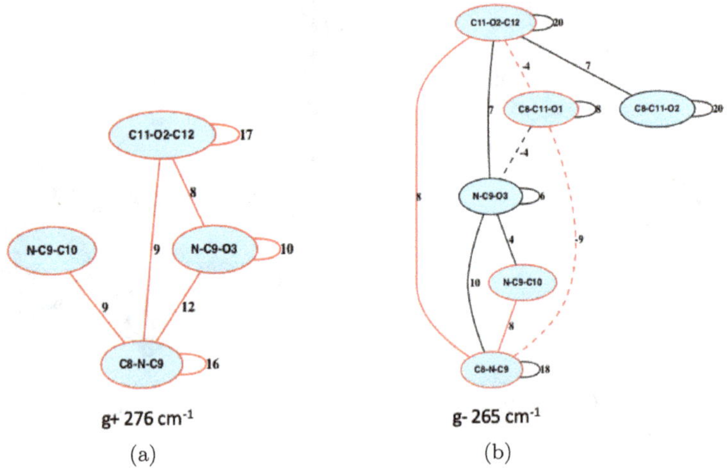

Figure 11.8. Illustration of the similarities of Vib-Graphs associated to (a) the 276 cm^{-1} mode of the g$^+$ conformer of gas phase Ac-Phe-OMe and (b) the 265 cm^{-1} mode of the g$^-$ monomer of Ac-Phe-OMe. Each Vib-Graph reports the quantitative decomposition of the associated IR mode (vertices: ICs; edges: couplings between ICs, see text), here made of bending motions in the THz domain. The vertices and edges highlighted in red are all the similarities measured between the two modes.

dimer, one can now measure how much the 276 cm^{-1} (g^+) and 265 cm^{-1} (g^-) modes of each monomer are preserved in the, *e.g.*, 257 cm^{-1} mode of the $g^+ \cdots g^-$ dimer. Figure 11.9 reports the Vib-Graphs with the MCIS sub-graphs highlighted in red. As is immediately obvious, the common sub-graphs obtained for each mode of the two monomers are very small in the Vib-Graph of the mode of the dimer, resulting in very low degrees of similarity, *i.e.*, respectively 0.12 and 0.14 for the modes of g^+ and g^-.

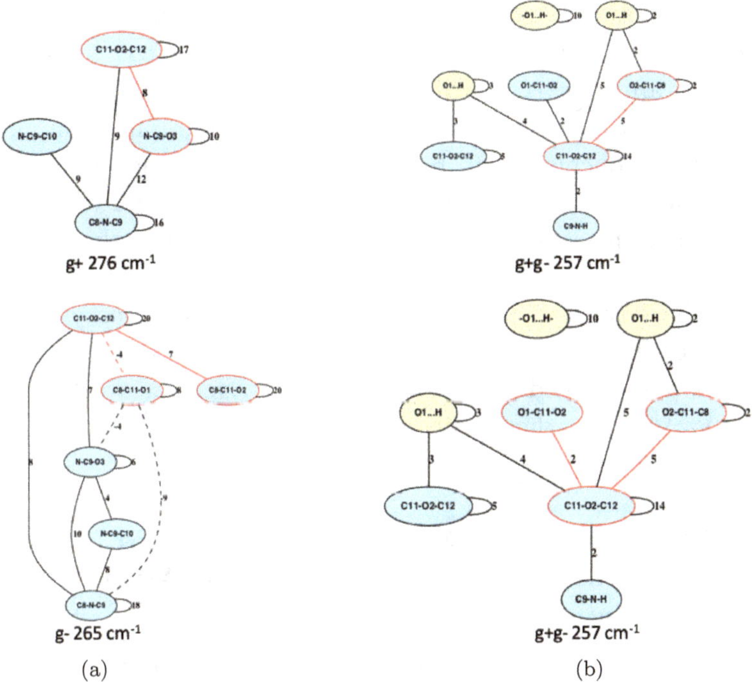

Figure 11.9. Illustration of the similarities of Vib-Graphs associated to the 276 cm^{-1} mode of g$^+$ (a, top) and the 265 cm^{-1} mode of g$^-$ (a, bottom) into the 257 cm^{-1} mode of the H-Bonded $g^+ \cdots g^-$ dimer (b). Each Vib-Graph reports the quantitative decomposition of the associated IR mode (vertices: ICs; edges: couplings between ICs, see text), here made of bending (blue vertices) and H-bonded (yellow) motions in the THz domain. The vertices and edges highlighted in red are all the similarities measured between the modes in the left side and the mode in the right side.

An alternate method for comparing graphs is to apply oriented similarity, which score is $\mathrm{sim}(G_i \rightarrow G_j) = \frac{|V_{ij}|+|E_{ij}|}{|V_i|+|E_i|}$. Applied to the comparison between the graphs presented in Figure 11.8, this measure leads to the Vib-Graph of g^+ being entirely included in the Vib-Graph of g^- with a score equal to 1, which means that the 276 cm^{-1} vibrational mode of g^+ is entirely included in the 265 cm^{-1} mode of g^-. This latter mode, as can be seen in the figure, has other internal components than the ones appearing in the mode of the g^+ conformer, which could be measured by the reverse calculation of similarity. If we apply to the graphs in Figure 11.9, values of 0.37 (similarity between the 276 cm^{-1} mode of g^+ and the 257 cm^{-1} mode of $g^+ \cdots g^-$) and 0.33 (mode of g^- compared to mode in $g^+ \cdots g^-$) are now obtained. This shows that the choice of the definition for the measure of similarity between graphs is important for the final conclusions to be drawn. One could then go one step further and include the weights known on the edges and vertices into the similarity measure. This has not been done yet in the presented works.

We have also developed algorithmic graph theory for the automatic recognition of structures over time in MD simulations, which is not detailed in this chapter. We refer the reader to our papers[11,114,115] for the details and applications in the gas phase, in liquids and at aqueous interfaces.

11.8. Illustration 1: IR-MPD Spectroscopy and Conformational Dynamics of Floppy Ala$_n$H$^+$ Protonated Peptides

The vibrational spectroscopy of gas phase protonated alanine peptides Ala$_n$H$^+$ is our first example that illustrates the crucial need to take into account the conformational dynamics of molecules into theoretical vibrational spectroscopy. See Refs. 67, 116–118 for more details on the theoretical spectroscopy of Ala$_2$H$^+$, Ala$_3$H$^+$ and Ala$_7$H$^+$, certain aspects of these works are highlighted hereafter. These papers also provide all details about the DFT-MD set-up.

Note that the theoretical dynamical spectra have been systematically averaged over several DFT-MD trajectories, in order to take into account the flexibility (described below) of the peptide into the spectroscopic signal.

Our room temperature trajectories on Ala_2H^+ showed a highly floppy molecule with a continuous isomerization dynamics between transA1 and transA2 conformers (see Figure 11.10(c)). This isomerization can be followed through the evolution with time of the dihedral angle $\Phi = C_2 - N - C_3 - C_4$ (top of Figure 11.10): this angle indeed distinguishes the two lower energy conformers of Ala_2H^+, *i.e.*, transA1 with $\Phi =198.1°$ and transA2 with $\Phi =284.2°$ (these values are from geometry optimisations at the BLYP/6-31G* level[118]). As observed in Figure 11.10, the continuous time evolution of Φ between $\sim 180°$ and $\sim 300°$ reveals the continuous exploration of the

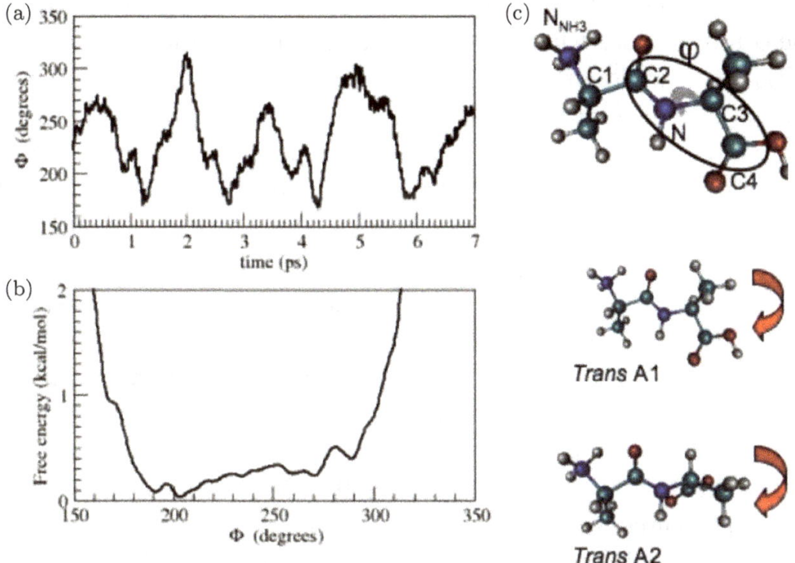

Figure 11.10. Ala_2H^+ gas phase dynamics from Ref. 118. (a) Time evolution of the dihedral angle $\Phi = C_2 - N - C_3 - C_4$ for a typical trajectory of the trans isomer at 300 K; (b) free energy profile along the Φ dihedral angle averaged over all trans trajectories. (c) atom labeling for Ala_2H^+ and schematic representations of transA1/transA2 conformers. *Source*: Reproduced with permission from Ref. 118.

two wells associated to transA1 and transA2 on the PES at room temperature. The free-energy curve along the dihedral coordinate Φ extracted from the dynamics ($G(\Phi) = -k_B T \ln P(\Phi)$, where $P(\Phi)$ is the probability of occurrence of a certain angle Φ in the course of the dynamics) presented in Figure 11.10(b) shows a rather flat well for angles between 190–300°, with an overall free energy difference of ~0.3 kcal/mol between the transA1 and transA2 conformations. There is thus no free energy barrier between transA1 and transA2 conformers of Ala_2H^+ at room temperature. It is interesting to note that the ~0.3 kcal/mol energy difference obtained in the dynamics is quite away from the ~2 kcal/mol energy difference between the two optimized conformers, and from the ~2.5 kcal/mol[119] energy difference between transA1 and the transA1–transA2 transition state, all computed at 0 K. This shows the importance of entropic contributions in structural equilibria, that are naturally taken into account in the dynamics. Beyond this isomerization, we found that the conformations where the N-terminal of the peptide is protonated (NH_3^+) are predominantly populated, with a strong hydrogen bond that is formed between NH_3^+ and the neighboring carbonyl group. Moreover, there is enough energy at room temperature for the NH_3^+ group to rotate and exchange the hydrogen atom that can be hydrogen bonded to the neighboring carbonyl. This is done a few times over the $10'$ ps trajectories achieved in that work. Back and forth proton transfers between NH_3^+ and C=O are also observed a few times. All of these results show the dynamicity of this rather small peptide at room temperature, which is the estimated temperature at which the IR-MPD spectroscopic experiments are performed.

The dynamical anharmonic infrared spectra of gas phase Ala_2H^+ in the 2800–4000 cm^{-1} and 1000–2000 cm^{-1} domains are respectively presented in Figures 11.11(a) and 11.11(b), and compared to the infrared multi-photon dissociation (IR-MPD) experiments from Refs. 119, 120. See Refs. 67, 118 for detailed description of the results and assignments. In the 3000–4000 cm^{-1} high frequency domain, we can see that our dynamical spectrum reproduces very well the IR-MPD spectrum. The band located at 3560 cm^{-1} is the O–H stretch of the C-terminal COOH group of the peptide, the

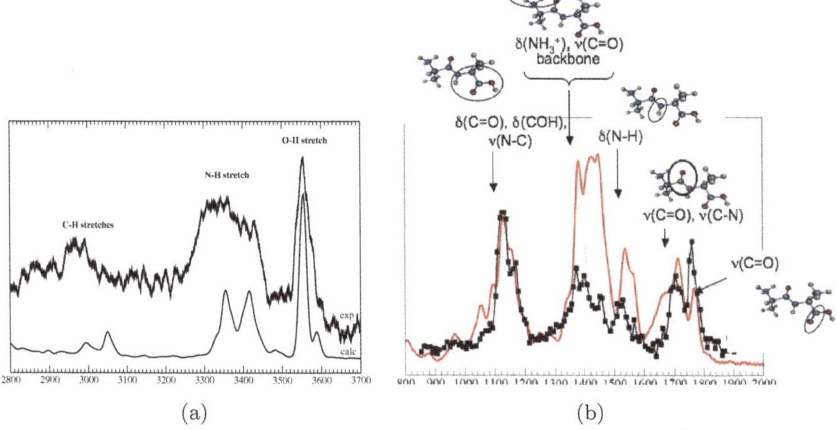

Figure 11.11. (a) Infrared spectrum of gas phase Ala_2H^+ in the trans form in the 2800–4000 cm^{-1} vibrational domain. Taken from Ref. 67. IR-MPD experiment (top), theoretical calculation from DFT-MD (bottom). (b) Infrared spectrum of gas phase Ala_2H^+ in the trans form in the 800–2000 cm^{-1} vibrational domain. Taken from Ref. 118. Squares: IR-MPD experiment taken from Ref. 119, Solid curve: dynamical spectrum from DFT-based molecular dynamics at ∼300 K. The band assignment deduced from the vibrational density of state (VDOS) analysis is illustrated with schemes on top of the bands. *Source*: Reproduced with permission from Refs. 67, 118.

calculation matches the experimental band in terms of band-position and band-shape. The shoulder located at ∼3590 cm^{-1} nicely reflects the feature also present in the experiment. The OH group is free of hydrogen bonding throughout the dynamics, thus the IR band located at high frequency for a free O–H. The broad band located between 3300 and 3500 cm^{-1} (including the small feature at ∼3490 cm^{-1}) is due to the N–H stretches from the amide N–H group and from the free N–H groups of the NH_3^+ N-terminal. The symmetric and antisymmetric stretches of the free N–H of the NH_3^+ can be seen at 3415 and 3370 cm^{-1} respectively. The backbone amide N–H stretch also appears on the higher frequency part of the band, around 3415 cm^{-1}. Although the calculated band nicely reflects the two parts which can be observed in the experimental band, it is nonetheless not broad enough in comparison to the experiment. As highlighted earlier in this chapter, the breadth of IR bands from molecular dynamics arises from a combination of temperature, conformational dynamics of the molecule, and anharmonicities (mode couplings, dipole anharmonicity, PES anharmonicity). The difference observed here is likely due to the temperature of the simulations not high enough to sample all these properties within the time-scale of the DFT-MD simulations.

Interestingly, the IR signature of the NH_3^+ hydrogen atom involved in the $N–H^+ \cdots O=C$ H-bond with the neighboring $C=O$ carbonyl throughout the trajectories is spread over the broad 2500–2850 cm^{-1} region of the spectrum, thus strongly down-shifted from the symmetric and antisymmetric modes of the free N–Hs of NH_3^+. Strong anharmonicity of the $N–H^+ \cdots O=C$ hydrogen bond are responsible for the displacement of this band to such lower frequencies, whereas mode-couplings and the intrinsic dynamics of the hydrogen bond at finite temperature are responsible for its ~350 cm^{-1} spread. The $N–H^+ \cdots O=C$ H-bond distance is found to fluctuate over time by 0.3–0.4 Å around its mean value, and there is enough energy for the NH_3^+ group to rotate and exchange the hydrogen atom which can be hydrogen bonded to the neighboring carbonyl. These H-bond dynamics therefore lead to vibrational signatures which are spread over a large frequency domain. At the time of our publications,[67, 118] there was no experimental signal recorded in the 2500–2850 cm^{-1} lower frequency domain in order to validate our theoretical spectral data. This has been later recorded by the group of Johnson[121] with tagging IR-PD experiments, confirming our dynamical spectral assignments. The double band around 3000–3100 cm^{-1} is due to combined C_α-H and C–H stretch modes of the methyl groups of the peptide. The dynamical bands are up-shifted by ~40 cm^{-1} from experiment, and the band spacing (~50 cm^{-1}) is slightly bigger than the experimental one (~30 cm^{-1}). It is nonetheless remarkable that our calculation predicts an intensity of the band so close to the experiment. The harmonic calculation[122] does not give such intensity to this mode.

In the 1000–2000 cm^{-1} mid-IR domain, the dynamical spectrum matches again extremely well the IR-MPD experiment, without the application of any scaling factor, see Figure 11.11(b).

Both experiment and calculation display three separate absorption domains, which are reproduced with a very good accuracy by the (average) dynamical spectrum in terms of both relative positions of the bands and band-widths. The relative positions of the different bands as well as fine details such as the two sub-bands of the 1100–1200 cm^{-1} domain or the two distinct IR active bands of the 1300–1600 cm^{-1} domain are very well reproduced too. The two peaks located at 1130 and 1150 cm^{-1} are due to stretching and bending movements of the N- and C-terminals of the peptide. The rather large band between 1340 and 1500 cm^{-1} corresponds to vibrations on the N-terminal side of the peptide, namely δ(N$-$H) bending motions of the NH$_3^+$ group coupled with the amide ν(C $=$ O) and skeletal ν(C$-$C) stretchings. The narrow band extending between 1500 and 1600 cm^{-1} is composed of the δ(N$-$H) Amide II motion coupled with ν(N$-$C$_2$) stretching of the backbone. The 1620–1800 cm^{-1} domain is composed of three separate peaks in the dynamical spectrum: the 1670 cm^{-1} peak comes from the Amide I movement of the C$_2$=O$_2$ carbonyl group, the 1720 cm^{-1} is due to the superposition of the two ν(C$_2$=O$_2$) and ν(C$_4$=O$_4$) stretchings, while the peak at 1770 cm^{-1} is due to C$_4$=O$_4$. The Amide I domain is clearly composed of three bands of high intensity in our calculation, while the experiment offers only two such bands, although a third sub-peak of far lower intensity may be distinguished in the 1660 cm^{-1} tail of this domain. Discrepancies in intensities between IR-MPD experiment and simulated dynamical spectrum can be traced back to the difference in the signals experimentally recorded (fragmentation yield) and calculated (linear IR absorption). See more discussions in Refs. 67, 118.

A further discussion on the sampling of the two lower energy conformers of Ala$_2$H$^+$ (transA1 and transA2 seen in Figure 11.10) becomes crucial for further understanding of the very good match of the two dynamical IR spectra with the experiments for this floppy peptide. As can be seen in Figure 11.10, the actual sampling over the two conformers can be assessed through the number of times the Φ dihedral angle value corresponding to each conformer (roughly Φ

around 198° for transA1 and 280° for transA2) has been obtained over the length of the trajectory. One can see that the majority of conformations sampled over time have angle values different from these two targets: in other words, the room temperature dynamics sample conformations outside the minima on the PES, which is also clear from the free energy profile in Figure 11.10 with the rather flat transA1–transA2 well. The direct consequence of this fact is that a Boltzmann-weighted spectrum of the harmonic spectra of transA1 and transA2 will not be useful for the interpretation of the IR-MPD experiment. These two conformers are indeed clearly not the ones of interest once temperature is included to the system, but rather all conformers explored during the continuous transA1–transA2 conformational dynamics are the ones that are participating to the final vibrational features. This property is naturally taken into account in MD simulations for vibrational spectra calculations. The weight of each conformation explored during the dynamics is indeed naturally taken into account in the final vibrational dynamical spectrum, by construction of the time-correlation of the dipole moment in, *e.g.*, Eq. (11.9) for the IR dynamical signal.

We continue the exploration of the dynamical vibrational signatures of protonated alanine peptides with the protonated Ala_3H^+. Prior to our dynamical investigations, three main structural families had been identified by Vaden *et al.*[120] and McMahon *et al.*[123] with geometry optimizations. These structures constitute (a) a "NH_2 family" in which the N-terminal of the tripeptide is NH_2 and COH, see Figure 11.12(a, top); (b) an "elongated NH_3^+ family" where the proton is located on the N-terminal NH_3^+ with and elongated peptide chain (similar to transA1/transA2 structures for Ala_2H^+), see Figure 11.12(a, middle); (c) and a "folded NH_3^+ family" in which the peptide chain is folded through a $NH_3^+ \cdots O{=}C$ H-bond, see Figure 11.12(a, bottom). Considering the initial photochemical scheme used for the production of the ions in the IR-MPD experimental set-up,[124] our DFT-MD trajectories have been accumulated at temperatures in the 380–530 K range, and have all been used for the final statistical average of the calculated

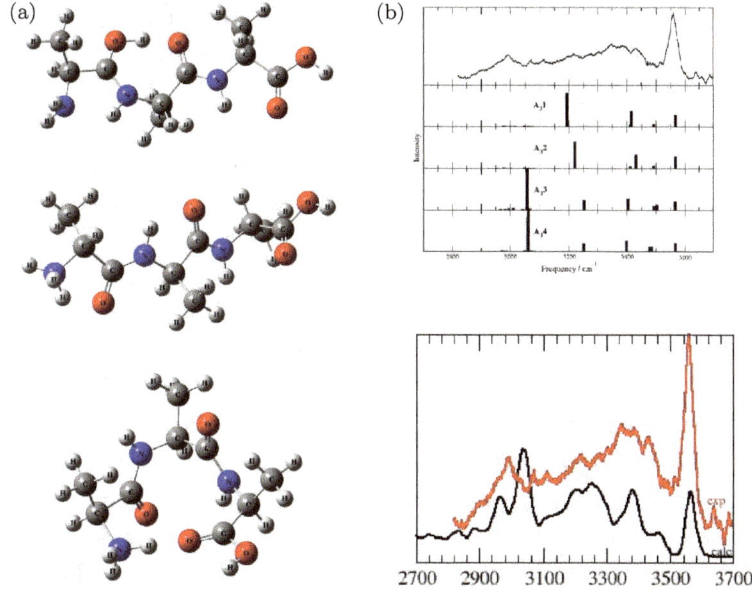

Figure 11.12. Infrared spectrum of gas phase Ala_3H^+ for the 2500–4000 cm^{-1}. Taken from Ref. 67. (a) Conformational families of Ala_3H^+ taken into account for the dynamics; (top) harmonic calculations from Ref. 120 and (bottom) comparison of IR-MPD experiment (exp) and dynamical spectrum from DFT-MD simulations (calc).[67] *Source*: Reproduced with permission from Ref. 67.

dynamical IR spectrum of Ala_3H^+ (each trajectory counts with the same weight in the final average). The production method of the ions in these experiments certainly does not follow a thermalized Boltzmann distribution, so that the "simple" average in our calculation is reasonable. Energy barriers between these conformational families had not been characterized in the optimization works.[120,123] Considering the changes into the conformations for these three families, we expect that these barriers are rather high with respect to 380–530 K temperatures (internal energies) of relevance for the IR-MPD experiments. We did not observe spontaneous conformational interconversion/isomerization in the dynamics at the lower temperatures, while this occurred for ∼500 K dynamics. The main dynamical behaviors seen in the trajectories are H-bonds dynamics and local geometrical reorganizations, including rotations of NH_3^+, CH_3 and COOH groups, which influence on the final vibrational features of the peptide are detailed below. See also Ref. 67 for details.

The dynamical spectrum of the protonated alanine tripeptide obtained from DFT-MD is presented in Figure 11.12(b, bottom) and compared to the IR-MPD experiment. It agrees remarkably well with the experimental spectrum in terms of band-positions and band-shapes, regardless of band intensities that are expected to give discrepancies with respect to the IR-MPD experimental band intensities, as previously discussed for Ala_2H^+.

The band located at 3560 cm^{-1} is due to the O–H stretch of the C-terminus, and is located identically for Ala_2H^+ and Ala_3H^+, corresponding to conformations with a similar free O–H group at the C-terminal of the peptide. Notably, the slight asymmetry of this band is correctly reproduced by the dynamical spectrum. The 3100–3500 cm^{-1} region can be separated into two parts. The 3300–3500 cm^{-1} frequency region is due to the stretches of the N–H groups that are not involved in hydrogen bonds along the dynamics. Depending on the tripeptide family, they are the symmetric and antisymmetric stretch of the N-terminal amine NH_2, the stretches of the free N–H of the NH_3^+, and the stretches of the C-terminal amide N–H. These motions give rise to similar vibrational features in this region, regardless of the underlying 3D-conformations of the peptide. The 3100–3300 cm^{-1} domain is uniquely due to the stretching motion of the N-terminal amide N–H group. Our trajectories show that this amide group can be weakly hydrogen bonded to the amine N-terminal in the NH_2 tripeptide family (through a distorted H-bond), and also weakly hydrogen bonded to the C-terminal carbonyl of the peptide in the NH_3^+ family (again, with a distorted H-bond). Despite being weak H-bonds, they are strong enough to induce the shift towards lower energy of the N-terminal N–H stretch in comparison to the positions of the free N–Hs, as very nicely captured boy the DFT-MD simulations. The overall shape of the N–H band is thus very well reproduced by our dynamical spectrum: the complex vibrational patterns of this large band are therefore a result of the local dynamics of the N–H groups in the various Ala_3H^+ families. Identically to Ala_2H^+, the 2800–3100 cm^{-1} domain of the spectrum of Ala_3H^+ is assigned to the C–H stretches arising from the methyls and the C_α–H groups of the tripeptide. All spectra calculated for all Ala_3H^+ families display two main intense bands located between 2900 and 3100 cm^{-1}, followed by a low intensity tail at lower frequency. Note that these bands are up-shifted by $\sim$40 cm^{-1} with respect to the experiment, as already observed for Ala_2H^+.

Also similar to the Ala_2H^+ peptide, the vibrational signatures of N–H$^+\cdots$O=C in the NH_3^+ families are strongly red-shifted to lower frequency compared to the other N–H stretches, and hence appear over the extended 2000–2800 cm^{-1} domain. Such a large 800 cm^{-1} spread is entirely due to the vibrational anharmonicities, mode-couplings and dynamics of the N–H$^+\cdots$O=C H-bond at finite temperature. Remarkably, the stretching of the protonated skeleton C–O–H in the NH_2 family is superimposed with the N–H$^+$ stretches of the NH_3^+ families. It is consequently not possible to distinguish both families using the stretching patterns of the H-bonded N–H$^+$ or C–O–H groups in the high frequency domain alone.

It is interesting to compare the scaled harmonic spectra of the three Ala_3H^+ peptides families with the dynamical anharmonic spectrum discussed above. This is done in Figure 11.12(b, top), where the scaled harmonic spectra are taken from Ref. 120 for the four lowest optimised geometries. As can be immediately observed, the N–H broad band of the experimental spectrum is systematically associated with only two main intense harmonic sharp bands, grossly separated by 200 and 150 cm^{-1}, respectively, for the NH_2 and NH_3^+ families; hints of a third band located close to the $\sim$3490 cm^{-1} experimental band can also be seen, with a very low intensity. With these harmonic normal modes, the highest frequencies are related to free N–H groups and the lowest frequencies to the hydrogen-bonded backbone N–H. Though these interpretations roughly agree with what has been obtained in our work from molecular dynamics simulations, one should fairly admit that the dynamical IR spectrum extracted from MD does offer vibrational details and comprehension of the 3100–3500 cm^{-1} N–H domain that are completely missed by the harmonic calculations. The interpretation given here from MD simulations, that the broad and complex N–H vibrational band comes from the intrinsic local dynamics of the N–H groups in the different conformers/isomers of Ala_3H^+, and the breaking/forming of these N–H hydrogen bonds, can only be achieved when performing molecular dynamics simulations. Scaled harmonic spectra, possibly including a Boltzmann-weighted final average, are not able to give such a broad dynamical band. Last but not least, as already mentioned for Ala_2H^+, the C–H harmonic modes predicted around 3000 cm^{-1} have no intensity in the harmonic spectra, while the anharmonic spectrum extracted from MD correctly predicts the intensity in this region.

We now turn to the larger Ala_7H^+ peptide. The initial structures for the trajectories were taken from Vaden *et al.*'s geometry optimisations.[120] The three lower energy conformers lie within less than 20 kJ/mol of energy, depending on the level of calculation and whether entropic effects are included.[120] These are globular and folded conformations of the peptide, with the N-terminal NH_3^+ being the central element for the folding of the peptide chain. The dynamical spectrum of Ala_7H^+ will be compared to the IR-MPD experimental signal from Ref. 120, that applies the same initial photochemical scheme for the production of Ala_7H^+ protonated peptides as the one for Ala_3H^+ in the previous section. It is thus more than likely that the three lower energy conformers of Ala_7H^+ would be equally produced and all participate to the final IR-MPD spectrum. We thus performed DFT-MD simulations, starting from these three conformers, similarly to the theoretical scheme employed for Ala_3H^+ above. Several trajectories are accumulated, all trajectories are included with the same weight in the final average theoretical dynamical IR spectrum that we present in Figure 11.13. See all details in Ref. 116.

The three lower energy optimized structures of Ala_7H^+ contain charge-solvating $NH^+ \rightarrow O{=}C$ hydrogen bonds involving one (denoted A_71), two (A_72) and three (A_73) NH^+ groups. A_73 has three $NH^+ \cdots O{=}C$ H-bonds in which two carbonyls of the chain and the carbonyl of the C-terminal (note the free hydroxyl) are

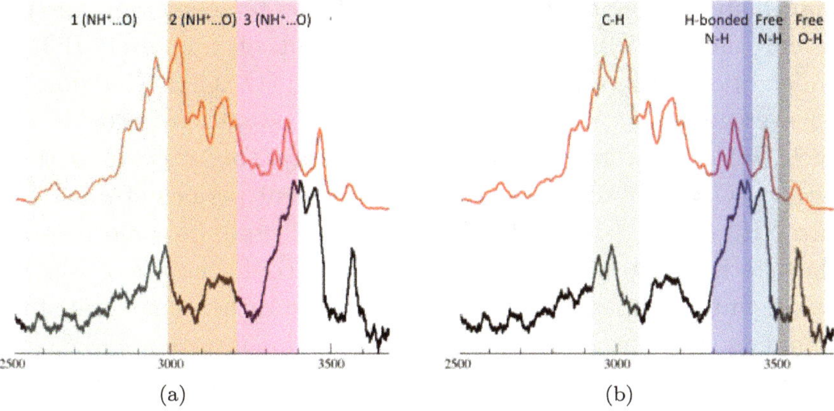

Figure 11.13. Experimental IR-MPD spectrum from Ref. 120 (bottom, black line) and dynamical spectrum (top, red line) obtained as an average over seven trajectories of Ala_7H^+ protonated peptide. Taken from Ref. 116. (a) has color-codings and labels explaining the assignments of the three vibrational domains in terms of average number of $NH^+ \cdots O{=}C$ H-bonds formed in Ala_7H^+ peptide. (b) displays the assignments in terms of stretchings of the O–H, N–H, N–H$^+$, and C–H (C_α–H and CH_3) motions. *Source*: Reproduced with permission from Ref. 116.

involved, $A_7 2$ has two $NH^+ \cdots O{=}C$ H-bonds in which one carbonyl of the chain and the carbonyl of the C-terminal are involved (note that the hydroxyl is H-bonded to a carbonyl of the chain), and $A_7 1$ has one single $NH^+ \cdots O{=}C$ H-bond in which one carbonyl of the chain is involved while the carboxylic acid is involved in two H-bonds with another part of the chain. Note that only $A_7 3$ has a free COOH hydroxyl. In total, seven trajectories have been accumulated at $\sim$350 K, two were initiated from $A_7 3$ optimized structure, two from $A_7 2$, and three from $A_7 1$, each one is accumulated for 3–5 ps time-lengths. The final dynamical spectrum is calculated as the "simple" average (*i.e.*, no weights applied) of the dynamical spectra generated from each trajectory.

Not unexpectedly, as already seen for Ala_2H^+ and Ala_3H^+ peptides, the average conformational view obtained for Ala_7H^+ at finite temperature differs from the frozen 0 K well-defined structures. While the 0 K temperature drives the formation of simultaneous multiple hydrogen bonds between the NH^+ charged groups and their immediate surroundings, dynamics at finite temperature does not favour such multiple H-bonds. Indeed, out of the seven BOMD trajectories, two provide one single $NH^+ \cdots O{=}C$ H-bond formed on average as the most relevant statistical event, five provide two such H-bonds on average, and only one trajectory provides a three $NH^+ \cdots O{=}C$ H-bonds situation over transient periods of time. The two $NH^+ \cdots O{=}C$ H-bonded situation ($A_7 2$ types of conformers) is thus statistically the most probable for the conformation of Ala_7H^+ at the finite temperature of 350 K (over the rather small statistical sampling that was achieved at the time of this investigation[116]).

Furthermore, these H-bonds can easily form and break. It is not too much surprising that maintaining three $NH_3^+ \cdots O{=}C$ H-bonds at finite temperature is rather difficult, as such a $A_7 3$ type of configuration provides a huge constraint over the whole geometry of the peptide. Once temperature and entropic effects are taken into account, the peptide conformation somehow does not comply easily to such constraint. Also, the (CO)OH C-terminal hydroxyl of Ala_7H^+ is on average not hydrogen bonded to its surrounding. Over the rather limited periods of time of the dynamics (5 ps) sampled for the spectroscopy, no remarkable structural reorganization of the peptide chain could be seen. The dynamical events of forming and breaking H-bonds are the sole events observed over such short timescales, which are in any case the sole events probed in the IR-MPD spectroscopy performed in the 2500–400 cm^{-1} high frequency domain.

Figure 11.13 presents the comparison between the IR-MPD experimental data extracted from Ref. 120 and the IR dynamical spectrum extracted from the seven trajectories of Ala_7H^+ peptide discussed above. It is clear that all experimental spectral features are obtained in the dynamical spectrum. As already emphasized in this text, when comparing IR-MPD and calculated spectra (harmonic or anharmonic spectra), one has to be cautious about discussing IR intensities, as they are not comparable between experiments and simulations. The band shapes and positions agree extremely well between the dynamical spectrum and the IR-MPD experiment. In particular, the experimental feature appearing at 3100–3230 cm^{-1}

for Ala_7H^+, that was not observed for the smaller peptidic chains, is immediately seen in the finite temperature dynamical spectrum of Ala_7H^+.

In Figure 11.13(b), bands have been colored according to the nature of the stretching assignments. These assignments are very similar to the ones obtained for the shorter peptides. The band located around 3560 cm^{-1} in the experiment and simulation and corresponding to the free O–H stretching match nicely, including the shoulder at ~3590 cm^{-1}. The broad band roughly extending between 3300 and 3480 cm^{-1} due to the N–H stretchings match nicely between experiment and simulation. The experimental band is clearly composed of at least two sub-bands, one between 3300 and 3400 cm^{-1}, the other between 3400 and 3480 cm^{-1}. The calculated spectrum shows two such sub-domains, with a more definite separation between these bands at 3400 cm^{-1}. These two spectral sub-domains are respectively due to non-H-bonded N–H (higher frequency part) and to H-bonded N–H (lower frequency part). As previously observed for Ala_3H^+, the broadness of this latter 3300–3400 cm^{-1} band as well as the supplementary sub-structures that can be seen within that band, arise from the H-bond dynamics, with sequences of breaking and reforming of H-bonds that occur at finite temperature. N–H$^+$ stretching signatures also show up within the tail of the 3300–3400 cm^{-1} band, arising from Ala_7H^+ conformations where the NH_3^+ group can be involved into *three* H-bonds on average. This participates to the broadening and "shoulder" of the 3300–3500 cm^{-1} band around 3300 cm^{-1}.

Figure 11.13(a) shows the assignment of the peaks in terms of the number of H-bonds made by the terminal NH_3^+ group in the conformations of Ala_7H^+ being explored over time. Very nicely, one can see that the supplementary band observed in the 3030–3300 cm^{-1} spectral domain of Ala_7H^+, which was absent for the smaller peptide chain lengths, is remarkably reproduced by the present dynamical spectrum, though slightly less broad than in the experiment. This band is entirely due to N–H$^+$ stretches arising from conformations where the NH_3^+ N-terminal is involved in *two* hydrogen bonds on average. Note that H-bonded O–H stretches also participate to this spectral domain. Although we have found that Ala_7H^+ peptide conformations with such H-bonded O–H are not statistically representative, they do exist, and these highly delocalized and anharmonic vibrational motions definitely participate to the intensity of the observed 3030–3300 cm^{-1} band. Remarkably, the two peaks seen at 3035 and 3065 cm^{-1} in the experiment are also present in the simulation, though with higher intensities and slightly blue-shifted from the experiment (roughly $+40$ cm^{-1}). These peaks are also due to $N-H^+$ and H-bonded O–H stretches, due to singly and doubly H-bonded NH_3^+ peptidic conformations. The 2500–3000 cm^{-1} experimental part of the spectrum is also remarkably reproduced by the dynamical spectrum. C–H stretchings from CH_3 and $C_\alpha H$ groups do participate to the experimental 2920–3000 cm^{-1} double band, as was already observed for the shorter peptide chains. This region is however more complex for Ala_7H^+, as the highly anharmonic vibrations of the *singly* H-bonded hydronium NH_3^+ conformations and H-bonded hydroxyl groups do also provide their vibrational signatures in that domain.

One can go one step further in the discussion of the $N-H^+ \cdots O=C$ vibrational signatures in the Ala_nH^+ series. Ala_2H^+, Ala_3H^+ and some of Ala_7H^+ conformers display one such H-bond on average at room temperature, which vibrational signature roughly appears in the 2000–3000 cm^{-1} domain. The $N-H^+ \cdots O=C$ H-bond is highly non-harmonic, its vibrational signature reflects its strength and degree of anharmonicity. Hence, we find that the strongest $N-H^+ \cdots O=C$ H-bonds are formed in the Ala_3H^+ peptide, where vibrational signatures down to 2000 cm^{-1} have been obtained, while the two other peptides have signatures above 2500 cm^{-1}. The IR spectrum of the Ala_7H^+ peptide very nicely shows that the strength of the $N-H^+ \cdots O=C$ H-bond is decreasing with the increase in the number of such H-bonds simultaneously formed. Hence, 1-H-bond conformers have $N-H^+ \cdots O = C$ signatures in the 2500–3000 cm^{-1}, while 2-H-bonds conformers have signatures in the 3000–3300 cm^{-1}, and 3-H-bonds conformers in the 3300–3400 cm^{-1}. In this later case, the $N-H^+ \cdots O=C$ vibrational signatures overlap with the ones arising from the amide N–H (neutral) H-bonded groups.

It is also clear from the analyses of the dynamics of Ala_nH^+ peptides that the knowledge of finite temperature properties is mandatory in order to characterize vibrational signatures of floppy molecules. Entropic effects are naturally taken into account in the dynamics, with consequences on conformational equilibria. Anharmonic vibrational effects such as anharmonic dipoles, anharmonicities from H-bonds, mode couplings, anharmonic motions on the PES, are naturally taken into account in the dynamics and therefore in the dynamical spectroscopy calculation, without *a priori* knowledge. Of further relevance for finite temperature dynamical spectroscopy is the formation of simultaneous H-bonds. They indeed might not be statistically relevant as shown here for the Ala_7H^+ peptide. While the 0 K frozen picture provides conformers in which the NH_3^+ group is involved into three simultaneous H-bonds as the most stable structures, the finite temperature shows that only two- and one-H-bonded structures are statistically relevant. Once this conformational property is taken into account, the final IR dynamical vibrational spectrum is in very good agreement with the experiment, and can be used in order to provide definitive answers on the structure-vibrational features relationships. Without this step, one can be left without definitive conformational assignments.

As was seen with these examples of di-, tri- and hepta-peptides the high frequency and mid-IR domains are directly informative on the H-bonded/non-H-bonded contributions to 3D-structures and on the dynamics of these H-bonds. For the peptides investigated above, the N–H, N–H$^+$ and O–H finite temperature dynamics were giving information on the 3D-structures adopted by these gas phase molecules. However, as could be seen with Ala_7H^+, this information is not sufficient for getting a definitive knowledge on the actual structure of the chain. The information on the globular organization of the Ala_7H^+ peptidic chain could be gained only by the geometry optimizations and energy ranking between the conformations (trusting that the conformational sampling has been complete). The dynamics confirmed that there were no changes in this globular state at the chosen temperature (at least, over the time-scale of the DFT-MD).

To get a direct knowledge on the 3D-structure of the peptide chain, one has to probe another spectral domain that directly provides information on the motions of the chain. The signatures of dihedral motions of the chain as well as the ones of H-bonds are in the THz domain below 800 cm^{-1}. We illustrate below how DFT-MD dynamical spectroscopy and THz-IR match extremely well and how, combined together, they can give relevant information on the conformation adopted by the chains of peptides.

11.9. Illustration 2: THz-IR Spectroscopy of Peptides and Mapping their Vibrational Motions

The 3D-structure of a series of neutral dipeptides have been characterized by THz-IR spectroscopy, combining dynamical DFT-MD spectroscopy and experimental IR-UV ion-dip spectroscopy. The principles for the experiments are detailed in Refs. 44,68, more details on our investigations on the dipeptides are reported in Refs. 60,61, other molecular systems can be found in Refs. 59, 62, and some reviews[44, 68] summarize our efforts in combining theoretical and experimental spectroscopy in the far IR/THz domain. These papers contain all details about the DFT-MD set-ups. We just emphasize here the low temperature of the spectroscopic experiments, therefore our DFT-MD simulations are performed at $\sim$50 K of temperature (see a discussion on this topic in Ref. 44), and all dynamical data are averaged over a series of DFT-MD trajectories to account for statistical sampling. We illustrate below some of the main results for the series of Ac-Phe-AA-NH$_2$ dipeptides, where 'Phe' is the phenylalanine that is used as the UV chromophore for the selection of the conformers in the IR-UV ion-dip experiments prior to the IR probing, "AA" stands for one of the following amino acids Gly, Ala, Pro, Cys, Ser, Val, and "Ac"/NH$_2$' are the two capped extremities. Figure 11.14 reports the nomenclature employed for the γ- and β-turns of these peptides. The dipeptide series presented here with THz-IR spectroscopy probing had been chosen because their 3D-structures had been well characterized in the literature using the "classical" 1000–4000 cm^{-1} spectral signatures.[125–127] Our goal in

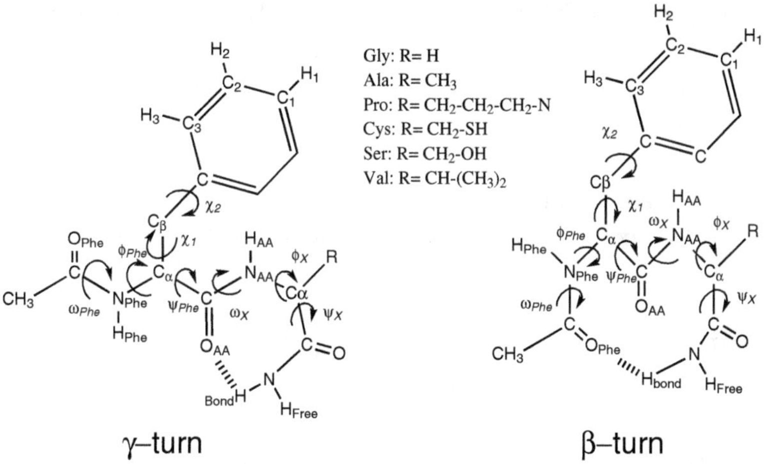

Figure 11.14.   Schemes of Ac-Phe-AA-NH$_2$ dipeptide for γ and β conformations with the labels used in the text for the atoms and dihedral angles (ϕ, ψ, ω and χ). Residue "AA" stands for Glycine (Gly), Alanine (Ala), Proline (Pro), Cysteine (Cys), Serine (Ser) and Valine (Val). *Source*: Reproduced with permission from Ref. 60.

using THz spectroscopy was to confirm these structural assignments by adding up spectral analyses in an uncharted domain.

Our analyses hence show how the THz-IR signatures carry direct information on the 3D-structure of peptidic chains, *i.e.*, their folding and the specificities of the folding into a γ- *vs* β-turn with, respectively, C$_7$ or C$_{10}$ H-bonded rings for the turn. In the particular case of Ac-Phe-Cys-NH$_2$ peptide, literature could not give a definitive conformational assignment,[126, 127] which was achieved thanks to the THz-IR signatures. The further strength of the theoretical analyses is the capability to provide a map of the motions that are responsible for the spectroscopic signatures recorded in the THz domain. This map can then be used to assign 3D-structures without the need for direct DFT-MD simulations (or for any calculation in general). The map of vibrational assignments is obtained from DFT-MD dynamical non-harmonic spectra, thus taking into account temperature effects, anharmonicities, including modes couplings, anharmonic dipole moments. Experiments are exploring the 100–800 cm^{-1} region (3–24 THz), the simulations also provide the signatures below 100 cm^{-1}.

All far-infrared/THz spectra measured experimentally and calculated from the DFT-MD simulations are reported in Figure 11.15 for the dipeptide series. The dipeptides spectra are organized from top to bottom, with the order of the γ-turn A1 conformation of Ac-Phe-Val-NH$_2$, the β-turn of Ac-Phe-Cys-NH$_2$, γ-turns of Ac-Phe-Gly-NH$_2$, Ac-Phe-Ala-NH$_2$, (A2) Ac-Phe-Val-NH$_2$, Ac-Phe-Ser-NH$_2$, Ac-Phe-Cys-NH$_2$, Ac-Phe-Pro-NH$_2$ and the β-turn of Ac-Phe-Pro-NH$_2$. We report the 0–800 cm^{-1} THz spectra of these peptides, split in two spectral regions for clarity reasons, left for 0–400 cm^{-1} (with a scaling of intensity), and right for 400–800 cm^{-1}. Experimental spectra are plotted in color while the theoretical anharmonic dynamical DFT-MD spectra are plotted in black, just below each experiment.

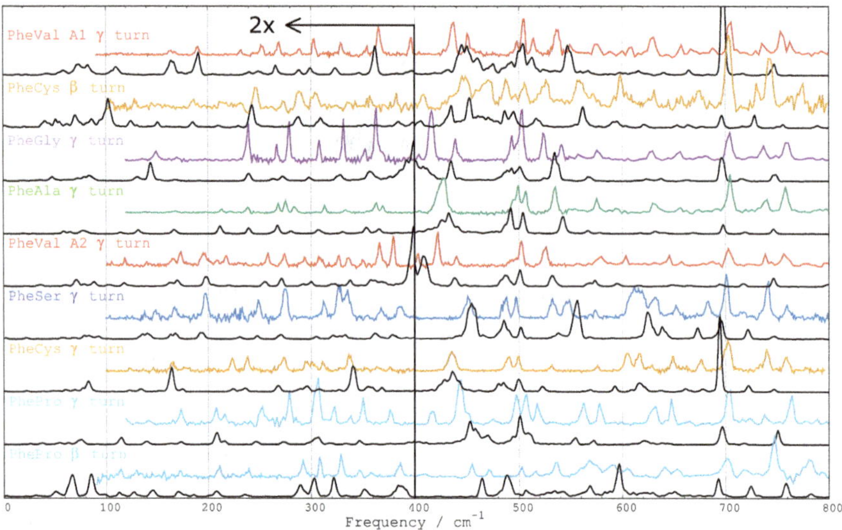

Figure 11.15. Experimental IR-UV ion dip spectra of the Ac-Phe-AA-NH$_2$ dipeptide series (in color) compared with the calculated dynamical DFT-MD IR spectra (in black). All the spectra intensities have been multiplied by a factor of 2 below 400 cm^{-1} for the sake of clarity. See Figure 11.14 for the nomenclature. *Source*: Reproduced with permission from Ref. 60.

The first remarkable result is the average 6 cm^{-1} difference between experimental and theoretical peaks positions, calculated over all nine dipeptides and over all peaks in the 90–800 cm^{-1} range. A maximum deviation from experiment of about 20 cm^{-1} is systematically obtained (in all systems) for the peak experimentally located at 740 cm^{-1} and found around 720 cm^{-1} in the theoretical spectra. This peak is part of a triplet that is not conformational selective, as it is present for all 3D-structures. This excellent match between theory and experiment validates the previous 3D-structural assignments[125–127] from literature using other spectral domains.

It is also remarkable that the theoretical dynamical THz-IR bands are found as well resolved as the experimental ones, hence showing the same number of absorption peaks, the same narrow band-widths and the same band-shapes, also including subtle shoulders in some broader peaks. There are some exceptions in the 400–500 cm^{-1} with less well resolved peaks in the calculations. Although experimental band-intensities are globally rather well reproduced in our dynamical spectra (without any adjustable parameter and/or model), some peaks however display either too low or too high intensities in comparison to the experiments. See for instance the peaks located around 700 cm^{-1} for the A1 conformation of the Ac-Phe-Val-NH$_2$ dipeptide or the γ-turn of Ac-Phe-Cys-NH$_2$. As already emphasized in this chapter, band-intensities from MD trajectories are sensitive to temperature, length of the trajectories and number of trajectories, and presumably some (or all) of these elements participate here to such deviations from experiments. They, however, do not prevent a fair assignment and interpretation of the experimental spectra. Note also that relative intensities between peaks are usually well reproduced by the dynamical spectra.

The great advantage of DFT-MD spectroscopy is to be able to provide the molecular assignment of the vibrational peaks. Using the dipeptide series together with other molecular systems characterized in the THz domain (including phenol derivatives, a capped alanine hexa-peptide, Ac-Phe-OMe and H-bonded dimers,

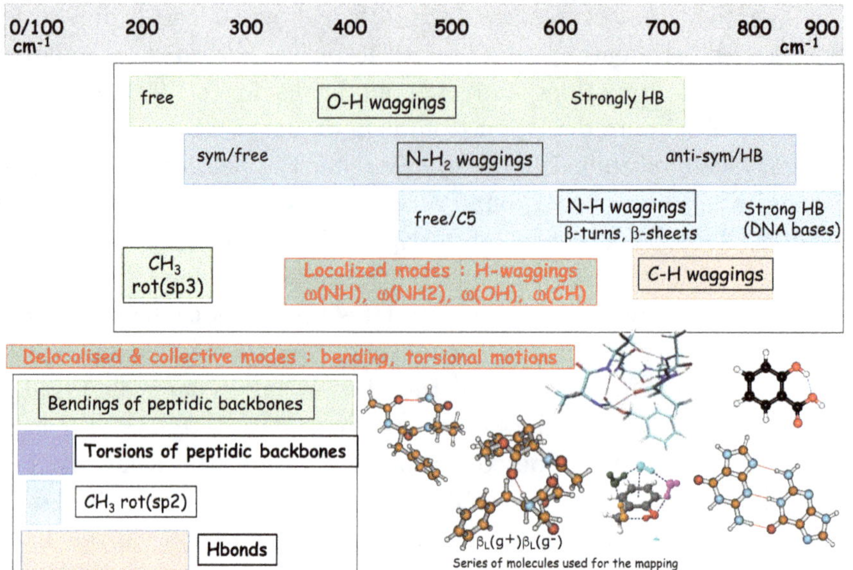

Figure 11.16.   Summary of the mapping of the vibrational motions that are found active in the THz domain below 900 cm^{-1} obtained from DFT-MD-IR spectra calculations and their decomposition over the series of molecules illustrated at the bottom-right of the figure.

DNA base-pairs),[44, 59, 62, 68] we have established a general map of the vibrational motions in the THz domain. This map is summarized in Figure 11.16. At the time of this work, the mapping was obtained with the ICDOS method described earlier in this chapter. Using the newly developed Vibrational-Graph method described in this chapter, one could obtain better interpretation of the delocalized and collective modes. It has not been done yet. There are localized and delocalized/collective vibrational motions in the THz spectral domain. The localized motions are waggings (out-of-plane movements) of the C–H, N–H, N–H$_2$, O–H and S–H groups in the gas phase molecules that we used for the vibrational mapping. Rotational and rocking motions of, *e.g.*, CH$_3$ groups are also catalogued into localized motions. The collective and delocalized vibrational motions are involving torsional motions of the molecular backbones as well as H-bond motions.

Whenever there are wagging motions of C–H groups as it occurs in the phenylalanine ring of the Ac-Phe-AA-NH$_2$ dipeptides, $\omega(C-H)_{\text{Phe}}$ have their vibrational signatures in the 700–800 cm^{-1} domain, with three absorption bands respectively located at 700, 730 and 740 cm^{-1}. Despite the fact that the phenylalanine aromatic ring can be found in three different environments within the Ac-Phe-AA-NH$_2$ dipeptides investigated by us (*i.e.*, interacting with the amide N–H group of the "AA" amino acid in some of the γ-turn geometries, or with a weak interaction in the β-turn conformation of Ac-Phe-Cys-NH$_2$, or with a π-interaction with proline in the γ- and β-turn conformations of Ac-Phe-Pro-NH$_2$), the same spectral signatures for the aromatic C–H wagging motions are found. They are therefore non-sensitive to the environmental details. Note that equivalent spectral signatures are also found in systems containing a phenol ring with equivalent substitution.[128]

Presumably $\omega(N-H)$ wagging motions are the most interesting to map. They are showing up in the 400–900 cm^{-1} domain with the signature of free $\omega(N-H)$ appearing in the lower 400 cm^{-1} range while $\omega(N-H)$ for strongly H-bonded N–H groups appear at the bluest side around 800–900 cm^{-1} (obtained for DNA base-pairs in the sample of biomolecules that we were mapping). For wagging motions in the THz, the weaker the H-bond of the group the lower the frequency, the stronger the H-bond the more blue-shifted position of the frequency of the motion. This is opposite to the stretching motions in the 3000–4000 cm^{-1} domain. The $\omega(N-H)$ of the H-bonded amide N–H group of the Ac-Phe-AA-NH$_2$ dipeptide series (Amide V motion) are showing up roughly between 500 and 700 cm^{-1}, for weak to medium H-bonded groups. In all the series of the dipeptides, the $\omega(N-H)_{\text{Phe}}$ Amide V is systematically located between 470 and 510 cm^{-1}, *i.e.*, with a common signature within 40 cm^{-1} interval. This is not too surprising as this N–H$_{\text{Phe}}$ is always involved in a weak C5 interaction with the C=O$_{\text{Phe}}$. The 470–510 cm^{-1} region can therefore be used as a reference for $\omega(N-H)$ wagging signatures of N–H amide groups engaged in a C5 backbone interaction. When this N–H$_{\text{Phe}}$ is free of interaction, as is found in the β-turn of the Ac-Phe-Cys-NH$_2$, ω(N$-$H)$_{\text{Phe}}$ is found around 410–490 cm^{-1}. This certainly provides an absolute reference for ω(N$-$H) of backbone free N$-$H groups.

ω(N$-$H)$_{AA}$ has various signatures between 400 and 640 cm^{-1} depending on the chemical composition of the residue and interactions of this backbone N$-$H with its surrounding. For Ac-Phe-Gly-NH$_2$, Ac-Phe-Ala-NH$_2$, and the two Ac-Phe-Val-NH$_2$ γ-turn structures, there is for instance an interaction between N$-$H$_{AA}$ and the Phe aromatic ring. The associated wagging signatures are therefore located between 530 and 550 cm^{-1} for these dipeptides. This represents up to a maximum of $\simeq$90 cm^{-1} blue-shift from the free ω(N$-$H)$_{\text{Phe}}$ wagging signature. For the Ac-Phe-Ser-NH$_2$ and Ac-Phe-Cys-NH$_2$ γ-turns, the same weak phenylalanine/N$-$H$_{AA}$ interaction is present. However, due to the different chemical nature of the side chain containing OH or SH groups respectively, a supplementary blue-shift of ω(N$-$H) is observable, which is now located at 555 and 620 cm^{-1} for Ac-Phe-Ser-NH$_2$ and Ac-Phe-Cys-NH$_2$, respectively. In the β-turn conformation of Ac-Phe-Cys-NH$_2$, N$-$H$_{AA}$ is only weakly interacting with the aromatic ring, the corresponding wagging ω(N$-$H)$_{AA}$ is found around 565 cm^{-1}.

Very nicely, the Amide V mode is hence found to be diagnosis for the local environment around the N$-$H backbone group, and can be used as a valuable tool for conformational assignment of gas phase peptides, thus providing information that are equivalent to the ones given in the 3000–4000 cm^{-1} N$-$H stretching range. As seen above, each backbone amide N$-$H group gives rise to a distinct wagging signature in the 400–700 cm^{-1} domain, reflecting the backbone interactions with the immediate surrounding. This is reminiscent of the separate ν(N$-$H) stretching signatures observed in other works[126, 129–131] in the 3000–4000 cm^{-1} domain under cold and isolated conditions in either a cold ion trap or a molecular beam environment. Although here the separate ω(N$-$H) peaks are slightly broader than the ones observed for ν(N$-$H), one can still distinguish one peak per N$-$H group without any ambiguity. One would therefore be able to experimentally separate ω(N$-$H) signatures by applying isotopic substitution, as done in the $\nu(N-H)$ stretching region[129] or more recently for the equivalent $\omega(O-H)$ wagging.[128]

The NH_2 terminal moiety in the Ac-Phe-AA-NH_2 dipeptide series is very specific of these neutral peptides, inducing their folding:[132] one N–H of this moiety is involved in one H-bond with one backbone carbonyl group in the β-turn geometries, while one N–H H–bonds to the $C=O_{AA}$ in the γ-turn structures. The second N–H of NH_2 is always free of H-bond. The NH_2 moiety has a symmetric and an antisymmetric $\omega(N-H)$ out-of-plane wagging motion. A strong correlation is observed between the $NH_2 \cdots O=C$ H-bond length and the frequency of the asymmetric out-of-plane wagging motion. The stronger the hydrogen bond the more blue-shifted the wagging signature. Such correlation was also found for the N–H backbone Amide V mode dissected above, and similar blue-shifts have also been reported for OH$\cdots$OH H-bonds.[133] We found asymmetric $\omega(NH_2)$ wagging in between $\sim$620 cm^{-1} for the weakest hydrogen bond formed in the γ-turn conformer A1 of Ac-Phe-Val-NH_2 and $\sim$695 cm^{-1} for the stronger H-bonds formed in the γ-turn and β-turn structures of Ac-Phe-Pro-NH_2. Such correlation between $\omega(NH_2)$ and $NH_2 \cdots O=C$ H-bond strength is not observed for the symmetric wagging motion. Signatures of this motion are systematically found in the 400–510 cm^{-1} interval. $\omega_{sym}(NH_2)$ is however found at 595 cm^{-1} for the β-turn conformation of Ac-Phe-Pro-NH_2.

$\omega(O-H)$ wagging motions were found between 200 cm^{-1} for free O–H groups (phenol) and 700 cm^{-1} for strongly H-bonded O–Hs in phenol derivatives. Again, the stronger the H-bond, the more blue-shifted frequency of the associated motion. For the dipeptide series, only Ac-Phe-Ser-NH_2 has one O–H group with $\omega(O-H) = 557$ cm^{-1}, which is hence measuring a medium strength H-bond. We found two spectral features for the CH_3 hindered rotational motions related to the two methyl populations in the dipeptide series. Signatures of the rotation of the terminal CH_3 group are found below 100 cm^{-1}, and are systematically coupled with delocalized modes that will be described afterwards. For residues alanine and valine in Ac-Phe-Ala-NH$_2$ and Ac-Phe-Val-NH$_2$, rotation of the methyl group is found at 220 and 242 cm^{-1} (Ala) and at 221, 239, 257 cm^{-1} (Val). Distinct rotational signatures for side-chain methyl groups and backbone methyl groups have already been observed for peptides in the condensed phase[134] within the same vibrational range. Interestingly, we clearly observe that the terminal methyl group of the dipeptide series has more amplitude in its rotational motion because of its sp^2 carbon hybridization. When the methyl is part of the peptide residues, its carbon being now sp^3 hybridized, a different environment is provided to the hydrogen atoms when the methyl group is rotating. The methyl is thus more constrained in its rotation leading to a rotational motion at higher frequency.

Collective vibrational modes delocalized over the peptide backbone were found over the full far-IR/THz range. However, their intensity dominates the range 0–400 cm^{-1}. The most prominent motions are the bending (100–400 cm^{-1}) and torsional (0–100 cm^{-1}) large amplitude and collective motions, which are coupled and delocalized over the peptide backbone. For the dipeptides presented in Ref. 60, these motions were followed by the ICDOS analysis of backbone dihedral angles. Multiple signatures of each internal coordinate were observed in the spectra, with several dihedral angles providing overlapping spectral signatures (they were observed within the same frequency ranges). This nicely indicates that the modes are collective and delocalized over the peptide backbone. Although it is not possible to determine the percentage of participation of each internal motion to the final mode from the ICDOS method, as emphasized earlier in this chapter, the ICDOS do reveal how many angles and dihedral angles are involved in a given vibrational peak, which is a measure of the collective nature of the vibrational mode. Furthermore, principle component analysis (PCA) was used to indicate which internal coordinates are connected and which do contribute to specific observed peaks in the experimental IR spectrum.[60] Both methods are however limited as they do not provide a quantitative assessment of each of the motions into the final vibrational delocalized mode. With the recently developed

Vib-Graph method described earlier in this chapter, delocalized collective vibrational modes can now be fully quantified, including the measure of the couplings between the participating internal motions. This analysis has not been done yet for the vibrational mapping presented here.

Hydrogen bond signatures are at the core of the THz signatures, see for instance the works of the Havenith and Marx groups for THz spectroscopy in the aqueous phase.[135–137] Based on ICDOS analyses, H-bond signatures were found in the 0–250 cm^{-1} domain. H-bond motions are highly coupled to other collective motions. To fully characterize the H-bond modes and their couplings to other motions, there again, the newly developed Vib-Graph method should be employed. This is where our current spectral investigations in the THz domain are now heading.

11.10. Illustration 3: THz-IR Spectroscopy and Conformational Assignment of the (Ac-Phe-OMe)$_2$ β-Sheet Model — How Vib-Graphs Provide a Quantitative View of Collective Anharmonic Modes

The next illustration of finite temperature DFT-MD simulations for anharmonic IR spectroscopy in the gas phase is taken from our work[44] where we have revealed the (elusive) 3D-structure of the (Ac-Phe-OMe)$_2$ dimer structure, which is a minimal model for β-sheet structures of peptides in general. As demonstrated in this work, definitely finding the 3D-structure of the (Ac-Phe-OMe)$_2$ H-bonded dimer requires the spectroscopic IR signatures in the THz domain, and especially the signatures that are probed below 400 cm^{-1}. Figure 11.17 reports the IR-UV ion dip THz-IR experimental signal and the six different DFT-MD THz-IR spectra calculated for the six possible dimeric forms of (Ac-Phe-OMe)$_2$. These structures are constructed based on the combination between the g^+ and g^- 3D-conformations of the Ac-Phe-OMe monomer. Despite the fact that we showed in Ref. 44 that the 3D-structure adopted by the monomer in the experimental conditions

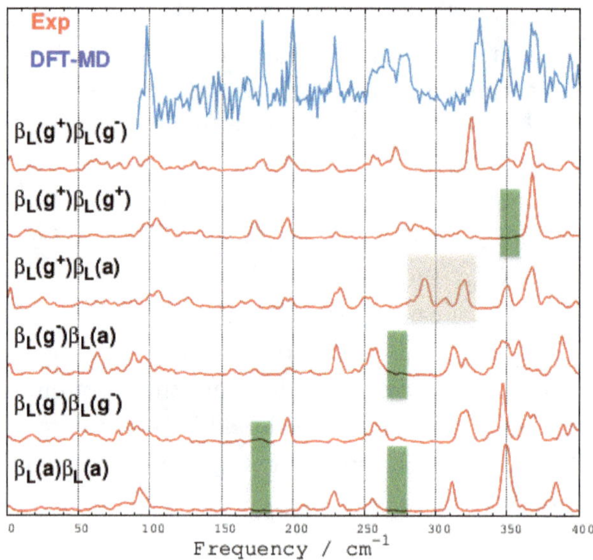

Figure 11.17. Experimental IR-UV ion dip spectrum of the gas phase (Ac-Phe-OMe)$_2$ β-sheet (in blue) compared with the calculated dynamical DFT-MD IR spectra (in red) of six possible isomers of the gas phase (Ac-Phe-OMe)$_2$. See Figure 11.7 for the chemical description of (Ac-Phe-OMe)$_{1,2}$. The green boxes highlight missing spectral features in the theoretical spectra that are however present in the experimental spectrum. The orange boxes highlight extra spectral features that are present in the theoretical spectra but are absent in the experimental spectrum. *Source*: Reproduced with permission from Ref. 44.

of the ion-dip experiment done by our collaborators is the g^+ (see Figure 11.7), the monomers could isomerize while forming the dimer into a different β_L orientation as a result of intermolecular interactions. Alternatively, all possible combinations of dimers might be produced during the initial stage of the experiment and only the lowest free energy dimer will remain once passing through the supersonic expansion, following thermodynamic rules. Looking at the relative electronic and free energies of the six possible (Ac-Phe-OMe)$_2$ β-sheet conformations built on the β_L orientations of the monomers, one could also see that $\beta_L(g^+)$-$\beta_L(g^-)$ and $\beta_L(g^+)$-$\beta_L(g^+)$ conformers are quasi-isoenergetic, while the $\beta_L(g^+)$-$\beta_L(a)$ conformer is only $\sim$1.1–1.5 kcal/mol higher in energy. All six possible conformers were thus found within less than 3 kcal/mol of energy.[44]

No search for energy barriers has been conducted. Therefore, all these dimeric conformations could be relevant.

Only vibrational spectroscopy will be able to decipher which 3D-structure(s) is(are) produced and probed in the experimental conditions. Note that the pioneering work by Gerhards and coworkers[138–140] took only the symmetrical $\beta_L(g^+)$-$\beta_L(g^+)$ and $\beta_L(a)$-$\beta_L(a)$ structures into account, because the $\beta_L(g^+)$ and $\beta_L(a)$ monomeric conformations were found lower in energy in these works. We thus investigated the six possible β_L-β_L structures through DFT-MD simulations and Figure 11.17 compares the six dynamical far IR/THz spectra (red lines) with the IR-UV ion dip experiment (blue line). Each theoretical spectrum is generated as an average over three separate trajectories (each one with different initial conditions for the dynamics). We focus here only on the 100–400 cm^{-1} spectral signatures, as the monomer spectroscopy showed that this is the domain that is the most conformational selective below 800 cm^{-1}.[44]

Analyzing the number of bands in the 100–400 cm^{-1} domain, their absolute and relative positions, their band-shapes and intensities with respect to the well resolved bands in the experimental spectrum, one can conclude that the DFT-MD theoretical spectrum of the $\beta_L(g^+)$-$\beta_L(g^-)$ isomer of the gas phase (Ac-Phe-OMe)$_2$ β-sheet structure provides a precise account of the experimental features. The 300–400 cm^{-1} triple-bands are in excellent agreement with the experiment (one however notes that the middle-band intensity appears too low), the 250–300 cm^{-1} double-band is not present with such accuracy in any other isomer (band-positions, shapes and intensities), the 150–250 cm^{-1} triple-bands are just located on-spot (although too low in absolute intensities), isomer $\beta_L(g^+)$-$\beta_L(g^+)$ being the only other isomer displaying the same triple-bands. The 100 cm^{-1} band seems less conformer-selective as all six isomers spectra display this particular band, within a few cm^{-1}.

The bands highlighted in colors in Figure 11.17 are the ones that are either missing in the theoretical spectra in order to match the experiment (green boxes) or the ones that are appearing in the theoretical spectra but are not present at these particular frequencies in the experiment (orange boxes). Strikingly, only the theoretical spectrum of $\beta_L(\mathrm{g}^+)$-$\beta_L(\mathrm{g}^-)$ does not have any of these boxes highlighted. We can hence unambiguously assign isomer $\beta_L(\mathrm{g}^+)$-$\beta_L(\mathrm{g}^-)$ to the (Ac-Phe-OMe)$_2$ β-sheet. This is also the lowest energy dimeric conformer.

Similarly to the Ac-Phe-OMe monomer,[44] the 100–400 cm^{-1} spectral range was found the most conformational selective. This is due to the nature of the vibrational modes in this domain. Delocalised modes over the whole backbone and side chain residue are found in this domain, mainly due to the combination of bendings and dihedrals motions. In particular, we found the dihedral angle χ_1 (Figure 11.7) involved in most of the modes below 200 cm^{-1}. This internal coordinate directly probes the rotation of the phenyl ring (rotation that would induce conformational transitions between the g^+, a and g^- orientations of AC-Phe-OMe) and we hence expect the signatures of this internal coordinate to be indeed relevant for conformational identification.

We would like to illustrate the strength of the Vib-Graphs (described earlier in this chapter) for extracting the quantitative

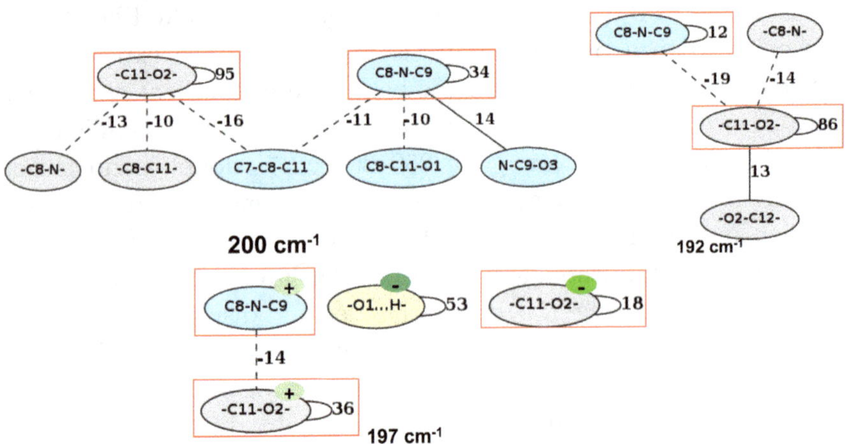

Figure 11.18. Vib-Graphs for the modes at 200 cm^{-1} (top, left) and 192 cm^{-1} (top, right), respectively for g^+ and g^- monomers, and 197 cm^{-1} (bottom) for $g^+ - g^-$ dimer. *Source*: Reproduced with permission from Ref. 44.

knowledge of the anharmonic vibrational modes of complex molecules such as the $(Ac\text{-}Phe\text{-}OMe)_2$ β-sheet model in its $g^+ \cdots g^-$ 3D-conformation just revealed by the THz-IR spectroscopy. See more discussions in Refs. 44, 68. The Vib-Graph vibrational analysis is illustrated here for the torsional modes of the peptides' backbones at $\sim$200 cm^{-1}, to show how these large amplitude, delocalized modes that involve several ICs are easily captured and quantified into anharmonic vibrational modes. See Figure 11.18 for the Vib-Graphs of the $\sim$200 and $\sim$192 cm^{-1} modes for each of the monomers obtained from 50 K DFT-MD trajectories of each isolated monomer (top-left and top-right), and the graph of the mode at 197 cm^{-1} for the $g^+ - g^-$ dimeric form (bottom of the figure), also obtained from 50 K DFT-MD. Please refer to Ref. 44 to see the excellent match between the dynamical DFT-MD-APT-IR spectra of the monomer (and dimer in Figure 11.17) with the experimental IR-UV ion dip action spectroscopies. See Figure 11.7 for the nomenclature of the atoms in the monomers for the following description of the vibrational modes.

The vibrational graph of the 200 cm^{-1} mode for the g^+ monomer nicely illustrates a highly delocalized torsional mode, where two motions dominate, *i.e.*, the torsion around the C-terminal $-$C11-O2$-$ backbone bond and the bending around the "central" backbone CNC. Each of these two motions is mechanically coupled to several torsions and bendings, these coupled motions however do not participate to the final IR activity of the mode (there are no values on the associated vertices). Note in the graphs the systematic out-of-phase couplings to the C–O torsion, and the mixing of in-/out-of-phase couplings to the CNC bending. Nicely, the 200 cm^{-1} IR mode of g$^+$ shows the mechanical coupling of the backbone motions to the bending of the side-chain, with the χ_1 (C7–C8–C11 in the graph) torsion coupled to both the dominant CO torsion and CNC bending. However, this is not IR active. The same motions are responsible for the 192 cm^{-1} mode of g^-. One can see that the Vib-Graph of this mode is simpler than the one for g^+, with only two motions being dominant, *i.e.*, the CO torsion and the CNC backbone bending.

The similarities between the two modes in the two isomers of Ac-Phe-OMe are highlighted with the red rectangles in the plot.

The IR mode at 197 cm^{-1} in the Ac-Phe-OMe$_2$ $g^+ \cdots g^-$ dimer (bottom of Figure 11.18) is nicely composed of the two CO torsions that were activated in the $\sim$200 cm^{-1} individual mode of each monomer strand (g^+ and g^- strands are recognizable by the $+$ and $-$ symbols in the vertices in the figure), however uncoupled to each other as the disconnected graphs show. The CO torsion on the g^+ strand is mechanically coupled to the CNC bending on the same strand, this latter is not participating to the final IR activity (no value on the associated vertex). Added to the overall IR activity of the 197 cm^{-1} mode of the dimer, one can see that the torsion around one of the intermolecular hydrogen bonds between the two monomers of the dimer has a large contribution to the anharmonic mode, as revealed by the largest value on the $-O1 \cdots H-$ vertex. The Vib-Graph of the 197 cm^{-1} mode in the dimer thus very nicely shows how some features of the equivalent mode in each monomer are conserved and how the new features arising from the H-bonding between the two monomers are appearing. The actual quantification of participation of each internal coordinate into the "intermolecular" mode is very important to obtain, as well as the mechanical and electronic couplings between the internal coordinates that are participating to the mode.

11.11. Illustration 4: Versatility of DFT-MD for Vibrational Spectroscopy, Some Illustrations on Aqueous Interfaces and SFG Spectroscopy

To illustrate further the versatility of DFT-MD for vibrational spectroscopy, we take one last example related to the condensed phase. We use again the aqueous silica interface as example, shown earlier in this chapter as illustration for the SFG spectroscopic signal calculation using the hybrid formalism that we have developed and showing the excellent agreement between this hybrid formalism and the SFG experimental spectrum. See Figure 11.2.

Beyond the excellent match between the experimental HD-SFG signal and the DFT-MD dynamical one seen in Figure 11.2, we illustrate in Figure 11.19 the peaks' assignments in the more complex condensed phase. In Figure 11.19(a) the theoretical SFG spectrum of the amorphous silica–water interface is deconvolved into distinct contributions arising from the most probable structural motifs identified from the MD simulations at the interface between liquid water and the top-surface of silica. Note that these motifs can be extracted from the MD simulations through the molecular graphs that we have developed.[114] This illustrates further the strength of theoretical spectroscopy from MD simulations, which allows deconvolution of the total spectrum of a complex molecular system, here composed of two phases, the solid and the liquid at its interface, into individual molecular contributions. Such decomposition is not possible from experiments alone. Only the MD simulations and the spectral decomposition as shown here can establish a one-to-one relationship between structural and spectroscopic properties. Please note that the decomposition shown in Figure 11.19 is done directly in terms of SFG deconvolve contributions, *i.e.*, keeping the exact activity of the signal in terms of SFG. This is not a VDOS decomposition that was shown earlier in this chapter to not include such activity. Being able to achieve such decomposition in keeping the SFG activity requires the hybrid formalism that we have developed for the calculation of any spectroscopic signal, *i.e.*, based on APT and Raman tensors as pre-factors. See the theory at the start of this chapter.

Four different O–H populations are found to contribute to the total SFG spectrum in the interfacial layer of aqueous silica, three of them are O–H groups of interfacial water molecules and one of them is the out-of-plane oriented O–H groups of the SiOH silanol terminations at the top-surface. It is interesting to note that water molecules further than 3 Å from the silica top-surface do not contribute to the SFG spectrum, because of the bulk-like centrosymmetry that is recovered beyond the topmost interfacial water layer. Note also that O–H groups (from water or silanols) that are oriented parallel to the silica surface do not contribute to

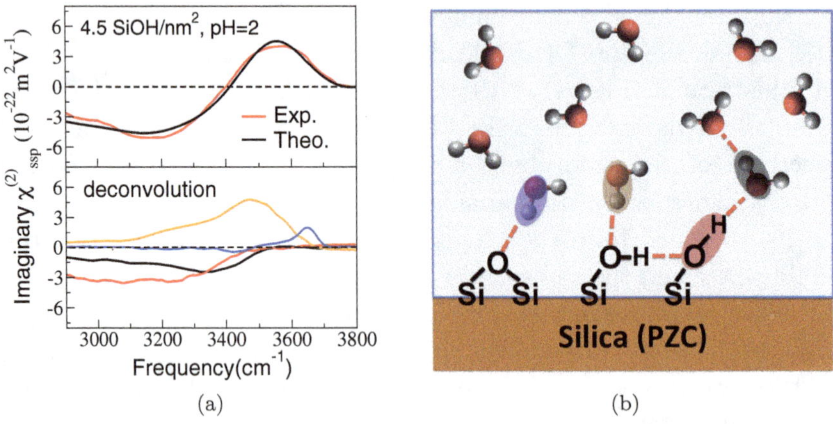

Figure 11.19. (a) Comparison between theoretical (black) and experimental[58] (red) $\Im(\chi^{(2)}(\omega))$ SFG-spectra of the fused/amorphous silica–water interface with an average silanol density of 4.5 SiOH/nm^2 (top). Theoretical deconvolution of the SFG spectrum into the individual contributions of the four SFG-active OH-populations in the BIL for isoelectric silica–water interfaces (bottom). See text for details. The color-coding of the curves is identical to the color-coding of the highlighted O–H groups in panel-B. (b) Schematic representation of the four SFG-active OH-populations. *Source*: Reproduced with permission from Ref. 15.

SFG, because of this orientation. The SFG signatures associated to each of the four O–H populations are reported in the lower panel of Figure 11.19(a), and a Schematic representation of the SFG-active O–H groups can be found in Figure 11.19(b). The color coding used for the deconvolved plots in Figure 11.19(a) and for the O–H groups highlighted in Figure 11.19(b) match. As illustrated in the figure, the four interfacial O–H populations that solely contribute to the SFG-response of the silica–water interface (with an average degree of hydroxylation of 4.5 SiOH/nm^2) at isoelectric conditions (pH$\sim$2) are as follows:

(1) O–H groups of interfacial water molecules donating H-bonds to in-plane oriented silanols (marked in orange in the figure). Such O–H groups provide a positive $\Im(\chi^{(2)}(\omega))$ SFG-signature (centered at 3450 cm^{-1}) because of their orientation from the liquid to the solid phase (along the normal to the surface as defined by convention in SFG). They are the major contributors

to the SFG spectrum of the silica–water interface in the high frequency range, *i.e.*, ≥ 3400 cm^{-1}, as can be seen by the intensity of the deconvolved peak.

(2) O–H groups of interfacial water molecules weakly interacting with siloxane bridges exposed at the amorphous surface (blue in the figure). Such O–H groups provide an $\Im(\chi^{(2)}(\omega))$ SFG-signature with the same positive sign as population 1 (since they both point towards the silica surface), but at a blue-shifted position (3660 cm^{-1} versus 3450 cm^{-1}). This is due to the nature of the Si-O-Si groups being exposed by the amorphous silica surface: they are indeed hydrophobic sites and hence far less good HB-acceptors than the hydrophilic silanol terminations.[56,141] The high frequency of these features in the SFG spectrum indeed indicates the quasi-free nature of the water O–H groups pointing to Si–O–Si sites. As we have shown in Ref. 56, the dynamics of such water quasi-free OH groups at the silica-water interface is accordingly found quantitatively comparable to the free OH at the air-water interface, both regarding its reorientational and vibrational dynamics. As concerns the lower intensity of the band associated to population two with respect to population one, it is due to the lower number of water molecules interacting with Si–O–Si groups than with SiOH groups at the fused 4.5 SiOH/nm^2 silica–water interface (all detailed in Ref. 56). The balance between these two water O–H populations is however strongly dependent on the surface hydroxylation state, and the intensities of the relative SFG signatures provide a direct SFG marker for molecular hydrophilicity/hydrophobicity at these heterogeneous surfaces.[142]

(3) O–H groups of out-of-plane oriented silanols donating strong HBs to interfacial water molecules (red in the figure). These O–H groups have opposite orientation as compared to the previous two populations, and thus provide a negative $\Im(\chi^{(2)}(\omega))$ SFG-signature. The red-shifted position of their SFG-band as compared to the one arising from population one (3150 versus 3450 cm^{-1}) is the spectroscopic evidence of the asymmetric H-bonding

properties of surface silanols, with the out-of-plane SiOH sites being able to donate stronger HBs than the ones accepted by in-plane SiOH sites, coherently with their pK_a values.[143,144] Among the four populations, this third one provides the most intense signal in the low frequency region, *i.e.*, below 3400 cm^{-1}, indicating how essential it is to consider the signature coming from the surface sites for a correct rationalization of the SFG spectra of aqueous interfaces.

(4) O–H groups of water molecules in the topmost interfacial layer donating HBs to water molecules located in the subsequent bulk layers (black in the figure). The water molecules that receive HBs from surface out-of-plane silanols (see population three) in turn donate water-water HBs preferentially to water molecules in the subsequent bulk layer, which are thus predominantly oriented toward the bulk of water and consequently provide a negative $\Im(\chi^{(2)}(\omega))$ SFG-signature centered at 3350 cm^{-1}.

11.12. Where the Field of MD-based Vibrational Spectroscopy is Going To: Some Perspectives That We Want to Highlight

We finish this chapter by providing some of the perspectives that we believe are highly promising for the field of MD-based vibrational spectroscopy. The perspectives highlighted below reflect our own opinions of where the field is evolving to. Others might have different opinions. We encompass the fields of spectroscopy in the gas phase as well as in condensed phases (liquids and interfaces).

11.12.1. *Machine learning*

The first item we would like to highlight concerns MD-based dynamical spectroscopy using force fields (FFs) for the calculation of the interactions, as FF-MD would allow us to address the vibrational spectroscopy of very large molecular systems (gas phase as well as condensed phase). As discussed earlier in this chapter, the current FFs of the literature are not accurate enough for spectroscopy.

Their parameterization for spectroscopy represents a huge amount of work, without warranty of success. Presumably one of the most exciting routes for improving FF-MD-based dynamical vibrational spectroscopy is machine learning (ML). One can either machine learn a classical FF or machine learn the properties that enter into the time-correlation functions for spectroscopy. For this latter, one has to machine learn molecular dipole moments for IR spectroscopy, machine learn molecular polarizability tensors for Raman spectroscopy, and both for SFG spectroscopy. The learning has to be done over relevant conformations that will be explored at the finite temperature of the MD simulations used for the spectroscopy calculations.

In the case of machine learning force fields, the goal is to obtain a highly accurate FF that would bring FF-MD at the level of accuracy of first principles MD. See Refs. 145,146 for general discussions on the topic. ML-FF-MD can hence be performed on very large molecular systems, over long time-scales, both non-accessible to first principles MD simulations, while maintaining first principles accuracy. See also some works by Rossi *et al.*[72] or Marx *et al.*[147] for ML-FF-MD that include a quantum treatment of the nuclei. Pioneers and well advanced research groups in ML-FF within the recent years include the groups of, *e.g.*, Behler *et al.*,[23, 70, 145, 146, 148–153] with applications mostly done for the condensed phases. Energies and forces are the central workhorse of the ML-FF techniques, ML-FFs express the PES as a sum of atomic energies that are a function of the local environment around each atom. They differ in the description of these local environments and in the ML approach or functional expressions to map the descriptors of the PES. See, *e.g.*, Ref. 146 for a review, see also Ref. 154 for alternative routes for ML-FF based on global representations.

To date, these ML-FF-MD simulations have been very seldom applied for vibrational spectroscopy calculations. Marquetand *et al.*[155] have pushed forward the ML to infrared spectra. In their pioneering work, they have coupled ML-FF dynamics with the simultaneous machine learning of dipole moments for the IR spectroscopy of gas phase molecules (methanol, *n*-alkanes, a tripeptide in their work). Molecular dipole moments were machine trained on DFT-MD simulations with techniques and algorithms similar to the training for energies and forces for the ML-FF. Excellent agreements of the ML-FF-dipole-MD dynamical spectra were hence obtained with the reference DFT-MD spectra (and experiments). This opened a highly promising route for vibrational spectroscopy, *i.e.*, accurate theoretical spectra at low computational cost. More recently, Rossi *et al.*[70] have continued on the same route for the vibrational spectroscopy of the porphycene gas phase molecule, *i.e.*, not only machine learning the PES of porphycene but also machine learning its molecular dipole moment. This investigation furthermore shows that, with such methods, the quantum dynamics of the nuclei can be included at very low cost, allowing us to take the quantum nature of the nuclei into account not only for the dynamics but ultimately for the vibrational spectra, which is essential in systems containing light atoms and possible proton transfers like the porphycene molecule there investigated. All together, such a successful investigation opens the route to ML-FF-MD and ML-spectroscopic properties that directly enter the time-correlation functions required in the vibrational spectra calculations for large size- and time-scale simulations of gas phase molecules, including nuclei quantum effects and ZPE effects on the final spectra.

Instead of machine learning molecular dipole moments, Ceriotti *et al.*[156] have machine learned atomic charges or/and atomic dipole moments in order to construct and predict the molecular dipole moments of a large variety of gas phase molecules extracted from databases. This "muML" model gives excellent agreements on molecular dipole moments (0.07 D in MAE) but the paper does not show any further application to IR spectroscopy. A similarly "AlphaML" model for polarizability tensors can be seen in Ref. 157. Machine learned polarizability tensors have also been obtained in Ref. 23 for molecular crystals and used for Raman spectroscopy calculations. Skinner and Corcelli[158] have also used machine learning for the transition frequencies and dipole derivatives of the O–H oscillators of water used in their vibrational maps for the vibrational spectroscopy of liquid water, furthermore noticeably increasing the accuracy of the spectroscopic-map approach.

This short review on the recent developments of machine learning for MD simulations and for theoretical vibrational spectroscopy shows where the field is moving. Such ML-strategies will allow us to generate long time-length MD trajectories that can sample conformational dynamics at finite temperature in a better way than first principles MD can do. The longer accessible time-lengths will in turn allow to address the elusive equipartition, that is especially crucial for gas phase molecules, as seen in this chapter. Including quantum nuclei and ZPE effects at low computational cost through these ML approaches will also change the theoretical spectroscopic field. Ultimately, these ML-strategies will allow us to reach better convergence of band-intensities of vibrational spectra, at low computational cost, improving further the interpretation of spectroscopic experiments.

11.12.2. *Hybrid formalism for dynamical MD-based spectroscopy based on pre-computed APT and Raman tensors*

Another route for addressing larger molecular systems together with longer time-scales for dynamical spectroscopy is the one that we have presented in this chapter in which the "computationally cheap" VDOS/ICDOS is modulated by APTs for IR spectroscopy, by Raman tensors for Raman spectroscopy, and by both in SFG spectroscopy. With an accurate ML-FF in hand, as described above, the MD trajectories for the VDOS/ICDOS could be generated at the FF-MD level with a low computational cost, while the APT/Raman tensors to be used for the spectroscopic signals can be obtained at a very high quantum level for the low number of reference geometries that are needed (as seen earlier in this chapter). We believe this is a very good alternative to ML-FF-ML-dipole-MD-IR (or to ML-FF-ML-polarizability-MD-Raman) discussed above, as it reduces the number of ML stages to go through. One could also imagine to further ML APT and Raman tensors.

11.12.3. *Algorithmic graph theory*

We believe that algorithmic graph theory is a key approach for the computational chemistry field. We have shown in this chapter one usage of algorithmic graph theory with vibrational graphs and the strength of the graphs in, *e.g.*, measures of similarities for comparing (anharmonic) vibrational modes and their couplings. Other applications of the Vib-Graphs can be thought of. In Ref. 114, we also developed algorithmic graph theory for the automatic recognition of structures over time in MD simulations. Applications in molecular dynamics simulations of gas phase molecules and clusters as well as aqueous interfaces were shown in Refs. 11, 114, 115. One possible dream could be to develop machine learning approaches, *e.g.*, for the calculation of spectra, for finding the molecular structures that are responsible for the experimental spectral fingerprints, without resorting to computational chemistry methods, but rather by solely resorting to graphs. This is where our research is typically heading.

Acknowledgments

I would like to acknowledge my collaborators and past/present students who worked with me to advance the field of MD-based vibrational spectroscopy over the past almost two decades. I mention by name only a few of these people. Over the years, I have had wonderful collaborations with spectroscopy experimentalists from the gas phase community, among whom the late Prof Jean-Pierre Schermann and Prof James M. Lisy who were both great mentors in my professional career and who trusted the perspectives offered by DFT-MD simulations for vibrational spectroscopy. Many thanks also to Prof Anouk M. Rijs who trusted me with DFT-MD for THz theoretical spectroscopy of gas phase molecules. I would also like to mention some of my more recent collaborations with spectroscopy experimentalists in the condensed phase, Prof Y. Ron Shen, Prof Martina Havenith, Prof Eric Borguet, Prof Ellen Backcus, Prof M. Bonn, and Prof Weitao Liu. I want to thank Prof Dominique Barth for the exciting and unique works that we share in coupling algorithmic Graph Theory and MD simulations, our recent collaboration has opened a new era in my research. I had the privilege to enroll gifted PhD/Post-Doc students in my group, who shared/share my passion for gas phase spectroscopy, among whom are Dr. Daria R. Galimberti, Dr. Sana Bougueroua, Dr. Jérôme Mahé, and PhD Vladimir Chantitch. Many thanks to them. Many thanks also to my other students and collaborators in the condensed phase for all the works done together on MD-based spectroscopies of liquids and aqueous interfaces. Special thanks to Dr. Simone Pezzotti, Dr. Flavio Siro Brigiano, Dr. Fabrizio Creazzo, Dr. Louis Potier, and PhD Wanlin Chen, for the most recent students in my group. Acknowledgments should also go to the French funding agencies that brought financial support to our research, Agence Nationale de la Recherche (ANR), LABEX CHARM3AT (Laboratory of Excellence) at the University Paris-Saclay, PICS/IEA-CNRS France-The Netherlands, 2MIB 'Chemical Sciences' Doctoral School at the University Paris-Saclay, Génopole Evry, the Institut Universitaire de France (IUF), and several research funds from the University of Evry val d'Essonne that encourages and supports collaborations and collaborative works. Last but not least, none of our research could be done without HPC resources, I therefore acknowledge the continuous support from GENCI-France under Grant 072484 (CINES/IDRIS/TGCC).

References

1. Frenkel, D.; Smit, B. *Understanding Molecular Dynamics*, 2nd ed. Academic Press, 2002.
2. Marx, D.; Hutter, J. Ab Initio Molecular Dynamics: Theory and Implementation. In *Modern Methods and Algorithms of Quantum Chemistry*; Grotendorst, J., Ed.; NIC: Forschungszentrum Julich, 2000; Chapter 13, pp. 301–449.
3. van Mourik, T.; Buhl, M.; Gaigeot, M. Density functional theory across chemistry, physics and biology. *Phil. Trans. Royal Soc.* **2014**, *372*, 20120488.
4. Leach, A. R. *Molecular Modelling: Principles and Applications*. Pearson Education, 2001.
5. Dykstra, C.; Frenking, G.; Kim, K.; Scuseria, G. *Theory and Applications of Computational Chemistry: The First Forty Years*. Elsevier, 2011.

6. VandeVondele, J.; Krack, M.; Mohamed, F.; Parrinello, M.; Chassaing, T.; Hutter, J. Quickstep: fast and accurate density functional calculations using a mixed gaussian and plane waves approach. *Comput. Phys. Commun.* **2005**, *167*, 103–128.

7. Thomas, M.; Brehm, M.; Fligg, R.; Vohringer, P.; Kirchner, B. Computing vibrational spectra from ab initio molecular dynamics. *Phys. Chem. Chem. Phys.* **2013**, *15*, 6608–6622.

8. Kirchner, B.; Blasius, J.; Esser, L.; Reckien, W. Predicting vibrational spectroscopy for flexible molecules and molecules with non-idle environents. *Adv. Theory Simul.* **2020**, *4*, 2000223–2000237.

9. Gaigeot, M.-P. Theoretical spectroscopy of floppy peptides at room temperature. A DFTMD perspective: gas and aqueous phase. *Phys. Chem. Chem. Phys.* **2010**, *12*, 3336–3359.

10. Gaigeot, M.-P.; Spezia, R. Theoretical Methods for Vibrational Spectroscopy and Collision Induced Dissociation in the Gas Phase. In *Gas-Phase IR Spectroscopy and Structure of Biological Molecules*; Rijs, A. M., Oomens, J., Eds.; Springer International Publishing, 2015; pp. 99–151.

11. Gaigeot, M.-P. Some opinions on MD-based vibrational spectroscopy of gas phase molecules and their assembly: an overview of what has been achieved and where to go. *Spectrochim. Acta A* **2021**, *260*, 119864–119883.

12. Gaigeot, M. P.; Sprik, M. Ab initio molecular dynamics computation of the infrared spectrum of aqueous Uracil. *J. Phys. Chem. B* **2003**, *107*, 10344.

13. Paesani, F. Getting the right answers for the right reasons: toward predictive molecular simulations of water with many-body potential energy functions. *J. Chem. Theory Comput.* **2016**, *49*, 1844.

14. Pezzotti, S.; Galimberti, D.; Gaigeot, M.-P. 2D H-bond network as the topmost skin to the air-water interface. *J. Phys. Chem. Lett.* **2017**, *8*, 3133–3141.

15. Pezzotti, S.; Galimberti, D.; Gaigeot, M.-P. Deconvolution of BIL-SFG and DL-SFG spectroscopic signals reveal order/disorder of water at the elusive aqueous silica interface. *Phys. Chem. Chem. Phys.* **2019**, *21*, 22188–22202.

16. Tang, F.; Ohto, T.; Sun, S.; Rouxel, J. R.; Imoto, S.; Backus, E. H. G.; Mukamel, S.; Bonn, M.; Nagata, Y. Molecular structure and modeling of water-air and ice-air interfaces monitored by sum-frequency generation. *Chem. Rev.* **2020**, *120*, 3633–3667.

17. Khatib, R.; Backus, E.; Bonn, M.; Perez-Haro, M.; Gaigeot, M.; Sulpizi, M. *Sci. Rep.* **2016**, *6*, 24287.

18. Medders, G. R.; Paesani, F. Dissecting the molecular structure of the air/water interface from quantum simulations of the sum-frequency generation spectrum. *J. Am. Chem. Soc.* **2016**, *138*, 3912–3919.

19. McQuarrie, D. A. *Statistical Mechanics*. University Science Books, 2000.

20. Kubo, R.; Toda, M.; Hashitsume, N. *Statistical Physics II*. Springer Series in Solid-State Sciences; Springer Berlin Heidelberg, 1991. Vol. 31.

21. Gaigeot, M. P.; Spezia, R. Theoretical methods for vibrational spectroscopy and collision induced dissociation in the gas phase. *Topics Curr. Chem.* **2015**, *364*, 99.

22. Pezzotti, S.; Galimberti, D.; Shen, Y.; Gaigeot, M.-P. Structural definition of the BIL and DL: a new universal methodology to rationalize non-linear $\chi^{(2)}(\omega)$ SFG signals at charged interfaces, including $\chi^{(3)}(\omega)$ contributions. *Phys. Chem. Chem. Phys.* **2018**, *20*, 5190–5199.

23. Raimbault, N.; Grisafi, A.; Ceriotti, M.; Rossi, M. Using Gaussian process regression to simulate the vibrational Raman spectra of molecular crystals. *New J. Phys.* **2019**, *21*, 105001–105014.

24. Shen, Y. R.; Ostroverkhov, V. Sum-frequency vibrational spectroscopy on water interfaces: polar orientation of water molecules at interfaces. *Chem. Rev.* **2006**, *106*, 1140–1154.

25. Shen, Y. Basic theory of surface sum-frequency generation. *J. Phys. Chem. C* **2012**, *116*, 15505.

26. Shen, Y. Phase-sensitive sum-frequency spectroscopy. *Annu. Rev. Phys. Chem.* **2013**, *64*, 129.

27. Geiger, F. Second harmonic generation, sum frequency generation, and $\chi(3)$: Dissecting environmental interfaces with a nonlinear optical Swiss army knife. *Annu. Rev. Phys. Chem.* **2009**, *60*, 61.

28. Morita, A.; Hynes, J. T. A theoretical analysis of the SFG spectrum of the water surface. II time dependent approach. *J. Phys. Chem. B* **2002**, *106*, 673.

29. Morita, A.; Ishiyama, T. Recent progress in theoretical analysis of vibrational sum frequency generation spectroscopy. *Phys. Chem. Chem. Phys.* **2008**, *10*, 5801.

30. Nagata, Y.; Mukamel, S. Vibrational sum-frequency generation spectroscopy at the water/lipid interface: molecular dynamics simulation study. *J. Am. Chem. Soc.* **2010**, *132*, 6434–6442.

31. Kaliannan, N.; Aristizabal, A.; Wiebeler, H.; Zysk, F.; Ohto, T.; Nagata, Y.; Kuhne, T. Impact of intermolecular vibrational coupling effects on the sum-frequency generation spectra of the water/air interface. *Mol. Phys.* **2020**, *118*, e1620358.

32. Sulpizi, M.; Salanne, M.; Sprik, M.; Gaigeot, M. Vibrational sum frequency generation spectroscopy of the water liquid–vapor interface from density functional theory-based molecular dynamics simulations. *J. Phys. Chem. Lett.* **2013**, *4*, 83.

33. Salanne, M.; Vuilleumier, R.; Madden, P.; Simon, C.; Turq, P.; Guillot, B. Polarizabilities of individual molecules and ions in liquids from first principles. *J. Phys. Condens. Matter* **2008**, *20*, 494207.

34. Partovi-Azar, P.; Kuhne, T. Efficient "on-the-fly" calculation of Raman spectra from ab-initio molecular dynamics: application to hydrophobic/hydrophilic solutes in bulk water. *J. Comput. Chem.* **2015**, *36*, 2188–2192.

35. Ishiyama, T.; Morita, A. Computational analysis of vibrational sum frequency generation spectroscopy. *Annu. Rev. Phys. Chem.* **2017**, *68*, 355–377.

36. Ishiyama, T.; Imamura, T.; Morita, A. Theoretical studies of structures and vibrational sum frequency generation spectra at aqueous interfaces. *Chem. Rev.* **2014**, *114*, 8447–8470.

37. Nihonyanagi, S.; Ishiyama, T.; Lee, T.; Yamaguchi, S.; Bonn, M.; Morita, A.; Tahara, T. Unified molecular view of the air/water interface based on experimental and theoretical spectra of an isotopically diluted water surface. *J. Am. Chem. Soc.* **2011**, *133*, 16875.

38. Marzari, N.; Vanderbilt, D. Maximally localized generalized Wannier functions for composite energy bands. *Phys. Rev. B* **1997**, *56*, 12847–12865.

39. Silvestrelli, P. L.; Parrinello, M. Water molecule dipole in the gas and in the liquid phase. *Phys. Rev. Lett.* **1999**, *82*, 3308–3311.

40. Silvestrelli, P. L.; Parrinello, M. Structural, electronic, and bonding properties of liquid water from first principles. *J. Chem. Phys.* **1999**, *111*, 3572–3580.

41. Thomas, M.; Brehm, M.; Kirchner, B. Voronoi dipole moments for the simulation of bulk phase vibrational spectra. *Phys. Chem. Chem. Phys.* **2015**, *17*, 3207–3213.

42. Luber, S. Local electric dipole moments for periodic systems via density functional theory embedding. *J. Chem. Phys.* **2014**, *141*, 234110.

43. Galimberti, D. R.; Milani, A.; Tommasini, M.; Castiglioni, C.; Gaigeot, M.-P. Combining static and dynamical approaches for infrared spectra calculations of gas phase molecules and clusters. *J. Chem. Theory Comput.* **2017**, *13*, 3802–3813.

44. Bakels, S.; Meijer, E. M.; Greuell, M.; Porskamp, S.; Rouwhorst, G.; Mahé, J.; Gaigeot, M.-P.; Rijs, A. M. Interactions of aggregating peptides probed by IR-UV action spectroscopy. *Faraday Discuss.* **2019**, *217*, 322–341.

45. Wilson, J. E.; Decius, J. C.; Cross, P. C. *Molecular Vibrations: The Theory of Infrared and Raman Vibrational Spectra*. Dover Publications, 1980.

46. Frisch, M. J.; Trucks, G. W.; Schlegel, H. B.; Scuseria, G. E.; Robb, M. A.; Cheeseman, J. R.; Scalmani, G.; Barone, V.; Mennucci, B.; Petersson, G. A. *et al.* Gaussian 09 and Revision C.02 and Gaussian and Inc. and Wallingford CT and 2009.

47. Cyran, J.; Donovan, M.; Vollmer, D.; Siro-Brigiano, F.; Pezzotti, S.; Galimberti, D.; Gaigeot, M.-P.; Bonn, M.; Backus, E. Molecular hydrophobicity at a hydrophilic surface. *Proc. Nat. Acad. Sci.* **2019**, *116*, 1520–1525.

48. Tuladhar, A.; Dewan, S.; Pezzotti, S.; Siro-Brigiano, F.; Creazzo, F.; Gaigeot, M.-P. Ions tune interfacial water structure and modulate hydrophobic interactions at silica surfaces. *J. Am. Chem. Soc.* **2020**, *142*, 6991–7000.

49. Galimberti, D. R.; Milano, A.; Tommasini, M.; Castiglioni, C.; Gaigeot, M.-P. Combining static and dynamical approaches for infrared spectra calculations of gas phase molecules and clusters. *J. Chem. Theory Comput.* **2017**, *13*, 3802.

50. Khatib, R.; Backus, E. H. G.; Bonn, M.; Perez-Haro, M.-J.; Gaigeot, M.-P.; Sulpizi, M. Water orientation and hydrogen-bond structure at the fluorite/water interface. *Sci. Rep.* **2016**, *6*, 24287.

51. Skinner, S. A. C. J. L. Infrared and Raman line shapes of dilute HOD in liquid H_2O and D_2O from 10 to 90 C. *J. Phys. Chem. A* **2005**, *109*, 6154–6165.

52. Pezzotti, S.; Galimberti, D. R.; Shen, Y. R.; Gaigeot, M.-P. Structural definition of the BIL and DL: a new universal methodology to rationalize non-linear chi(2)(omega) SFG signals at charged interfaces, including chi(3)(omega) contributions. *Phys. Chem. Chem. Phys.* **2018**, *20*, 5190–5199.

53. Pezzotti, S.; Galimberti, D. R.; Shen, Y. R.; Gaigeot, M.-P. What the diffuse layer (DL) reveals in non-linear SFG spectroscopy. *Minerals* **2018**, *8*, 305.

54. Pezzotti, S.; Gaigeot, M.-P. Spectroscopic BIL-SFG invariance hides the chaotropic effect of protons at the air-water interface. *Atmosphere* **2018**, *9*, 396.

55. Creazzo, F.; Galimberti, D. R.; Pezzotti, S.; Gaigeot, M.-P. DFT-MD of the (110)-Co_3O_4 cobalt oxide semiconductor in contact with liquid water, preliminary chemical and physical insights into the electrochemical environment. *J. Chem. Phys.* **2019**, *150*, 041721.

56. Cyran, J.; Donovan, M.; Vollmer, D.; Brigiano, F. S.; Pezzotti, S.; Galimberti, D.; Gaigeot, M.-P.; Bonn, M.; Backus, E. Molecular

hydrophobicity at a macroscopically hydrophilic surface. *Proc. Nat. Acad. Sci.* **2019**, *116*, 1520–1525.

57. Nihonyanagi, S.; Kusaka, R.; Inoue, K.; Adhikari, A.; Yamaguchi, S.; Tahara, T. Accurate Determination of Complex $\chi^{(2)}$ Spectrum of the Air/Water Interface. *J. Chem. Phys.* **2015**, *143*, 124707.

58. Myalitsin, A.; Urashima, S.-h.; Nihonyanagi, S.; Yamaguchi, S.; Tahara, T. Water structure at the buried silica/aqueous interface studied by heterodyne-detected vibrational sum-frequency generation. *J. Phys. Chem. C* **2016**, *120*, 9357–9363.

59. Bakker, D.; Dey, A.; Tabor, D.; Ong, Q.; Mahé, J.; Gaigeot, M.-P.; III, E. S.; Rijs, A. Fingerprints of inter- and intra-molecular hydrogen bonding in saligenin-water clusters revealed by mid- and far- infrared spectroscopy. *Phys. Chem. Chem. Phys.* **2017**, *19*, 20343.

60. Mahé, J.; Bakker, D.; Jaeqx, S.; Rijs, A.; Gaigeot, M.-P. Mapping gas phase dipeptides motions in the far-infrared and terahertz domain. *Phys. Chem. Chem. Phys.* **2017**, *19*, 13778.

61. Mahé, J.; Jaeqx, S.; Rijs, A.; Gaigeot, M. Can far-IR action spectroscopy and BOMD simulations be conformation selective? *Phys. Chem. Chem. Phys.* **2015**, *17*, 25905.

62. Bakker, D.; Ong, Q.; Dey, A.; Mahé, J.; Gaigeot, M.-P.; Rijs, A. Anharmonic, dynamic and functional level effects in far-infrared spectroscopy: phenol derivatives. *J. Mol. Spectrosc.* **2017**, *342*, 4.

63. Brites, V.; Lisy, J. M.; Gaigeot, M. P. Infrared predissociation vibrational spectroscopy of Li + (H2O)3 − 4Ar0,1 reanalyzed using Density Functional Theory Molecular Dynamics. *J. Phys. Chem. A* **2015**, *119*, 2468.

64. Beck, J. P.; Gaigeot, M.-P.; Lisy, J. M. Anharmonic vibrations from the chloride...amide ionic hydrogen bonds in $Cl^-(N$-methylacetamide$)_1(H_2O)_{0-2}Ar_2$ cluster ions. Combined IRPD experiments and AIMD simulations. *Phys. Chem. Chem. Phys.* **2013**, *15*, 16736–16745.

65. Beck, J. P.; Gaigeot, M.-P.; Lisy, J. M. O-H anharmonic vibrational motions in $Cl^- \cdots (CH_3OH)_{1-2}$ ionic clusters. Combined IRPD experiments and AIMD simulations. *Spectrochim. Acta A* **2014**, *119*, 12.

66. Marinica, C.; Grégoire, G.; Desfrançois, C.; Schermann, J. P.; Borgis, D.; Gaigeot, M. P. Ab-initio molecular dynamics of protonated dialanine and comparison to infrared multiphoton dissociation experiments. *J. Phys. Chem. A* **2006**, *110*, 8802.

67. Cimas, A.; Vaden, T. D.; de Boer, T. S. J. A.; Snoek, L. C.; Gaigeot, M. P. Vibrational spectra of small protonated peptides from finite temperature MD simulations and IRMPD spectroscopy. *J. Chem. Theory Comput.* **2009**, *5*, 1068.

68. Bakels, S.; Gaigeot, M.-P.; Rijs, A. Gas phase spectroscopy of neutral peptides: Insights from the far IR domain. *Chem. Rev.* **2020**, *120*, 3233–3260.

69. Ivanov, S. D.; Witt, A.; Marx, D. Theoretical spectroscopy using molecular dynamics: theory and application to CH5+ and its isotopologues. *Phys. Chem. Chem. Phys.* **2013**, *15*, 10270.

70. Litman, Y.; Behler, J.; Rossi, M. Temperature dependence of the vibrational spectrum of porphycene: a qualitative failure of classical-nuclei molecular dynamics. *Faraday Discuss.* **2020**, *221*, 526–546.

71. Qu, C.; Bowman, J. Quantum approaches to vibrational dynamics and spectroscopy: is ease of interpretation sacrificed as rigor increases? *Phys. Chem. Chem. Phys.* **2019**, *21*, 3397–3413.

72. Rossi, M.; Liu, H.; Pasesani, F.; Bowman, J.; Ceriotti, M. Temperature dependence of the vibrational spectrum of porphycene: a qualitative failure of classical-nuclei molecular dynamics. *J. Chem. Phys.* **2014**, *141*, 181101.

73. Thomas, D.; Mucha, E.; Lettow, M.; Meijer, G.; Rossi, M.; von Helden, G. Characterization of a trans!trans carbonic acid-fluoride complex by infrared action spectroscopy in helium nanodroplets. *J. Am. Chem. Soc.* **2019**, *141*, 5815–5823.

74. Rossi, M.; Kapil, V.; Ceriotti, M. Fine tuning classical and quantum molecular dynamics using a generalized Langevin equation. *J. Chem. Phys.* **2018**, *148*, 102301.

75. Poltavsky, I.; Kapil, V.; Ceriotti, M.; Kim, K.; Tkatchenko, A. Accurate description of nuclear quantum Effects with High-Order Perturbed Path Integrals (HOPPI). *J. Chem. Theory Comput.* **2020**, *16*, 1128–1135.

76. Oanh, N.-T. V.; Falvo, C.; Calvo, F.; Lauvergnat, D.; Basire, M.; Gaigeot, M.-P.; Parneix, P. Improving anharmonic infrared spectra using semi-classically prepared molecular dynamics simulations. *Phys. Chem. Chem. Phys.* **2012**, *14*, 2381.

77. Esser, T.; Knorke, H.; Asmis, K. R.; Schollkopf, W.; Yu, Q.; Qu, C.; Bowman, J. M.; Kaledin, M. Deconstructing prominent bands in the terahertz spectra of H_7O_{3+} and H_9O_{4+}: Intermolecular modes in eigen clusters. *J. Phys. Chem. Lett.* **2018**, *9*, 798–803.

78. Peslherbe, G.; Wang, H.; Hase, W. Analysis and extension of a model for constraining zero-point energy flow in classical trajectory simulations. *J. Chem. Phys.* **1993**, *100*, 1179.

79. Zhang, X.; Rheinecker, J.; Bowman, J. Quasiclassical trajectory study of formaldehyde unimolecular dissociation: $H_2CO \rightarrow H_2 + COH_2CO \rightarrow H_2 + CO$, $H + HCO$. *J. Chem. Phys.* **2005**, *122*, 114313.

80. Bowman, J.; Gazdy, B.; Sun, Q. A method to constrain vibrational energy in quasiclassical trajectory calculations. *J. Chem. Phys.* **1989**, *91*, 2859–2862.

81. Miller, W.; Hase, W.; Darling, C. A simple-model for correcting the zero-point energy problem in classical trajectory simulations of polyatomic-molecules. *J. Phys. Chem.* **1989**, *91*, 2863–2868.

82. Xie, Z.; Bowman, J. Quasiclassical trajectory study of the reaction of fast H atoms with C-H stretch excited CHD_3. *Chem. Phys. Lett.* **2006**, *429*, 355–359.

83. Farag, M.; Ruiz-Lopez, M.; Bastida, A.; Monard, G.; Ingrosso, F. Hydration effect on amide I infrared bands in water: An interpretation based on an interaction energy decomposition scheme. *J. Phys. Chem. B* **2015**, *119*, 9056–9067.

84. Bistafa, C.; Kitamura, Y.; Martins-Costa, M.; Nagaoka, M.; Ruiz-Lopez, M. Vibrational spectroscopy in solution through perturbative ab initio molecular dynamics simulations. *J. Chem. Theory Comput.* **2019**, *15*, 4615–4622.

85. Dubosq, C.; Falvo, C.; Calvo, F.; Rapacioli, M.; Parneix, P.; Pino, T.; Simon, A. Mapping the structural diversity of C60 carbon clusters and their infrared spectra. *A&A* **2019**, *625*, L11.

86. Dubosq, C.; Calvo, F.; Rapacioli, M.; Dartois, E.; Pino, T.; Falvo, C.; Simon, A. Quantum modeling of the optical spectra of carbon cluster structural families and relation to the interstellar extinction UV bump. *A&A* **2020**, *634*, A62.

87. Simon, A.; Rapacioli, M.; Mascetti, J.; Spiegelman, F. Vibrational spectroscopy and molecular dynamics of water monomers and dimers adsorbed on polycyclic aromatic hydrocarbons. *Phys. Chem. Chem. Phys.* **2012**, *14*, 6771–6786.

88. McGill, C.; Forsuelo, M.; Guan, Y.; Green, W. Predicting infrared spectra with message passing neural networks. *J. Chem. Inf. Model.* **2021**, *61*, 2594–2609.

89. Pracht, P.; Grant, D.; Grimme, S. Comprehensive assessment of GFN tight-binding and composite density functional theory methods for calculating gas-phase infrared spectra. *J. Chem. Theory Comput.* **2020**, *16*, 7044–7060.

90. Palmo, K.; Krimm, S. Electrostatic model for IR intensities in a spectroscopically determined molecular mechanics force field. *J. Comput. Chem.* **1998**, *19*, 754–768.

91. Galimberti, D.; Milani, A.; Castiglioni, C. Charge mobility in molecules: Charge fluxes from second derivatives of the molecular dipole. *J. Chem. Phys.* **2013**, *138*, 164115.

92. Galimberti, D.; Milani, A.; Castiglioni, C. Infrared intensities and charge mobility in hydrogen bonded complexes. *J. Chem. Phys.* **2013**, *139*, 074304.

93. Semrouni, D.; Sharma, A.; Dognon, J.; Ohanessian, G.; Clavaguera, C. Finite temperature infrared spectra from polarizable molecular dynamics simulations. *J. Chem. Theory Comput.* **2014**, *10*, 3190–3199.

94. Kratz, E.; Walker, A.; Lagardere, L.; Lipparini, F.; Piquemal, J.; Cisneros, G. LICHEM: A QM/MM program for simulations with multipolar and polarizable force fields. *J. Comput. Chem.* **2016**, *37*, 1019–1029.

95. Kroutil, O.; Pezzotti, S.; Gaigeot, M.-P.; Predota, M. Phase-sensitive vibrational SFG spectra from simple classical force-fields molecular dynamics simulations. *J. Phys. Chem. C* **2020**, *124*, 15253–15263.

96. Thaunay, F.; Dognon, J.-P.; Ohanessian, G.; Clavaguera, C. Vibrational mode assignment of finite temperature infrared spectra using the AMOEBA polarizable force field. *Phys. Chem. Chem. Phys.* **2015**, *17*, 25968–25977.

97. Martinez, M.; Gaigeot, M.-P.; Borgis, D.; Vuilleumier, R. Extracting effective normal modes from equilibrium dynamics at finite temperature. *J. Chem. Phys.* **2006**, *125*, 144106.

98. Gaigeot, M.-P.; Martinez, M.; Vuilleumier, R. Infrared spectroscopy in the gas and liquid phase from first principle molecular dynamics simulations: Application to small peptides. *Mol. Phys.* **2007**, *105*, 2857–2878.

99. Mathias, G.; Ivanov, S. D.; Witt, A.; Baer, M. D.; Marx, D. Infrared spectroscopy of fluxional molecules from (ab initio) molecular dynamics: Resolving large-amplitude motion, multiple conformations, and permutational symmetries. *J. Chem. Theory Comput.* **2012**, *8*, 224–234.

100. Schmitz, M.; Tavan, P. Vibrational spectra from atomic fluctuations in dynamics simulations. I. Theory, limitations, and a sample application. *J. Chem. Phys.* **2004**, *121*, 12233–12246.

101. Schmitz, M.; Tavan, P. Vibrational spectra from atomic fluctuations in dynamics simulations. II. Solvent-induced frequency fluctuations at femtosecond time resolution. *J. Chem. Phys.* **2004**, *121*, 12247–12258.

102. Bowman, J.; Zhang, X.; Brown, A. Normal-mode analysis without the Hessian: A driven molecular-dynamics approach. *J. Chem. Phys.* **2003**, *119*, 646.

103. Kaledin, M.; Brown, A.; Kaledin, A.; Bowman, J. Normal mode analysis using the driven molecular dynamics method. II. An application to biological macromolecules. *J. Chem. Phys.* **2004**, *121*, 5646.

104. Nonella, M.; Mathias, G.; Tavan, P. Infrared spectrum of p-benzoquinone in water obtained from a QM/MM hybrid molecular dynamics simulation. *J. Phys. Chem. A* **2003**, *107*, 8638.

105. Gaigeot, M.-P.; Martinez, M.; Vuilleumier, R. Infrared spectroscopy in the gas and liquid phase from first principle molecular dynamics simulations: Application to small peptides. *Mol. Phys.* **2007**, *105*, 2857–2878.

106. Martinez, M.; Gaigeot, M.-P.; Borgis, D.; Vuilleumier, R. Extracting effective normal modes from equilibrium dynamics at finite temperature. *J. Chem. Phys.* **2006**, *125*, 144106.

107. Mathias, G.; Ivanov, S.; Witt, A.; Baer, M.; Marx, D. Infrared spectroscopy of fluxional molecules from (*ab initio*) molecular dynamics: resolving large-amplitude motion, multiple conformations, and permutational symmetries. *J. Chem. Theory Comput.* **2012**, *8*, 224.

108. Galimberti, D.; Bougueroua, S.; Mahe, J.; Tommasini, M.; Rijs, A. M.; Gaigeot, M.-P. Conformational assignment of gas phase peptides and their H-bonded complexes using far-IR/THz: IR-UV ion dip experiment, DFT-MD spectroscopy, and graph theory for mode assignment. *Faraday Discuss.* **2019**, *217*, 67–97.

109. Bougueroua, S.; Spezia, R.; Pezzotti, S.; Vial, S.; Quessette, F.; Barth, D.; Gaigeot, M.-P. Graph theory for automatic structural recognition in molecular dynamics simulations. *J. Chem. Phys.* **2018**, *149*, 184102–184115.

110. Raymond, J. W.; Willett, P. Maximum common subgraph isomorphism algorithms for the matching of chemical structures. *J. Comput. Aided Mol.* **2002**, *16*, 521–533.

111. Vismara, P.; Valery, B. Finding maximum common connected subgraphs using clique detection or constraint satisfaction algorithms. International Conference on Modelling, Computation and Optimization in Information Systems and Management Sciences. 2008; pp. 358–368.

112. Depolli, M.; Szabó, S.; Zaválnij, B. An improved maximum common induced subgraph solver. *Match* **2020**, *84*, 7–28.

113. Ramachandran, G.; Ramakrishnan, C.; Sasisekharan, V. Stereochemistry of polypeptide chain configurations. *J. Mol. Biol.* **1963**, *7*, 95–99.

114. Bougueroua, S.; Spezia, R.; Pezzotti, S.; Vial, S.; Quessette, F.; Barth, D.; Gaigeot, M.-P. Graph theory for automatic structural recognition in molecular dynamics simulations. *J. Chem. Phys.* **2018**, *149*, 184102.

115. Serva, A.; Pezzotti, S.; Bougueroua, S.; Galimberti, D. R.; Gaigeot, M.-P. Combining ab-initio and classical molecular dynamics

simulations to unravel the structure of the 2D-HB-network at the air-water interface. *J. Mol. Struc.* **2018**, *1165*, 71–78.

116. Sediki, A.; Snoek, L. C.; Gaigeot, M.-P. N-H$_+$ vibrational anharmonicities directly revealed from DFT-based molecular dynamics simulations on the Ala(7)H(+) protonated peptide. *Int. J. Mass Spectrom.* **2011**, *308*, 281.

117. Grégoire, G.; Gaigeot, M. P.; Marinica, D. C.; Lemaire, J.; Schermann, J. P.; Desfrançois, C. Resonant infrared multiphoton dissociation spectroscopy of gas-phase protonated peptides. Experiments and Car-Parrinello dynamics at 300 K. *Phys. Chem. Chem. Phys.* **2007**, *9*, 3082.

118. Marinica, C.; Grégoire, G.; Desfrancois, C.; Schermann, J.-P.; Borgis, D.; Gaigeot, M.-P. Ab-initio molecular dynamics of protonated dialanine and comparison to infrared multiphoton dissociation experiments. *J. Chem. Phys. A* **2006**, *110*, 8802–8810.

119. Lucas, B.; Gregoire, G.; Lemaire, J.; Maitre, P.; Ortega, J. M.; Rupenyan, A.; Reimann, B.; Schermann, J. P.; Desfrancois, C. Investigation of the protonation site in the dialanine peptide by infrared multiphoton dissociation spectroscopy. *Phys. Chem. Chem. Phys.* **2004**, *6*, 2659.

120. Vaden, T. D.; de Boer, T. S. J. A.; Simons, J. P.; Snoek, L. C.; Suhai, S.; Paizs, B. Vibrational spectroscopy and conformational structure of protonated polyalanine peptides isolated in the gas phase. *J. Phys. Chem. A* **2008**, *112*, 4608.

121. Leavitt, C. M.; DeBlase, A. F.; Johnson, C. J.; van Stipdonk, M.; McCoy, A. B.; Johnson, M. A. Hiding in plain sight: unmasking the diffuse spectral signatures of the protonated n-terminus in isolated dipeptides cooled in a cryogenic ion trap. *J. Phys. Chem. Lett.* **2013**, *4*, 3450.

122. Vaden, T. D.; de Boer, T. S. J. A.; MacLeod, N. A.; Marzluff, E. M.; Simons, J. P.; Snoek, L. C. Infrared spectroscopy and structure of photochemically protonated biomolecules in the gas phase: a noradrenaline analogue, lysine and alanyl alanine. *Phys. Chem. Chem. Phys.* **2007**, *9*, 2549.

123. Wu, R.; McMahon, T. B. Infrared multiple photon dissociation spectroscopy as structural confirmation for GlyGlyGlyH(+) and AlaAlaAlaH(+) in the gas phase. Evidence for amide oxygen as the protonation site. *J. Am. Chem. Soc.* **2007**, *129*, 11312.

124. MacLeod, N.; Simons, J. Infrared photodissociation spectroscopy of protonated neurotransmitters in the gas phase. *Mol. Phys.* **2007**, *105*, 689.

125. Chin, W.; Piuzzi, F.; Dimicoli, I.; Mons, M. Probing the competition between secondary structures and local preferences in gas phase isolated peptide backbones. *Phys. Chem. Chem. Phys.* **2006**, *8*, 1033–1048.

126. Yan, B.; Jaeqx, S.; van der Zande, W. J.; Rijs, A. M. A conformation-selective IR-UV study of the dipeptides Ac-Phe-Ser-NH2 and Ac-Phe-Cys-NH2: probing the SH[three dots, centered]O and OH[three dots, centered]O hydrogen bond interactions. *Phys. Chem. Chem. Phys.* **2014**, *16*, 10770–10778.

127. Alauddin, M.; Biswal, H. S.; Gloaguen, E.; Mons, M. Intra-residue interactions in proteins: interplay between serine or cysteine side chains and backbone conformations, revealed by laser spectroscopy of isolated model peptides. *Phys. Chem. Chem. Phys.* **2015**, *17*, 2169–2178.

128. Bakker, D. J.; Ong, Q.; Dey, A.; Mahé, J.; Gaigeot, M.-P.; Rijs, A. M. Anharmonic, dynamic and functional level effects in far-infrared spectroscopy: Phenol derivatives. *J. Mol. Spectrosc.* **2017**, *342*, 4–16.

129. Stearns, J. A.; Boyarkin, O. V.; Rizzo, T. R. Spectroscopic signatures of gas-phase helices: Ac-Phe-(Ala)5-Lys-H+ and Ac-Phe-(Ala)10-Lys-H+. *J. Am. Chem. Soc.* **2007**, *129*, 13820.

130. Gloaguen, E.; Mons, M. Isolated neutral peptides. *Topics Curr. Chem.* **2015**, *364*, 225.

131. Abo-Riziq, A.; Grace, L.; Nir, E.; Kabelac, M.; Hobza, P.; de Vries, M. S. Photochemical selectivity in guanine–cytosine base-pair structures. *Proc. Nat. Acad. Sci.* **2005**, *102*, 20–23.

132. Chin, W.; Piuzzi, F.; Dognon, J.-P.; Dimicoli, I.; Tardivel, B.; Mons, M. Gas phase formation of a 310-helix in a three-residue peptide chain: Role of side chain-backbone interactions as evidenced by IRUV double resonance experiments. *J. Am. Chem. Soc.* **2005**, *127*, 11900–11901.

133. Bakker, D. J.; Peters, A.; Yatsyna, V.; Zhaunerchyk, V.; Rijs, A. M. Far-infrared signatures of hydrogen bonding in phenol derivatives. *J. Phys. Chem. Lett.* **2016**, *7*, 1238.

134. Perticaroli, S.; Russo, D.; Paolantoni, M.; Gonzalez, M.; Sassi, P.; Nickels, J.; Ehlers, G.; Comez, L.; Pellegrini, E.; Fioretto, D. *et al.* Painting biological low-frequency vibrational modes from small peptides to proteins. *Phys. Chem. Chem. Phys.* **2015**, *17*, 11423–11431.

135. Heyden, M.; Sun, J.; Funkner, S.; Mathias, G.; Forbert, H.; Havenith, M.; Marx, D. Dissecting the THz spectrum of liquid water from first principles via correlations in time and space. *Proc. Natl. Acad. Sci. U. S. A.* **2010**, *107*, 12068–12073.

136. Bohm, F.; Schwaab, G.; Havenith, M. Mapping hydration water around alcohol chains by THz calorimetry. *Angew. Chem. Int. Ed.* **2017**, *56*, 9981–9985.

137. Schwaab, G.; Sebastiani, F.; Havenith, M. Ion hydration and ion pairing as probed by THz spectroscopy. *Angew. Chem. Int. Ed.* **2019**, *58*, 3000–3013.

138. Gerhards, M.; Unterberg, C. Structures of the protected amino acid Ac-Phe-OMe and its dimer: A [small beta]-sheet model system in the gas phase. *Phys. Chem. Chem. Phys.* **2002**, *4*, 1760–1765.

139. Gerhards, M.; Unterberg, C.; Gerlach, A. Structure of a [small beta]-sheet model system in the gas phase: Analysis of the C[double bond, length half m-dash]O stretching vibrations. *Phys. Chem. Chem. Phys.* **2002**, *4*, 5563–5565.

140. Fricke, H.; Gerlach, A.; Gerhards, M. Structure of a [small beta]-sheet model system in the gas phase: Analysis of the fingerprint region up to 10 [small mu]m. *Phys. Chem. Chem. Phys.* **2006**, *8*, 1660–1662.

141. Isaienko, O.; Borguet, E. Hydrophobicity of hydroxylated amorphous fused silica surfaces. *Langmuir* **2013**, *29*, 7885–7895.

142. Pezzotti, S.; Serva, A.; Sebastiani, F.; Brigiano, F. S.; Galimberti, D.; Potier, L.; Alfarano, S.; Schwaab, G.; Havenith, M.; Gaigeot, M.-P. Molecular fingerprints of hydrophobicity at aqueous interfaces from theory and vibrational spectroscopies. *J. Phys. Chem. Lett.* **2021**, *12*, 3827–3836.

143. Gaigeot, M.-P.; Sprik, M.; Sulpizi, M. Oxide/water interfaces: How the surface chemistry modifies interfacial water properties. *J. Phys. Condens. Matter* **2012**, *24*, 124106.

144. Sulpizi, M.; Gaigeot, M.-P.; Sprik, M. The silica-water interface: How the silanols determine the surface acidity and modulate the water properties. *J. Chem. Theory Comput.* **2012**, *8*, 1037–1047.

145. Behler, J. First principles neural network potentials for reactive simulations of large molecular and condensed systems. *Angew. Chem. Int. Ed.* **2017**, *56*, 12828–12840.

146. Zuo, Y.; Chen, C.; Li, X.; Deng, Z.; Chen, Y.; Behler, J.; Czanyi, G.; Shapeev, A.; Thompson, A.; Wood, M. *et al.* Performance and cost assessment of machine learning interatomic potentials. *J. Phys. Chem. A* **2020**, *124*, 731–745.

147. Schran, C.; Behler, J.; Marx, D. Automated fitting of neural network potentials at coupled cluster accuracy: protonated water clusters as testing ground. *J. Chem. Theory Comput.* **2020**, *16*, 88–99.

148. Imbalzano, G.; Anelli, A.; Giofre, D.; Klees, S.; Behler, J.; Ceriotti, M. Automatic selection of atomic fingerprints and reference configurations for machine-learning potentials. *J. Chem. Phys.* **2018**, *148*, 241730.

149. Cole, D.; Mones, L.; Czanyi, G. A machine learning based intramolecular potential for a flexible organic molecule. *Faraday Discuss.* **2020**, *224*, 247–265.

150. Deringer, V.; Caro, M.; Czanyi, G. Machine learning interatomic potentials as emerging tools for materials science. *Adv. Mater.* **2019**, *31*, 1902765–1902780.

151. Dral, P.; Owens, A.; Dral, A.; Czanyi, G. Hierarchical machine learning of potential energy surfaces. *J. Chem. Phys.* **2020**, *152*, 204110.

152. Cheng, B.; Griffiths, R.; Wengert, S.; Kunkel, C.; Stenczel, T.; Zhu, B.; Deringer, V.; Bernstein, N.; Margraf, J.; Reuter, K. *et al.* Mapping materials and molecules. *Acc. Chem. Res.* **2020**, *53*, 1981–1991.

153. Ceriotti, M. Unsupervised machine learning in atomistic simulations, between predictions and understanding. *J. Chem. Phys.* **2019**, *150*, 150901.

154. Chmiela, S.; Tkatchenko, A.; Sauceda, H.; Poltavsky, I.; Schutt, K.; Muller, K.-R. Machine Learning of accurate energy-conserving molecular force fields. *Sci. Adv.* **2017**, *3*, e1603015.

155. Gastegger, M.; Behler, J.; Marquetand, P. Machine learning molecular dynamics for the simulation of infrared spectra. *Chem. Sci.* **2017**, *8*, 6924–35.

156. Veit, M.; Wilkins, D.; Yang, Y.; DiStasio, R.; Ceriotti, M. Predicting molecular dipole moments by combining atomic partial charges and atomic dipoles. *J. Chem. Phys.* **2020**, *153*, 024113.

157. Yang, Y.; Lao, K. U.; Wilkins, D.; Grisafi, A.; Ceriotti, M.; DiStasio, R. Quantum mechanical static dipole polarizabilities in the QM7b and AlphaML showcase databases. *Sci. Data* **2019**, *6*, 152–161.

158. Kananenka, A.; Yao, K.; Corcelli, S.; Skinner, J. Machine learning for vibrational spectroscopic maps. *J. Chem. Theory Comput.* **2019**, *15*, 6850–6858.

Chapter 12

Introduction to Vibropolaritons: Spectroscopy, Relaxation and Chemical Reactions

Raphael F. Ribeiro[*,**] and Joel Yuen-Zhou[†]

*Department of Chemistry and Cherry Emerson Center for Scientific Computation, Emory University
†Department of Chemistry and Biochemistry, University of California, San Diego
**rfloren@emory.edu

12.1. Introduction

Recent advances in the fabrication and patterning of nano and microscopic materials have enabled the discovery of novel chemical phenomena in optical cavities and other devices that confine electromagnetic fields to distances on the scale of an optical or infrared wavelength (Figure 12.2).[1-3] This subfield of chemistry gained widespread attention especially after the pioneering observations by the Ebbesen group that molecular photochemistry,[4] chemical reactivity[5,6] and energy transport[7] could be controlled with optical microcavities.

There remain many open questions on the mechanisms underlying the reported dramatic microcavity effects on chemical phenomena such as intermolecular energy transfer,[8] chemical equilibria[9] and reaction dynamics.[5,6,10] Nevertheless, it is understood that the *strong light–matter interactions* promoted by photonic devices are the fundamental features that enable the control of molecular materials. These interactions lead to the formation of hybrid quantum states denoted *molecular polaritons* consisting of a superposition of stationary waves (resonant modes) of the optical cavity and excited molecular states delocalized across extended regions of size of the order of a μm.

Most studies of polariton chemistry investigated the strong coupling of molecular systems with optical cavities resonant with transitions in the visible or infrared regions of the electromagnetic spectrum. Cavities with resonances in the visible region may undergo strong coupling with bright (optically allowed) electronic transitions of organic aggregates or chromophores with large absorbance,[4,11,12] whereas infrared cavities are generally employed to modulate the optical response or chemical properties of molecular systems with bright vibrational modes.[13–15] The hybrid excitations of organic materials under strong coupling with microcavities are denoted *organic polaritons*, whereas excitations emergent from strong interactions between molecular vibrations and the modes of infrared cavities are denoted *vibropolaritons*.

In this chapter, we present a detailed description of the theory of strong light–matter interactions in infrared cavities and the main consequences of vibropolariton formation. We explain the basics properties of polaritons and their emergence from strong light-matter interactions in optical cavities in Section 12.2, describe vibropolariton dynamics, nonequilibrium optical response and cavity-assisted inter-molecular energy transfer in Section 12.3, and provide our perspective on vibropolariton effects on chemical reactions in Section 12.4. Our goal is to provide a pedagogical description of the theory guided by selected experiments that in our opinion show either universal or striking features of molecular polaritons. Our considerations span the *collective strong coupling regime* where a large collection of molecules (estimated to be between 10^6 and 10^{12}) interact with near-resonant stationary states of an optical cavity (Figure 12.2). This choice is motivated by the vast majority of polariton chemistry experiments which require a macroscopic molecular ensemble to achieve vibrational strong coupling.

12.2. Fundamentals of Strong Light-Matter Interactions and Polariton Chemistry

12.2.1. *Review and phenomenology of weak light–matter interactions*

Light is a fundamental tool for the investigation of chemical systems. By characterizing the light emitted, absorbed, reflected or transmitted through a sample, we can characterize its equilibrium (*e.g.*, composition) and non-equilibrium (*e.g.*, how fast it relaxes back to the thermal state after application of stress) properties. Light is also a driver of physical and chemical transformations as evidenced, for example, by its fundamental role as source of energy in photosynthesis,[16] and by the various methods to control the chemical and physical properties of materials with sufficiently strong electromagnetic fields.[17–19]

Spectroscopy and photochemistry typically proceed in free space under conditions where the interaction between light and matter is extremely *weak* relative to the internal electrostatic fields holding atoms and molecules together.[20] Therefore, the action of light on matter can often be assumed to be *perturbative*: light induces transitions between the quantum states of a molecular system, but, the material states are insignificantly affected by the interaction with the electromagnetic field. In other words, the weak disruptions caused by most light sources are insufficient to appreciably change the observable properties of atoms and molecules (except for potentially provoking effectively irreversible changes to the state of a material as is sometimes the case in photochemistry). The weakness of free space light–matter interactions allows spectroscopy to be a useful technique to identify the energy levels of a system and to track its dynamics while disturbing the system only by changing its quantum state.

Note also that, typically, the energy exchange processes between light and matter are effectively *irreversible*: the energy initially carried by a photon and subsequently absorbed by a molecule is ultimately converted into heat or emitted via fluorescence. In either case, the excitation energy is quickly dissipated (lost) to a thermal reservoir.

These considerations justify the application of first-order time-dependent perturbation theory in the form of Fermi's Golden Rule (FGR)[20] for the computation of the rate of photon absorption (absorption spectrum) by a molecular system in the presence of a light beam. In particular, assuming an isotropic sample in a volume V containing N molecules of a size ($\sim$1–10 nm) much smaller than the wavelength λ_M of light irradiated into the system ($\sim$400–1000 nm), the rate of single-photon absorption events by the molecular ensemble takes the following form[21]:

$$\Gamma_A = \frac{2\pi}{3\hbar} \frac{\langle n(\omega_M)\rangle \hbar \omega_M}{2\epsilon_0 V} N |\mu_M|^2 \rho(\omega_M), \qquad (12.1)$$

where $n(\omega_M)$ is the mean number of photons in the incident beam in resonance with the molecular transition frequency ω_M, ϵ_0 is the electrical permittivity of free space, μ_M is the single-molecule dipole matrix element between the initial and final states (corresponding to a transition with frequency ω_M) and $\rho(\omega_M)$ is the single-molecule density of states at frequency ω_M.

A similar perturbative treatment of light–matter interaction leads to the following expression for the rate of spontaneous photon emission by a molecule in free space[21]

$$\Gamma_E = \frac{2\pi}{3\hbar} \frac{\hbar \omega_M}{2\epsilon_0 V} |\mu_M|^2 \rho_F(\omega_M), \qquad (12.2)$$

where $\rho_F(\omega_M) = 4\pi \omega_M^2 V / [\hbar(2\pi c)^3]$ is the density of free space electromagnetic field (photon) modes with frequency ω_M. Note the essential role played in the last equation by the photon density of states ρ_F. This quantity arises from a summation over all possible final states satisfying the energy conservation condition that any spontaneously emitted light must carry the energy released by the molecular system upon excited state decay.

Before concluding this subsection, we remind the reader that we have assumed throughout that the molecular sample is hosted by a

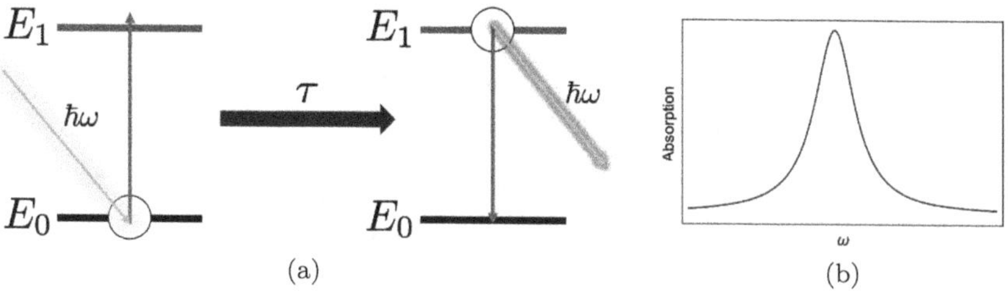

Figure 12.1. (a) Excitation of a molecular system via absorption of light with frequency $\omega = (E_1 - E_0)/\hbar$ and subsequent energy dissipation (after waiting time τ) by light emission or non-radiative decay (*e.g.*, conversion into heat). (b) Example of molecular absorption spectrum in a frequency region where the where the molecule has a single available optical transition. Spectroscopy is a useful tool for the characterization of molecules because the perturbation induced by light on the molecular system is typically weak enough that its energy levels are not significantly disturbed by the electromagnetic field. Therefore, the absorption spectrum of a molecular system reveals intrinsic information about its energy levels and quantum state transitions.

macroscopic medium where light propagation is almost indistinguishable from that in free space (neglecting absorption and refraction index changes due to the presence of a material system). This assumption is acceptable in many instances, but may break down severely in microscopic regions of space confined by moderate quality mirrors as we discuss next.

12.2.2. *Light–matter interactions in optical cavities*

Consider an optical cavity consisting of two planar mirrors (Figure 12.2) separated by a distance L of the order of a micron (10^{-6} m). It can be shown that this device supports electromagnetic standing waves satisfying the condition[1]

$$\frac{n\lambda}{2} = Lcos(\phi), \tag{12.3}$$

where $n \geq 1$ is any natural number and ϕ corresponds to the angle between the cavity longitudinal axis (z axis, Figure 12.2) and any direction parallel to the mirrors (also known as incidence angle). These standing waves are denoted *cavity modes* and their frequency

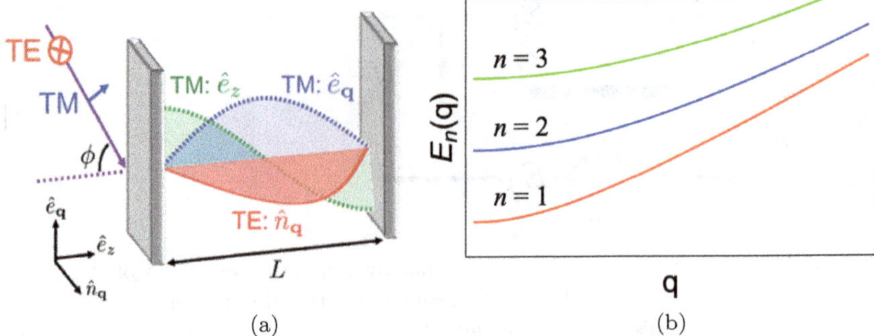

Figure 12.2. (a) Planar optical cavity obtained from the introduction of two planar mirrors placed perpendicular to the z-axis and separated by the longitudinal length L. (b) Resonance spectrum for a planar optical cavity given as energies versus in-plane wave-vector magnitude $q = \sqrt{q_x^2 + q_y^2}$ for each cavity band represented by the integer $n \geq 1$. *Source*: Adapted from Ref. 2.

is given by

$$\omega_n(\mathbf{q}_\parallel) = \frac{c}{\sqrt{\epsilon}} \sqrt{\frac{n^2 \pi^2}{L^2} + q_x^2 + q_y^2}, \tag{12.4}$$

where ϵ is the frequency-independent dielectric constant of the intracavity medium (from now on we take $\epsilon = 1$ for simplicity), and q_x and q_y are real numbers corresponding to the momentum carried by the standing wave along the directions parallel to the mirrors (in-plane directions). Each cavity mode $(n, \mathbf{q}_\parallel)$ also has a specific direction for the electric field: transverse-electric (the electric field at each point is orthogonal to the cavity axis) or transverse magnetic (the magnetic field is orthogonal to the cavity axis). The quantum number n is simply related to the number of nodes $n-1$ of the cavity mode profile along the z axis, whereas $\mathbf{q}_\parallel$ is a continuous 2D vector because we limit our considerations to Fabry–Perot (FP) cavities employing mirrors with macroscopic dimensions (roughly 10^{-3} m or larger). This feature implies that energy quantization is irrelevant along the cavity in-plane directions and, in principle, q_x and q_y are continuous unbounded variables.

In clear contrast to free space, an optical microcavity has a low-energy cutoff in its wave spectrum: the lowest energy standing wave is that with no nodes ($n = 1$) and zero in-plane momentum ($q_x = q_y = \phi = 0$). From Eq. (12.4), we find the cavity low-energy cutoff to be given by

$$E_c \equiv \hbar\omega_1(\mathbf{0}) = \frac{\hbar c\pi}{L}. \tag{12.5}$$

It follows that photons with energy $E < E_c$ are not supported by a FP cavity with length L. This fact has outstanding consequences: emission of photons with $E < E_c$ would be highly suppressed in these devices! This is our first example of the fundamental fact that the dynamics of molecular systems may be sharply dependent on its electromagnetic environment. For instance, if fluorescence is the dominant mechanism for decay of a particular molecular excited state, then in an optical cavity that suppresses the fluorescent decay, the molecular excited state will have longer lifetime and energy dissipation would proceed via a new microscopic mechanism (although in reality the suppression is far from perfect due to various material imperfections).

Purcell showed in fact that the rate of spontaneous emission of light by a material system is a function of its environment.[22] The *Purcell effect* denotes the enhancement or suppression of light emission by a material in an optical cavity. Spontaneous emission acceleration may happen because the vacuum fluctuations of the cavity electromagnetic field have greater intensity relative to free space, thus leading to faster radiative decay. These enhanced field fluctuations are associated to a greater *local* density of modes of the confined electromagnetic field in an optical cavity. In fact, the density of states at the resonance energies of a suitable optical cavity (with significant separation between modes with equal in-plane momentum but distinct longitudinal quantum numbers $n \geq 1$) is enhanced relative to that in free space by a linear function of the cavity quality factor per unit length $Q/L = \omega_C/(L\kappa)$, where ω_C is the frequency of a representative cavity mode and κ is the photon escape (leakage) rate.[1] Observations of mild suppression and enhancement

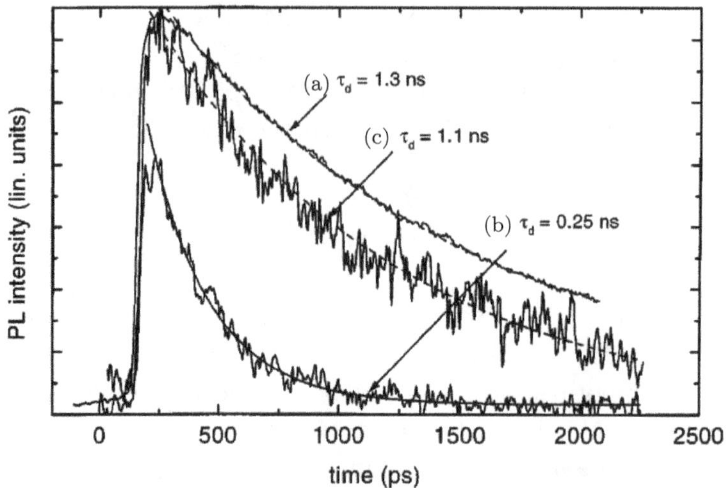

Figure 12.3. Light emission dynamics of excited state population of semiconductor (GaAs) quantum dots in (a) bulk environment, (b) in a pillar optical cavity resonant with the corresponding quantum dot transition, and (c) in an optical cavity slightly off-resonant with the ground to excited state transition of the quantum dot. A shorter decay time τ_d indicates faster light emission by the quantum dot. Therefore, introduction of the sample into a resonant optical cavity leads to a factor of five enhancement of the light emission by a quantum dot relative to its bulk emission rate. *Source*: From Ref. 23.

of spontaneous emission of semiconductor quantum dots in optical cavities were reported, *e.g.*, in Ref. 23 (Figure 12.3).

The Purcell effect consists of the change in the rates of radiative processes in optical cavities when compared to free space. Nevertheless, although distinct in magnitude, the interactions between light and matter remain weak and effectively irreversible.

12.2.3. *Strong light–matter interactions and hybrid modes in optical cavities*

In addition to the Purcell effect, optical microcavities also present the opportunity for reversible exchange of energy between the cavity electromagnetic field and matter. To see this, note that if the cavity mirrors have high reflectivity, any photon emitted by a molecule may be reabsorbed after one or more round-trips through the optical device. This analysis therefore suggests that a new regime of interaction between radiation and matter with qualitatively new phenomena emerges when the light–matter interaction is stronger than dissipative interactions (*i.e.*, those leading to photon leakage or heat conversion) acting on the molecular and photonic subsystems. In this regime, light and matter are said to be under *strong coupling*. Figure 12.4 shows experimental observation of the transition from the weak (Purcell) to the strong light–matter interaction regime as observed by Simpkins *et al.*[24] for the organic polymer PMMA in an infrared optical cavity resonant with its carbonyl stretch modes.

The evolution of the spectra presented in Figure 12.4 with molecular concentration clearly shows new phenomenology in the presence of a sufficiently high density of molecules with large oscillator strength at a transition resonant with an optical cavity. In order to explain this result, we introduce the *Tavis-Cummings* (TC) model for the description of light–matter interactions in an optical cavity.[25, 26] In the TC model, the cavity is represented by a single mode (*e.g.*, that with $n = 1$ and $\mathbf{q}_{\parallel} = 0$ and arbitrary polarization) with frequency ω_c, while all other cavity degrees of freedom are ignored (we provide comments on this and other assumptions at the end of this section). The molecular system contains N non-interacting molecules (it is assumed that the intermolecular interactions are negligible relative to the effects of strong light–matter interactions). Each molecule is represented by a two-level system with transition frequency ω_M (Figure 12.1), so the measurement of a molecule's quantum state may yield ground (0) or excited (1). The non-interacting part of the total Hamiltonian of light and matter (restricted to the subspace of the light–matter Hilbert space containing a single excitation of the molecular *or* cavity mode subsystems) can be written as

$$H_0 = \sum_{i=1}^{N} \hbar\omega_M |1_i\rangle\langle 1_i| + \hbar\omega_c |1_c\rangle\langle 1_c|, \tag{12.6}$$

where we assume that the global ground state $|0\rangle$ (where all molecules are in the ground state and the cavity has no photons) has zero energy, $|1_c\rangle$ is the quantum state with all molecules in the ground

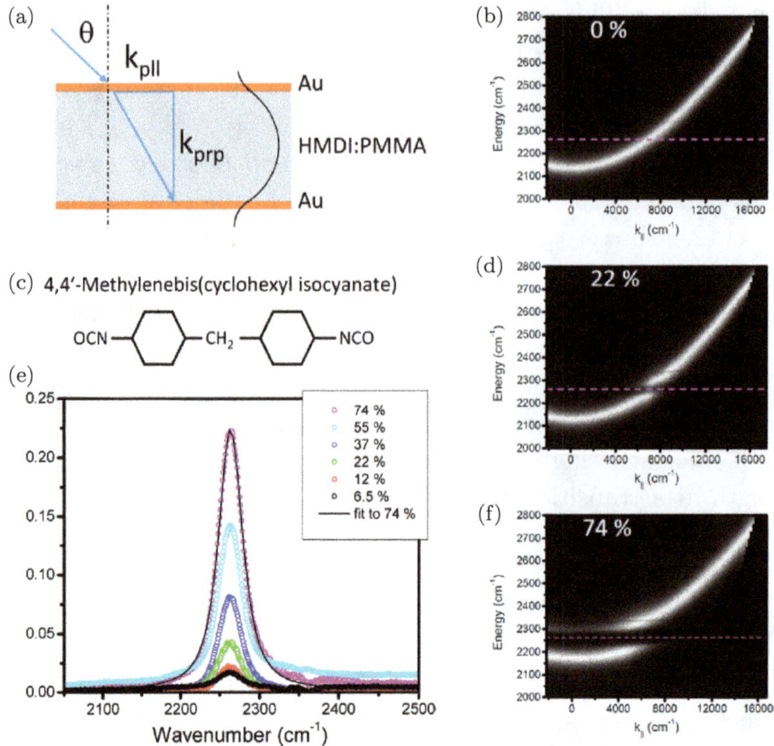

Figure 12.4. Left: (a) A planar cavity obtained from two parallel gold planar mirrors separated by a polymer spacer (HMDI:PMMA) where a liquid solution of (c) 4,4′-methylenebis(cyclohexyl isocyanate) is inserted. (e) Absorption spectrum of the compound in free space as a function of its concentration. Right: (b) Transmission spectrum of an *empty* infrared cavity as a function of the in-plane wave-vector magnitude $k_\parallel$ corresponding to incidence angle θ. The bright regions of the spectrum correspond to the resonant modes of the cavity. The dashed pink horizontal line lies at the peak of the absorbance spectrum of the molecular system in free space. This picture shows that the molecules are resonant with the cavity at a particular in-plane wave-vector where the white and pink curves cross. (d) When the molecular concentration is 22% a visible transmission loss can be observed in the cavity transmission for incidence angles θ corresponding to stationary waves of the optical cavity nearly resonant with the molecular transition. A band gap is formed. (f) At 74% concentration, the band gap is appreciably larger, and we can clearly see the existence of two new energy bands in clear contrast to the empty cavity spectrum. This indicates the formation of new normal modes (or alternatively) new excited states for the optical cavity in the presence of a sufficiently dense molecular solution. *Source*: Adapted from Ref. 24.

state, while the cavity is in its first excited state (*i.e.*, a single photon is in the system), and $|1_i\rangle$ denotes the state with the ith molecule in its excited state while all other molecules and the cavity are in the ground state.

The cavity electromagnetic field may promote quantum transitions between the molecular states via emission or absorption of photons according to the interaction Hamiltonian[a]

$$H_{\mathrm{LM}} = -i \sum_{i=1}^{N} \sqrt{\frac{\hbar\omega_c}{2\epsilon_0 LS}} \mu_M \left(|1_i\rangle\langle 1_c| - |1_c\rangle\langle 1_i| \right), \qquad (12.7)$$

where μ_M is the magnitude of the transition dipole moment of each molecule. It is assumed that all of these transition dipoles are aligned with the electric field of the selected cavity mode. The operator $|1_i\rangle\langle 1_c|$ promotes photon annihilation with simultaneous excitation of the ith molecule, whereas $|1_c\rangle\langle 1_i|$ leads to the reverse process where an excited molecule goes to the ground state and a cavity photon is created.

The light–matter interaction Hamiltonian takes a particularly simple form when we introduce the definition of the molecular bright state

$$|1_B\rangle = \sum_{i=1}^{N} \frac{1}{\sqrt{N}} |1_i\rangle. \qquad (12.8)$$

Specifically, this expression can be employed in the definition of H_{LM} to obtain the following expression for the interaction between a single cavity mode (with $q = 0$) and a molecular ensemble with N molecules:

$$H_{\mathrm{LM}} = -i\sqrt{\frac{\hbar\omega_c N}{2\epsilon_0 V}} \mu_M \left(|1_B\rangle\langle 1_c| - |1_c\rangle\langle 1_B| \right). \qquad (12.9)$$

---

[a]This Hamiltonian is given in the multipolar gauge[21] after several key approximations. In particular, we employ the electrical dipole and rotating-wave approximations,[27] and also ignore the self-polarization term.[28] All of these approximations are deemed reasonable for the description of systems operating in the regime of collective strong light–matter interactions.[28,29]

Thus, we see that the interaction of the cavity mode and the material system gains a $\sqrt{N}$ boost when matter consists of an ensemble with N absorbers. Importantly, by writing the light–matter interaction in this form we see that the cavity mode interacts with a single *delocalized* molecular excited state. Given that a change of basis for the singly excited molecular states is a unitary transformation in the manifold of molecular states with a single excitation, there are $N-1$ states $|D_\mu\rangle, \mu = 1, 2, ..., N-1$ orthogonal to $|1_B\rangle$ which are unaffected by light and remain degenerate with energy $\hbar\omega_M$. Using the delocalized (symmetry-adapted) molecular basis formed by $\{|1_B\rangle, |D_1\rangle, ..., |D_{N-1}\rangle\}$, the total Hamiltonian can be expressed as

$$H = H_0 + H_{\mathrm{LM}}, \tag{12.10}$$

$$H = \sum_{\mu=1}^{N-1} \hbar\omega_M |D_\mu\rangle \langle D_\mu| + \hbar\omega_M |1_B\rangle \langle 1_B|$$

$$+ \hbar\omega_c |1_c\rangle \langle 1_c| - i\frac{\Omega_R}{2} \left(|1_B\rangle \langle 1_c| - |1_c\rangle \langle 1_B|\right), \tag{12.11}$$

where $\Omega_R/2 = \mu_M \sqrt{\hbar\omega_c N/2\epsilon_0 V}$ is the *collective light–matter interaction strength* and Ω_R is the *Rabi splitting*. In a matrix representation, the Hamiltonian can be written in the following block-diagonal form

$$H = \begin{pmatrix} [\hbar\omega_M]_{N-1\times N-1} & 0 & 0 \\ 0 & \hbar\omega_M & i\Omega_R/2 \\ 0 & -i\Omega_R/2 & \hbar\omega_c \end{pmatrix}, \tag{12.12}$$

where the notation $[\hbar\omega_M]_{N-1\times N-1}$ denotes a *diagonal* $N-1 \times N-1$ matrix where each non-vanishing element is equal to $\hbar\omega_M$. The Hamiltonian matrix indices $i = 1, ..., N-1$ correspond to the non-interacting dark states $|D_\mu\rangle$ (Eq. (12.11)), while the indices N and $N+1$ correspond to the molecular bright mode $|1_B\rangle$ (Eq. (12.8)) and the cavity photon $|1_c\rangle$, respectively. As indicated in Eq. (12.11), the bright molecular state and the cavity mode interact *exclusively* with each other in the TC model.

Our choice of symmetry-adapted basis states for the molecular system has allowed the reduction of the initial $N+1$-body Hamiltonian into the significantly simpler two-state Hamiltonian

$$H_B = \begin{pmatrix} \hbar\omega_M & -i\Omega_R/2 \\ i\Omega_R/2 & \hbar\omega_c \end{pmatrix}.$$
(12.13)

The eigenmodes of H_B are denoted lower and upper polariton $|\text{LP}\rangle$ and $|\text{UP}\rangle$, respectively. They are hybrid modes of light and matter with energies E_{LP} and E_{UP}, respectively. These eigenvalues and eigenstates can be written as

$$E_{\text{LP}} = \frac{\hbar\omega_c + \hbar\omega_M}{2} - \frac{1}{2}\sqrt{\Omega_R^2 + (\omega_c - \omega_M)^2}, \quad |\text{LP}\rangle$$

$$= -\sin(\theta/2)e^{-i\phi/2}|1_B\rangle + \cos(\theta/2)e^{i\phi/2}|1_c\rangle,$$
(12.14)

$$E_{\text{UP}} = \frac{\hbar\omega_c + \hbar\omega_M}{2} + \frac{1}{2}\sqrt{\Omega_R^2 + (\omega_c - \omega_M)^2}, \quad |\text{UP}\rangle$$

$$= \cos(\theta/2)e^{-i\phi/2}|1_B\rangle + \sin(\theta/2)e^{-i\phi/2}|1_c\rangle,$$
(12.15)

where $\tan(\theta) = 2\Omega_R/(\hbar\omega_M - \hbar\omega_c)$, $0 \le \theta < \pi$ and $\phi = \pi/4$.[30] Figure 12.5 presents a pictorial description of the formation of the polaritons and dark excitations in two different situations: $\omega_M = \omega_c$ (on-resonance or zero detuning) and $\omega_M \gg \omega_c$ (red detuning, since the cavity mode has lower frequency than the molecular transition) — the blue-detuned case where $\omega_c > \omega_M$ is analogous to the red-shifted case (with a reversal of roles for the molecular system and the cavity). The on-resonance scenario provides equal molecular and photonic character to each polariton. Conversely, when the cavity is detuned from the molecular system ($\omega_M \ne \omega_c$), the UP is dominated by the mode with higher energy, whereas the LP has greater content of the subsystem with lower energy. This expected feature of the polariton excitations may also be directly obtained from Eqs. (12.14) and (12.15), *e.g.*, when $\omega_c - \omega_M \gg \Omega_R$, we may ignore Ω_R in Eqs. (12.14) and (12.15) to obtain $E_{\text{LP}} \approx \hbar\omega_M$, $|\text{LP}\rangle \approx |1_B\rangle$, and $E_{\text{UP}} \approx \hbar\omega_c$, $|\text{UP}\rangle \approx |1_c\rangle$. Note that the dark states are completely insensitive to variations of the cavity detuning

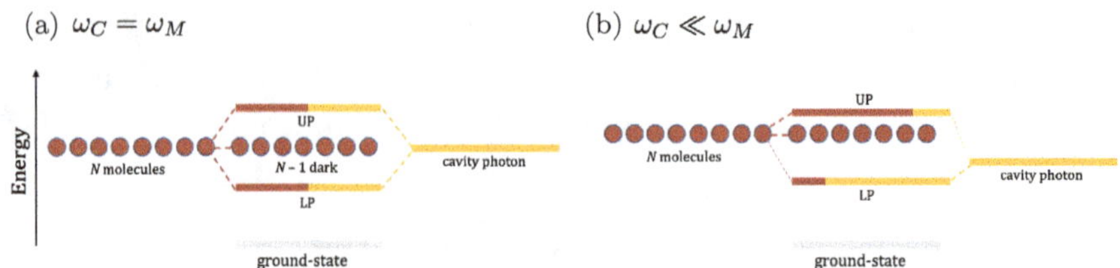

Figure 12.5. State correlation diagrams corresponding to the formation of polaritons and dark modes according to the TC model for the interaction between a single cavity mode and a collection of molecular systems. (a) The energy spectrum of the hybrid system for the case where the molecules are on-resonance ($\omega_c = \omega_M$) with the optical cavity indicates the formation of hybrid polaritonic excitations with equal contributions from light and matter degrees of freedom. $N - 1$ dark states ("non-bonding orbitals") are shown with the same energy as the initial molecular states. (b) Polariton formation diagram for the case where the cavity is red-shifted ($\omega_c < \omega_M$) relative to the molecular system. This off-resonance condition creates an asymmetry: the lower energy hybrid mode is now dominated by the cavity, whereas the UP excitation is dominated by its molecular content. The ground state and the dark states are not shifted in energy relative to the on-resonance case shown left since these states are unaffected by the light–matter interaction in the TC model.

as expected given that in the TC model, the cavity only interacts with the bright superposition of molecular excited states and there are always $N - 1$ molecular dark states.

Despite their simplicity, the results provided in this section may already be employed to explain a variety of experimental observations associated to the behavior of molecules in optical cavities. For instance, the transition from weak to strong coupling in the data presented by Simpkins *et al.*[24] (Figure 12.4) can be explained as follows: when the concentration of the molecular system is small, the collective light–matter interaction strength $\Omega_R/2 \propto \sqrt{N}$ is small enough that dissipative processes acting on the molecular bright state and/or the photonic subsystem prevent polariton formation and their signatures are absent from the optical spectra of the composite system. In this case, the light–matter interaction is effectively irreversible and may be treated perturbatively. When N is sufficiently large that $\Omega_R/2$ is greater than the interaction strength of the cavity and molecular bright state with their respective environments, the hybrid excitations of the composite material become sufficiently long-lived and markedly distinct from independent molecular and cavity excitations. In particular, the emergence of polaritons is observed with energies sufficiently distinct from the empty cavity and molecular subsystem at low concentration. Indeed, the energy gap between the polariton modes has been repeatedly observed to be proportional to the square root of the molecular density in agreement with the predictions of the TC model. Therefore, polaritonic signatures become visible at sufficiently high molecular concentrations as indicated in Figure 12.4.

A quantitative criterion for the emergence of strong light-matter interactions and polaritons in optical cavities following from this analysis is that $\Omega_R > \gamma_M + \kappa$, where γ_M is the molecular absorption band linewidth and κ is the rate of cavity photon escape (leakage). This condition finds support from the fact that the quantities γ_M and κ provide rough measures for the strength of dissipative interactions acting on the molecular and photonic systems, respectively.

Our discussion of the TC model incorporating a single-cavity mode under strong interaction with an N-molecule ensemble employed various approximations. For instance, the TC model assumes that a single photon mode interacts with a large number N of molecules and this leads to two polariton states and $N - 1$ dark states. However, in optical cavities where the electromagnetic field is confined along one (FP cavities) or two directions (photonic wires), there exists a continuous set of cavity standing waves labeled by the in-plane wave-vector $\mathbf{q}_\parallel$ corresponding to the various incidence directions for input radiation (Figures 12.2 and 12.4). Polaritons formed with cavity modes that are highly off-resonant with the molecular system are likely to be indistinguishable from pure molecular or photonic quantum states (see, *e.g.*, the diagram on the r.h.s of Figure 12.5 — in the case where $|\omega_c - \omega_M| \gg \omega_R/2$, we can essentially ignore the hybridization of cavity and matter modes) and may be ignored. Nevertheless, cavity modes which are nearly resonant with the molecular ensemble *do* form hybrid polaritonic excitations. Indeed, Figure 12.4 shows experimental data that supports the view that cavity photons with energy close to the mean molecular transition frequency hybridize with molecular excited states to form lower and upper polariton states with well-defined values of $q_\parallel$ indicating the in-plane wave-vector may remain a nearly conserved quantity even in the presence of a disordered molecular system. These observations suggest a simple multimode generalization of the TC model that would retain the simplicity of Figure 12.5. In particular, we may view the polaritons with distinct wave-vectors $\mathbf{q}_\parallel$ to be non-interacting modes originating from the exclusive interaction between cavity modes with $\mathbf{q}_\parallel$ and corresponding symmetry-adapted molecular bright states $|1_B(\mathbf{q}_\parallel)\rangle$. The energies and eigenstates of the polaritons with distinct wave-vectors would be obtained from Eqs. (12.14) and (12.15). In this way, the complete polariton spectra shown in Figure 12.4 would be reproduced.

In this simple multimode generalization of the TC model, a number $N_D = N - N_c$ of dark states would also be formed. Like in the single-mode case (Figure 12.5), these dark states are purely molecular and completely unaffected by the cavity. Note that we have implicitly assumed $N > N_c$. This holds because the density of molecular excitations ρ_M with transition energy $\hbar\omega_M$ is generally much larger than the density of cavity photon states at this same frequency.[31] This ensures the reliability of computational models adopting a number of molecules greater than cavity modes. To see that the number of molecular states per unit area is much greater than the number of photon states, recall that a polyatomic molecule occupies a spatial region of size on the order of 1 nm^3, whereas optical radiation of wavelength $\lambda = 10^2 - 10^4$ nm occupies a much larger volume λ^3. Furthermore, only the small fraction of cavity modes nearly resonant with the molecules interact effectively with the latter. Therefore, the number of dark reservoir states is always much larger than the number of polaritons, and it is often productive (especially when analyzing optical response data) to view polaritons as consisting of independent hybrid excitations emergent from the strong interaction of a single cavity photon mode with specific $\mathbf{q}_\parallel$ and a large ensemble of molecules. A formal verification of the relative density of states estimates given above may be found in Refs. 31 and 32.

Another essential approximation of the TC model is that it assumes the molecules have equal transition frequencies ω_M and transition dipole matrix elements μ_M parallel to the cavity mode electric field. These assumptions lead to a many-body Hamiltonian that is symmetric under permutation of the molecular labels. Therefore, symmetry-adapted molecular delocalized states can be constructed to obtain the polariton and dark eigenstates of H_{TC} (and likewise, the same idea can be employed with the discussed multimode generalization explained above). This convenience is not the only reason that it is often reasonable to take the molecules to have the same excitation frequency and transition dipole moment. Generally, although molecular excitation frequencies undergo fluctuations induced by the solvent environment, these variations are small enough that at any moment, a vast majority of molecules have transition frequencies nearly equal to the mean. The approximation that the molecular transition dipole moments are equal is justified for isotropic bulk samples undergoing strong light–matter coupling on the basis that there will always be a large number of molecules in the sample with significant transition dipole moment projection along any cavity electric field direction. In the TC model, we include only those molecules with transition dipole aligned or nearly aligned with the electric field of the chosen cavity mode. Molecules that do not satisfy this condition will contribute negligibly to the considered polariton modes and are therefore disregarded without much consequence for the description of the polariton signatures in optical response spectra such as those visible from Figure 12.4. Nevertheless, except for the very small fraction of molecules that either have transition frequencies highly off-resonant with all cavity modes, or are located in negligible regions of space where the cavity electric field vanishes (*e.g.*, at nodes or near the mirrors), every molecule contributes appreciably to *some* polaritonic wave function.

In the next section, we will specialize our discussion of molecular polaritons to the case of vibropolaritons emergent from vibrational strong coupling with infrared cavities. We will describe a simple open quantum system generalization of the TC model (Eq. (12.11)) to explain essential aspects of polariton and dark state non-equilibrium dynamics (relaxation kinetics) and the optical signatures of the non-equilibrium excited state population obtained from pump-probe spectroscopy experiments.[15,33,34] This analysis will further corroborate the reliability of the standard polariton picture presented in this subsection.

Nevertheless, we conclude this section by pointing out that although the TC model (and its simple generalizations that *e.g.*, add an additional quantum degree of freedom to each molecule,[31,35,36] introduce another molecular ensemble with a distinct transition frequency,[37] describe the molecular system as a three-level system,[38] *etc.*) has been successful to a large extent in rationalizing and predicting *optical response data from experiments in the infrared and UV-visible*, the theories described in this Section fail[39–41] to explain the observations of large effects of optical cavities on chemical thermodynamics, reaction kinetics,[5,6,10] and intermolecular vibrational energy transfer.[8] We revisit this point in the context of vibropolariton effects on intermolecular energy transfer and chemical reactions briefly in Section 12.3.5, and in detail in Section 12.4, respectively.

12.3. Vibropolariton Dynamics and Spectroscopy

12.3.1. *Infrared strong coupling and the formation of vibropolaritons*

Light–matter strong coupling is more easily achieved when the photonic subsystem is nearly resonant with large oscillator strength electronic excited state transitions of a high-density molecular system leading to giant Rabi splittings Ω_R. In fact, the first experiments showing molecular cavity-polariton formation were performed with organic J-aggregates in solution.[11] Molecular vibrational transitions generally have much weaker oscillator strength relative to electronic partly due to the smaller transition dipole moments associated to infrared excitations. From this perspective, it would seem challenging for strong interactions to exist between infrared optical cavities of moderate quality and molecular vibrations (note that infrared wavelengths are also larger by a factor of 10–1000 relative to

UV-visible, so infrared cavities tend to have significantly longer L and reduced local density of photon states relative to UV-visible cavities). However, vibrational excitations have weaker interactions with their environment relative to electronic (as evidenced, *e.g.*, by the typically much smaller linewidths of strong infrared transitions). Therefore, the magnitude of Ω_R required to form vibropolaritons is much smaller than that needed for molecular electronic transitions to undergo strong interactions with optical cavities.

In fact, in 2015, Long and Simpkins,[13] and George *et al.*[14] provided the first reports of infrared cavity strong coupling and (cavity)-vibropolariton emergence. These experiments showed optical vibropolaritonic spectra in agreement with the picture given by the TC model as detailed in Section 12.2.3. Given the significant interest in employing vibrational strong coupling to control inter and intramolecular vibrational energy flow, and chemical reactivity (Section 12.4), Dunkelberger *et al.*[15] and subsequently Xiang *et al.*[33] examined the dynamics and non-equilibrium optical properties of vibropolaritons via pump-probe and 2D infrared spectroscopy, respectively. These experiments showed intriguing aspects of infrared strong coupling dynamics that we examine from a theoretical perspective below.

In Section 12.3.2, we examine the most salient features of the pump–probe transmission spectra obtained under vibrational strong coupling in Refs. 15 and 33. This analysis motivates the examination in Section 12.3.3 of the elementary energy loss processes and corresponding timescales relevant to the dynamics of vibropolaritonic and dark excitations. In Section 12.3.4, we introduce a theory of vibropolariton pump–probe optical response that allows us to connect the experimental data with the dynamics described in Section 12.3.3. We conclude our discussion of polariton relaxation kinetics and spectroscopy with an application of the main ideas discussed in Sections 12.3.2– 12.3.4 to analyze the recent observations of enhanced intermolecular vibrational energy transfer under infrared cavity strong coupling.[8] This discussion will highlight an important open problem in the field of polariton chemistry.

12.3.2. *Vibropolariton pump–probe spectroscopy: Experiments*

In pump–probe spectroscopy, a sample is irradiated with a strong light source ("pump") at time $t = 0$ and is later probed at time t with a weak light beam ("probe"). The time elapsed between the action of the pump and probe fields on the material sample is called the pump–probe delay time Δt_{pp}. The difference between the time-dependent probe transmission spectrum $T_{pp}(t)$ of a material and its transmission spectrum in the absence of any pumping T_0 provides information about the excited state dynamics and high-energy excitations of molecular systems. For instance, the (average) time τ where $T_{pp}(\tau) \approx T_0$ is decay time of the excited state population created by the pump. Thus, the differential pump–probe transmission spectrum

$$\Delta T_{pp}(t) = T_{pp}(t) - T_0, \tag{12.16}$$

provides fundamental information about dynamics of non-equilibrium distributions of molecular excited states in materials.

Given the exotic features of polaritons as hybrid light–matter excitations, the puzzling effects of vibrational strong coupling on thermal reactivity first observed by Thomas *et al.*[5] and the promises of vibropolariton applications in enhanced sensing and quantum information science, it was not long after the first demonstration of strong interactions between optical cavities and vibrational excitations that the nonequilibrium dynamics of vibropolaritons was studied via pump–probe spectroscopy by Dunkelberger *et al.*[15] These authors obtained pump–probe spectra of vibropolaritons consisting of a superposition of carbonyl asymmetric stretch modes of 10 mM $W(CO)_6$ (in a hexane solution) and infrared optical cavity modes.

Figure 12.6 contrasts the pump–probe spectra reported for the $W(CO)_6$-cavity vibropolaritons obtained at $\Delta t_{pp} = 3$ and 100 ps to the corresponding spectra for the same system outside of a cavity. Clear distinctions can be seen between these two spectra, as expected given that the optical response is controlled by vibropolaritons in the experiments that occurred under strong light–matter interactions, whereas outside of a cavity, the equilibrium and non-equilibrium

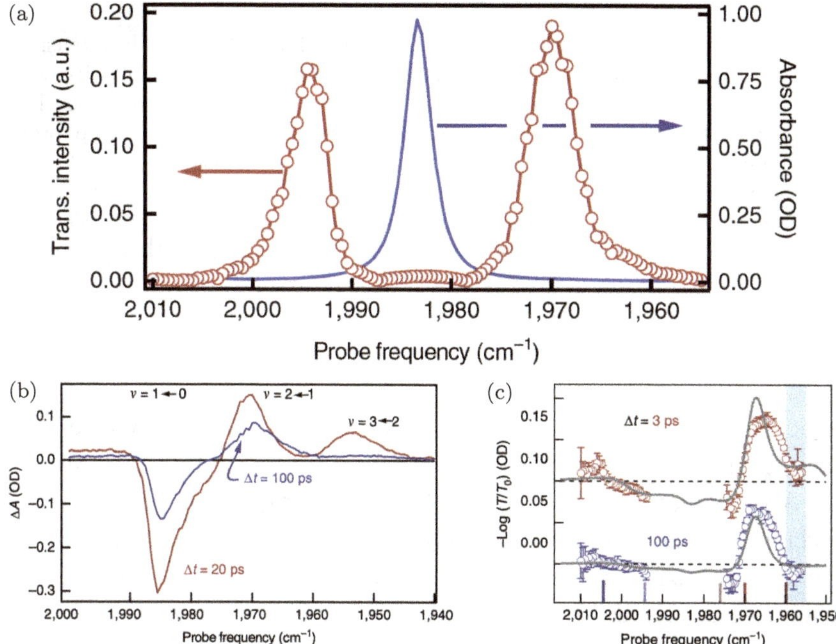

Figure 12.6. (a) Absorption spectrum (blue) of 10 mM $W(CO)_6$ solution in hexane and transmission spectrum (red) of the same solution in a resonant optical cavity with $L = 25\ \mu m$ filled by a spacer and a 10-mM solution of $W(CO)_6$ in hexane. The incidence angle is chosen to match photon mode in resonance with the carbonyl asymmetric stretch vibrational mode of $W(CO)_6$. (b) Pump–probe differential absorption spectra ($\Delta A = A_{pp} - A_0$ where A_0 is the absorption spectrum of the sample before pumping) of the $W(CO)_6$ solution outside of an optical cavity obtained 20 and 100 ps after pumping the system with an incident pulse containing photons in the frequency region where the asymmetric carbonyl stretch of the molecule absorbs radiation. In this figure, a negative value of ΔA implies that less light is absorbed at the corresponding frequencies, whereas a positive ΔA indicates an increased photon absorption rate relative to the system absorption before pumping. In fact, if the molecular system is excited by a pump field resonant with $0 \to 1$ transition of the carbonyl asymmetric stretch, then a population of molecules will be driven into vibrational excited states. Therefore, *less* light will be absorbed by the molecular system at the frequency v_{01} and $\Delta A < 0$. The excited state transitions $1 \to 2$, $2 \to 3$ all display $\Delta A > 0$ because the excited state population generated by the pumping process may now absorb light from the probe pulse thus leading to reduced light transmission at the associated excited state transitions. Note that stimulated emission provoked by the probe pulse on the excited state population leads to more photons emerging from the pumped sample. Therefore stimulated emission processes provide an

Figure 12.6 (*Continued*).   additional negative contribution to ΔA. The intensities of the pump–probe signals at the various frequencies show simple decay as the molecular excited state populations decay from their excited states into the ground state taking A_{pp} closer to A_0 as time increases. (c) Pump–probe transmission spectrum obtained for a 40-mM hexane solution of $W(CO)_6$ in a resonant optical microcavity. Both at 3 and 100 ps, the spectrum shows a significantly reduced transmission and blue shift of the LP mode. The UP resonance undergoes a redshift, but displays much weaker signals than the region around the LP. *Source*: Adapted from Ref. 15.

optical response of the sample is governed by localized vibrational excitations.

A stark feature of the vibropolaritonic pump–probe transmission spectrum is the strong asymmetry between the optical responses at the LP and UP frequencies. In particular, LP shows large $|\Delta T_{pp}|$, whereas the signals around the UP frequency are much weaker at both 3 and 100 ps pump-probe delay times. This was initially a surprising feature because the TC model predicts that the linear optical response of LP and UP is essentially symmetric (see top of Figure 12.6) when the incidence angle (of the external pump and probe radiation) is that corresponding to the cavity mode which is resonant with the molecular system. Another intriguing result provided by the polaritonic pump–probe measurements was that the peak near the LP frequency underwent a blueshift upon pumping the material, whereas the UP resonance shows an inverse redshift to lower frequencies thus indicating a contraction of the Rabi splitting.

In order to decode the presented vibropolariton pump–probe spectra and understand exactly what physical or chemical phenomena are responsible for the described pump–probe signals, it is necessary to analyze what happens to polaritonic excitations after they are pumped by an external source, as ultimately, the vibropolariton pump–probe spectra are manifestations of the non-equilibrium state of the cavity-matter system (initially generated by the pump) around the time where the probe irradiates the strongly coupled sample.

12.3.3. *Vibropolariton dynamics*

Our discussion of strong light-matter interactions in optical cavities in Section 12.2.3 assumed that both the optical cavity and the molecular system were *lossless, i.e.*, that these two sub-systems form a closed system isolated from the rest of the universe. In reality, optical cavities are leaky, since the reflectivity of metallic or dielectric mirrors is always less than 100%.[1] Similarly, the molecular excited states that generally undergo infrared strong coupling interact and may exchange energy with a thermal bath consisting of intramolecular (*e.g.*, vibrational or rotational modes) or external degrees of freedom (*e.g.*, solvent modes). Thus, both the cavity modes and the molecular bright states have finite lifetimes, and the interactions of each subsystem with their corresponding environment provide multiple pathways for polaritons to decay.

Photon leakage due to mirror imperfections leads to vibropolariton decay into the ground state by emission of an external (output) photon with the same energy as the initial state. Similarly, molecular excited state population relaxation (with heat generation) may also drive vibropolariton annihilation. However, the lifetime of vibrational modes with high oscillator strength is generally much longer than infrared cavity photon lifetimes.[2] This would imply that, in the absence of any other decay processes, energy stored in polaritonic degrees of freedom would predominantly decay via photon emission and the cavity–matter system would return to the ground state with a rate approximately equal to that of cavity leakage. However, as we discuss below, experiments contradict this picture. In fact, it has been observed repeatedly for cavities with lifetimes on the order of $1 - 10$ ps and molecular vibrations with population relaxation time $T_1 = 100 - 200$ ps, that after a period of $\Delta t_{pp} \approx 10 - 30$ ps, the cavity–matter system behaves as if its excited-state population consists of modes with properties indistinguishable from pure molecular excitations under normal conditions in free space.[15, 33] Recalling from Section 12.2.3 that according to the TC model every polaritonic system in the collective strong coupling regime also contains a large number of "dark" states with energy equal to that of the bare molecules, we are led to conclude that polaritons likely undergo efficient decay accompanied by excitation of a molecular dark state.

It follows that, in addition to cavity leakage and molecular excited state population relaxation, there exist additional important pathways for polaritons to decay with simultaneous excitation of a dark state. Energy conservation requires that any deficiency or excess of energy of these molecular states relative to the initial vibropolariton must be accounted by either borrowing energy from the molecular thermal bath (if the initial state is the LP, see Figure 12.5) or transferring energy to the environment (if the initial state is the UP). Thus, a precondition for *efficient* polariton decay into the dark manifold is that the environment has motions with characteristic frequency matching the frequency gap between polaritons and dark modes (so they can be either excited or deexcited as needed for energy to be conserved in the polariton decay process). If the LP is the initial state, another requirement of efficient polariton decay into the manifold of dark states is that $\Omega_R/(2kT)$ (where k is the Boltzmann constant and T is the temperature) is not larger than 1, as in this case, the donation of energy from the environment to the polaritonic system would be highly unlikely. Both of these preconditions are generally satisfied for vibropolaritonic systems: the achieved values of $\Omega_R/2$ have almost always been much smaller than kT. Additionally, the molecular system is generally in a solution phase where low-frequency intermolecular modes are abundant.

Yet another fundamental reason for the efficient decay of vibropolaritons into dark molecular modes comes from the discussion in Section 12.2.3 where we explain that the number of polaritonic excitations with significant photonic content is orders of magnitude smaller than the number of dark states. Note the same argument implies that direct conversion of UP into LP (or vice-versa) is much less likely relative to polariton decay into dark states. Likewise, the small values of typical vibropolaritonic $\Omega_R/2$ relative to kT and the extremely larger dark molecular density of states relative to the polaritonic suggest that environment-assisted conversion of dark excitations into vibropolaritons is almost always suppressed.

The arguments provided above can be made quantitative by employing the FGR[20] to compute transition rates between any pair of initial or final state described above. According to FGR, the rate of transition from an *individual* initial state i with energy E_i into a *specific* final state f with energy E_f is given by

$$k_{if} = \frac{2\pi}{\hbar}|V_{if}|^2 \delta(E_f - E_i), \qquad (12.17)$$

where $V_{if} = \{f|V_{\mathrm{SB}}|i\}$ is the matrix element for the system-bath interaction (see below) that mediates the conversion of the state i into the state f, $|i\rangle = |i_S, i_B\rangle$ is the initial state consisting of the i_S system state (*e.g.*, polariton or dark excitation) and i_B is a particular bath state, and analogous definitions apply for $|f\rangle = |f_S, f_B\rangle$. The delta function enforces energy conservation. Assuming a thermal probability distribution $\{p_{\{i_B\}}\}$ for the initial state of the environment and summing over all of the possible final bath states we obtain the quantum transition rate between the various states of the cavity–matter system.

$$k_{i_S f_S} = \frac{2\pi}{\hbar} \sum_{i_B} \sum_{f_B} p_{i_B}|V_{i_S i_B, f_S f_B}|^2 \delta(E_f - E_i). \qquad (12.18)$$

In order to obtain the polariton decay rate into the manifold of dark modes we must sum over all of the latter. In the TC model with N molecules, we have $N - 1$ dark states $|D_\mu\rangle$, therefore,

$$k_{P\to D} = \frac{2\pi}{\hbar} \sum_{\mu=1}^{N-1} \sum_{i_B} \sum_{f_B} p_{i_B}|V_{P i_B, D_\mu f_B}|^2 \delta(E_f - E_i). \qquad (12.19)$$

Assuming each molecule has an independent thermal bath (as would be the case if we assumed the molecules strongly coupled to the cavity correspond to the solute of an ideal, dilute solution), the molecular part of the system-bath interaction Hamiltonian is given by a sum of independent interactions between each molecule and its environment. In this case, we find $k_{P\to D}$ can be written as

$$k_{P\to D} = \frac{2\pi}{\hbar} \sum_{m_A=1}^{N} \sum_{\mu=1}^{N-1} |c_{Pm_a}|^2 |c_{D_\mu m_a}|^2 W_{m_a}(E_P - E_D),$$

(12.20)

$$W_m(E_P - E_D) = \sum_{i_B} \sum_{f_B} p_{i_B} |V_{m_a i_B, m_a f_B}|^2$$
$$\times \delta(E_D + E_{f_B} - E_P - E_{i_B}),$$

(12.21)

where c_{Pm_a} is the amplitude for the molecule m_a to be in the excited state given that the system is in the polariton state denoted by P and $c_{D_\mu m_a}$ is the analogous amplitude for m_a to be excited when the μ dark state is present. Because the dark states are orthogonal to the molecular bright state and form a degenerate subspace of the TC model eigenstates, it is possible to write a basis of dark states of maximally delocalized pure molecular states with equal probability $1/N$ for an excitation to be detected in each molecule. Thus, each factor in Eq. (12.20) can be taken to be $1/N$. Similarly, $|c_{Pm_a}|^2$ is of order $1/N$, since polaritons in the TC model are equally delocalized over all molecules in the cavity. The sum over all molecules and dark states has $N(N-1)$ terms which for $N \gg 1$ essentially cancel the $1/N^2$ factor coming from the polariton and dark state amplitudes (another way to extract these results is via the identity $\sum_{\mu=0}^{N-1} |D_\mu\rangle \langle D_\mu| = \sum_{i=1}^{N} |i\rangle \langle i| - |B\rangle \langle B|$). This results in $k_{P\to D}$ being of the same order of magnitude of the single-molecule quantity $W_m(E_P - E_D) \equiv W_{m_a}(E_P - E_D)$ (we omit the molecular label a as the molecules and their thermal baths are assumed to be identical) which characterizes the interaction between any molecule and the degrees of freedom of its environment with natural frequency $|E_P - E_D|/\hbar$. When same steps are applied in the reverse direction, we find the rate of environment-assisted dark $\to$ polariton transition is given by:

$$k_{D\to P} = \frac{2\pi}{\hbar(N-1)} \sum_{m_A=1}^{N} \sum_{\mu=1}^{N-1} |c_{Pm_a}|^2 |c_{D_\mu m_a}|^2 W_{m_a}(E_D - E_P).$$

(12.22)

Note that the final state now consists of a single polariton mode, but the rate is obtained by averaging over the initial dark states. This provides the essential difference between Eqs. (12.20) and (12.22) which are related by the detailed balance condition

$$k_{D \to P} = k_{P \to D} \frac{W_m(E_D - E_P)}{W_m(E_P - E_D)} \frac{1}{N - 1}. \qquad (12.23)$$

Limiting our considerations to the case where $\Omega_R/2$ is smaller than kT (as in almost all vibropolariton experiments), $W_m(E_D - E_P)/W_m(E_P - E_D)$ is much less than $N - 1$ (see discussion in Section 12.2.3). It follows that the dark state $\to$ polariton transition rate is an infinitesimal fraction of the polariton $\to$ dark state decay rate due to the large density of states of dark modes. This same reasoning leads to polariton $\to$ polariton environment-assisted inter-conversion rates proportional to $1/N$ of the polariton $\to$ dark-state transition.

The above arguments suggest a picture for vibropolariton dynamics summarized in Figure 12.7 where we employ numbers suggested by the pioneering experiments in Refs. 15 and 33. Without any external pumping, we may assume the system is in the ground state since the Boltzmann or Bose–Einstein distribution at ambient temperatures give very low probability for excited states to be thermally occupied. Upon irradiation with an external field, polaritons are created. They have a relatively short lifetime due to efficient decay via cavity leakage (acting on the photonic part of the polaritonic wave function) or by ultrafast conversion into dark states (with simultaneous exchange of energy with the molecular bath). The dark states are much more numerous than the polaritons, and therefore, the former may be taken as a thermal bath relative to the latter, and the polariton $\to$ dark state energy transfer process is almost irreversible (in Section 12.3.5 we discuss a potential exception to the scenario discussed here). According to this view, the last step in the relaxation of a polaritonic system into the global ground state is the release of the energy stored in the dark manifold into the molecular environment via heat generation.

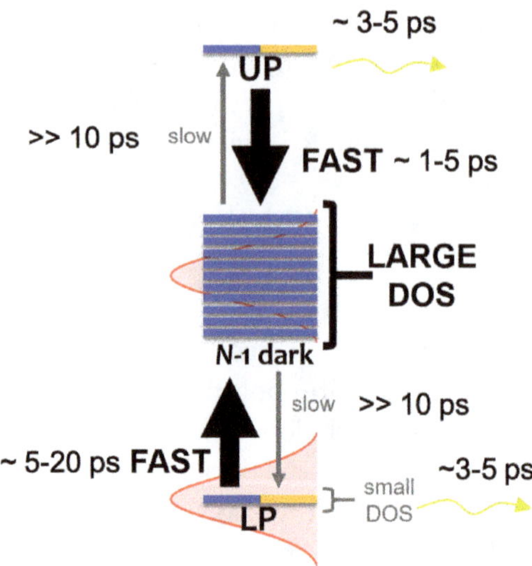

Figure 12.7. Relaxation kinetics diagram for polaritons and dark excitations of an infrared optical cavity under strong coupling with a molecular ensemble containing a bright vibrational mode with significantly large absorption cross-section. The wavy arrows correspond to cavity leakage processes where polaritons decay by light transmitted through the mirrors into the exterior world. The provided estimated timescales were obtained from Refs. 2, 15, 33, 38. *Source*: Adapted from Ref. 2.

12.3.4. *Vibropolariton pump–probe spectroscopy: Theory*

The polariton dynamics discussed in the previous section suggests that the pump-probe spectrum of vibropolaritons carry information about complex many-body phenomena involving polaritons, dark states and their thermal environments. Nevertheless, although the polaritonic ΔT_{pp} spectra in Figure 12.6 is puzzling when viewed in light of the TC model (Section 12.2.3), the Rabi splitting contraction and giant reduction of light transmission via the LP mode are unambiguous and were subsequently verified to arise in several other experiments.[8,42] The simplicity and universality of these optical signatures suggest that, notwithstanding the complexity of the detailed microscopic processes governing polariton relaxation kinetics (Section 12.3.2) and the quantum wave function evolution, vibropolariton pump–probe spectra can be interpreted with a simple effective model. In this subsection, we explain the physical origins of the main features observed in Figure 12.6 with the theory first given in Ref. 38.

In order to describe the transient polariton pump–probe response, we must invoke a mathematical framework that allows us to obtain the probe transmission spectrum as a function of the macroscopic state of the cavity–matter system at the delay time Δt_{pp}. We consider here the simpler scenario where Δt_{pp} is long enough (20–50 ps for the system of Figure 12.6) that fast polariton relaxation into molecular excited states has occurred so the pump–probe response is a function of the number of purely molecular excited states, and that the energy stored in polaritons is negligible (this idea is confirmed by the fact that after a initial ultrafast relaxation time, the spectral features of ΔT_{pp} cease to undergo oscillations and decay in intensity with a rate essentially equal to the population relaxation rate of the molecular excited states outside of a microcavity[15, 33, 43]). We also employ a "maximal entropy" assumption where we assign the excited state energy in the system to be equally distributed between all molecules interacting with the cavity mode. As we will see below, these assumptions lead to results in excellent agreement with experimental data.

Assuming the cavity has an excited state population of dark modes provides an initial state for the probe measurement that is drastically different relative to the global ground state of the system. This has important consequences: upon interaction with a photon, molecules in the $\nu = 1$ state can absorb energy and go to $\nu = 2$. This process will be important if the cavity is irradiated with photons with frequency matching the energy gap $E_{12} = \hbar\omega_{12} - \hbar\omega_{01}$ between the $\nu = 2$ and $\nu = 1$ vibrational states (note in this section, the molecular vibrational frequency of interest is denoted ω_{01}). For the asymmetric carbonyl stretch of $W(CO)_6$ in hexane, $\hbar\omega_{01} = 1,983\,\mathrm{cm}^{-1}$, while $\hbar\omega_{12} = \hbar\omega_{01} - 2\Delta$, where $\Delta = 7$ cm^{-1} is the vibrational anharmonicity parameter. Interestingly, there is a near match between the LP energy and the energy required for the $\nu = 1 \to 2$ transition of $W(CO)_6$ in hexane. On the other hand, the

UP is highly off-resonant with the molecular excited state absorption transition. This provides a strong hint that the transmission of photons with frequency around the LP is much weaker after pumping because the weakly coupled molecules in the first excited state efficiently absorb LP energy while undergoing a transition into the molecular $\nu = 2$ state. No such excited state absorption is likely to happen when the matter–cavity system is illuminated with photons that are resonant with the UP because $E_{\mathrm{UP}} \gg E_{1 \to 2}$.

The other puzzling spectral feature of the vibropolariton pump–probe response was the Rabi splitting contraction. It may also be understood by analyzing the elementary processes that light can induce on the considered molecular vibrational states. First, note that in the TC model we obtained a Rabi splitting Ω_R proportional to $\sqrt{N}$ because we assumed all molecules that participate in polaritons were in the ground state and could undergo a transition into the excited state by absorption of a cavity photon (Section 12.2.3). If the number of molecules in the ground state were $N_0 = N - N_1$ (where N_1 is the number of molecules in the first excited state) and transitions from the first excited state were highly off-resonant with the cavity mode frequency, the observed Rabi splitting would *decrease*. According to this argument, the Rabi splitting for a system with N_0 molecules in the ground state would be approximately equal to $\Omega_R\sqrt{1 - f_{\mathrm{pu}}}$, where $f_{\mathrm{pu}} = N_1/N$. However, this argument is incomplete. In particular, cavity photons may also induce stimulated emission of molecular excited state population. It turns out that when the stimulated emission process is considered, the Rabi splitting observed after pumping the system is found to be[38]

$$\Omega_R^{pp}(t) \approx \Omega_R\sqrt{1 - 2f_{\mathrm{pu}}(t)}. \tag{12.24}$$

Indeed, when simple estimates of $f_{\mathrm{pu}}(t)$ from polariton relaxation theory are inserted into Eq. (12.24), the theoretical Rabi splitting contraction agrees with the experimental to a reasonable extent.[38]

The arguments presented to this point suggest that indeed a simple microscopic treatment of the vibropolariton pump–probe optical response can rationalize the main observed trends. The complete

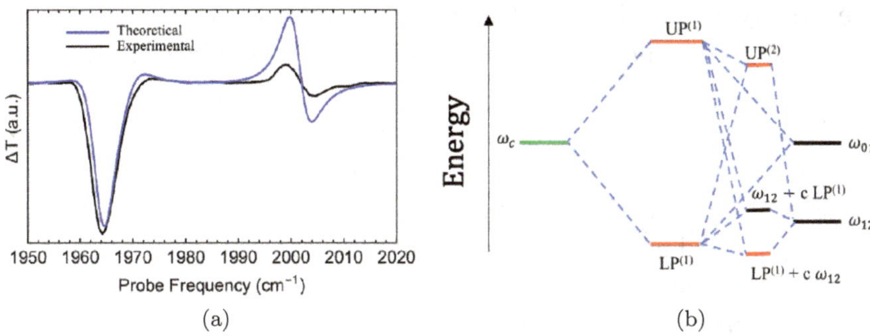

Figure 12.8. (a) Theoretical and experimental pump–probe differential transmission spectrum ($\Delta T \equiv \Delta T_{pp} = T_{pp} - T_0$ where T_0 is the transmission spectrum of the sample in equilibrium) obtained 25 ps after irradiating vibropolaritons in an infrared cavity resonant with the asymmetric carbonyl stretch of a 40-mM solution of $W(CO)_6$ in hexane. (b) State correlation diagrams for interpretation of the vibropolariton pump–probe spectrum. Here, the cavity photon mode has frequency ω_c in resonance with the molecular $\nu = 0 \to \nu = 1$ vibrational frequency $\omega_M = \omega_{01}$. Strong light–matter coupling leads to the formation of polaritons $UP^{(1)}$ and $LP^{(1)}$ separated by a Rabi splitting $\Omega_R \propto \sqrt{N}$ where N is the number of molecules in the system. Upon irradiation of the sample with a strong pump, polaritons are generated, and a significant fraction of those subsequently decay into purely molecular modes ("dark states") creating a finite population of molecules in the first excited state. Upon interaction with photons the molecular excited state population may undergo absorption into the doubly excited molecular state (with $\nu = 2$). This transition is nearly resonant with the LP, and therefore, probe light at frequencies near $\omega_{LP}^{(1)}$ are efficiently absorbed by the molecular excited state population. This process leads to the strong negative ΔT feature in the spectra shown left. There is some weak mixing between the $LP^{(1)}$ state and the molecular excited state absorption transition represented by the parameter $c \to 0$ in the diagram. The UP state undergoes a redshift after pumping due to the Rabi splitting contraction explained in the text. *Source:* (a) From Refs. 33 and 34. (b) From Ref. 33 (Supplementary Material).

detailed quantum-optical treatment was presented in Ref. 38. Its predictions are given in Figure 12.8.

In Figure 12.8, we compare the theoretical and experimental vibropolariton $\Delta T_{pp}(\omega)$ obtained 25 ps after pumping for 40 mM $W(CO)_6$ in hexane assuming that an average of 10% of the molecules are excited when the probe illuminates the sample (this number was justified by numerical estimates presented in the Supporting Information of Ref. 34). The accuracy of the theoretical results reflect the fidelity of the key implemented assumptions discussed above. In other words, the qualitative considerations provided above were indeed corroborated by our mathematical theory.[38] In particular, the large negative peak near the LP frequency is due to an overlap between the LP resonance and the molecular $1 \to 2$ excited state absorption which significantly reduces the transmission of light at the LP frequency. The UP undergoes a redshift largely driven by a Rabi splitting contraction induced by the reduced number of molecules in the ground state and by stimulated emission processes acting on molecular excited states (a similar shift is harder to recognize for the LP due to the broad lineshape of the feature around the LP frequency). In fact, we showed in Ref. 34 that the transient polariton resonances present in a system with incoherent molecular excited state population can be obtained from diagonalization of a 3×3 non-Hermitian Hamiltonian (*i.e.*, including the effects of dissipation in the molecular and photonic subsystems) consisting of the cavity mode, a molecular bright state consisting of a superposition of the $N \times (1 - f_{\mathrm{pu}})$ molecules in the ground state and a molecular bright state consisting of a superposition of states containing only those molecules that are in the first excited state after Δt_{pp}. In this reduced description, the light–matter coupling constants are given by $\Omega_R \sqrt{1 - 2N_D(t)/N}$ and $\Omega_R \sqrt{2N_D(t)/N}$ for the cavity interaction with the $0 \to 1$ and $1 \to 2$ transitions, respectively.

We conclude this subsection by mentioning that our treatment focused entirely on the effects of the molecular reservoir excitations on the polaritonic optical response since in this case the Rabi splitting contraction and large LP differential transmission peaks are well understood and follow naturally from the theory in Ref. 38. When vibropolariton pump–probe spectrum is measured with delay time Δt_{pp} shorter than the polariton decay time (by cavity leakage or by relaxation into the dark manifold), different phenomenology recently reported in Ref. 44 ensues. We present no such discussion here as the physical origin of these ultrafast spectral features remains open.

12.3.5. *Cavity-assisted vibrational energy transfer*

Before concluding our discussion of vibropolariton dynamics and spectroscopy, we briefly comment on another application of the theory discussed above as well as outline an aspect of our discussion of dynamics that is in contradiction with recent experiments.

In Ref. 8, the theory described in this section was generalized to analyze polariton-mediated energy exchange between equally concentrated $W(CO)_6$ and $W\left(C^{13}O\right)_6$ in a hexane/dichloromethane solvent mixture. The carbonyl asymmetric stretch of $W(CO)_6$ (hereafter denoted as the energy "donor") and $W\left(C^{13}O\right)_6$ (the energy "acceptor" species) are different by about 40 cm^{-1}, but their oscillator strength and absorption linewidth are essentially the same. Thus, both species undergo strong coupling with a suitable infrared optical cavity. This system has three polariton bands: LP, MP (middle polariton) and UP (Figure 12.9). Two-dimensional pump–probe spectroscopy was performed in this system to investigate cavity-assisted vibrational energy transfer between the donor and acceptor molecules. The choice of donor and acceptor subsystems

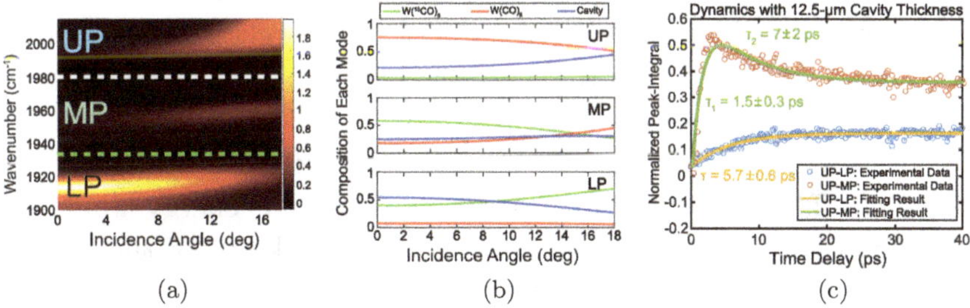

Figure 12.9. (a) Transmission spectrum obtained for an infrared cavity under strong coupling with the asymmetric stretch modes of $W(CO)_6$ (donor) and $W\left(C^{13}O\right)_6$ (acceptor) in a dichloromethane/hexane mixture. This spectrum shows the formation of LP, MP and UP resonances (compare to Figure 12.4). (b) Composition of each polariton excitation in the LP, MP and UP bands at low incidence angles. (c) Time-dependent variation of spectral signatures corresponding to pumping the UP mode at $\theta = \pi/12$ rad and probing the transmission of probe light at the MP and LP modes with the same incidence angle. The intensity of the UP-MP and UP-LP peaks is proportional to the time-dependent excited state population of donor and acceptor molecules, respectively. *Source*: Adapted from Ref. 8.

was suitable as under standard conditions in free space, no significant energy exchange happens between the donor and the acceptor molecules. This indicates that direct intermolecular energy transfer between the CO asymmetric stretch of donors and acceptors is slower than the lifetimes of their vibrational excited states.

In the experiments performed by Xiang *et al.*, UP states with minimal acceptor contribution (the ratio of acceptor to donor component was 1:14) were initially pumped, and energy relaxation was measured with a probe pulse. The population of acceptor and donor excited states was probed following the strategy outlined in Section 12.3.4: from the reduction in transmitted intensity of light with a given polaritonic frequency nearly resonant with a molecular excited state absorption, the population of donor and acceptor excited states was inferred. This idea was shown to work for the particular pair of donor and acceptor molecules studied by Xiang *et al.* because the MP states were nearly resonant with the $1 \rightarrow 2$ vibrational transition of donor, and similarly the $1 \rightarrow 2$ acceptor transition was nearly resonant with the LP (at the same incidence angle as the pumped UP). Hence, by examining the time evolution of the differential probe transmission spectrum at the LP and MP frequencies, it was possible to track the energy transfer relaxation dynamics and the ratio of donor to acceptor population at steady state (Figure 12.9).

According to our previous discussion of vibropolariton dynamics (Section 12.3.3), the majority of the pumped UP energy would be quickly deposited into donor dark states, or leak out of the optical cavity since the acceptor contribution to the UP modes is much smaller. Instead, it was observed that the ratio of donor to acceptor excited state population at steady state was approximately 2.5:1, a much smaller number than the 14:1 that would follow from statistical considerations based solely on the vibropolaritonic composition of the excited UP mode.

In analogy with studies of polariton-assisted energy transfer in organic microcavities, Ref. 8 investigated the possibility that perhaps donor dark states decayed into MP states subsequently converted into acceptor excitations. This hypothesis was also motivated by measurements of the time-dependent pump-probe signals at the LP and MP frequencies (Figure 12.9, right). In particular, it was observed that the differential transmission signals for MP (proportional to the donor excited state population) initially increased at an ultrafast rate close to the theoretical rate of UP $\to$ dark donor decay. Within a few ps after pumping, the MP signal started to decrease indicating a relatively slower depletion of the dark donor excited state population until a steady state was reached. This depletion of the MP signal cannot be attributed to population relaxation of dark donor excited states as their lifetime is greater than 100 ps. Furthermore, the differential transmission signal at the LP showed a rise time that was nearly equal to the second component of the MP dynamics indicating almost that as the excited state population of donors decayed, the number of excited dark acceptors increased.

Despite all of this evidence, a computation of the dark donor $\to$ MP transition rate according to the FGR and the theory outlined in Section 12.3.3 reveals that the conversion of dark donors into polaritons is slower by several orders of magnitude relative to what it would have to be in order to match the energy transfer timescales reported in Ref. 8. This seems to rule out the simplest mechanism where MP modes are intermediate states in the donor $\to$ acceptor energy transfer process. The essential reason this mechanism fails is that infrared cavity lengths are generally larger than UV-visible microcavities by one or two orders of magnitude, so the number of molecules interacting with any infrared cavity mode is likely much larger than is the case with organic microcavities. This makes dark $\to$ polariton transition rates extremely slow as discussed in Section 12.3.3. Therefore, while polariton-enhanced intermolecular energy transfer in organic donor–acceptor pairs via MP intermediate states is a plausible mechanism for electronic excitation energy transfer in optical cavities, the same idea is unable to explain experimentally observed ultrafast cavity-assisted vibrational energy transfer.

12.4. Vibropolariton Chemistry: Reaction Effects

In 2016, the Ebbesen group in Strasbourg pioneered a series of remarkable experiments showing that the rates of thermally-activated (TA) chemical reactions can be altered by collective vibrational strong coupling (VSC).[5] What is puzzling about the reported phenomena (hereafter referred to as vibropolaritonic chemistry, VPC) is that it occurs in the absence of external optical pumping (such as that from a laser source), thus relying solely on thermal fluctuations. These experiments are interesting in that they seem to open doors to a revival of the "mode-selective chemistry" (MSC) program,[45,46] a research effort that received great attention especially in the 1980–1990s where selective breaking of chemical bonds was intended using intense infrared (IR) laser light tuned to resonance with the corresponding normal mode frequency. However, MSC proved to be difficult to achieve in general, with the deposited laser energy quickly dissipating into other modes of the molecule through anharmonic couplings, in a process known as intramolecular vibrational redistribution (IVR).[47,48] Thus, only a handful of examples exist where MSC is successful, largely pertaining to small molecules in gas phase.

VPC seems to circumvent the drawbacks of the original version of MSC, while bearing its spirit of targeting specific chemical bonds by simply dialing the frequency of light. Contrary to the original MSC program, no intense optical pump is needed in VPC, which seems to only depend on the small occupation of IR photon modes of the optical microcavity driven by thermal fluctuations. Yet, the fine tuning of the frequencies of the optical modes is readily achievable using microfabrication techniques and piezoelectric components.

Despite the great promise of VPC, there are at present many unknowns about its mechanisms. As we shall explain below, standard reaction-rate theories in combination with standard treatments of cavity–matter interactions seem to be incapable or limited in explaining several of the results that have been obtained experimentally. For VPC to reach a broad synthetic chemistry audience, it is essential that a solid foundation be developed. Before we describe

the theoretical efforts along these lines, we will give a brief overview
of the experimental observations.

12.4.1. *Experimental observations*

The first VPC experiment[5] consisted of the deprotection of 1-phenyl-
2-trimethyl-silylacetylene (PTA) with tetra-n-butylammonium fluoride (TBAF) in methanol solution (see Figure 12.10(a)). In this
reaction, the fluoride nucleophile of TBAF attacks the silicon center
in PTA, displacing the leaving carbanion group. The reaction
proceeds in an S_N2 fashion under normal circumstances. In their
experimental report, Thomas and coworkers inserted the microfluidic

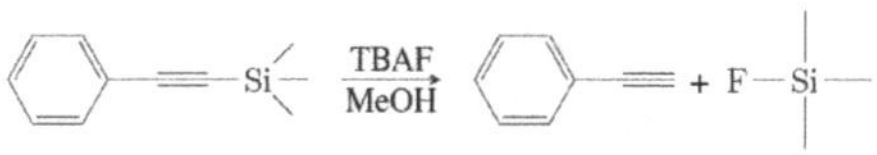

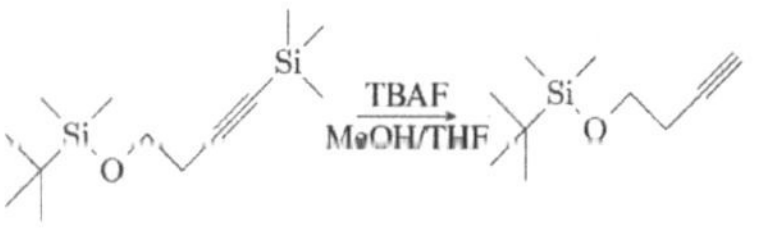

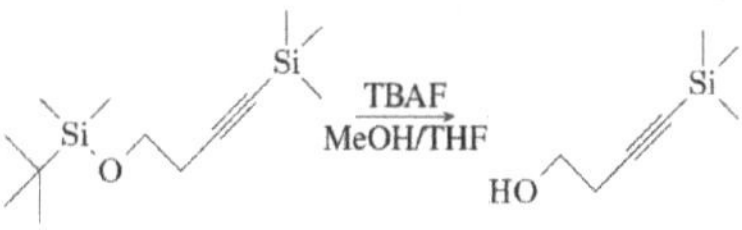

Coupled vibrational mode	Ω/ω	$k_{\mathrm{VSC}}/k_{\mathrm{bare}}$
(a)		
Si—C	0.081	0.25
(b)		
Si—CH$_3$	0.034	
Si—C	0.083	0.27
Si—O	0.077	0.25
Si—CH$_3$	0.034	
Si—C	0.083	0.65
Si—O	0.077	0.49

Figure 12.10. The first examples of vibropolaritonic chemistry (VPC) were
reported in the literature by the Ebbesen group. In (a), suppression of the Si—C
bond breaking rate is achieved under vibrational strong coupling (VSC).[5] In (b)
suppression of both Si—C and Si—O bonds is attained, but a reversal of the yields
is obtained under VSC with a number of reactive and spectator modes.[6]

solution of reactants inside of a cavity supporting a $q_\parallel = 0$ photon mode at $\omega_0 = 860$ cm^{-1}, which is resonant with a normal mode of PTA with significant C-Si stretch character. The reaction has rate constant on the order of $k \sim 10^{-3}$ s^{-1} at room temperature, corresponding to a half lifetime of $\tau = \ln 2/k \sim 10$ min. With a very high concentration of reactant, [PTA] $= 2.53$ M, these researchers were able to demonstrate collective VSC, with a Rabi splitting of $\Omega_{\mathrm{R}} = 98$ cm^{-1} at normal incidence ($q_\parallel = 0$). More interestingly, they observed a reduction in the reaction rate by a factor of approximately 4 under those circumstances. In a rather puzzling observation, these researchers report reduced changes in kinetics when compared to the outside of the cavity case when the $q_\parallel = 0$ cavity mode is detuned from the targeted molecular resonance (by varying the cavity longitudinal length), despite the presence of other $q_\parallel \neq 0$ modes that are resonant with the molecular resonance, and thus able to engage in VSC. Another interesting observation is that as the concentration of reactants was increased and the Rabi splitting was consequently varied from $\Omega = 58$ to 114 cm^{-1}, the reaction became even slower.

In a more elaborate experiment reported by the same group (see Figure 12.10(b)),[6] the researchers were able to achieve something even more dramatic than the first time. They considered the cleavage of the tert-butyldimethyl{[4-(trimethylsilyl)but-3-yn-1-yl]oxy}silane molecule along the Si$-$O and Si$-$C (with normal mode frequencies of 1,110 and 842 cm^{-1}, respectively) bonds using TBAF as the nucleophile in a methanol/tetrahydrofurane solution. Typically, cleavage of the Si$-$O bond is faster than that of the Si$-$C bond, offering a kinetic control over the product branching ratio. However, VSC with $q = 0$ cavity mode resonant with either the Si$-$O, Si$-$C, and even with a spectator Si$-$CH$_3$ bending mode (at 1,250 cm^{-1}) yields a global reduction of both product formation rates; yet, this reduction is stronger for the Si$-$O bond than for the Si$-$C bond, thus resulting in an inversion of branching ratios in the yield. An analogous VSC situation with the C$=$O absorption band at 1,040–1,060 cm^{-1} does not yield large changes to the reaction rates or product branching ratio, and neither is the case for a cavity where $q = 0$ cavity mode is not resonant with any molecular modes.

Further experiments by Ebbesen and coworkers validate similar traits of VPC. Hirai and coworkers demonstrated reduction in the rate of Prins cyclization upon VSC with a ketone C=O mode,[10] while proteolytic activity of pepsin was also decelerated upon VSC with the O−H stretch of water, which acts as reactant and solvent.[49] The systematic observation of rate deceleration prompted researchers to inquire whether VSC can lead to the opposite trend, that is, rate acceleration or catalysis. In fact, reports by Hiura[50] show that this can indeed be the case in a series of carbonyl addition reactions. Finally, it is worth mentioning an important class of VPC experiments that operate under so-called "cooperative" VSC,[51] where VSC occurs at the expense of having a high concentration of solvent rather than reactant, although polaritons are formed with both near-resonant solvent and reactant vibrational modes. This was demonstrated with the solvolysis deceleration of a dilute sample of para-nitrophenyl acetate in ethyl acetate solvent, upon VSC with the C=O bonds of reactant and product. This modality of VPC seems quite useful given that it is often the case that the reactant must be in dilute concentration, therefore precluding the onset of VSC in the absence of solvent.

Let us summarize some of the key phenomenology reported for the VPC experiments so far. The experimental conditions are as follows:

C1 They occur under collective VSC, where the ratio of dark modes per polariton has been estimated to be between $N \approx 10^6 - 10^{12}$.
C2 Optical pumping is not needed, VPC thus relying on thermal fluctuations.

The experimental observations are as follows:

O1 Both enhancement and suppression of reactions can occur, although the latter seems to be more common.
O2 Resonance (and the consequent VSC) must ensue between the optical cavity and the reactant, spectator, and/or solvent modes.
O3 Stronger effects on reaction rates seem to require VSC with a $q = 0$ mode.

12.4.2. *General theoretical considerations*

A successful theory must account for the experimental conditions and observations above. However, each of them brings a highly non-trivial challenge. For instance, O1 is puzzling in light of C1 and C2: Boltzmann statistics at room temperature entropically favors the occupation of dark rather than polariton modes, rendering VPC presumably indistinguishable from chemistry outside of a cavity. O3 is puzzling in light of O2: resonance between the molecular system and the $q_\parallel = 0$ photon mode seems to be required for the cavity to induce large changes to thermal reaction rates, but polariton formation and VSC also clearly occur when the molecular vibrations are resonant with $q_\parallel$ that can be significantly different from zero (see *e.g*, Figure 12.4). In fact, there is substantial evidence from numerical simulations[52] and theoretical arguments[52,53] that the effects of a photonic cavity on kinetic and thermal properties of a disordered molecular system are strongest when the molecules are resonant with photon modes satisfying $q_\parallel \neq 0$.[52]

In the following sections, we will review elementary theories developed in the last few years, which start with the conditions C1 and C2 while trying to explain VPC. As we shall explain, none of them are fully capable of addressing all of the three observations above. Yet, we shall present them because it provides a pedagogical way of thinking about extensions of reaction kinetics theory that incorporate the effects on ensemble reaction rates of a delocalized harmonic mode (photon) under strong interaction with a collective variable of the molecular system. In addition, the correct theory may arise from a generalization of the results presented below.

Before we dive into the details, let us review a simplified theoretical treatment of a chemical reaction. Labeling the nuclear displacement, corresponding generalized momenta, and mass by q, p, and m, respectively, we write the total Hamiltonian for a molecule undergoing a reaction as

$$H = \frac{p^2}{2m} + \begin{pmatrix} V_\mathrm{R}(q) & J_\mathrm{RP}(q) \\ J_\mathrm{RP}(q) & V_\mathrm{P}(q) \end{pmatrix}. \tag{12.25}$$

Here, $V_R(q)$ and $V_P(q)$ are potential energy surfaces (PESs) corresponding to the diabatic electronic states of reactant $|R\rangle$ and product $|P\rangle$, while $V_{RP}(q)$ is the electronic coupling between such states. The corresponding adiabatic PESs are given by the eigenvalues of the electronic Hamiltonian in Eq. (12.25)

$$V_\pm(q) = \frac{V_R(q) + V_P(q)}{2} \pm \sqrt{\left(\frac{V_R(q) - V_P(q)}{2}\right)^2 + J_{RP}(q)^2}.$$

$$(12.26)$$

Let us denote by $\{\omega_i\}$ the set of characteristic frequencies of harmonic motion associated with the V_R potential. If $\min|V_+(q) - V_-(q)| \gg \max \omega_i$, then the adiabatic PESs are well separated in energy and the chemical reaction can be reasonably described by focusing on the lowest PES $V_-(q)$. This occurs typically when $J_{RP}(q)$ is large, corresponding to so-called *adiabatic* reactions, which describe heavy-atom rearrangements. In its simplest incarnation, $V_-(q)$ features two minima, corresponding to reactant and product, respectively, separated by an energy barrier at q^* (transition state, TS) that must be surmounted for the reaction to occur (see Figure 12.11(a)).

In the opposite limit, $\min|V_+(q) - V_-(q)| \ll \min \bar{\omega}_i$ is typically the consequence of $J_{RP}(q)$ being small, as occurs in tunneling reactions of light particles such as electrons and protons. Those reactions are known as *non-adiabatic* and involve an interplay of nuclear motion in two different PESs. The most convenient description is a diabatic one, where the light particle tunnels between the reactant and product wells at the nuclear configuration q^* where they cross (*i.e.*, at the TS). The tunneling probability is proportional to $|J_{RP}(q^*)|^2$. An important qualitative difference with respect to the adiabatic reactions is that, in non-adiabatic ones, a given nuclear configuration q does not correspond uniquely to a chemical species (reactant, product), given that the electron has a non-negligible probability of "forgetting" to tunnel to the product state despite

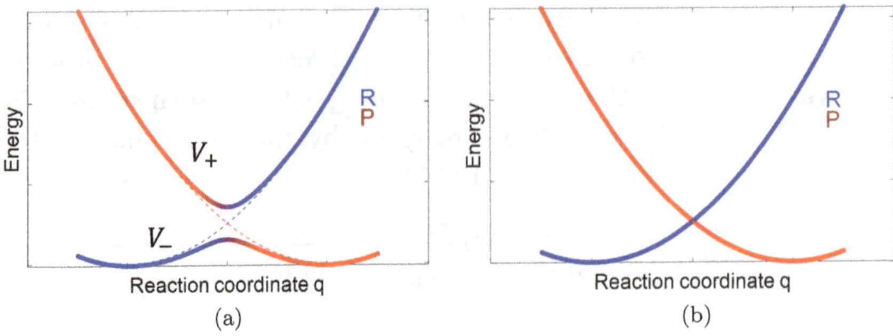

Figure 12.11.　The coupling between two electronic states with different chemical character (R and P, representing reactant and product) gives rise to adiabatic potential energy surfaces (PESs) $V_\pm$. (a) Adiabatic reaction: if such coupling is very large, the kinetic energy of the reaction coordinate q has little probability to induce transitions into the excited adiabatic PES V_+ and the reaction dynamics reduces to barrier crossing along the ground adiabatic PES V_-. The formalism describing this dynamics is TST. (b) Non-adiabatic reaction: when the coupling is small, the dynamics is dictated essentially by the diabatic PESs (V_R and V_P), except at the vicinity of their crossing, where tunneling can occur. The rate of transfer between these PESs is described by Marcus theory.

it being lower in energy than the reactant, given the smallness of $|J_{\mathrm{RP}}(q^*)|^2$ (see Figure 12.11(b)).

In practice, most organic synthesis transformations involve a series of adiabatic and non-adiabatic reaction steps. Thus, in the next subsections, we will discuss adiabatic and non-adiabatic theories for VPC.

12.4.3.　*Adiabatic reactions: Cavity transition state theory*

In this section, we denote $V \equiv V_-$ when referring to the ground adiabatic PES. TST[54–58] is the most commonly used formalism to model adiabatic chemical reactions. It predicts that the reaction rate is given by

$$k_{\mathrm{TST}} = \frac{k_{\mathrm{B}}T}{2\pi\hbar}\frac{Z_\ddagger}{Z_{\mathrm{eq}}}e^{-\frac{E_a}{k_BT}}. \tag{12.27}$$

We use the labels $\alpha = \mathrm{eq}, \ddagger$ to denote the reactant equilibrium and TS nuclear configurations, which correspond to local minima and first-order saddle points in V, respectively. $E_a = V_\ddagger + \frac{1}{2}\sum_i \hbar\omega_{i,\ddagger} - V_{\mathrm{eq}} - \frac{1}{2}\sum_j \hbar\omega_{j,\mathrm{eq}}$ is the activation energy to surmount. V_α and Z_α are the values of the potential and partition functions at the respective locations, while $\omega_{k,\alpha}$ correspond to the normal mode frequencies obtained from diagonalizing the Hessian of V around them (while the summation over j goes through all modes, that over i skips the reactive one; similarly, the partition function Z_{eq} includes all modes, while $Z_\ddagger$ skips the reactive one). At the core of TST is the assumption of equilibrium between the TS and the reactant well. Hence, it is no surprise that the rate is proportional to an "equilibrium constant" $\frac{Z^\ddagger}{Z_{\mathrm{eq}}}$ relating those two macrostates.

Galego and co-workers were first to develop a TST for ensembles of molecules in optical cavities.[39] Their theory was subsequently reexamined by Li,[41] Zhdanov,[59] and Campos-Gonzalez-Angulo.[40] In particular, the latter work recast Galego's work in the very compact formalism of polariton normal modes, making its salient features rather easy to understand. We summarize this framework in the following paragraphs, while referring the reader to the original publication for further details.

We first consider N identical molecules in the reactant configuration coupled to a photon mode through their dipole moment, yielding a potential energy of,

$$V(\mathbf{R} \approx \mathbf{R}_{\mathrm{eq}}, q_0) = \sum_{i=1}^{N} V_{\mathrm{nuc}}(\mathbf{R}_{i,\mathrm{eq}}) + \frac{\omega_{\mathrm{eq}}^2}{2}\sum_{i=1}^{N} q_i^2$$

$$+ \frac{\omega_0^2}{2} q_0^2 + \omega_0 g q_0 \sum_{i=1}^{N}\left(\mu_{i,\mathrm{eq}} + \mu'_{i,\mathrm{eq}} q_i\right), \quad (12.28)$$

where ω_0 and q_0 are the frequency and displacement of the photonic mode, q_i is the nuclear displacement of the ith molecule with respect to the equilibrium geometry of the reactant, $\omega_{\mathrm{eq}}^2 = \left.\frac{\partial^2 V_{\mathrm{nuc}}^{(i)}}{\partial q_i^2}\right|_0$ is the squared reactant vibrational frequency, and the dipole moment of the ith molecule affords an $O(q_i)$ Taylor expansion $\mu_i = \mu_{i,\mathrm{eq}} + \mu'_{i,\mathrm{eq}} q_i$, where $\mu_{i,\mathrm{eq}}$ and $\mu'_{i,\mathrm{eq}}$ are permanent and transition dipole moment contributions.

We wish to evaluate the TST rate of a single molecule in the ensemble undergoing a chemical reaction. This is a reasonable assumption given that reactions are rare events: the probability of two or more molecules simultaneously undergoing a reaction is much smaller than that of a single one. Without loss of generality, we take the Nth molecule to be the one reacting; the potential energy corresponding to all molecules remaining in the equilibrium configuration of the reactants except for the Nth molecule reaching the TS is

$$V(\mathbf{R} \approx \mathbf{R}_\ddagger, q_0)$$

$$= \sum_{i=1}^{N-1} V_{\mathrm{nuc}}(\mathbf{R}_{i,\mathrm{eq}}) + V_{\mathrm{nuc}}(\mathbf{R}_{N,\ddagger}) + \frac{\omega_{\mathrm{eq}}^2}{2} \sum_{i=1}^{N-1} q_i^2 + \frac{\omega_0^2}{2} q_0^2 + \frac{\omega_\ddagger^2}{2} q_N^2$$

$$+ \omega_0 g q_0 \left[\sum_{i=1}^{N-1} \left(\mu_{i,\mathrm{eq}} + \mu'_{i,\mathrm{eq}} q_i \right) + \mu_\ddagger + \mu'_\ddagger (q_N - q_{N,\ddagger}) \right]. \quad (12.29)$$

Here, $\omega_\ddagger^2 = \left. \frac{\partial^2 V_{\mathrm{nuc}}^{(N)}}{\partial q_N^2} \right|_{q_\ddagger} < 0$ is the squared frequency of the unstable mode at the TS.

Eqs. (12.28) and (12.29) contain all the information needed to evaluate the TST rate according to Eq. (12.27). In fact, all we need to do is find the minimum of $V(\mathbf{R} \approx \mathbf{R}_{\mathrm{eq}}, q_0)$ and the saddle-point of $V(\mathbf{R} \approx \mathbf{R}_\ddagger, q_0)$, after which we solve for the normal mode frequencies at each location. When $g = 0$ (outside of the cavity), the rate reduces to

$$k_{\mathrm{TST}} = \frac{k_{\mathrm{B}}T}{\pi\hbar} \sinh\left(\frac{\hbar\omega_{\mathrm{eq}}}{2k_B T} \right) \exp$$

$$\times \left(-\frac{V_{\mathrm{nuc}}(\mathbf{R}_{N,\ddagger}) - V_{\mathrm{nuc}}(\mathbf{R}_{N,\mathrm{eq}}) - \frac{1}{2}\hbar\omega_{\mathrm{eq}}}{k_B T} \right), \quad (12.30)$$

while inside of the cavity, we obtain

$$k_{\mathrm{TST}}^{\mathrm{VSC}} = \kappa_N k_{\mathrm{TST}}.$$

The ratio of rate constants inside and outside of the cavity reads

$$\kappa_N = A_{\text{VSC}}(T) \exp\left(-\frac{\Delta V_{\text{VSC}} + \Delta E_0^{\text{VSC}}}{k_{\text{B}}T}\right), \qquad (12.31)$$

with an entropic contribution

$$A_{\text{VSC}}(T) = \frac{\sinh\left(\hbar\omega_{+(N)}/2k_{\text{B}}T\right)\sinh\left(\hbar\omega_{-(N)}/2k_{\text{B}}T\right)}{\sinh\left(\hbar\omega_{+(N-1)}/2k_{\text{B}}T\right)\sinh\left(\hbar\omega_{-(N-1)}/2k_{\text{B}}T\right)}, \qquad (12.32\text{a})$$

a cavity-induced potential energy renormalization

$$\Delta V_{\text{VSC}} = \omega_0^2 \omega_{\text{eq}}^2 g^2 \left[\left(\frac{N\langle\mu_{\text{eq}}\rangle_N}{\omega_{+(N)}\omega_{-(N)}}\right)^2 - \left(\frac{(N-1)\langle\mu_{\text{eq}}\rangle_{N-1} + \mu_{\ddagger}}{\omega_{+(N-1)}\omega_{-(N-1)}}\right)^2\right], \qquad (12.32\text{b})$$

and an additional zero-point-energy shift

$$\Delta E_0^{\text{VSC}} = \frac{\hbar\omega_{+(N-1)} + \hbar\omega_{-(N-1)} - \hbar\omega_{+(N)} - \hbar\omega_{-(N)}}{2}. \qquad (12.32\text{c})$$

Here, $\langle X\rangle_N = \frac{1}{N}\sum_i X_i$ is the molecular ensemble average of property X and

$$\omega_{\pm(N)}^2 = \frac{1}{2}\left[\omega_0^2 + \omega_{\text{eq}}^2 \pm \sqrt{4\omega_0^2 g^2 N\langle\mu_{\text{eq}}'^2\rangle_N + (\omega_0^2 - \omega_{\text{eq}}^2)^2}\right] \qquad (12.33)$$

are the upper and lower polariton frequencies squared. As expected, Eq. (12.32) depends on differences and ratios of terms featuring N and $N-1$ molecules undergoing VSC with the cavity in the reactant configuration. Since we are interested in the collective VSC regime, we shall examine the regime where $N \gg 1$. Under these circumstances, $A_{\text{VSC}}(T) \approx 1$, $\Delta E_0^{\text{VSC}} \approx 0$, and the ratio of rate constants becomes

$$\kappa_N \approx \exp\left[\frac{(\omega_{\text{eq}}g\mu_{\ddagger})^2}{\left(\omega_{\text{eq}}^2 - g^2 N\langle\mu_{\text{eq}}'^2\rangle_N\right)k_{\text{B}}T}\right], \qquad (12.34)$$

where we assumed that, for typical reactions in liquid solution, the molecular dipoles are isotropically distributed; therefore, $\langle \mu_{\text{eq}} \rangle_N = 0$. At first glance, Eq. (12.34) is puzzling because it does not recognize the polaritonic normal modes, and hence, is independent of light-matter resonance, let alone VSC for that matter. In fact, since the collective light-matter coupling is typically a small fraction of the optical gap $\omega_{\text{eq}}^2 - g^2 N \langle \mu_{\text{eq}}'^2 \rangle \approx \omega_{\text{eq}}^2$, and the predicted cavity effect on the rate reduces essentially to the electrostatic shift of a single molecule in the TS, $\kappa_N \approx \exp\left[\dfrac{g^2 \mu_{\ddagger}^2}{k_B T}\right]$, which turns to be negligible for all practical purposes when $N \gg 1$ (see Figure 12.12). Thus, underwhelmingly, TST under conditions C1 and C2 cannot predict any of the observations O1–O3.

12.4.4. *Non-adiabatic reactions: Cavity Marcus–Levich–Jortner theory*

Marcus was the first to derive non-adiabatic reaction rate theory for the particular case of electron transfer processes.[60] His treatment

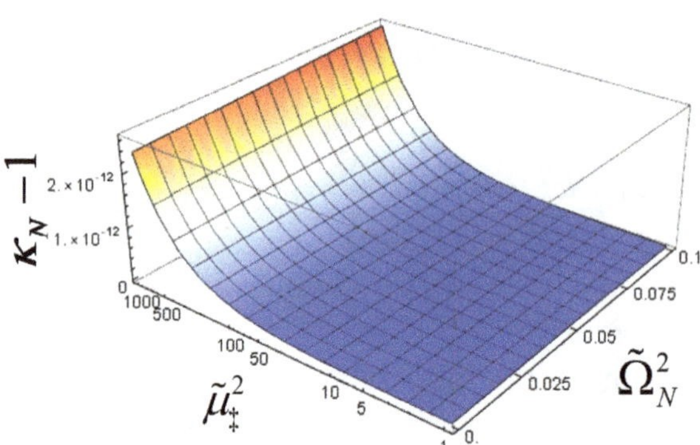

Figure 12.12. Ratio of in- and out-of-cavity reaction rate constants κ_N as a function of the TS permanent dipole and the collective light-matter coupling. In particular, $\tilde{\mu}_{\ddagger} = |\mu_{\ddagger}/\langle 1|\mu_{\text{eq}}'|0\rangle|$ is the ratio of the permanent dipole moment in the TS to the transition dipole moment at the equilibrium configuration, and $\tilde{\Omega}_N = g\sqrt{N \langle \mu_{\text{eq}}^2 \rangle}/\omega_{\text{eq}}$ is the ratio of light-matter coupling to the frequency in the same configuration. Throughout all explored values, $\kappa_N \approx 1$, a consequence of the number of molecules being large, $N = 10^9$. Other parameters in the calculation are $\omega_{eq} = 2000$ cm^{-1} and $k_B T = 208.5$ cm^{-1}. *Source*: From Ref. 40.

assumes that the diabatic potentials are parabolas centered at $q = d_R$ and $q = d_P$, respectively,

$$V_R(q) = \frac{m\omega_S^2(q - d_R)^2}{2},$$

$$V_P(q) = \frac{m\omega_S^2(q - d_P)^2}{2} + \Delta E.$$

Here, ΔE is the difference in energy between the equilibrium geometries of the product and reactant. q is an effective "slow" classical coordinate describing the orientation of solvent dipoles, and it is associated with an effective mass m and frequency ω_S. The reorganization energy $\lambda_S = V_P(d_R) - V_P(d_P) = \frac{1}{2}m\omega_S^2(d_P - d_R)^2$ is a measure of the displacement between the R and P parabolas. Assuming that J_{RP} is very small and q independent, the TS occurs at $q = q^*$ where $V_P(q^*) = V_R(q^*)$, which yields $q^* = \frac{\Delta E}{2\lambda}(d_P - d_R) + \frac{d_R + d_P}{2}$ with activation energy $\frac{(\Delta E + \lambda_S)^2}{4\lambda_S}$ starting from $q = q_R$. In fact, assuming $\hbar\omega_s \ll k_B T$, an FGR computation of the reaction rate mediated by the diabatic coupling yields Marcus' celebrated expression,[61]

$$k_{R\to P} = \sqrt{\frac{\pi}{\lambda_S k_B T}} \frac{|J_{RP}|^2}{\hbar} \exp\left(-\frac{(\Delta E + \lambda_S)^2}{4\lambda_S k_B T}\right). \tag{12.35}$$

Levich and Jortner[62, 63] generalized Marcus' expression to account for the participation of an additional high-frequency "quantum" vibrational mode of frequency ω_P in the electron transfer process,

$$k_{R\to P} = \sqrt{\frac{\pi}{\lambda_S k_B T}} \frac{|J_{RP}|^2}{\hbar} e^{-S} \sum_{v=0}^{\infty} \frac{S^v}{v!} \exp\left(-\frac{(\Delta E + \lambda_S + v\hbar\omega_P)^2}{4\lambda_S k_B T}\right). \tag{12.36}$$

This latter expression (known as the Marcus-Levich-Jortner (MLJ) formula) can be interpreted as a sum of Marcus rates over channels where v excitations are deposited in the high-frequency mode of the product electronic state (an additional average over Boltzmann populations in the reactant side is unnecessary due to the negligible equilibrium occupation of the excitations of the quantum modes at room temperature). $S = \lambda_P/\hbar\omega_P$ is a Huang–Rhys parameter with λ_P being the reorganization energy of the quantum mode.

Let us now consider the case where N high-frequency vibrational modes are coupled to a single optical cavity mode, each with amplitude g. This coupling results in $N + 1$ normal modes out of which $N - 1$ are purely vibrational dark modes with the bare frequency ω_P, as well as an upper polariton (UP$_N$) and a lower polariton (LP$_N$) with respective frequencies $\omega_{\pm(N)} = \frac{\omega_0+\omega_P}{2} \pm \frac{\Omega_N}{2}$, where the Rabi frequency is given by $\Omega_N = \sqrt{4g^2 N + \Delta^2}$ and $\Delta = \omega_0 - \omega_P$ is the light–matter detuning.

In Ref. 64, we studied the case where only the vibrational modes in the product electronic state couple to the cavity (extensions to cases where both reactant and product molecules are coupled to the cavity yield similar conclusions, as discussed below). Assume there are a total of M molecules in the cavity. Consider the point of the reaction progression where there are $M - N + 1$ reactant and $N - 1$ product molecules. We now consider the reactive event where (without loss of generality), the Nth molecule converts from reactant to product. Outside of the cavity, this event would merely involve two vibrational modes: the classical and quantum vibrational modes of the Nth molecule itself. In principle, VSC converts this problem into one involving $N + 2$ vibrational modes: the classical one of the Nth molecule, and the $N + 1$ (polariton and dark) modes that are in principle delocalized across all product molecules. However, given the large degeneracy of the dark-mode manifold (Sections 12.2.3 and 12.3.3), a judicious quasi-localized basis can be chosen such that only four modes are involved. Obviating the participation of the classical coordinate, the reactive event can be described as a transfer of vibrational amplitude between 3D parabolic PESs that are displaced, rotated, and squeezed with respect to each other,

$$R + UP_{N-1} + LP_{N-1} \longrightarrow UP_N + LP_N + D_N^{(N)}. \tag{12.37}$$

Here, R, UP$_N$, LP$_N$ and D$_N^{(N)}$ represent the localized quantum vibrational mode in the Nth molecule, upper and lower polariton modes arising from VSC with N product molecules, and D$_N^{(N)}$ is a highly localized dark mode in the product of the Nth molecule. The interpretation is that when the Nth molecule reacts, the polariton modes change from involving $N-1$ to N product molecules in VSC, and an extra dark mode appears. Thus, the VSC generalization of the MLJ expression in Eq. (12.36) involves a sum over possible quanta $\{v_+, v_-, v_\mathrm{D}\}$ that can be deposited into the product modes UP_N, LP_N and $\mathrm{D}_N^{(N)}$, respectively, upon the reactive event,

$$k_{\mathrm{R}\to\mathrm{P}}^{\mathrm{VSC}} = \sqrt{\frac{\pi}{\lambda_\mathrm{S} k_\mathrm{B} T}} \frac{|J_{\mathrm{RP}}|^2}{\hbar} \sum_{v_+=0}^{\infty} \sum_{v_-=0}^{\infty} \sum_{v_\mathrm{D}=0}^{\infty} W_{v_+,v_-,v_\mathrm{D}}, \qquad (12.38)$$

with $W_{v_+,v_-,v_\mathrm{D}} = |F_{v_+,v_-,v_\mathrm{D}}|^2 \exp\left(-\frac{E_{v_+,v_-,v_\mathrm{D}}^\ddagger}{k_\mathrm{B} T}\right)$, and

$$
\begin{aligned}
|F_{v_+,v_-,v_\mathrm{D}}|^2 &= |\langle 0_{+(N-1)} 0_{-(N-1)} 0_\mathrm{R} | v_+ v_- v_\mathrm{D} \rangle|^2 \\
&= \left(\frac{\sin^2 \theta_N}{N}\right)^{v_+} \left(\frac{\cos^2 \theta_N}{N}\right)^{v_-} \left(\frac{N-1}{N}\right)^{v_\mathrm{D}} \\
&\quad \times \binom{v_+ + v_- + v_\mathrm{D}}{v_+, v_-, v_\mathrm{D}} |\langle 0' | v_+ + v_- + v_\mathrm{D} \rangle|^2, \quad (12.39)
\end{aligned}
$$

being a Franck–Condon (FC) factor between the global ground state in the reactants and an excited vibrational configuration in the product.[65] Via a change of basis, in the second line of Eq. (12.39), we have related the three-dimensional FC factor to a one-dimensional one involving only the quantum mode of the Nth molecule. $\theta_N = \frac{1}{2} \arctan \frac{2g\sqrt{N}}{\Delta}$ is the mixing angle due to VSC. $|0'\rangle$ is the vibrational ground state of the Nth molecule in the reactant electronic state and $|v_+ + v_- + v_\mathrm{D}\rangle$ is the vibrational state of the Nth molecule with $v_+ + v_- + v_\mathrm{D}$ in the product electronic state. The activation energy for each of the channels is

$$E_{v_+,v_-,v_\mathrm{D}}^\ddagger = \frac{(E_\mathrm{P}^{v_+,v_-,v_\mathrm{D}} - E_\mathrm{R}^0 + \lambda_\mathrm{S})^2}{4\lambda_\mathrm{S}}, \qquad (12.40)$$

with $E_\mathrm{R}^0 = \frac{\hbar}{2}\left(\omega_{+(N-1)} + \omega_{-(N-1)} + \omega_\mathrm{R}\right)$ being the zero-point energy contributions of the reactant state and $E_\mathrm{P}^{v_+,v_-,v_\mathrm{D}} = \Delta E + \hbar\left[\omega_{+(N)}\left(v_+ + \frac{1}{2}\right) + \omega_{-(N)}\left(v_- + \frac{1}{2}\right) + \omega_\mathrm{P}\left(v_\mathrm{D} + \frac{1}{2}\right)\right]$ is the energy corresponding to the polaritonic Fock state that is accessed in the product upon the reactive event. Eq. (12.39) affords a physical interpretation: the state $|v_+ v_- v_D\rangle$ is accessed by creating $v_+ + v_- + v_D$ excitations in the quantum mode of the Nth product molecule; there are $\binom{v_+ + v_- + v_D}{v_+, v_-, v_D}$ ways to do this; $\left(\frac{\sin^2 \theta_N}{N}\right)$, $\left(\frac{\cos^2 \theta_N}{N}\right)$, and $\left(\frac{N-1}{N}\right)$ are prefactors that result from overlaps of the product normal modes and the quantum mode of the Nth product molecule. Importantly, these prefactors indicate that from an entropic standpoint, polariton modes are unlikely to partake in the reaction compared to the dark mode when $N \gg 1$. However, polariton modes offer activation energy barriers that are different from those of the bare quantum mode of the molecule, a feature that the dark modes do not (see Figure 12.13). Thus, there are certain geometric circumstances under which the

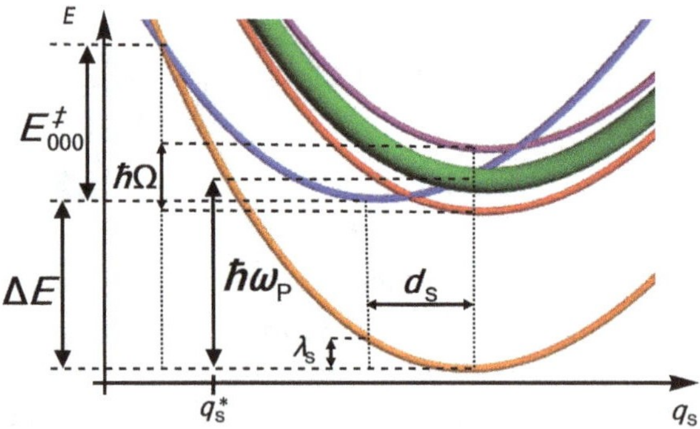

Figure 12.13. An ensemble of molecules that undergo a non-adiabatic reaction and whose "quantum" modes at frequency ω_P in the product electronic state are strongly coupled to an optical cavity mode. Under this VSC with N product vibrations, the new normal modes are upper and lower polaritons (separated by Rabi frequency Ω), and $N-1$ dark modes parked at the same frequency as the bare modes. The reactant and product diabatic PESs as a function of the slow coordinate q_S in the absence of excitations in the quantum modes are depicted in blue and orange, respectively. The difference in the energies of their equilibrium geometries is ΔE, while their intersection at q_S^* denotes the transition state (TS) with an activation energy $E_{000}^\ddagger$. The channels due to the lower polariton, upper polariton, and dark mode PESs are labeled in red, purple, and green. The lower-polariton channel features a lower activation energy than the dark modes, potentially catalyzing the reaction under VSC compared to the bare case; however, this is only possible if the decrease in activation energy can outcompete the entropically favorable dark modes, of which there are many. *Source*: From Ref. 64.

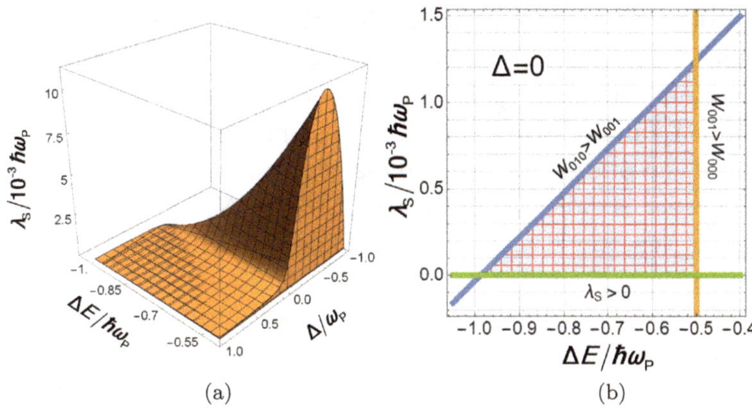

Figure 12.14. (a) Non-adiabatic reaction parameters for catalytic behavior under VSC. The lower polariton channel (due to the W_{010} term in Eq. (12.38)) dominates the kinetics over the many dark (D) channels certain values of ΔE (the energy difference between P and R), λ_S (the classical mode reorganization energy), and Δ (the detuning between the cavity and the high-frequency mode). (b) shows a cross section of the plot in (a) when $\Delta = 0$. The chosen parameters are $\omega_R = \omega_P$, $k_B T = 0.2\hbar\omega_P$, $N = 10^{10}$, $S = 1$, and $\hbar\Omega = 5 \times 10^{-2}\hbar\omega_P$. *Source:* From Ref. 64.

polariton channels outcompete the dark-mode channels, giving rise to catalysis. For instance, Figure 12.14 shows the region in parameter space where the term W_{010} due to one single excitation deposited in the lower polariton in the products dominates the summation in Eq. (12.38).

Unlike in the cavity TST case, the cavity MLJ theory under conditions C1 and C2 can explain some of the observations. O1 is only partially accounted for, as the aforementioned mechanism can catalyze reactions under VSC, but not suppress them, for if the polariton channels do not offer energetic benefits to be utilized, the dark states will still be present, rendering reaction rates that are identical to outside of the cavity case. However, there could be a possibility that VSC catalyzes competing reactions, rendering an apparent suppression of the reaction of interest. O2 is definitely accounted for: under highly off-resonant circumstances (see Figure 12.5), one of the polariton states is dominated by its molecular content (with barely no energy difference compared to the bare molecular excitation) while the other has barely no molecular content (in which case, there is no matrix element coupling it to the reactant channel), giving rise to no changes in reaction rate in either case. Finally, given that this is a single cavity-mode theory, O3 cannot be accounted for. Regardless of the promise of this theory, a big drawback is indicated in Figure 12.14: for $N = 10^6 - 10^{10}$, it only works for very small values of λ_S, while typical values of λ_S much larger, in the 0.1–1 eV range.[66]

Extensions of this cavity MLJ theory to account for non-equilibrium effects of "vibrationally hot" products or intermediates as well as cavity leakage do not seem to ameliorate the problems of the cavity-assisted charge transfer theory described above.[67] However, a generalization that accounts for dark-state semi-localization in the presence of disorder seems to offer a potential way to solve these issues, despite O3 still not being addressed.[68] Cavity effects on dynamical corrections to TST[69, 70] were also investigated recently by the Huo group.[71, 72] However, their work focused on the regime of single-molecule strong light-matter interactions with strength that is several orders of magnitude greater than is observed in the FP cavities employed in the experiments discussed in this section (where a macroscopic number of molecules interact with the confined electromagnetic field, see Section 12.2.3).

To conclude, it is interesting to inquire why the cavity MLJ theory does feature collective and resonant effects, while the cavity TST theory does not. In the MLJ case, the collective VSC of the macroscopic number of product molecules gives rise to energetic shifts of product channels that translate into changes in activation energies and concomitant catalysis. The fact that the energetics (and the wave functions) of these product channels matter is a consequence of quantum tunneling; this contrasts with TST, where whatever happens after the TS is immaterial.

12.5. Conclusions

In this chapter, we aimed to convey to the reader a sample of the intriguing, rich new chemical phenomenology that happens when molecules are placed in resonant infrared microcavities.

Despite significant theoretical and experimental progress, we have also shown that there remain important open questions and much is yet to be discovered about chemistry under vibrational strong coupling. In addition, given the several recent reports of cavity control of chemical equilibria,[9] intra[73] and intermolecular energy flow,[8] and reaction dynamics,[5,6,10,50] it seems we have just scratched the surface of polariton chemistry. Therefore, it is paramount that careful new experiments and theories are designed to help elucidate and predict the operating mechanisms for the reported effects of cavities on molecular systems. We expect this field will continue to progress at a fast rate propelled by a productive collaboration between theory and experiments with the ultimate goal of unraveling the microscopic mechanisms allowing for control of chemical phenomena with infrared microcavities.

Acknowledgments

R.F.R. acknowledges generous start-up funds from Emory University. J.Y.Z. acknowledges support from AFOSR award FA9550-18-1-0289. The theoretical work described on vibropolaritonic chemistry: reaction effects was largely carried out by Dr. Jorge A. Campos-Gonzalez-Angulo.

References

1. Kavokin, A. V.; Baumberg, J. J.; Malpuech, G.; Laussy, F. P. *Microcavities*, Vol. 21. Oxford University Press, 2017.
2. Ribeiro, R. F.; Martínez-Martínez, L. A.; Du, M. Campos-Gonzalez-Angulo, J.; Yuen-Zhou, J. Polariton chemistry: controlling molecular dynamics with optical cavities. *Chem. Sci.* **2018**, *9*, 6325–6339.
3. Ebbesen, T. W. Hybrid light–matter states in a molecular and material science perspective. *Acc. Chem. Res.* **2016**, *49*, 2403–2412.
4. Hutchison, J. A.; Schwartz, T.; Genet, C.; Devaux, E.; Ebbesen, T. W. Modifying chemical landscapes by coupling to vacuum fields. *Angew. Chem. Int. Ed.* **2012**, *51*, 1592–1596.
5. Thomas, A.; George, J.; Shalabney, A.; Dryzhakov, M.; Varma, S. J.; Moran, J.; Chervy, T.; Zhong, X.; Devaux, E.; Genet, C.; Hutchison, J. A.; Ebbesen, T. W. Ground-state chemical reactivity under vibrational coupling to the vacuum electromagnetic field. *Angew. Chem. Int. Ed.* **2016**, *55*, 11462–11466.

6. Thomas, A.; Lethuillier-Karl, L.; Nagarajan, K.; Vergauwe, R. M. A.; George, J.; Chervy, T.; Shalabney, A.; Devaux, E.; Genet, C.; Moran, J.; Ebbesen, T. W. Tilting a ground-state reactivity landscape by vibrational strong coupling. *Science* **2019**, *363*, 615–619.

7. Zhong, X.; Chervy, T.; Zhang, L.; Thomas, A.; George, J.; Genet, C.; Hutchison, J. A.; Ebbesen, T. W. Energy transfer between spatially separated entangled molecules. *Angew. Chem. Int. Ed.* **2017**, *56*, 9034–9038.

8. Xiang, B.; Ribeiro, R. F.; Du, M.; Chen, L.; Yang, Z.; Wang, J.; Yuen-Zhou, J.; Xiong, W. Intermolecular vibrational energy transfer enabled by microcavity strong Light–Matter coupling. *Science* **2020**, *368*, 665–667.

9. Pang, Y.; Thomas, A.; Nagarajan, K.; Vergauwe, R. M. A.; Joseph, K.; Patrahau, B.; Wang, K.; Genet, C.; Ebbesen, T. W. On the role of symmetry in vibrational strong coupling: the case of charge-transfer complexation. *Angew. Chem. Int. Ed.* **2020**, *132*, 10522–10526.

10. Hirai, K.; Takeda, R.; Hutchison, J. A.; Uji-i, H. Modulation of prins cyclization by vibrational strong coupling. *Angew. Chem. Int. Ed.* **2020**, *59*, 5332–5335.

11. Lidzey, D. G.; Bradley, D. D. C.; Skolnick, M. S.; Virgili, T.; Whittaker, S.; Whittaker, D. M. Strong Exciton–Photon coupling in an organic semiconductor microcavity. *Nature* **1998**, *395*, 53.

12. Tartakovskii, A. I.; Emam-Ismail, M.; Lidzey, D. G.; Skolnick, M. S.; Bradley, D. D. C.; Walker, S.; Agranovich, V. M. Raman scattering in strongly coupled organic semiconductor microcavities. *Phys. Rev. B* **2001**, *63*, 121302.

13. Long, J. P.; Simpkins, B. S. Coherent coupling between a molecular vibration and Fabry–Perot optical cavity to give hybridized states in the strong coupling limit. *ACS Photonics* **2015**, *2*, 130–136.

14. George, J.; Shalabney, A.; Hutchison, J. A.; Genet, C.; Ebbesen, T. W. Liquid-phase vibrational strong coupling. *J. Phys. Chem. Lett.* **2015**, *6*, 1027–1031.

15. Dunkelberger, A. D.; Spann, B. T.; Fears, K. P.; Simpkins, B. S.; Owrutsky, J. C. Modified relaxation dynamics and coherent energy exchange in coupled vibration-cavity polaritons. *Nat. Commun.* **2016**, *7*, 13504.

16. Nelson, P. *From Photon to Neuron: Light, Imaging, Vision*. Princeton University Press, 2017.

17. Shapiro, M.; Brumer, P. (Eds.). *Principles of the Quantum Control of Molecular Processes*. Wiley-VCH: Weinheim, 2003.

18. Frei, H.; Pimentel, G. C. Infrared induced photochemical processes in matrices. *Annu. Rev. Phys. Chem.* **1985**, *36*, 491–524.

19. Mitrano, M.; Cantaluppi, A.; Nicoletti, D.; Kaiser, S.; Perucchi, A.; Lupi, S.; Di Pietro, P.; Pontiroli, D.; Riccò, M.; Clark, S. R.; Jaksch, D.; Cavalleri, A. Possible light-induced superconductivity in K_3C_{60} at High temperature. *Nature* **2016**, *530*, 461.

20. Commins, E. D. *Quantum Mechanics*. Cambridge University Press, 2014.

21. Craig, D. P.; Thirunamachandran, T. *Molecular Quantum Electrodynamics: An Introduction to Radiation-Molecule Interactions*. Dover Books on Chemistry Series; Dover Publications, 1998.

22. Purcell, E. M. Minutes of the spring meeting at Cambridge. *Phys. Rev.* **1946**, *69*, 674–674.

23. Gérard, J.-M.; Sermage, B.; Gayral, B.; Legrand, B.; Costard, E.; Thierry-Mieg, V. Enhanced spontaneous emission by quantum boxes in a monolithic optical microcavity. *Phys. Rev. Lett.* **1998**, *81*, 1110.

24. Simpkins, B. S.; Fears, K. P.; Dressick, W. J.; Spann, B. T.; Dunkelberger, A. D.; Owrutsky, J. C. Spanning strong to weak normal mode coupling between vibrational and Fabry–Pérot cavity modes through tuning of vibrational absorption strength. *ACS Photonics* **2015**, *2*, 1460–1467.

25. Tavis, M.; Cummings, F. W. Exact solution for an N-molecule-radiation-field Hamiltonian. *Phys. Rev.* **1968**, *170*, 379–384.

26. Dicke, R. H. Coherence in spontaneous radiation processes. *Phys. Rev.* **1954**, *93*, 99–110.

27. Steck, D. A. *Quantum and Atom Optics*, revision 0.12.0 ed. (available online at https://atomoptics.uoregon.edu/~dsteck/teaching/quantum-optics/quantum-optics-notes.pdf, 2017).

28. Stokes, A.; Nazir, A. Implications of gauge-freedom for nonrelativistic quantum electrodynamics. 2020, arXiv:2009.10662.

29. Ribeiro, R. F.; Martínez-Martínez, L. A.; Du, M.; Campos-Gonzalez-Angulo, J.; Yuen-Zhou, J. Polariton chemistry: Controlling molecular dynamics with optical cavities. *Chem. Sci.* **2018**, *9*, 6325–6339.

30. Cohen-Tannoudji, C.; Diu, B.; Laloë, F. *Quantum Mechanics, Volume 1: Basic Concepts, Tools, and Applications*. John Wiley & Sons, 2019.

31. del Pino, J.; Feist, J.; Garcia-Vidal, F. J. Quantum theory of collective strong coupling of molecular vibrations with a microcavity mode. *New J. Phys.* **2015**, *17*, 053040.

32. Daskalakis, K. S.; Maier, S. A.; Kéna-Cohen, S. Polariton Condensation in Organic Semiconductors. In *Quantum Plasmonics*; Bozhevolnyi, S. I., Martin-Moreno, L., Garcia-Vidal, F., Eds.; Springer International Publishing, Cham, 2017; pp. 151–163.

33. Xiang, B.; Ribeiro, R. F.; Dunkelberger, A. D.; Wang, J.; Li, Y.; Simpkins, B. S.; Owrutsky, J. C.; Yuen-Zhou, J.; Xiong, W.

Two-dimensional infrared spectroscopy of vibrational polaritons. *Proc. Natl. Acad. Sci.* **2018**, *115*, 4845–4850.

34. Ribeiro, R. F.; Dunkelberger, A. D.; Xiong, B.; Xiang, W.; Simpkins, B. S.; Owrutsky, J. C.; Yuen-Zhou, J. Theory for nonlinear spectroscopy of vibrational polaritons. *J. Phys. Chem. Lett.* **2018**, *9*, 3766–3771.

35. Ćwik, J. A.; Reja, S.; Littlewood, P. B.; Keeling, J. Polariton condensation with saturable molecules dressed by vibrational modes. *EPL* **2014**, *105*, 47009.

36. Herrera, F.; Spano, F. C. Cavity-controlled chemistry in molecular ensembles. *Phys. Rev. Lett.* **2016**, *116*, 238301.

37. Du, M.; Martínez-Martínez, L. A.; Ribeiro, R. F.; Hu, Z.; Menon, V. M.; Yuen-Zhou, J. Theory for polariton-assited remote energy transfer. *Chem. Sci.* **2018**, *9*, 6659–6669.

38. Ribeiro, R. F.; Dunkelberger, A. D.; Xiong, B.; Xiang, W.; Simpkins, B. S.; Owrutsky, J. C.; Yuen-Zhou, J. Theory for nonlinear spectroscopy of vibrational polaritons. *J. Phys. Chem. Lett.* **2018**, *9*, 3766–3771.

39. Galego, J.; Climent, C.; Garcia-Vidal, F. J.; Feist, J. Cavity Casimir-Polder forces and their effects in ground-state chemical reactivity. *Phys. Rev. X* **2019**, *9*, 021057.

40. Campos-Gonzalez-Angulo, J. A.; Yuen-Zhou, J. Polaritonic normal modes in transition state theory. *J. Chem. Phys.* **2020**, *152*, 161101.

41. Li, T. E.; Nitzan, A.; Subotnik, J. E. On the origin of ground-state vacuum-field catalysis: equilibrium consideration. *J. Chem. Phys.* **2020**, *152*, 234107.

42. Dunkelberger, A. D.; Grafton, A. B.; Vurgaftman, I.; Soykal, Ö. O.; Reinecke, T. L.; Davidson, R. B.; Simpkins, B. S.; Owrutsky, J. C. Saturable absorption in solution-phase and cavity-coupled Tungsten hexacarbonyl. *ACS Photonics* **2019**, *6*, 2719–2725.

43. Dunkelberger, A. D.; Davidson, R. B.; Ahn, W.; Simpkins, B. S.; Owrutsky, J. C. Ultrafast transmission modulation and recovery via vibrational strong coupling. *J. Phys. Chem. A* **2018**, *122*, 965–971.

44. Xiang, B.; Ribeiro, R. F.; Li, Y.; Dunkelberger, A. D.; Simpkins, B. B.; Yuen-Zhou, J.; Xiong, W. Manipulating optical nonlinearities of molecular polaritons by delocalization. *Sci. Adv.* **2019**, *5*, eaax5196.

45. Frei, H.; Pimentel, G. C. Infrared induced photochemical processes in matrices. *Annu. Rev. Phys. Chem.* **1985**, *36*, 491–524.

46. Levine, R. D.; Jortner, J. Mode Selective Chemistry. In *Mode Selective Chemistry*; Jortner, J., Levine, R. D., Puliman, B., Eds.; Springer, 1991; pp. 535–571.

47. Nesbitt, D. J.; Field, R. W. Vibrational energy flow in highly excited molecules: Role of intramolecular vibrational redistribution. *J. Phys. Chem.* **1996**, *100*, 12735–12756.

48. Gruebele, M.; Wolynes, P. G. Vibrational energy flow and chemical reactions. *Acc. Chem. Res.* **2004**, *37*, 261–267.

49. Vergauwe, R. M. A.; Thomas, A.; Nagarajan, K.; Shalabney, A.; George, J.; Chervy, T.; Seidel, M.; Devaux, E.; Torbeev, V.; Ebbesen, T. W. Modification of enzyme activity by vibrational strong coupling of water. *Angew. Chem. Int. Ed.* **2019**, *58*, 15324–15328.

50. Hiura, H.; Shalabney, A.; George, J. Vacuum-field catalysis: accelerated reactions by vibrational ultra strong coupling. 2019, 10.26434/chemrxiv.7234721.v4.

51. Lather, J.; Bhatt, P.; Thomas, A.; Ebbesen, T. W.; George, J. Cavity catalysis by cooperative vibrational strong coupling of reactant and solvent molecules. *Angew. Chem. Int. Ed.* **2019**, *131*, 10745–10748.

52. Ribeiro, R. F. Strong light-matter interaction effects on molecular ensembles. 2021, arXiv preprint arXiv:2107.07032.

53. Litinskaya, M.; Reineker, P. Loss of coherence of exciton polaritons in inhomogeneous organic microcavities. *Phys. Rev. B* **2006**, *74*, 165320.

54. Eyring, H. The activated complex in chemical reactions. *J. Chem. Phys.* **1935**, *3*, 107–115.

55. Wigner, E. The transition state method. *Trans. Faraday Soc.* **1938**, *34*, 29–41.

56. Hänggi, P.; Talkner, P.; Borkovec, M. Reaction-rate theory: fifty years after Kramers. *Rev. Modern Phys.* **1990**, *62*, 251–341.

57. Pollak, E.; Talkner, P. Reaction rate theory: what it was, where is it today, and where is it going? *Chaos* **2005**, *15*, 026116.

58. Arnaut, L.; Burrows, H. *Chemical Kinetics: From Molecular Structure to Chemical Reactivity*. Elsevier, 2006.

59. Zhdanov, V. P. Vacuum field in a cavity, light-mediated vibrational coupling, and chemical reactivity. *Chem. Phys.* **2020**, *535*, 110767.

60. Marcus, R. A. On the theory of oxidation-reduction reactions involving electron transfer. i. *J. Chem. Phys.* **1956**, *24*, 966–978.

61. Nitzan, A. *Chemical Dynamics in Condensed Phases: Relaxation, Transfer and Reactions in Condensed Molecular Systems*. Oxford University Press, 2006.

62. Levich, V. G. Present state of the theory of oxidation-reduction in solution (bulk and electrode reactions). *Adv. Electrochem. Electrochem. Eng.* **1966**, *4*, 249–371.

63. Jortner, J. Temperature dependent activation energy for electron transfer between biological molecules. *J. Chem. Phys.* **1976**, *64*, 4860–4867.

64. Campos-Gonzalez-Angulo, J. A.; Ribeiro, R. F.; Yuen-Zhou, J. Resonant catalysis of thermally activated chemical reactions with vibrational polaritons. *Nat. Commun.* **2019**, *10*, 1–8.

65. Strashko, A.; Keeling, J. Raman scattering with strongly coupled vibron-polaritons. *Phys. Rev. A* **2016**, *94*, 023843.

66. Chaudhuri, S.; Hedstrom, S.; Mendez-Hernandez, D. D.; Hendrickson, H. P.; Jung, K. A.; Ho, J.; Batista, V. S. Electron transfer assisted by vibronic coupling from multiple modes. *J. Chem. Theory Comput.* **2017**, *13*, 6000–6009.

67. Du, M.; Campos-Gonzalez-Angulo, J. A.; Yuen-Zhou, J. Nonequilibrium effects of cavity leakage and vibrational dissipation in thermally activated polariton chemistry. *J. Chem. Phys.* **2021**, *154*, 084108.

68. Du, M.; Yuen-Zhou, J. Can dark states explain vibropolaritonic chemistry? 2021, arXiv:2104.07214 [quant-ph].

69. Kramers, H. A. Brownian motion in a field of force and the diffusion model of chemical reactions. *Physica* **1940**, *7*, 284–304.

70. Grote, R. F.; Hynes, J. T. The stable states picture of chemical reactions. II. Rate constants for condensed and gas phase reaction models. *J. Chem. Phys.* **1980**, *73*, 2715–2732.

71. Li, X.; Mandal, A.; Huo, P. Cavity frequency-dependent theory for vibrational polariton chemistry. *Nat. Commun.* **2021**, *12*, 1315.

72. Li, X.; Mandal, A.; Huo, P. Theory of mode-selective chemistry through polaritonic vibrational strong coupling. *J. Phys. Chem. Lett.* **2021**, *12*, 6974–6982.

73. Erwin, J. D.; Wang, Y.; Bradley, R. C.; Coe, J. V. Changing vibration coupling strengths of liquid acetonitrile with an angle-tuned etalon. *J. Phys. Chem. B* **2021**, *125*, 8472–8483.

Index

CPSIA information can be obtained
at www.ICGtesting.com
Printed in the USA
JSHW050939010722
27433JS00001B/2